Inductive Devices in Power Electronics

Other related titles:

You may also like

- PBPO152 | Iannuzzo | Modern Power Electronic Devices: Physics, applications, and reliability | 2020
- PBPO074 | Trzynadlowski | Power Electronic Converters and Systems | 2015
- PBPO239| Bendaoud | Intelligent Control of Medium and High Power Converters | 2023

We also publish a wide range of books on the following topics:
Computing and Networks
Control, Robotics and Sensors
Electrical Regulations
Electromagnetics and Radar
Energy Engineering
Healthcare Technologies
History and Management of Technology
IET Codes and Guidance
Materials, Circuits and Devices
Model Forms
Nanomaterials and Nanotechnologies
Optics, Photonics and Lasers
Production, Design and Manufacturing
Security
Telecommunications
Transportation

All books are available in print via https://shop.theiet.org or as eBooks via our Digital Library https://digital-library.theiet.org.

IET ENERGY ENGINEERING SERIES 203

Inductive Devices in Power Electronics

Materials, measurement, design and applications

Edited by
Peter Zacharias

The Institution of Engineering and Technology

About the IET

This book is published by the Institution of Engineering and Technology (The IET).

We inspire, inform and influence the global engineering community to engineer a better world. As a diverse home across engineering and technology, we share knowledge that helps make better sense of the world, accelerate innovation and solve the global challenges that matter.

The IET is a not-for-profit organisation. The surplus we make from our books is used to support activities and products for the engineering community and promote the positive role of science, engineering and technology in the world. This includes education resources and outreach, scholarships and awards, events and courses, publications, professional development and mentoring, and advocacy for governments.

To discover more about the IET please visit https://www.theiet.org/.

About IET books

The IET publishes books across many engineering and technology disciplines. Our authors and editors offer fresh perspectives from universities and industry. Within our subject areas, we have several book series steered by editorial boards made up of leading subject experts.

We peer review each book at the proposal stage to ensure the quality and relevance of our publications.

Get involved

If you are interested in becoming an author, editor, series advisor, or peer reviewer, please visit https://www.theiet.org/publishing/publishing-with-iet-books/ or contact author_support@theiet.org.

Discovering our electronic content

All of our books are available online via the IET's Digital Library. Our Digital Library is the home of technical documents, eBooks, conference publications, real-life case studies and journal articles. To find out more, please visit https://digital-library.theiet.org.

In collaboration with the United Nations and the International Publishers Association, the IET is a Signatory member of the SDG Publishers Compact. The Compact aims to accelerate progress to achieve the Sustainable Development Goals (SDGs) by 2030. Signatories aspire to develop sustainable practices and act as champions of the SDGs during the Decade of Action (2020–30), publishing books and journals that will help inform, develop, and inspire action in that direction.

In line with our sustainable goals, our UK printing partner has FSC accreditation, which is reducing our environmental impact on the planet. We use a print-on-demand model to further reduce our carbon footprint.

Published by The Institution of Engineering and Technology, London, United Kingdom

The Institution of Engineering and Technology (the "**Publisher**") is registered as a Charity in England & Wales (no. 211014) and Scotland (no. SC038698).

First published 2024

The Institution of Engineering and Technology
Futures Place,
Kings Way,
Stevenage,
Herts, SG1 2UA
United Kingdom

www.theiet.org

British Library Cataloguing in Publication Data
A catalogue record for this product is available from the British Library

ISBN 978-1-83953-447-8 (hardback)
ISBN 978-1-83953-448-5 (PDF)

Typeset in India by MPS Limited

Cover image: Wilmar Martinez/Future of Magnetics.

Contents

Foreword

Modern society is inconceivable without the use of various forms of energy. Social life, the division of labour and technological development are not possible without the use of energy. Electrical energy is becoming increasingly important because it is easy to transport and can be easily converted into other forms of energy. However, electrical energy also needs to be processed for a wide variety of applications. The applications require very different quality parameters such as voltage, current, frequency or impedance of the energy supplier. Power electronic converters are used for this purpose. These consist of switched electrical networks that are able to adapt the characteristic parameters of the final electrical energy supplied to the respective requirements with virtually no loss. The electrical power is essentially controlled via the ratio of the switch-on and switch-off times (pwm) of the switches used or via the magnitude of the frequency used. The desired output characteristics are achieved by recharging capacitive and inductive storage devices in the converter.

The aim is to achieve conversion efficiencies that are as close as possible to 100%. Modern semiconductor switches based on silicon (Si), silicon carbide (SiC) or gallium nitride (GaN) offer a wide selection in terms of electrical parameters and switching characteristics. Theoretically, the generally quite voluminous storage units for electrical and magnetic energy can be made smaller and smaller if the cycle times for their recharging are reduced. One trend is therefore towards increasing the switching frequency.

However, this theoretical approach ignores the fact that both the components and the entire converter are the result of the energy exchange of electric and magnetic fields. Various properties of materials are used in this process. It is not only the semiconductor switches that cause higher losses for the same current at a higher frequency. The same applies to the storage elements and transformers used. The size of the capacitors used to store electrical energy depends primarily on the dielectric used for storage and generally has a high ratio of reactive power to power loss (quality factor Q). This is much less the case with inductive components.

The size of inductive components depends not only on the maximum inductance that can be used (limited by saturation) but also on the magnitude of power loss per volume and how effectively the heat generated can be dissipated. Without heat dissipation, the temperature in the components becomes too high at critical points and leads to their failure. In this way, inductive components lead to limits in the miniaturization of power electronic converters.

In recent decades, significant improvements have been made to magnetic materials used to conduct the magnetic field. These make it possible to increase the maximum operating frequency. At the same time, however, the trend towards ever-increasing power leads to ever-increasing currents to be switched. This trend comes up against other limiting factors. The electric current density field is always connected to a magnetic field. When the current changes, the induced effects of the magnetic field have revers influence on the electrical current density field causing the change and penetration of neighbouring conductive materials. These processes all lead to increased losses. However, unwanted interactions also occur in switched meshes of electrical networks.

As mentioned, every circuit is connected to a magnetic field in which energy is stored. This means that an inductance can also be assigned to this circuit. Rapid changes in current lead to changes in voltage at this inductance. These are proportional to the value of the acting inductance and the change in the current rise. If the maximum permissible values are exceeded, the semiconductor switches in particular may be destroyed. The switchable currents in a particular setup are therefore limited by characteristics that are not evident from the circuit diagram.

This book deals with the properties of materials used in inductive components and their measurement and modelling as well as with measures for component design and changes in circuit technology to push the physically set limits. The most commonly used method is that of pulse width modulation (pwm). How to modify the inductive components used through design measures is addressed in the same way as splitting the current over several magnetically coupled switched phases in a converter.

The higher the switching frequencies and power, the greater the effect of the relationships described above. Approaches for integrated magnetic components and a holistic converter design are therefore becoming increasingly important in these cases.

Part I deals with material-related issues. The most relevant types of material for the construction of inductive components are discussed, and their properties of interest to the user are presented. This means that both the magnetic materials for the cores of inductive components and questions relating to the design of windings and the effects of design measures are dealt with.

Part II deals with various aspects of modelling and its parameterization. The applications relate to the large-signal behaviour of the magnetic components. This means that non-linear models would actually be required. The relationships between small-signal and large-signal behaviour of the magnetic materials are explained, how the metrological characterization can be carried out and how the expected electrical properties and losses can be determined using the parameters obtained.

Part III deals with special questions of component design for single-phase and multi-phase applications. The focus here is on the targeted use of linear and non-linear core properties. For example, the combination of materials can be used to influence saturation or achieve an electrically controllable dependence on the inductance value. The combination of materials with very different permeabilities can lead to a separation of low-frequency flux and high-frequency flux.

Part IV presents design approaches for the design of electrical energy converters based on various applications. These are intended to help in our own considerations for solving design problems by adopting different perspectives.

Finally, Part V addresses the problems of electromagnetic compatibility. Due to the fast semiconductor switches used, conducted interference also occurs, e.g. when connecting to the public grid. This must be minimized so as not to interfere with the function of other connected devices. Passive filters and shielding measures as well as circuitry measures can be used to achieve the desired objectives.

The book is intended to support the development of ideas and the explanation of correlations in a way that is particularly helpful for developers of electronic power converters.

About the editor

Peter Zacharias is professor emeritus for Electric Power Systems at the University of Kassel, Germany. Prior affiliations include the University of Magdeburg, the Institute for Solar Energy Technology (ISET) in Kassel and eupec/Infineon in Warstein, with responsibility for power electronic assemblies. In 2009, he founded the 'Competence Center for Decentralized Electrical Power Systems' (KDEE). His main interest is inductive components in power electronic converters. He was honoured with the Outstanding Achievement Award at EPE 2022.

Part I

Magnetic materials and winding concepts

Chapter 1

Soft ferrite materials for high-frequency applications

Jonas Pfeiffer[1], *Michael Baumann*[1] *and Michael Schmidhuber*[1]

High-frequency applications are widely used, e.g. in information and communication technologies, power electronics or medical technologies. The most used magnetic material in that frequency area is soft ferrite. The first research into the magnetic properties of oxides was made in 1908 by the German chemist Siegfried Hilpert, whereas the first mixed ferrites made of zinc oxide and iron oxide were investigated by Takei and Kato in 1930. The Japanese electronics group TDK emerged from this.

Improvements for the later soft-magnetic ferrite cores were made by the Philips company under Snoek in the 1940s. The ferrite cores were marketed in 1946 with the trade name 'Ferroxcube' (Fe: iron, ox: oxide, cube: cubic crystal structure). This later became the company Ferroxcube.

According to the literature [1], the worldwide demand for soft ferrite is in the range of 350,000 metric tons per year.

This chapter describes the theoretical principles and the usage of ferrite materials with a focus on soft ferrites. After a brief overview of different types of magnetic materials, the physical and chemical principles of soft ferrites are explained. Finally, the selection of the appropriate soft ferrite material based on datasheets is described.

1.1 Different magnetic materials – a brief overview

There are different kinds of magnetic materials. Their use depends largely on the application's frequency. Figure 1.1 shows a selection of different magnetic materials, their range of initial permeability and the frequency range in which they are usually used.

In the following subsections, a selection of materials for high-frequency applications is discussed in more detail.

[1]SUMIDA Components & Modules GmbH, Obernzell, Germany

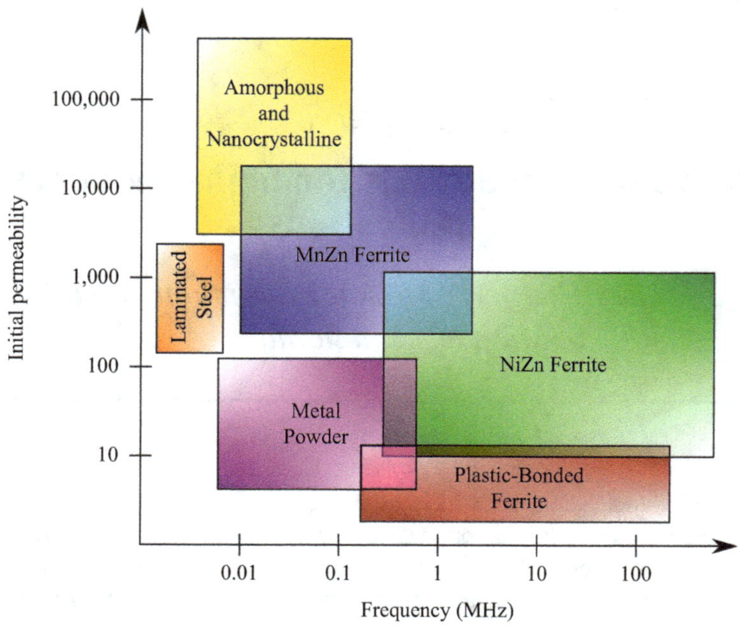

Figure 1.1 Various magnetic materials for power magnetic devices

1.1.1 Metal powder

The majority of metal powder cores are based on iron powder, which is based on carbonyl iron powder. Various additives are added to the metal powder, depending on the desired application. The mixed powder is provided with an insulating binder, pressed into shape and in some cases annealed to counteract mechanical distortion after pressing. To prevent electrical short circuits and metal oxidation, the core is coated with an insulating protective layer. The insulating binder within the core has the additional effect of a distributed air gap over the entire core geometry. This means that a magnetic field strength of several kiloamperes is required to drive the material into saturation. The saturation flux density is up to 1.5 T and in special cases beyond that. Due to defects in the insulation of the individual grains and the high conductivity of the metal powder particle, eddy currents can form within the core material, which have a negative effect on core losses.

There are different types of powder cores depending on the admixed main components. The so-called 'high-flux powder cores' are based on an alloy of almost equal proportions of iron and nickel. They are usually used if a high saturation flux density and high energy storage in combination with low core losses are requested. If the nickel proportion is significantly increased and molybdenum is also added, the powder core can be assigned to the so-called 'moly permalloy powder' (MPP) cores, which are characterized by a very small hysteresis and thus minimal core losses. Another alloy made of iron, silicon and aluminium is known as 'Sendust' powder cores. It is a low-cost material with a medium saturation flux density and medium core losses but also a low magnetostriction.

1.1.2 Ferrites

Ferrite cores represent another category of soft magnetic cores. The basic material is now iron oxide, in the range of 70 wt per cent. In addition, there are also numerous additives. Similar to metal powder cores, a binder is added to the mixed ferrite powder. The mixture is pressed in the requested core geometry and sintered afterwards. The duration and temperature have a great impact on the characteristics of the material. During the sintering process, the volume of the ferrite core is nearly halved. The cores are therefore shrinking in the range of a factor of 1.2. In contrast to metal powder cores, the binder in ferrite cores is burned off and due to compaction during the sintering process there is nearly no distributed air gap. So, a discrete air gap is needed in most applications. The grains formed by sintering are insulated from each other, so that the formation of eddy currents across individual grains is very low. Nevertheless, they are possible due to defects in this insulation and due to conductivity and parasitic capacitances – especially with large core geometries. Compared to metal powder cores, ferrites have lower core losses in the high-frequency range.

The additives used also have a major impact on the core material's characteristics. The most commonly used soft ferrite materials are manganese–zinc (MnZn) and nickel–zinc (NiZn) ferrites, which are explained in more detail below.

MnZn ferrites: The main components of this ferrite are the oxides of iron (Fe_2O_3), manganese (MnO) and zinc (ZnO). In addition, a variety of other oxides are used as additives. MnZn ferrites are used in low-signal applications, e.g. telecommunication. They are also used in power applications with a frequency between 10 kHz and 2 MHz. Their saturation flux density is usually between 400 and 550 mT at room temperature. Modern MnZn power ferrites therefore have very low core losses due to their high saturation flux density.

NiZn ferrites: In high-frequency applications with frequencies of up to 1 GHz, NiZn ferrites usually are used because of the material's smaller electrical conductivity. In comparison to MnZn ferrites, they have a smaller saturation flux density, which is usually between 250 and 430 mT, and higher losses in comparable frequency ranges.

For the sake of completeness, semi-hard and hard ferrites should be mentioned here. Hard ferrites have a high coercive field strength H_c and a high remanence B_r. They have a very low permeability and a medium flux density of up to 370 mT. Hard ferrites, e.g. barium-ferrite or strontium-ferrite, are usually used as permanent magnets. Ferrite materials in between soft and hard ferrites usually are referred to as semi-hard ferrites. Their applications include sensor technology, e.g. magnesium-ferrite or lithium-ferrite. However, these types of ferrites are not discussed further in the following.

1.1.3 Plastic-bonded soft ferrites and injection moulded soft ferrites

If complex shapes are needed, e.g. for housing, connectors, etc., the specialized plastic-bonded and injection moulded soft ferrites could be the best choice. A high content of nearly 50 vol per cent organic material is mixed via an extrusion process to the oxide material. This composite is called feedstock and can be injected with special machines into the desired shape.

Plastic-bonded soft ferrite belongs to the polymer-bonded soft magnetic composites (PBSMC) family. It is based on an oxide that is already magnetic (soft ferrite), and no further thermal process after the manufacturing process is needed. Therefore, the plastic-bonded ferrites have the best tolerances according to the values of the standard manufacturing process of the plastic materials.

Instead, the so-called 'injection moulded soft ferrites' are based on non-magnetic oxides. This process belongs to the family of the well-known ceramic injection moulding process. Before sintering the injection moulded green core, the organic content has to be removed, e.g. by thermal burning off or solvent debinding.

The injection moulding process generally enables the production of complex core geometries. However, the disadvantages are a deterioration in the magnetic properties and the cost. Plastic-bonded ferrites generally have very low initial permeabilities ($\mu_i = 5$–20), high losses (factor 10–50) by medium costs (factor 2) compared to standard ferrite materials. In the case of the injection moulded ferrite, the magnetic performance is in the range of the standard ferrite. However, the complicated process results in high costs (factor 3) compared to the standard ferrite.

1.2 Physical and chemical principles of ferrites

Ferrites belong to the oxide ceramic materials. These materials consist of crystallites and generally are polycrystalline. The crystallites in ceramic materials have a lattice structure in which the respective atoms have fixed places. Basically, the atomic nucleus and electrons are firmly connected to each other. Thus, unlike metals, there is no conduction band filled with freely moving electrons. The boundary line between adjacent crystallites is called the 'grain boundary'.

Many ceramics consist of elements of strongly different electronegativity, i.e. of different tendency to capture electrons from the shells of other atoms. This leads to the formation of ions. The ions are very tightly bound to each other by electrical forces. The crystal structure, i.e. the arrangement of the ions in the unit cell, forms in such a way that there is an optimum packing density, since this is energetically most favourable.

1.2.1 Ferrimagnetism

A distinction is made between diamagnetic, paramagnetic and ferromagnetic materials. Diamagnetic materials (zinc, water, superconductors) have a lower magnetic conductivity than vacuum ($\mu_i < 1$). Paramagnetic materials (aluminium, air) conduct the magnetic field lines slightly more than vacuum ($\mu_i > 1$). Ferromagnetic materials (iron, nickel, amorphous metals) conduct the field lines up to 10^5 times more than vacuum or air ($\mu_i >> 1$) and are magnetized in a magnetic field. If a clear magnetization remains after the magnetic field has been removed, the material is said to be hard magnetic. If, instead, the magnetization largely decays again, the material is soft magnetic.

In contrast to ferromagnetic metals, soft magnetic ferrites are ferrimagnetic. In metals, the resulting atomic dipoles are aligned parallel to each other. In ferrite materials, the two sublattices of the spinel crystal (see Section 1.2.2) result in partial compensation of the dipoles. This special magnetic order is called ferrimagnetism, which is often categorized as ferromagnetism.

In order to keep the scattered field energy low, Weiss domains are formed within a crystallite, a part of which are orientated in opposite directions (a domain contains 10^{12}–10^{15} atoms). The area in which the continuous change of orientation between two domains takes place is called the domain wall (Bloch wall). At low field energy (magnetization), the Bloch wall shift is reversible. However, if the field strength is increased further, the shift becomes partially irreversible due to crystal lattice defects. Reversible rotation processes occur at high field strengths [2].

1.2.2 The structure of soft ferrites

Soft ferrites are ferrimagnetic ceramic materials based on iron oxide compounds [3]. The structure is similar to an AM2X4 spinel. An ideal model of this spinel is shown in Figure 1.2. From the oxygen atoms of the oxides, the cubic surface-centred basic oxygen lattice of the ferrite is built up in the densest sphere packing [2]. This packing contains gaps, which are divided into tetrahedral gaps and octahedral gaps. If the trivalent iron ions (Fe^{3+}) are located in the octahedral gaps (B places), and if the divalent metal ions such as manganese (Mn^{2+}) and zinc ions (Zn^{2+}) are located in the tetrahedral gaps (A places) then the spinel is ideal. Otherwise, the spinel is inverse. The elementary cell is composed of 32 oxygen atoms. The cell contains 16 trivalent iron ions and 8 divalent metal ions.

In most cases, the spinel deviates from the previously presented ideal form. For example, trivalent iron ions are usually also located in the tetrahedral gaps, while

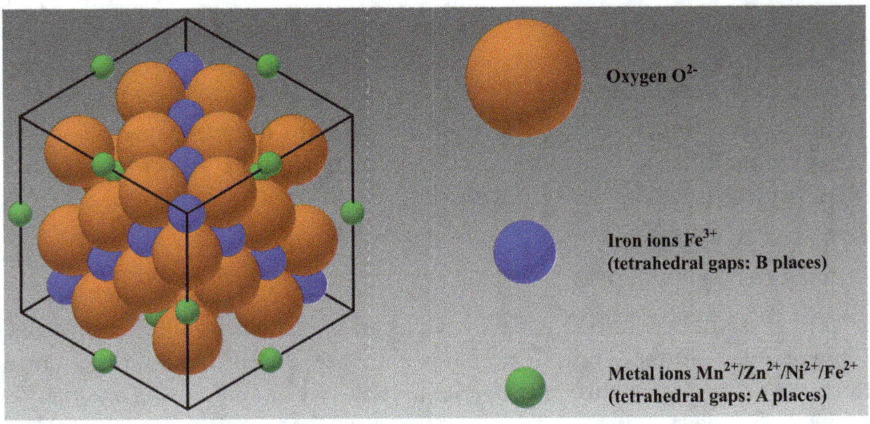

Figure 1.2 Model of an ideal AM2X4 spinel

divalent metal ions can also be found in the octahedral gaps. The degree of deviation of the spinel from its ideal form is called the degree of inversion. The degree of inversion is temperature dependent.

MnZn ferrite is unstable due to the instability of the phase MnZn and the phase magnetite (shown in the table of chemical reaction). To prevent decay during the sintering process, the unstable manganese ferrite has to be cooled under 200°C using low oxygen partial pressure, according to Figure 1.8, so that the decomposition proceeds very slowly [4]. Below 1,000°C, there should be no oxygen partial pressure.

With regard to the composition of ferrites, a differentiation is made regarding the ratio of iron oxide (Fe_2O_3). Compositions with 50 mol per cent iron oxide, in which each divalent metal ion has exactly two trivalent iron ions as partners, are called stoichiometric. Mixtures below 50 mol per cent are referred to as substoichiometric. These soft ferrites have a high resistivity. In this case, high permeability is achieved only for very high zinc contents, which, however, are responsible for a very low curie temperature. The saturation magnetostriction does not reach zero for this range of composition. Desaccommodation is very small in such ferrites because there is a low amount of cation vacancies within the material.

However, the properties of a stoichiometric soft ferrite are not useful for most applications. A mixture of more than 50 mol per cent iron oxide results in a hyperstoichiometric composition. Then, with suitable sintering, the excess ferric oxide is partly converted into magnetite (Fe_3O_4) or ferrous oxide (FeO), as is explained in Section 1.2.4, depending on the residual oxygen present.

In a hyperstoichiometric ferrite material, the amount of crystal anisotropy and magneto-crystalline energy decreases. Furthermore, the saturation flux density increases. In turn, cation vacancy concentration and desaccommodation increase due to hyperstoichiometry [4]. The high amount of iron oxide facilitates the so-called 'electron hopping', i.e. the release of electrons from the Fe^{2+} ions and their onward transmission through the crystal. The specific resistivity of the ferrite consequently decreases [5]

$$Fe^{2+} \leftrightarrow Fe^{3+} + e^- \quad (1.1)$$

1.2.3 Additives

Additives are materials or dopants that are added to the manganese–zinc–iron mixture or nickel–zinc–iron mixture to change its physical properties. The following two categories can be distinguished:

- Intercrystalline additives and
- Intracrystalline additives.

Intercrystalline additives, also known as grain boundary additives, often act as precipitates at the grain boundaries and are dosed in the ppm range. Microscopic homogeneous mixing is particularly important. This can be achieved by very fine particle sizes and long milling times. Many intercrystalline additives cause irregular grain growth when overdosed above all in the presence of impurities.

Standard intercrystalline additives are, e.g. silicon dioxide (SiO_2), which either provides, if combined with calcium oxide (CaO), an insulation layer at the grain boundaries or promotes crystal growth, depending on the quantity of its addition. The amount of additive added affects the losses of the soft ferrite material, as shown in Figure 1.3.

Another example of a standard intercrystalline additive is CaO or calcium carbonate ($CaCO_3$). As mentioned above, it forms a high-resistance phase of calcium silicate at the grain boundaries when combined with SiO_2, CaO or $CaCO_3$ suppresses grain growth.

In addition, there are pentavalent intercrystalline additives, such as niobium pentoxide (Nb_2O_5). This additive leads to irregular grain growth when overdosed. Other examples of pentavalent intercrystalline additives are vanadium pentoxide (V_2O_5) or tantalum pentoxide (Ta_2O_5).

Intracrystalline additives, also known as lattice additives, are incorporated into the spinel lattice. They are usually dosed in the per mill range. Technologically, they are therefore easier to incorporate, and the requirement for purity of the basic composition is therefore not as decisive as for intercrystalline additives. The fineness of the particles therefore not quite as important.

Intracrystalline divalent additives, such as cobaltous oxide (CoO), nickel monoxide (NiO), copper oxide (CuO), etc., take the place of the divalent Fe^{2+} ions in the ferrite crystal. Depending on which metal oxide is used, the characteristics of the ferrite and its physical properties can be modified.

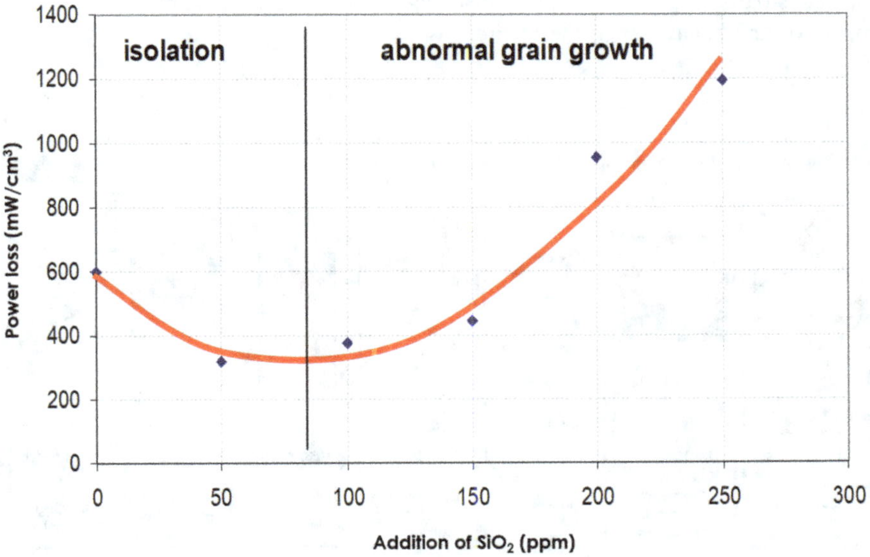

Figure 1.3 Power loss of a soft ferrite material depending on the addition of SiO_2

Intracrystalline tetravalent additives are supposed to reduce the so-called 'electron hopping' and thus increase the activation energy, resulting in a higher resistivity of the ferrite. A tetravalent additive often used is titan dioxide or tin dioxide.

In addition, there are trivalent 'additives' in the form of impurities that substitute the place of the trivalent iron Fe^{3+} ion. Such an impurity as chrome oxide strongly increases the power loss level.

1.2.4 Soft ferrite's manufacturing process

The manufacturing process of ferrite is a complicated operation, the result of this depends on numerous different parameters. It is shown as an example of MnZn ferrite in Figure 1.4 and is explained in general terms below.

In the first step, the ferrite powder is produced by mixing the main components. Depending on the desired properties of the ferrite, various additives are added. The highest accuracy is required because even small deviations from the desired mixing ratio can result in large deviations in the material properties of the ferrite.

The ferrite powder is then calcined in the range of 900–1,100°C. This serves to clean the powder because unwanted substances evaporate at high temperature. In addition, first crystallization nuclei are formed for subsequent sintering. Furthermore, this step promotes the homogenization of the mixture.

After the calcination is completed, the cooled calcinate is subjected to a milling process. The resulting median (the fifth percentile) of the particle size should be $d_{50} \approx$ 1–2 µm [7]. In addition, a narrow distribution is crucial for a homogeneous microstructure of the end product, as shown in Figure 1.5. Drum mills, among others, are used for this process. The calcinate is ground in a drum by numerous balls. The abrasion of the used aggregates and milling balls can have an impact on the behaviour of the ferrite.

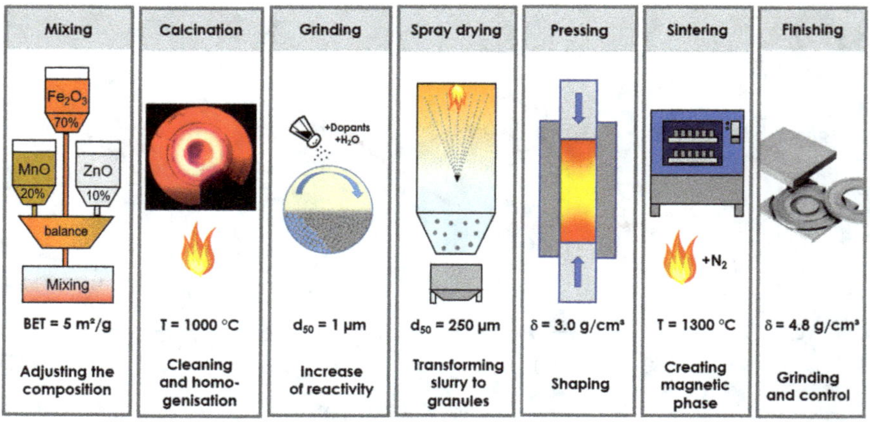

Figure 1.4 Standard manufacturing process of a MnZn ferrite (according to Ref. [6])

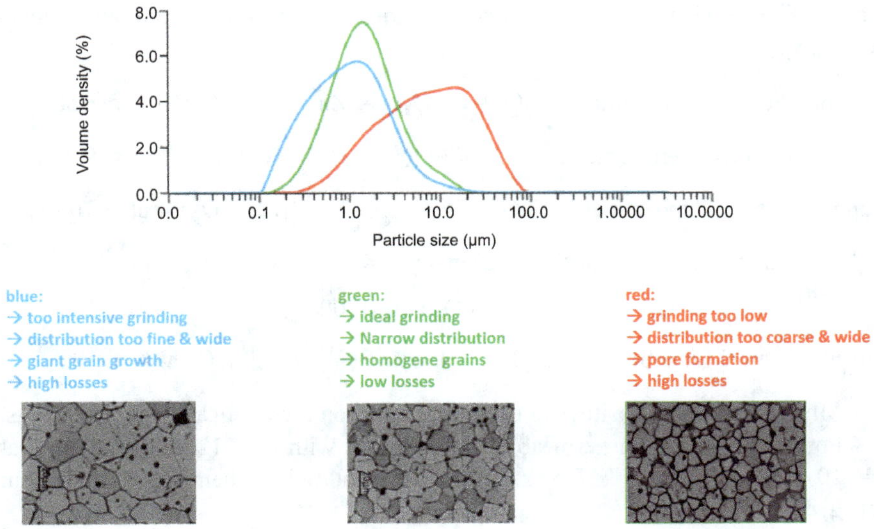

Figure 1.5 Impact of the particle size after the milling process on the microstructure of a typical MnZn power ferrite

After the completion of the spray drying process, the particle size distribution should be as small as possible (d_{50} = 80–350 μm, for example) to ensure a smooth course of further processing.

In the next step, the ferrite powder is pressed into shape. In this respect, the production of a large number of different core shapes is possible – even off the standard core geometries. A pressing tool must be manufactured for each core geometry. The specific pressing force in the production of ferrite materials is usually between 1,000 and 2,000 kg/cm^2. The pressed but not yet sintered ferrite core is called green core.

After pressing, the green cores are sintered. The shrinkage is about 20 per cent during the process. The driving force in the sintering process is the extraordinarily high surface energy of the pressed powder [3]. This means that the reactivity depends on the specific surface of the powder as well as the sintering temperature. The sintering process is divided into the following three phases, which are explained in more detail below:

- Heating phase,
- Holding phase and
- Cooling phase.

During the heating phase, the atmosphere in the sintering chamber is heated to up to the range of 1,350°C. Up to 500°C and with the addition of oxygen, a decarburization process of the green compacts takes place. As the temperature continues to rise from 600 to 800°C, the zinc ferrite is formed. Finally, at approx. 1,000°C, manganese- and iron-ferrites are formed, which diffuse into the zinc ferrite. The associated chemical reaction equations are simplified and shown below [8]:

$$\text{Zinc-Ferrite(Franklinite)}: ZnO + Fe_2O_3 \rightarrow Zn^{2+}\left[Fe_2{}^{3+}O_4{}^{2-}\right] \text{ at} \approx 750°C$$

$$\text{Nickel-Ferrite(Trevorite)}: NiO + Fe_2O_3 \rightarrow Ni^{2+}\left[Fe_2{}^{3+}O_4{}^{2-}\right] \text{ at} \approx 950°C$$

$$\text{Copper-Ferrite(Cuprospinel)}: CuO + Fe_2O_3 \rightarrow Cu^{2+}\left[Fe_2{}^{3+}O_4{}^{2-}\right] \text{ at} \approx 700°C$$

$$\text{Manganese-Ferrite(Jacobsite)}: 2Mn_2O_3 + 4Fe_2O_3 \leftrightarrow 4Mn^{2+}\left[Fe_2{}^{3+}O_4{}^{2-}\right] + O_2 \text{ at} \approx 1,000°C$$

$$\text{Iron-Ferrite(Magnetite)}: 6Fe_2O_3 \leftrightarrow 4Fe^{2+}\left[Fe_2{}^{3+}O_4{}^{2-}\right] + O_2 \text{ at} \approx 1,100°C$$

Mixtures of these ferrites are used depending on the application requirements.

For example, a high permeable MnZn-ferrite with μ_i = 12,000 (measured at f = 10 kHz; $B < 0.1$ mT; T = 25°C) has the following chemical composition: $Mn_{0.47}Zn_{0.47}Fe_{2.06}O_4$.

A typical MnZn-power-ferrite with specific losses of $P_v \approx 350$ mW/cm^3 (measured at 100 kHz; B = 200 mT; T = 100°C) has a typical composition (without additives): $Mn_{0.69}Zn_{0.23}Fe_{2.08}O_4$.

Furthermore, a typical high-permeability NiZnCu-Ferrite with μ_i = 1,000 (measured at f = 10 kHz; $B < 0.1$ mT; T = 25°C) has the following chemical mixture: $Ni_{0.3}Zn_{0.6}Cu_{0.1}Fe_2O_4$.

In the heating phase, the structure of the ferrite is thus formed. The oxygen partial pressure set during this phase can affect the density of the ferrite. The grain size is independent of this, as shown in Figure 1.6 qualitatively.

The holding phase serves to homogenize the ferrite grains. A relatively low oxygen content promotes the formation of iron-ferrite (magnetite Fe_3O_4). However, oxygen cannot be left out entirely because a certain amount of oxidation at the grain boundaries is necessary for this process, so that layers of high electrical resistance (insulation layers) form at the grain boundaries [4]. A high oxygen

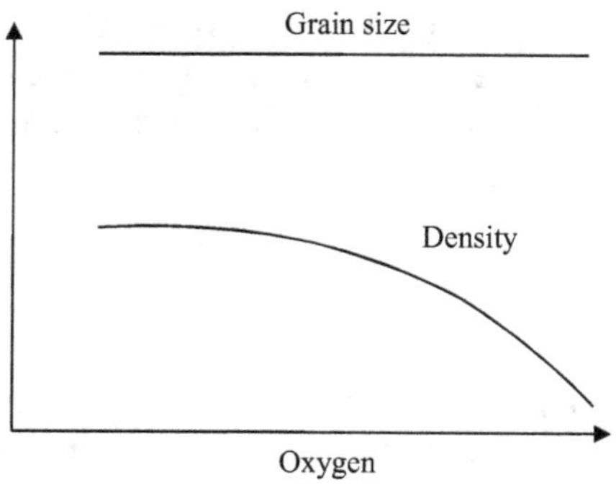

Figure 1.6 Grain size and ferrite density over the oxygen partial pressure during the heating phase (according to Ref. [8])

partial pressure has a significant effect on grain growth and, thus, the density of the ferrite material, which both increase, as shown in Figure 1.7 qualitatively. This is useful for manufacturing high-permeability MnZn-ferrites with $\mu_i > 10{,}000$.

In the cooling phase, the temperature is slowly reduced. As the temperature decreases, the reaction rate decreases [3]. The diffusion rate of the oxygen also decreases as the temperature falls [3]. In order not to obtain an oxygen gradient in the ferrite material, the oxygen partial pressure must be lowered analogously, as shown in Figure 1.8 qualitatively.

During the cooling process, a distinction is made between oxidizing cooling behaviour and reducing cooling behaviour.

An oxidizing cooling behaviour oxidizes ferro ions (Fe^{2+}) to ferric ions (Fe^{3+}). In this process, the electrical resistance in the material increases, which increases the hysteresis losses and desaccommodation. However, the eddy current losses are reduced. The boundary layers of high electrical resistance lower the permeability via the mechanism of the internal shearing. In case of a reduced cooling process,

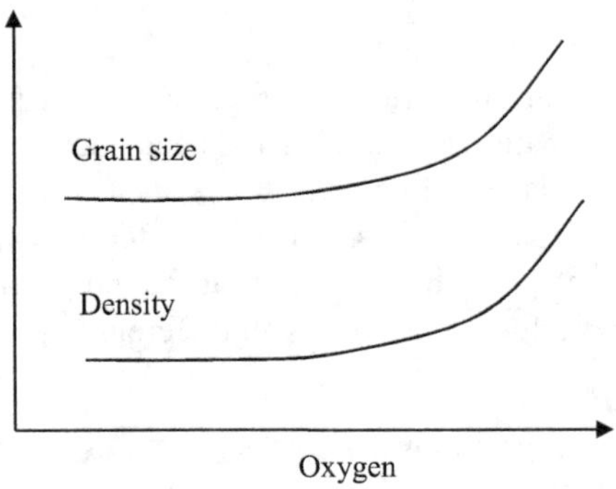

Figure 1.7 Grain size and ferrite density over the oxygen partial pressure during the holding phase (according to Ref. [8])

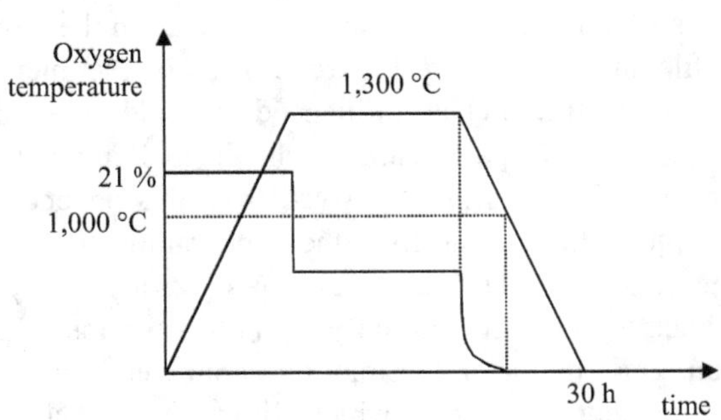

Figure 1.8 Temperature and oxygen profile during the sintering process (according to Ref. [8])

the lower oxygen content of the atmosphere prevents oxidation during the cooling period. The higher percentage of ferric oxide remains as magnetite (Fe_3O_4). The anisotropies and thus the hysteresis losses are thereby reduced. The eddy current losses are enhanced by the increased conductivity due to the ferro ions [7]. The Fe^{2+}/F^{3+} ratio can be subsequently adjusted with different cooling behaviour. This allows some electrical parameters to be fine-tuned in case of an inaccurate weighing of the raw materials.

The final step in the manufacturing process is finishing. The contact surfaces of the core halves get grinded to minimize the air gaps between them due to roughness. In some cases, the opposing surfaces of the contact faces are also grinded to simplify the grinding process. Otherwise, a defined air gap can be grinded in the centre leg in order to avoid saturation. Subsequently, the cores can still be individually machined, if necessary. For example, different shapes of the air gaps can be grinded in order to get a special characteristic of the permeability in dependence of the magnetic field strength. After quality control, the finished ferrite core can be incorporated into a magnetic component.

1.2.5 Physical characteristics

In this section, the physical properties of ferrites are discussed. In particular, various types of loss of the material as well as its mechanical stability and behaviour under thermal shock are discussed below.

The specific power losses P_L of soft ferrite materials can be divided into the following loss components, which all depend on the temperature T, the oscillating frequency f and the ripple flux density $\widehat{B}$ with different exponents:

- Hysteresis losses P_h,
- Eddy current losses at grain level P_e,
- Dielectric losses P_ε,
- DC bias losses P_{dc} and
- Residual losses P_r.

Hysteresis losses are defined as losses due to the change of magnetic moments within the ferrite crystal structure over time. They are the sum of crystal anisotropy losses and magnetostriction losses. They mainly depend on the basic composition, on the oxygen profile during sintering (Figure 1.8) and on the microstructure of the ferrite core. The microstructure is influenced by additives, milling process (Figure 1.5) and the sintering parameter, as shown in Figures 1.6 and 1.7. The parameter of an energy density $K(T,\sigma)$ derived from this depends on the internal and external mechanical stress σ and from the temperature [9].

Eddy current losses at the grain level are strongly dependent on the temperature-dependent specific conductivity $\sigma(T)$ of the grains. This is determined by the composition, grain diameter D, sintering parameter (Figure 1.8) and added intracrystalline tetravalent additives, which limit the ‘electron hopping’, as explained in Section 1.2.2. The eddy currents attributed to this type of loss are formed only within the grain boundaries of the ferrite material, i.e. they are limited by the insulating layer between the individual grains. These losses are therefore not dependent on the core geometry.

Dielectric losses occur mainly in large-scale cores. They are the sum of conductivity losses and polarization losses and can be considered as eddy current losses across the grain boundaries (Figure 1.3). These losses can be expressed with the loss modulus $\varepsilon''(T,f)$, which depends on the frequency and temperature. They are strongly dependent on the core geometry, in particular on the shape and size of the core cross section A_e. A detailed description is presented in Ref. [10].

The DC bias losses are caused by the pre-rotation of the dipoles caused by a constant external magnetic field. These losses are influenced by the type of material, the DC flux density B_{DC}, the ripple flux density, the frequency and the temperature [11].

The residual losses mainly have an effect at higher frequencies in the range of MHz. In some cases, the dielectric losses, which are shown here as separate losses, are counted as residual losses. Residual losses also include losses caused by a phase shift between the external magnetic field and the oscillating magnetic moments at high frequencies. This also includes the relaxation effect caused by the interaction of the oscillating magnetization with diffusing ions. The residual losses are also dependent on the ripple flux density, the frequency and the temperature [12,13].

In the following, the individual losses are each totalled with a proportionality constant k in order to show their physical character as a first approximation:

$$\begin{aligned} P_L &= P_h + P_e + P_\varepsilon + P_{DC} + P_r \\ &\approx k_h K \widehat{B}^3 f + k_e \sigma D^2 \widehat{B}^2 f^2 + k_\varepsilon \varepsilon'' A_e \widehat{B}^2 f^3 + k_{DC} B_{DC} \widehat{B}^{1..2} f + k_r \widehat{B}^{2...3} f^{3...5} \end{aligned} \quad (1.2)$$

Unfortunately, it has not yet been possible to describe the manufacturer's data with sufficient accuracy using such a simple relationship with a few material-specific constants. For this reason, the losses are calculated in practice using simple empirical equations (Steinmetz) with the help of defined material constants for each section. Alternatively, the losses can be calculated on the basis of digitized measurement data tables with the aid of an interpolation algorithm (mapping).

Ferrites belong to the oxide ceramic materials. Analogous to ceramics, they are very brittle and thus susceptible to mechanical stress. As a rule, the endurable tensile strength of soft ferrites is lower than their compressive strength by a factor of up to 20. However, it should be noted that a subcritical crack growth appears already below the breaking strength, which significantly reduces the mechanical strength. Ferrite cores under mechanical stress show deviating physical behaviour compared to an equivalent unloaded core. Thus, the permeability of the material may change in complicated way, as shown in Figure 1.9.

Furthermore, the core losses of MnZn ferrites can slightly decrease and increase with multiple factors in dependence of the direction and amount of mechanical stress, as shown in Figure 1.15.

As Figure 1.10 shows, the mechanical strength of soft ferrite can be significantly increased by changing some process parameters. However,

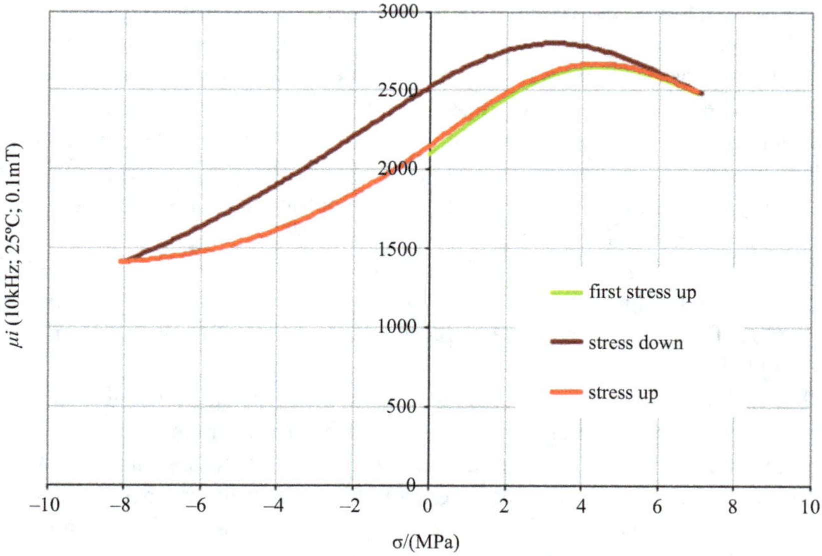

Figure 1.9 *μ(σ)-hysteresis of a typical power MnZn ferrite $Mn_{0.69}Zn_{0.23}Fe_{2.08}O_4$ (according to Section 1.2.4) with respect to an applied external force. The negative forces indicate compressive (tensile) stress which is oriented parallel (orthogonal) to the flux orientation. The positive forces indicate compressive (tensile) stress which is oriented orthogonal (parallel) to the flux orientation [14].*

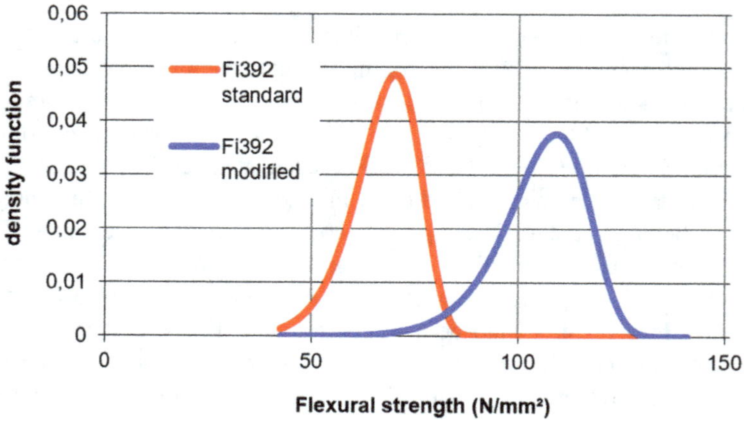

Figure 1.10 *Modified ferrite material for higher mechanical strength*

modifications of this kind are always associated with losses in other physical characteristics of the core material. Thus, in this specific example, the core losses of the modified Fi392 increase by approximately 50 per cent compared to the original material.

1.2.6 Differences between MnZn ferrites and NiZn ferrites

The statements made in the previous sections using the example of MnZn ferrites can be transferred to NiZn ferrites without difficulty. In this section, however, NiZn ferrites are examined in more detail and the differences to MnZn ferrites are worked out.

Figure 1.11 shows the characteristics of different core materials for a specified loss density. From this, the different applications of MnZn ferrites and NiZn ferrites can be derived. Due to their higher specific resistance and the associated reduced eddy current losses within the grains of the core material, NiZn ferrites are particularly suitable for high-frequency applications. In addition, the high resistivity ensures a certain insulation strength of the ferrite material, making it suitable for applications with high voltage values.

The increasing specific resistance is accompanied by a lower saturation flux density. Thus, NiZn ferrites can be used at high voltage values, but the voltage/current changes occurring in the application must not be too large in order to avoid high losses. Furthermore, NiZn ferrites exhibit hysteretic losses increased by a factor of up to three compared to MnZn ferrites.

The high resistivity of NiZn ferrites is mainly achieved by the absence of manganese ions and by nearly ideal stoichiometry. The latter significantly reduces the so-called 'electron hopping', which results in high resistivity and low eddy current losses at a grain level.

NiZn ferrites are a good basis for using the so-called 'perminvar effect'. Appropriate composition of the ferrite material results in a narrowing of the hysteresis, which is normally linear for small amplitude modulations. This constriction of the hysteresis, which is like a wasp's waist, significantly reduces hysteresis losses. This effect is achieved by a kind of fixation of the domain walls within the

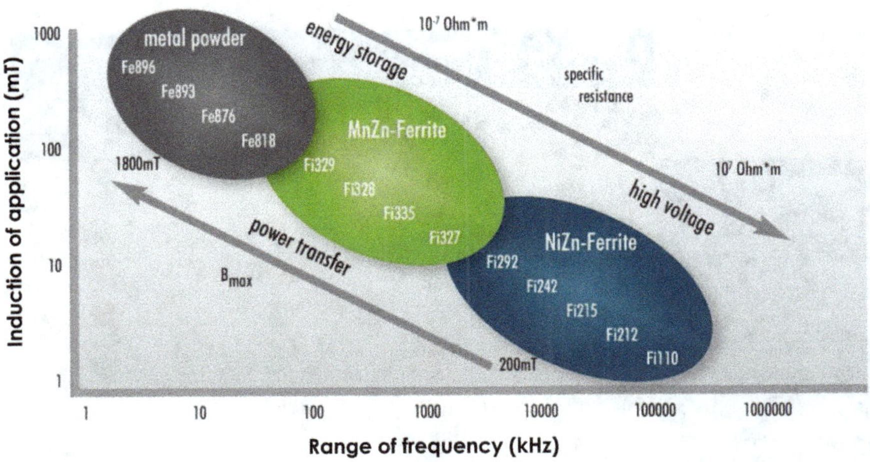

Figure 1.11 Characteristics of different core materials for a specified loss density [15]

ferrite. If the induction of the material becomes too large, this fixation is loosened, which leads to a considerable increase in hysteresis losses. Re-fixation of the domain walls in conventional operation is then no longer possible. The effect can only be restored by heating the ferrite material above its curie temperature or by a decreasing oscillating magnetic field strength.

Two mainly used additives used for NiZn ferrites are cobalt oxide (CoO) and CuO. CoO promotes the perminvar effect, but also lowers the permeability of the NiZn ferrite. Among other things, copper oxide promotes crystal growth in NiZn ferrites and thus acts as a sintering aid. This results in a lower top temperature during the sintering process.

1.3 Selection of the appropriate soft ferrite material in high-frequency applications

When designing magnetic components, numerous conditions and specifications have to be taken into account. Conversely, the developer has several degrees of freedom to meet the specification. However, various data and parameters have to be considered and compared. This section is intended to provide assistance in the selection of materials and the handling of datasheets. The contents presented are illustrated by means of an application-related example.

At the beginning of the design of a magnetic device, a clear specification should be available. In addition, it should be clear which component characteristics are to be prioritized. For example, is it a device for automotive applications, where the focus usually is on low costs, or is the application in the aerospace industry, where volume and weight tend to be more important?

Based on the given specification, a first basic selection of the magnetic material can be made. Figure 1.12 shows the characteristics of the different

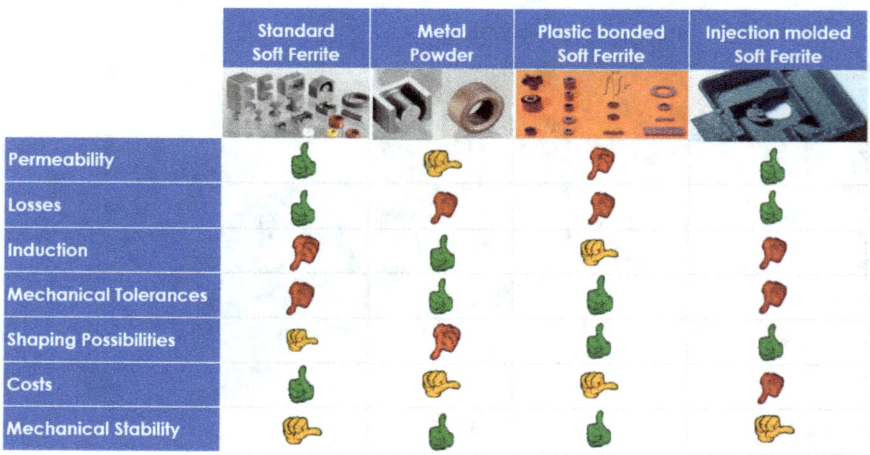

Figure 1.12 Characteristics of different high-frequency soft magnetic materials

magnetic materials for high-frequency applications. It can be clearly seen that the materials differ widely in their properties and thus also in their areas of application. This basic selection therefore points the way for the further design process.

Once the fundamental type of magnetic material is selected, a further restriction can be made based on the application of the magnetic device to be designed. Depending on whether this is to be a storage inductor, a filter inductor, a transformer, etc., some specific magnetic materials can be excluded in advance. Then, a comparison within the remaining group of materials and a selection of a specific material can be made. The following parameters in particular are decisive for this:

- Frequency,
- Current resp. current ripple,
- Available mounting space,
- Operating temperature and
- Inductivity.

For the sake of completeness, it has to be mentioned that the criteria of magnetostriction can result in an audible noise in some kHz-range applications. Furthermore, electromagnetic compatibility (EMC) usually plays an important role with regard to material selection in numerous applications. Since both topics are complex and would go beyond the scope of this section, they will not be discussed further in the following.

Now, the datasheets of the magnet materials have to be compared. The magnetic field strength can be determined from the specification via the current, the targeted number of turns and the effective length of the magnetic path. The maximum flux density can be determined from this in combination with the available mounting space and the derived magnetic cross section of the core. This should be well below a maximum of 80 per cent of the saturation flux density of the material specified in the datasheet. Depending on the core geometry, local increases in magnetic flux density may occur. These can only be detected to a limited extent by the simple calculation presented above. For this reason, the use of three-dimensional (3D) finite-element method (FEM) simulation software to support the design is highly recommended.

From the data obtained so far, the core losses can now be determined in conjunction with the frequency and the targeted core volume. The specific power loss functions in Figure 1.13a serve this purpose. Also, at this point, it should be said that support by means of 3D FEM simulation software is very helpful and useful in the calculation of core and winding losses. For example, the dielectric losses of the core material are usually not included in the power loss curves of the material datasheets because their measured values are determined on small toroidal cores. Thus, significant deviations between the calculation of the core losses from the datasheets and the results of the 3D FEM simulation can occur, especially when large-scale cores are used. That's why the guaranteed specific power loss values related to the desired core are 30 per cent up to 300 per cent higher than the values from the datasheets, depending on the core shape and manufacturer.

The temperature rise can now be calculated from the core losses and the additional winding losses. It should be noted that the temperature of the soft

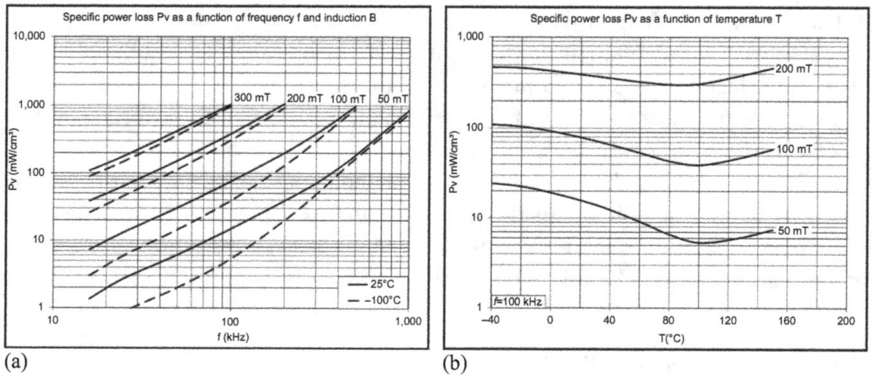

Figure 1.13 Specific power loss curves of the magnetic material Fi395 [16]: (a) depending on frequency and induction and (b) depending on temperature and induction

Table 1.1 Specifications of the storage inductor

Inductance (μH)	41
Maximum peak current (A)	44.5
Nominal peak current (A)	31
RMS current (A)	25.6
Frequency (kHz)	100
Input voltage (V)	375–400
Maximum volume (cm^3)	<40
Coolant temperature (°C)	60

magnetic material should not exceed its loss minimum in the case of ferrites (Figure 1.13b), because otherwise there is a risk of thermal runaway if cooling is insufficient.

The procedure described above is illustrated in the following application example.

A storage inductor for a bidirectional DC–DC converter is to be designed. Its specifications are listed in Table 1.1.

The design results in a PQ 50/50 core as the best core geometry. The core has an effective cross section of $A_e = 332\ \text{mm}^2$. The number of turns is set to $N = 18$. From this follows for the maximum flux density:

$$B_{\text{max}} = \frac{L * I_{\text{peak,max}}}{N * A_e} = \frac{41\ \mu\text{H} * 44.5\ \text{A}}{18 * 332\ \text{mm}^2} = 315\ \text{mT} \tag{1.3}$$

Based on the requirements, two materials are selected for the worst case (maximum flux density). Fi395 is a low loss power ferrite with a flat power loss curve for frequencies up to 400 kHz. It should be mentioned that the calculated

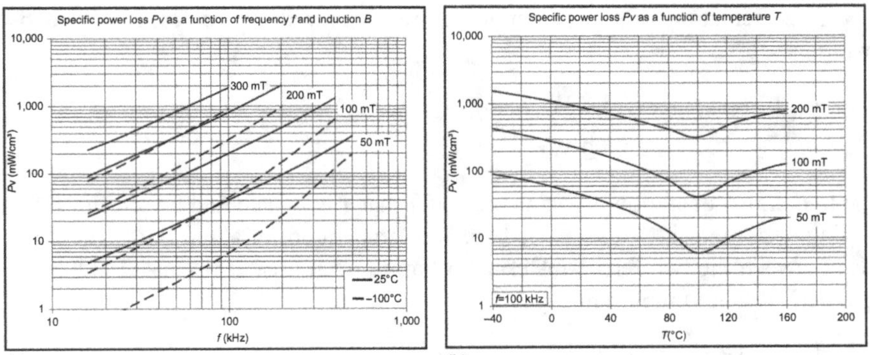

(a) (b)

Figure 1.14 Specific power loss functions of the magnetic material Fi329 [17]: (a) depending on frequency and induction and (b) depending on temperature and induction

maximum flux density in this case is above the 80 per cent value of the maximum flux density of the material. For this reason, the second material selected is Fi329, a high saturation power ferrite for frequencies up to 200 kHz. The specific power loss functions of Fi395 are shown in Figure 1.13, whereas the functions of Fi329 are shown in Figure 1.14.

Calculating the losses using the power loss curves of both materials (Figures 1.13a and 1.14 a)) gives the following results for the volume of the PQ 50/50 core ($V_e = 37.6 \ \ \mathrm{cm}^3$):

$$P_{Core,Fi395,100^\circ C} = 41.7\ W$$

$$P_{Core,Fi329,100^\circ C} = 40.8\ W$$

As already mentioned, the values are the core losses which the cooling system must at least be able to dissipate in the worst case in order to keep the core at a temperature of 100°C. It should also be noted that the copper losses of the component are not yet included in this value.

The comparison of the calculated losses shows that Fi329 is better suited for the worst case than Fi395. From Figure 1.13b, it can be seen that the loss minimum for Fi395 tends to be at 80°C at a flux density of 300 mT. For this temperature, the core losses decrease to $P_{Core,Fi395,80^\circ C} = 39.8$ W. The loss minimum of Fi329 is at 100°C, which is why the value does not decrease further.

In addition to the worst case, the nominal operation of the component should also be investigated, because it is operated around this operating point most of the time. The nominal ripple peak current is $I_{peak,nom} = 31\mathrm{A}$. That leads to the following flux density:

$$B_{nom} = \frac{L * I_{peak,nom}}{N * A_e} = \frac{41\ \mu H * 31\ \mathrm{A}}{18 * 332\ \mathrm{mm}^2} = 213\ \mathrm{mT} \tag{1.4}$$

A recalculation of the power losses results for the core losses of the two materials:

$$P_{Core,Fi395,100^{\circ}C} = 13.8 \text{ W}$$

$$P_{Core,Fi329,100^{\circ}C} = 14.3 \text{ W}$$

In nominal operation, Fi395 is the better material and should be selected. If the cooling system is able to keep the core at a temperature of about 80°C, the difference between the two ferrite materials becomes even greater because Fi329 is not operated at its loss minimum:

$$P_{Core,Fi395,80^{\circ}C} = 13.6 \text{ W}$$

$$P_{Core,Fi329,80^{\circ}C} = 18.8 \text{ W}$$

However, when selecting the ferrite material, there are some pitfalls that should be considered to ensure that the design of the magnetic device is successful. The most important ones are briefly listed and explained below:

An important point is the perminvar effect, already explained in Section 1.2.5, which tends to be used, but not exclusively, in NiZn ferrites. By constricting the hysteresis curve, hysteresis losses can be significantly reduced. However, if the material is steered out to the point where the domain walls become detached, this results in significantly larger core losses than estimated in the design. This can lead to a thermal runaway of the magnetic component if cooling is insufficient.

Another important impact factor that has to be taken into account when designing a magnetic component is mechanical stress that acts on the core material, e.g. due to the potting or the screw connection in combination with thermal expansion. Not only can thermal expansion alone cause the potting or bolted core to fracture, but the stress as mentioned above also affects the magnetic properties of the core material called inverse magnetostrictive effect, as shown in Figure 1.15. Multi-physical 3D FEM simulations can be used to test this impact.

Lastly, it is important to mention that the cooling system as a whole has to be considered. Thus, local hotspots can occur in the core as well as in the winding. The heat there has to be able to be dissipated by the cooling system – either by a correspondingly large thermal gradient or by direct cooling of these areas. Particularly in the case of cores with distributed air gaps, it is useful to connect the individual ferrite discs in the leg with a thermal connection in order to avoid local hotspots. Here, too, 3D FEM simulations can support this to a considerable extent.

Adequate cooling is important because especially some cobalt-doped MnZn-ferrite materials age thermally. The higher the temperature at which they are loaded, the faster the aging process takes place. As a result, core losses can increase over time, which leads to an increase in temperature and an acceleration of the ageing process if cooling is inadequate. The speed of the ageing process can be counteracted by appropriate doping. However, the effect of thermal ageing can only be included in the design to a limited extent because most manufacturers do not publish any data on this. Only on the basis of the datasheets, this effect is not calculable.

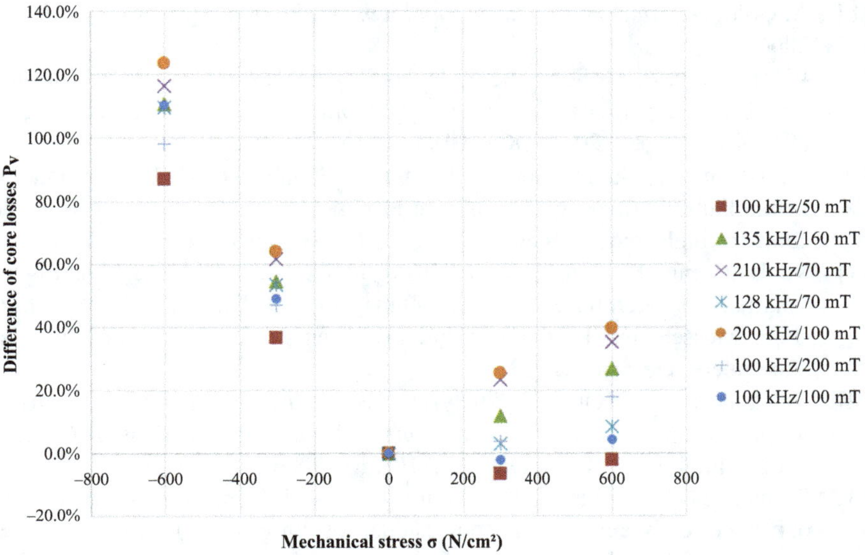

Figure 1.15 Relative power loss density of Fi335 measured at 25°C with respect to an applied external force. The negative forces indicate compressive (tensile) stress, which is oriented parallel (orthogonal) to the flux orientation. The positive forces indicate compressive (tensile) stress which is oriented orthogonal (parallel) to the flux orientation [18].

References

[1] M. J. Ruthner. "Desired development aspects for spray roasted iron oxides predominantly for the production of ferrites". *Journal of the Japan Society of Powder and Powder Metallurgy*, vol. 61, issue S1, pages: S183–S188. Japan. 2014. DOI: 10.2497/jjspm.61.S183.

[2] E. Ivers-Tiffée and W. von Münch. *Werkstoffe der Elektrotechnik*. 10th Edited and Extended Edition. B.G. Teubner. Wiesbaden. 2017.

[3] H.-J. Bargel and G. Schulze. *Werkstoffkunde*. 11th Edited Edition. Springer. Berlin. 2012. DOI: 10.1007/978-3-642-17717-0.

[4] W. Kampczyk and E. Röß. *Ferritkerne: Grundlagen, Dimensionierung, Anwendungen i. d. Nachrichtentechnik*. Siemens AG. Germany. 1978.

[5] E. P. Wohlfarth. *Ferromagnetic Materials*. 2nd Edition. North-Holland. Amsterdam. 1981. DOI: 10.1002/crat.19810160127.

[6] M. Baumann, C. Drexler and J. Schültzke. "Improved magnetic materials for high frequency applications and their limits". *ECPE Workshop: Magnetic Components in Power Electronics*. Grenoble, France. 2020.

[7] E. C. Snelling. *Soft Ferrites: Properties and Applications*. Butterworth-Heinemann. Oxford. 1989.

[8] A. Goldman. *Modern Ferrite Technology*. 2nd Edition. Springer. New York. 2006.

[9] W. H. Jeong, B. M. Song and Y. H. Han. "Analysis of power losses in Mn–Zn ferrites". *Japanese Journal of Applied Physics*, vol. 41, pages: 2912–2915. Japan. 2002. DOI: 10.1143/JJAP.41.2912.

[10] M. Baumann, Ch. Drexler, J. Pfeiffer, J. Schueltzke, E. Lorenz and M. Schmidhuber. "Investigation of core-loss mechanisms in large-scale ferrite cores for high-frequency applications." EPE. Hanover, Germany. 2022.

[11] M. Baumann, E. Lorenz, J. Schültzke and C. Drexler. "Relative Eigenschaften von Kernmaterialien (weichmagnetische Ferrite)". ECPE Workshop: Messen und Prüfen von Induktivitäten in der Leistungselektronik. Nuremberg, Germany. 2023.

[12] S. Yamada and E. Otsuki. "Analysis of eddy current loss in Mn-Zn ferrites for power supplies". *IEEE Transactions on Magnetics*, vol. 31, issue 6, pages 4062—4064. USA. 1995. DOI: 10.1109/20.489862

[13] S. Yamada, T. Chiba, K. Kondo and E. Otsuki. "Analysis of temperature dependent eddy current loss and residual loss in Mn–Zn ferrites for power supplies". *Proceedings of the 8th International Conference of Ferrites (ICF)*, pages 556–560. Tokyo, Japan. 2000.

[14] M. Baumann, S. Sortier and J. Weidinger. "A model of initial permeability of soft ferrite in dependence of biaxial stress". *Journal of the Jpn. Soc. of Powder and Powder Metallurgy*, vol. 61, issue S1, pages: S70–S74. Japan. 2014. DOI: 10.2497/jjspm.61.S70

[15] H. Jungwirth, M. Schmidhuber, M. Baumann and M. Schmeller. "A new high frequency ferrite material for GaN applications". PCIM. Nuremberg, Germany. 2016.

[16] SUMIDA Components & Modules GmbH. "Magnetic material Fi395". Datasheet. Obernzell, Germany. 2022.

[17] SUMIDA Components & Modules GmbH. "Magnetic material Fi329". Datasheet. Obernzell, Germany. 2018.

[18] SUMIDA Components & Modules GmbH. J. Schültzke. Internal Document Physical Laboratory.

Chapter 2

Metallic magnetic materials

Peter Zacharias[1], Jens Friebe[1] and Christian P. Dick[2]

2.1 Introduction

Metallic materials have long been used in inductive components. Almost the majority of the materials used are iron-based. Alloy components and manufacturing steps are used to give them the properties that make them interesting for electrotechnical applications [1]. The desired properties are

- high saturation induction,
- low magnetization losses, and
- low-temperature dependency.

In addition, for example, permeabilities that are as constant as possible across various parameters (e.g. temperature, excitation, etc.) are also sought. Figure 2.1 shows a comparison of different materials using selected examples for the highest possible frequencies at limited specific core losses. The saturation induction at 100 °C is shown on the horizontal axis of the diagram. On the vertical axis, the maximum achievable frequency is plotted at core losses of 300 mW/cm^3 with sinusoidal core excitation at B_{max} = 0.1 T, resulting in a core temperature of 100 °C. The black dots represent samples of power ferrites. These are characterized by high achievable frequencies. However, they also have disadvantages. In particular, the strong dependence of the saturation induction and the losses on the working temperature should be mentioned (Section 2.1). Metallic materials in sheet or powder form allow much higher saturation inductions to be achieved. However, the specific losses are higher with complete or partial magnetization than with ferrites. High saturation inductions enable low numbers of turns and, therefore, small components and more degree of freedom in the winding design. Because a further limit is imposed by the maximum temperature inside the components, high losses act in the direction of large component dimensions. The heat must be dissipated via parts of the surface. This usually leads to components with an enlarged surface area or special cooling measures.

Figure 2.1 shows that cores made of electrical steel are usually only used for low frequencies. There are still many applications for this. It should be remembered

[1]University of Kassel, Department of Power Electronics, Germany
[2]TH Köln – University of Applied Sciences, Institute of Automation Engineering, Germany

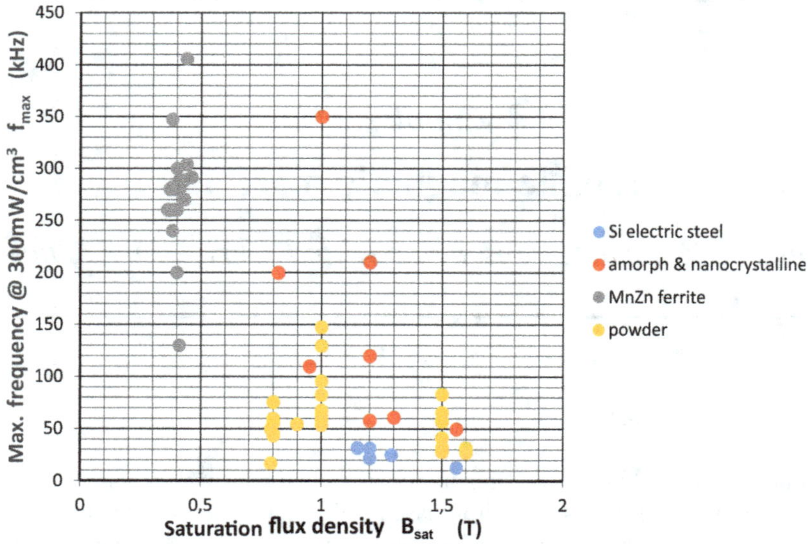

Figure 2.1 Comparison of different magnetic core materials with regard to their saturation induction B_{sat} at 100 °C and the achievable operating frequency at B_{max} = 0.1 T at 100 °C. (Data: TDK/EPCOS, Sumida, Ferroxcube, Vacuumschmelze, Magnetics, Micrometals, Hitachi, Thyssen, JFE Steel.)

that even an electric motor with a pole pair with 15 000 min^{-1} in a passenger car only has a rotating field of 250 Hz. The high switching frequencies used in a drivetrain only lead to partial magnetization with correspondingly low losses. A frequency of 400 Hz is used in aircraft, which also allows the use of laminated materials for both inductor cores and machine rotors and stators. However, significantly higher frequencies are often used in power electronic energy converters in order to keep device volume, weight and thus costs low. As the red dots in Figure 2.1 show, there are special metallic core materials that combine a high saturation induction with low magnetization losses. This means that they can also be used in the higher frequency range.

2.2 Use of electrical steel sheets

This section focuses on cores made from layered sheets or coiled strips. Sheets can be produced in certain minimum sheet thicknesses by rolling. There are also limits to the processing of sheet metal by punching. Thick sheets are stacked or coiled to achieve larger cross sections. The individual layers are insulated from each other such that induced currents cannot spread beyond the boundaries of a sheet layer. The thinner the material, the lower the induced voltage that the insulation between the layers has to withstand. Insulation by a thin oxide layer on one of the two surfaces is therefore usually sufficient for very thin strips. For thicker sheets, a

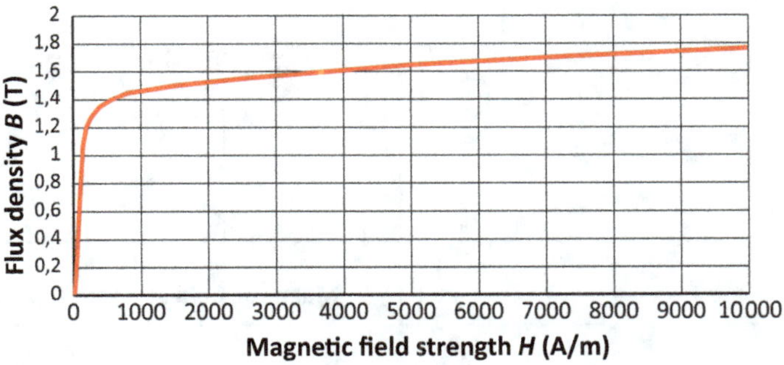

Figure 2.2 Flux density characteristic B(H) for a non-grain-oriented electrical steel sheet M400-50A (based on data from [2])

coating of insulating organic or inorganic materials is used. As a result, the geometrically available cross section for the core is only filled with a fill factor of around 95%. In pulse power technology applications, very high winding voltages up to the kilovolt range are present for a brief moment. The layer insulation of the tape cores must be able to withstand the resulting induced voltages. Thin Hostaphan films, for example, are used as layer insulation. Ferrites suffer an internal thermal breakdown in such applications featuring kilovolt winding voltages.

If the magnetic field strength in an electrical steel sheet is increased, the magnetic flux density initially increases rapidly (Figure 2.2). At high magnetic field strengths, the material is driven into saturation and the increase in the $B(H)$ curve drops to

$$\lim_{H\to\infty}\left(\frac{dB}{dH}\right) = \mu_0 = 4\pi\cdot 10^{-7}\frac{\mathrm{Vs}}{\mathrm{Am}} \tag{2.1}$$

This relationship also applies in a vacuum. Magnetization curves are therefore also given as a dependence of the magnetic polarization on the magnetic field strength. In this case, the proportion of material-free space is excluded. The relationship between the magnetic flux density $B(H)$ and the magnetic polarization $M(H)$ is as follows:

$$B(H) = M(H) + \mu_0 \cdot H \tag{2.2}$$

If the magnetization is measured periodically in the interval $\{-B_{\max}; B_{\max}\}$, an ambiguous curve is obtained, as shown in Figure 2.3, even if the magnetization speed is arbitrarily low. This is generally known as a hysteresis curve. The area of this curve has the unit $\mathrm{VAs/m^3}$. This means that a certain amount of energy is required per volume for a complete remagnetization process. This process can be interpreted as 'internal friction'. It leads to the heating of the core. This proportion of the power loss increases proportionally with the frequency f and volume V

$$P_{hys} = V\cdot f\cdot \oint B\cdot dH \tag{2.3}$$

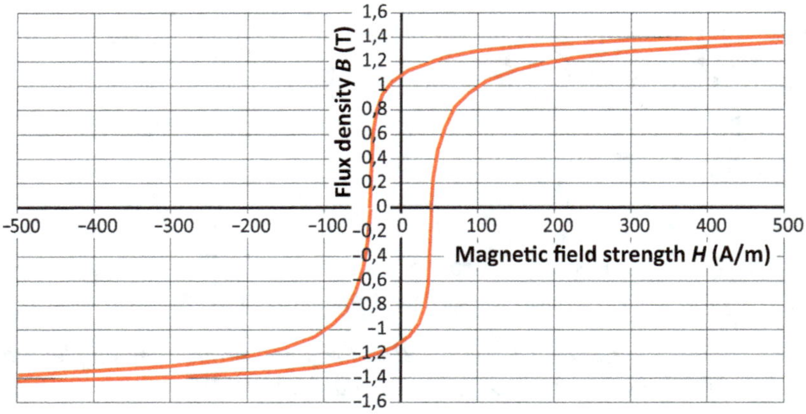

Figure 2.3 B(H) dependence of M400-50A electrical steel sheet (based on [3])

In addition to hysteresis losses, which occur when the magnetic field in the core changes and the magnetic domains in the material are realigned, there are eddy current losses in metallic cores. When the magnetic field in the core changes, eddy currents are induced. Since the induced voltage is proportional to the frequency, this loss component depends on the square of the frequency, i.e. $P_{eddy} \sim f^2$. In addition, there are so-called after-effect losses (also known as residual or viscosity losses). They arise due to the mechanical toughness of the microstructure. After switching off the excitation current responsible for the magnetic field, a change in polarization can still be observed for several seconds, for example. With high magnetization values, this proportion is almost negligible. A $B(H)$ dependence based on measurements [3] is shown in Figure 2.3.

A significant proportion of the losses that occur are eddy current losses. If only small propagation volumes are allowed for the eddy current fields, the corresponding eddy current losses are comparatively low. Technically, two approaches are used here. This is a low thickness for sheet metal and also a small particle size for metal powder. For metal strips, a thickness of 0.5 mil = 12.7 μm can be technically achieved. The magnetic properties of Si electrical steel sheets improve up to a Si content of around 6.5%. However, the material then becomes extremely brittle. It is then very difficult to machine. With layers produced using the chemical vapour desposition (CVD) process, however, 'sheets' with a thickness of 100 μm can be produced, which also have very low magnetization losses [4,5].

For powder material, the particle size is in the range of 10–70 μm, and for carbonyl iron even lower at less than 4 μm. Cores for inductive components are discussed in Section 2.3, so the topic will not be discussed further here. Powder cores represent magnetic cores with a very finely distributed air gap. This makes these cores particularly suitable for magnetic energy storage but not for transformers or the construction of a stator or rotor of electric machines. These require the highest possible effective permeability to concentrate the magnetic flux and enable high flux densities in the airgap between the stator and the rotor.

Figure 2.4(a) shows the formation of eddy current fields in the stacked sheets or strips of a core. Figure 2.4(b) shows the maximum frequencies that can be achieved for a power loss density of 0.1 W cm^{-3} in the core. In this case, a quadrupling of the material thickness leads to a halving of the maximum operating frequency. The material Orthonol used as a reference is an amorphous material with a nickel content of 50%. The manufacturing process can be used to produce very thin ribbons. Their application range is in the kHz range.

If all loss components are summarized, characteristic specific losses are obtained for the core losses of a material; typically, gravimetric or volumetric loss densities are plotted over the peak flux density of a sinusoidal excitation with frequency as a parameter. Figure 2.5 shows such a representation, for an example material. It is noticeable that the data sheet values in the diagram are grouped as a

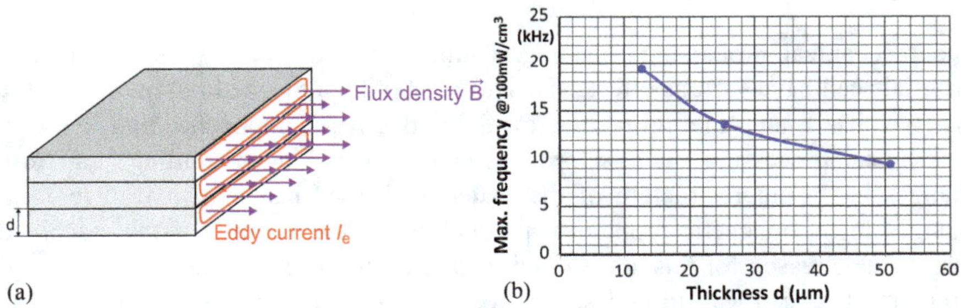

Figure 2.4 Eddy currents and their effects as a result of AC magnetization: (a) principle sketch of the formation of eddy currents, (b) achievable working frequency at a maximum induction of B_{max} = 0.1 T and a loss density of 100 mW/cm^3 for ORTHONOL (data from [6])

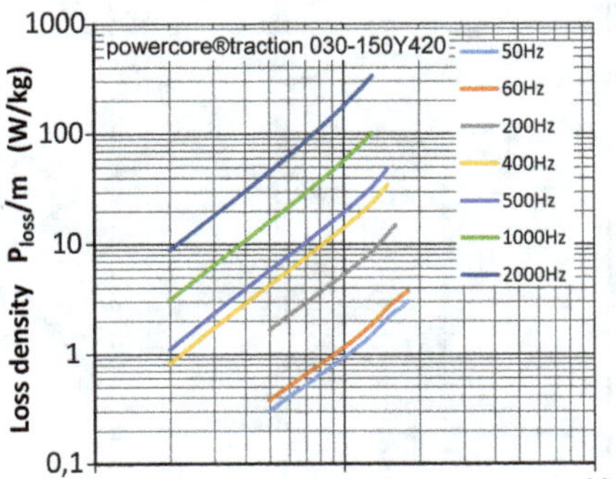

Figure 2.5 Specific power loss dependent on the maximum induction and the operating frequency for the powercore®traction 030-150Y420 electrical steel (0.3 mm thick, data: ThyssenKrupp AG)

set of parallel straight lines. As the representation is double logarithmic, this results in the form of the relationship known from the Steinmetz equation

$$p(f, B) = \frac{P}{V} = k_c \cdot \left(\frac{f}{Hz}\right)^{\alpha} \left(\frac{B_{\max}}{T}\right)^{\beta} \tag{2.4}$$

The characteristic exponents in the example of the material shown in Figure 2.5 are approximately $\alpha = 1.58$, $\beta = 1.94$. This makes it easy to estimate the losses in a component under a certain load. The electrical model of an inductive component is given here by a parallel connection of an ideal inductor and a loss resistance.

2.3 Changes in magnetic properties due to mechanical processing and heat treatment

The mechanical processing (bending, cutting, pressing, etc.) of soft magnetic materials can have various effects. This affects both the magnetic properties of the material itself and the properties of the inductive component built with it. Mechanical processing such as forming, cutting, punching or milling can cause changes in the microstructure of the material. These changes can lead to local changes in the magnetic properties, particularly in the vicinity of the machined surface. The reason for this is the modification in the state of order in the material structure. This can lead to undesirable losses or a reduction in permeability. The permeability of mu-metal/permalloy (Figure 2.6) is given as an example. It is clearly visible that cold-formed mu-metal loses a large part of its permeability. The desired state is therefore usually achieved by an annealing process.

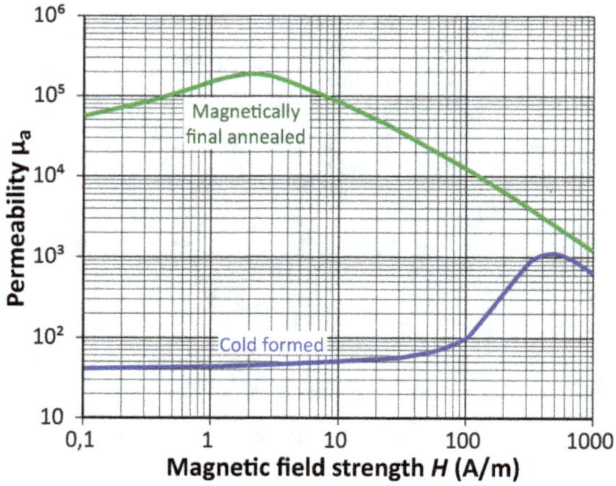

Figure 2.6 Influence of mechanical deformation on the permeability μ_a of mu-metal/permalloy (according to [7])

Since even the smallest air gaps lead to a noticeable change in the inductance in a core, a change in the magnetic properties can also be expected with each processing step to create a certain geometry. For example, certain mechanical processing methods can influence the magnetic anisotropy of the material, which affects the directional dependence of its magnetic properties, especially the permeability. This can be important if the material needs to be aligned in a specific magnetic field.

The seemingly most favourable core shape is that of a toroidal core with toroidal winding. With this core, the magnetic field in the core has approximately the same shape as it would have without a core. This results in a minimal magnetic stray field in the surroundings. Oval core shapes also have such advantageous properties. In areas where the magnetic field is forced to run around corners, the conditions are practically always different from those in the area where the magnetic field runs parallel to the edges.

Since the winding of closed toroidal cores can be difficult, so-called cut-strip cores have been developed. With these, a ferromagnetic tape is wound up. The tool for holding the core forms the subsequent winding window. Once winding is complete, the core is fixed in its shape. The core is then divided using a cutting process so that winding bodies with windings can be slid on (Figure 2.7).

M-cores can also be produced from the cut cores (C cores) for compact designs at higher power levels. The surfaces that have resulted from the cutting are particularly sensitive. Manufacturing defects and surface roughness lead to a deterioration in the conduction of the magnetic field. In addition, short circuits can occur between the layers, resulting in increased eddy current losses. Cutting is practically always associated with a loss of magnetic properties.

As described above, heat treatment has a major influence on the development of special magnetic properties of metallic materials. If cold-rolled or drawn material has to undergo further mechanical processing before the final heat treatment, soft annealing above the recrystallization temperature of approx. 700 °C–800 °C in a protective atmosphere or in a vacuum is advisable in order to facilitate processability.

A final heat treatment finally sets the optimum magnetic properties. Because all these processes take place at a certain rate of change, the duration, temperature and, in many cases, the cooling conditions must be maintained. The heat treatment processes are preferably carried out in hydrogen as a protective gas.

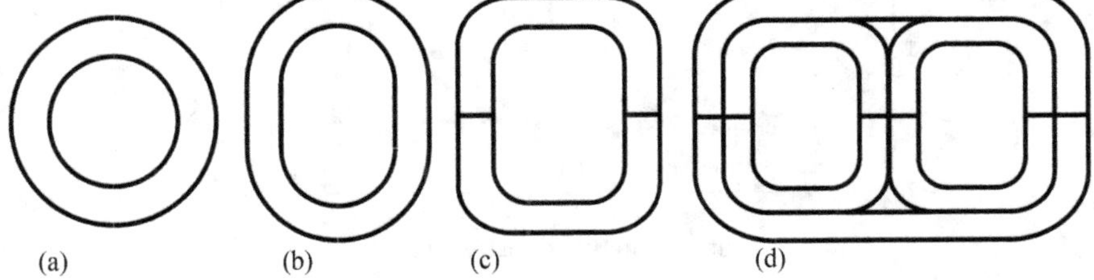

Figure 2.7 Core shapes based on metallic strips: (a) toroidal core, (b) oval core, (c) cut core (C core), M core based on cut strip cores

2.4 Non-grain-oriented electric steel (NGOES) and grain-oriented electric steel (GOES)

Grain-oriented sheets are a special type of electrical steel used for applications in electrical engineering and especially in the manufacture of transformers. These sheets are grain-oriented due to the process used in their manufacturing. This gives them special magnetic properties. Steel with low hysteresis and low eddy current losses is used for GOES sheets. It is brought to the sheet thickness required for further processing by hot rolling. In addition to the change in shape, this also results in a change in its microstructure. This makes it more malleable, among other things. After hot rolling, the material is cold rolled to further reduce its thickness. During the cold rolling process, the crystal structure begins to align in a preferred direction. An etching process for surface treatment is followed by a recrystallization phase. Recrystallization is a step in which the cold-rolled material is heated to regenerate or recrystallize the crystal structure. During this process, the internal stresses in the material are relieved, and the crystals are rearranged. This leads to a finer and more homogeneous crystal structure. As a result, the magnetic properties of the material can be optimized, leading to lower losses and better performance in applications such as transformers. In a final heat treatment, the orientation of the crystal structure is stabilized so that the magnetic properties are further improved [8].

In the rolling direction, both the saturation induction and the effective permeability are larger than in the transverse direction. As an example, please refer to Figure 2.8.

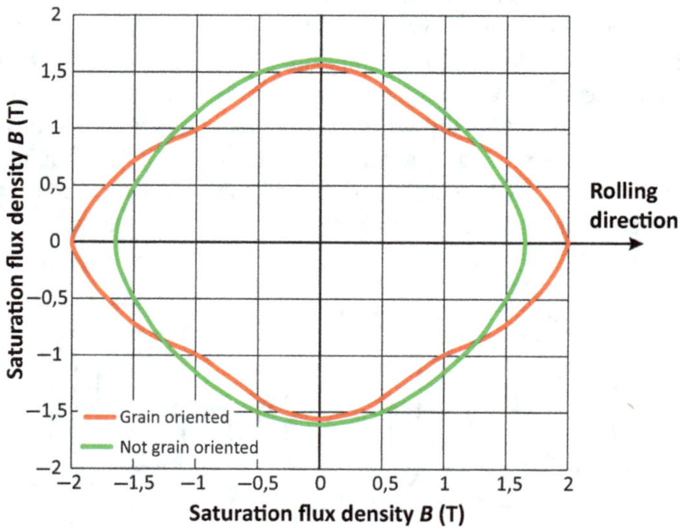

Figure 2.8 Exemplary comparison of the direction-dependent saturation induction of NGOES and GOES sheet steel (data from [9])

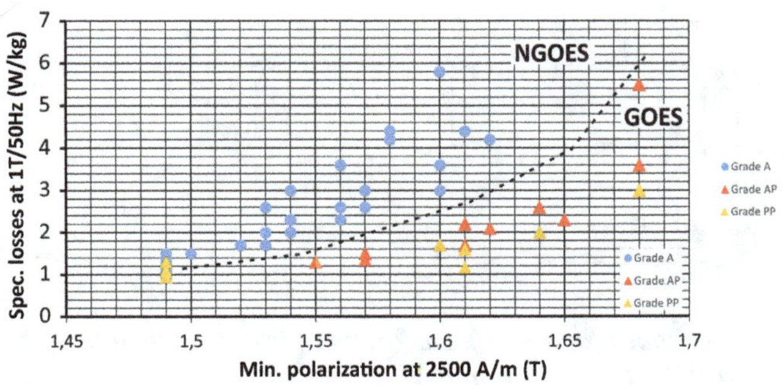

Figure 2.9 Comparison of the specific losses at (50 Hz, 1 T) and the minimum polarization at 2500 A/m) for the electrical steel sheets of one manufacturer (according to ThyssenKrupp AG datasheets)

The outlined material treatment enables the grain-oriented sheets to achieve a higher magnetic flux density and, at the same time, reduce losses, which makes them particularly efficient for applications in transformers (Figure 2.9). The use of grain-oriented sheets also helps to reduce the amount of material used in transformers.

The most important difference between GOES and NGOES lies in their magnetic properties. GOES has a magnetic anisotropy due to its controlled grain orientation, which leads to significantly lower core losses and a higher magnetic permeability in the rolling direction. NGOES, in contrast, has a magnetic isotropy that ensures relatively uniform magnetic properties in all directions.

The directional properties of GOES ensure that the magnetic flux is efficiently bundled, which increases the overall efficiency of electrical devices built based on it. Conversely, NGOES are used in electric motors, generators and smaller transformers where the direction of magnetic flux changes frequently and magnetic isotropy is an advantage.

As the explanations above show, the production of GOES is associated with a complex manufacturing process. Therefore, NGOES are generally associated with lower costs. When efficiency and performance are critical, GOES are preferred despite their higher cost and limited availability. Both GOES and NGOES play an important role in electrical engineering. GOES, with its magnetic anisotropy, is ideal for high-efficiency power transformers and large power distribution systems, while NGOES, with its isotropic magnetic properties, is suitable for electric machines and smaller transformers.

When the electrical steel sheet is thick enough, it is stamped into specific shapes. Certain basic shapes have been developed for small transformers. Figure 2.10 provides an overview. A distinction is made between UI, UU, LL, EI, EE and M cores.

The cores of high-power inductive components are assembled from strip-shaped sheets. They are stacked in an overlapping manner so that the magnetic field lines do not pass into the next sheet at the joints, but through the insulation between the sheets in the overlapping area. This prevents unwanted (parasitic) air gaps.

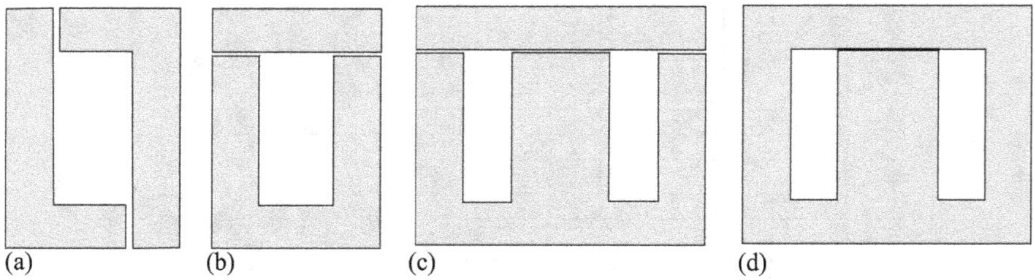

Figure 2.10 Cut shapes for small transformers for cores made of electrical sheet: UI, UU, LL, EI, EE and M cores

In transformers for the power supply network, the core laminations are layered in such a way that an approximately round cross-section is created. The main purpose of this measure is to absorb large forces in the event of a short circuit leading to high currents.

L, U and E cuts can be layered in such a way that a certain specified air gap is maintained. With the M cut, an air gap can in principle be provided when the shape is punched out of the sheet. However, subsequent adaptation is not possible. For transformers, high values are generally required for the magnetization inductances, so that the smallest possible air gap is aimed for. This is achieved by alternating layers of metal sheets. As a result, the next layer covers the air gap of the previous one, and the magnetization inductance is maximized.

2.5 Amorphous ferromagnetic materials

The amorphous 'structure' of a metallic material is created by cooling the alloy from a molten mass at a cooling rate of approximately -10^6 K/s. This rapid solidification process involves pouring the molten alloy onto a rotating water-cooled cylinder. In this one step, the resulting metal ribbon is cast from the melt into its final dimensions (thickness approx. 0.020 mm, width approx. 1–50 mm).

Due to their homogeneous and isotropic structure, the thin amorphous ribbon has good magnetic properties immediately after casting. This applies in particular to the magnetostriction-free Co-based alloys. Latter feature coercive field strengths of 1 A/m and below, for example. Heat treatment is also used to develop and optimize the final properties of amorphous magnetic materials. If the material properties are to be exploited to the maximum, this is primarily possible with toroidal core shapes.

For this purpose, the material is first annealed at low stress above the Curie temperature (approx. 400 °C) without exceeding the crystallization temperature. Otherwise, crystalline precipitation will start, which will lead to a deterioration of the magnetic properties.

The material is then annealed in an applied magnetic field in a range below the Curie temperature, i.e. approx. 200 °C–350 °C. Depending on the direction of the magnetic field, an F or Z loop can be set (Figure 2.11). The permeability level

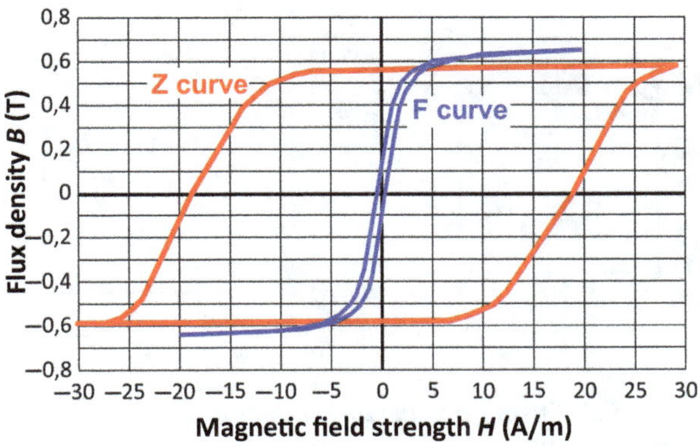

Figure 2.11 Principle comparison of the magnetization characteristics B(H) of Vitrovac 6025F and 6025Z, which can be adjusted by heat treatment under the influence of a magnetic field (based on [10])

required for the respective application is also set in this way. Toroidal cores are particularly suitable as core shapes because the amorphous metallic ribbons become brittle at annealing temperatures. This makes subsequent shape changes very difficult.

A typical application for materials with a flat (*F*) magnetization curve is the use of transformers. Due to the high permeability of up to μ_r = 150,000 (Figure 2.11), low magnetization currents are obtained even at low frequencies. As a result, the no-load losses of such transformers are very low. The material Vitrovac 6025 in Figure 2.11 is a cobalt-containing material, which will only be reserved for special applications due to the expensive cobalt component. Toroidal cores with *Z* characteristics also have a high permeability (in the case shown, μ_r = 50.400). However, they also have a high ratio of remanence induction relative to saturation induction (B_r/B_{sat} = 0.98). The magnetization losses (without the eddy current component) amount to 42.56 μWs/cm^3 per cycle. This makes this material interesting for use in magnetic amplifiers in the higher frequency range. Only a very low magnetomotive force is required to 'reset', i.e. 'open' the switches

$$MMF \geq H_c \cdot l_{Fe} = 19\ \text{A m}^{-1} \cdot l_{Fe} = I_{contr} \cdot N_{contr} \tag{2.5}$$

The function of a magnetic switch for energy transfer in a circuit is based on the fact that an unsaturated inductance is initially connected in the connection and load. The saturable inductance with a *Z* characteristic is finally driven into saturation by the increasing voltage difference between the energy source and the load. Due to the relatively sharp bend in the magnetization characteristic, the saturation process appears like switching between two effective inductors. The small increase in current, which is determined by the unsaturated inductance L_{20}, is followed by a "conduction" phase of the component. Its behaviour is determined by the residual inductance of L_{20} in the saturated state. A high effective permeability relative to the saturated inductance is therefore of particular interest. The main applications are in

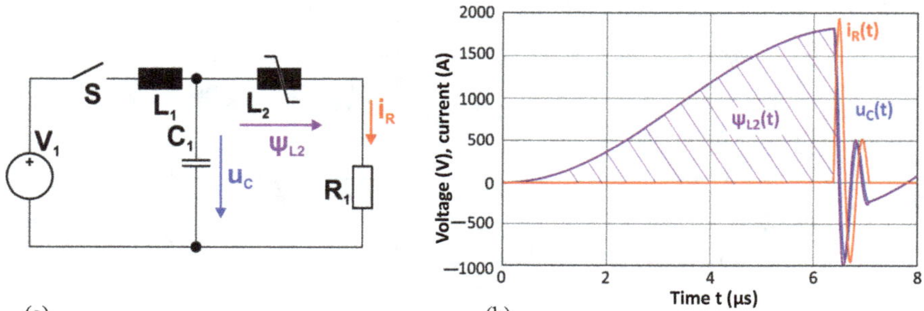

(a) (b)

Figure 2.12 Principle of using a saturable inductor as a flux-controlled switch in a pulse application: (a) circuit diagram and (b) current and voltage curves (V_1 = 1000 V, L_1 = 50 µH, C_1 = 200 nF, L_{20} = 585 µH, L_{2sat} = 50 nH, R_1 = 0.3 Ohm)

the field of pulsed lasers, radar technology and pulse power applications. Figure 2.12 shows the underlying principle schematically. Saturation of L_2 is reached at just over 6.5 µs. The load current then rises rapidly from 10 to almost 2000 A. The switch is only loaded with a maximum of 45 A. The current increase is driven by the intermediate energy storage capacitor C_1. When the recorded voltage–time area Ψ_{L2} reaches the saturation value, the component in the simulation switches from 585 µH to 50 nH. This starts a rapid increase in the current through the load resistor R_1. A DC-fed winding is placed on the core to actively reset the choke, which is used as a switch here.

However, it can also be used as a magnetic amplifier. The current through the additional DC winding is then used to control the saturation time via the flux linkage.

2.6 Nanocrystalline ferromagnetic materials

Ferromagnetic materials such as iron, cobalt, nickel and their alloys can also be produced in nanocrystalline form. Nanocrystalline magnetic materials exhibit improved magnetic properties such as magnetic saturation, coercivity and magnetic permeability and are used in applications such as data storage, sensor technology and magnetic recording.

Iron is a classic ferromagnetic material and can be produced in nanocrystalline form. Nanocrystalline iron shows improved magnetic properties such as high saturation induction and low hysteresis losses. Iron–nickel alloys such as permalloy (an alloy of about 80% nickel and 20% iron) can also be produced in nanocrystalline form. These alloys exhibit high magnetic permeability.

Alloys based on iron and cobalt, such as Fe–Co, can also be produced in nanocrystalline form. These alloys are characterized by relatively high magnetic saturation, low coercivity and low hysteresis losses. They are used in applications

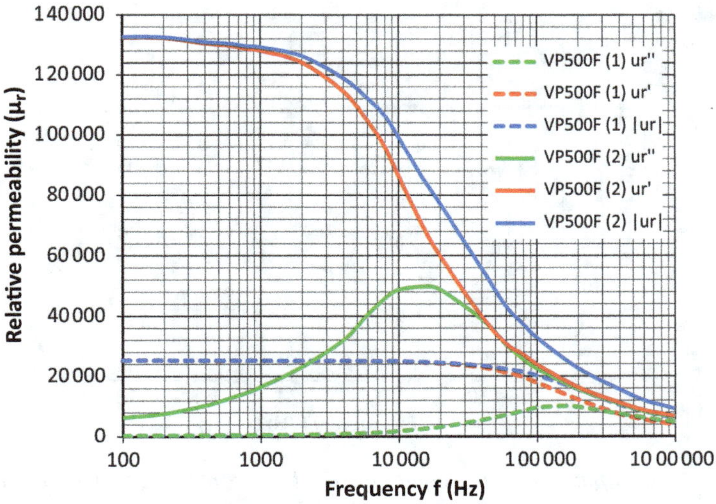

Figure 2.13 Moderate (1) and high (2) permeability for Vitroperm VP500F material (data from [11])

such as magnetic switches, actuators and current transformers. Due to the cobalt content alone, however, the material costs are significantly higher than for the aforementioned alloys.

Nanocrystalline materials (such as VITROPERM) require heat treatment of the initially amorphous base material (Section 2.5). To adjust the nanocrystalline structure and the associated soft magnetic properties, heat treatment at temperatures of typically 540 °C–600 °C must be carried out so that crystallization in the nanometre range occurs. If this temperature range is exceeded, hard magnetic phases may be precipitated, which severely impair the soft magnetic properties. After the transition to the nanocrystalline state, materials with very high permeabilities are obtained. Again, during the annealing process, either F or Z characteristics can be generated by means of an externally generated magnetic field. In addition, the permeabilities can be set in the range from 20,000 to 100,000 by selecting the parameters for the heat treatment in an F-loop (Figure 2.13). Here, achieving the desired magnetic properties is associated with a transition to significant brittleness. This prevents any real possibility of further processing. As a result, the material can only be heat-treated in its final form, i.e. usually as toroidal cores.

The core shapes (a) and (b) shown in Figure 2.7 therefore come into question here. Common mode chokes are widely used on the basis of such toroidal cores for EMI suppression purposes. Cutting the magnetic material always leads to a significant deterioration in the magnetic properties. The amount of work required to achieve a simpler winding or to introduce an air gap is considerable. The expected benefit must justify this effort.

Compared to the less expensive ferrite materials, significantly higher values of the real part of the permeability can be achieved with nanocrystalline materials at

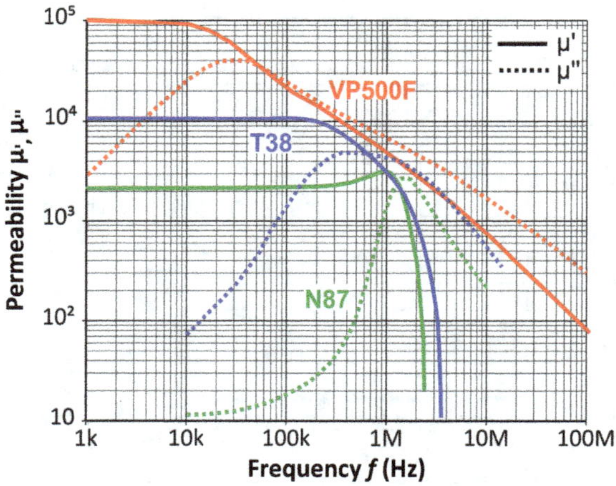

Figure 2.14 Comparison of the components of the complex permeability μ', μ'' of the nanocrystalline material VP500F ($B_{sat20°C}$ = 1.24 T), with those of two MnZn ferrites [T38 (power, low THD, R10 toroids, $B_{sat100°C}$ = 280 mT) and N87 (power, R34 toroids, $B_{sat100°C}$ = 390 mT)] (data from [6,11–13])

low frequencies and also very high frequencies (Figure 2.14). This means that inductive components based on these materials still behave as inductive components at very high frequencies. However, the winding capacitances play an increasingly important role at very high frequencies. Due to the high permeability, comparatively few turns are required for the same inductance. As a result, the winding capacitances are different, with winding to core capacitance usually being lower, but winding capacitance itself strongly depends on the distance of the single turns.

The components of the complex permeability of two typical MnZn ferrites are shown for comparison. In addition to the significantly lower saturation flux density, it is noticeable that the positive real part of the complex permeability disappears at around 2–3 MHz, while it is still significantly larger in the 10–100 MHz range for VP500F.

The storage of energy and the magnetic coupling of windings require – each in itself – cores with very different properties. The energy density of a magnetic storage device increases in inverse proportion to the effective permeability of the core material. In the practical design of storage cores, highly permeable core materials with a built-in air gap are usually used. If the required size is appropriate, this is divided into several partial air gaps. The magnetic field 'breaks' out of the core at the edges of the air gaps. A stray field forms around the core, which can interfere with other areas of the system and cause eddy currents and thus additional losses in metallic components.

Today nanocrystalline materials can also be produced with low-effective permeabilities, yielding in a quasi-distributed (like powder cores) airgap with the advantages of low proximity losses and significant energy storage capability. Nanocrystalline ribbon material is typically known for its high relative permeability. In a subsequent production step, mechanical stress is applied, resulting in a creep-induced anisotropy [4]. Relative permeabilities can be set as a design parameter down to a few hundred [5,9,14,15], maintaining low loss behaviour (Figure 2.5) and the high saturation flux density of approx. 1.2 T. It is therefore possible to significantly store magnetic energy in a quasi distributed airgap of the ribbon. As toroidal cores, these have almost no stray field. As a result, the proximity and fringing losses in these storage chokes are very low. This is particularly advantageous for applications with high currents. With multiphase interleaved operated chokes, the resulting current ripple at the output and also the current ripple in one phase can be significantly reduced. This effect is even more pronounced if magnetic coupling is present. For magnetic coupling, it is best to use cores made of highly permeable material with low core losses. If the cores are combined, as shown in Figure 2.15, a very compact integrated magnetic component for the coupling of four phases is obtained. The approach to realize such inductors is presented in detail in Chapter 12. The materials are primarily used according to their physical properties. The stray fields in the vicinity of the component are very low due to the uninterrupted core shapes.

There are also formal advantages of amorphous materials for main transformers for 50/60 Hz. These advantages are based on the significantly lower core losses due to lower hysteresis losses and lower eddy current losses. The reasons for the latter lie in the low strip thickness and the relatively high specific resistance of the amorphous core material. The higher manufacturing costs of amorphous core material certainly have to be considered.

Transformers with amorphous metal cores can achieve no-load losses that are up to 70%–75% lower than those of conventional transformers with silicon steel

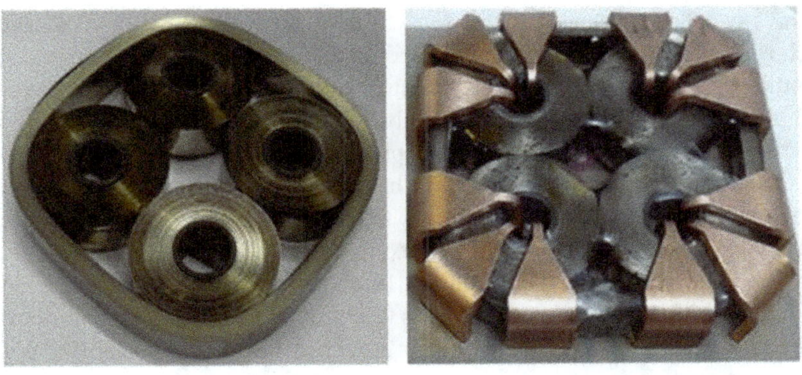

Figure 2.15 Coupling of four storage chokes made of low-permeability nanocrystalline tape-wound cores: inner four cores with $\mu_r = 160$, coupling core with $\mu_r = 1500$, current waveforms, see chapter 12 (photographs: Patrick Deck, TH Köln)

cores. This significant reduction in losses contributes to energy savings and lower operating costs over the life of the transformer. In [16], a reduction to 46.5% is given for a calculation example for the no-load losses of a 400 kVA oil transformer. Because the no-load losses occur over the entire service life (approx. 30 years), this results in energy losses reduced by 2 MWh per year for a 400 kVA transformer. This results in an amortization period of approx. 8.9 years for the case under consideration. If the foreseeable energy price increases in connection with CO_2 pricing are taken into account, the amortization period is even shorter [17]. Since power electronic energy converters for the use of solar and wind resources at the medium-voltage level are connected to the public grid via transformers, the increasing number of these transformers provides a very interesting opportunity to save energy by using such new materials [18–20].

Amorphous transformers have been used in common grid applications since the 1980s. In the mid-2000s, they were increasingly promoted and used due to growing demands for energy savings and concerns about CO_2 emissions. They have already been deployed in many countries, including Japan (410,000 units installed), the United States (420,000 units installed), China (385,000 units installed) and India (800,000 units installed). In Europe, trials are currently being carried out in several countries [18].

As a result of Directive 2009/125/EC of the European Parliament and of the Council of October 21, 2009, owners/users of transformers must comply with the ecodesign requirements as of 1 July 2015. In addition, Commission Regulation (EU) No. 548/2014 of 21 May 2014, applies to small, medium and large power transformers [19].

Moving on to power electronics applications, it is clear that switching frequencies are reduced at high power levels. On the one hand, this is due to the fact that it becomes very difficult to control fast switching processes at high currents. The inductances in the commutation loop cannot be reduced at will. Every extended conductor and every conductor loop has an inductance that leads to overvoltages when the current is switched off quickly. These easily reach a range where they can damage the switch itself. In addition, the skin effect and proximity effect mean that only a small proportion of the conductor cross section can be utilized at high switching frequencies. For this reason, switching frequencies of only 2–3 kHz are traditionally used for power converters of more than 500 kW. This applies in particular to transformers that are used to transmit energy via an insulating barrier. To minimize volume and weight, high frequencies are of interest here. Due to the problems mentioned above, the actual frequency at which the energy is transmitted remains well below the theoretical possibilities of the semiconductors. Figure 2.16 shows an overview of higher power transformers with the operating frequencies used from the literature [21]. In nearly all cases above around 200 kW, amorphous or nanocrystalline ribbon material is used as the core material. It is also clear here that the switching frequency drops to a few kilohertz with high power to be transmitted. Because heat dissipation from the windings is a particular problem at high power levels, the materials used are also specified.

Although the required size of medium-frequency transformers decreases with increasing frequency, the losses of the transformer are generated in a much smaller

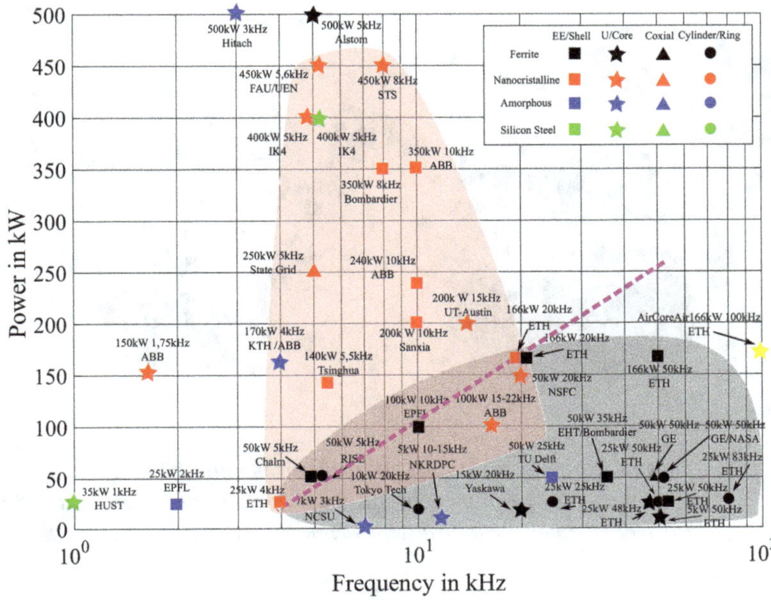

Figure 2.16 Literature examples for high-power transformers in the medium frequency range [21] with details of the rated power and the operating frequency

volume and must be dissipated. Effective cooling of the winding and core is therefore absolutely essential. Figure 2.16, therefore, also shows examples where the losses of the windings are dissipated via water cooling-based concepts. In the design approach in [21], both core and winding are cooled in an oil bath with forced convection. The core is fixed in this oil bath with primary and secondary windings. The weight and current forces are absorbed by insulating spacers, which are not shown in Figure 2.17. On the one hand, the width of the conductors is matched to the current to be carried. On the other hand, the width, spacing and position also determine the level of leakage inductance. The large oil-cooled surfaces of the conductors enable very effective cooling of the windings. The high-voltage winding (primary winding) is arranged on the outside, while the low-voltage winding is arranged coaxially inside the primary winding. The design was tested so that a power density of 22 kW/L including the oil tank can be determined, with a temperature increase of only about 20 K.

In summary, it can be stated that amorphous and nanocrystalline crystalline core materials have special and beneficial properties compared to conventional crystalline materials. These materials generally have lower magnetic hysteresis losses than crystalline materials. This means efficient energy transfer with lower losses. Their high magnetic permeability leads to high impedances even at low frequencies but also at high frequencies. This makes them particularly interesting for transformers and current-compensated chokes. Compared to ferrite materials,

(a) (b)

Figure 2.17 Design example for a 250 kW transformer for 50 kHz operating frequency: (a) primary winding with spacers on a toroidal core and (b) transformer structure with primary and secondary winding [21]

they are also characterized by high saturation inductances and significantly lower temperature dependencies of the parameters. Due to the manufacturing processes, the base materials of the inductive cores are thin layers. On the one hand, this means that the properties are characterized by anisotropy. On the other hand, the brittle tapes with their properties that depend on heat treatment and mechanical processing limit the design possibilities of cores.

References

[1] Herzer, G. Modern soft magnets: Amorphous and nanocrystalline materials. *Acta Materialia*, Volume 61, Issue 3, 2013, Pages 718–734.

[2] Devillers, Emile, Hecquet, M., Le Besnerais, J. A new hybrid method for the fast computation of airgap flux and magnetic forces in IPMSM. *12th International Conference on Ecological Vehicles and Renewable Energies (EVER)*, 2017, DOI: 10.1109/EVER.2017.7935904.

[3] https://doc.modelica.org/Modelica%203.2.3/Resources/helpMapleSim/Magnetic/FluxTubes/Material/HysteresisTableData.

[4] Hilzinger, R., Rodewald, W. Magnetic Materials: Fundamentals, Products, Properties, Applications, Vacuumschmelze GmbH & Co.KG, Hrsg., Publicis Publishing Erlangen, 2013.

[5] Vacuumschmelze GmbH & Co.KG. Material Information "Nanocrystalline Vitroperm" – relative permability down to 400 available. https://vacuumschmelze.com/products/soft-magnetic-materials-and-stamped-parts/nanocrystalline-material-vitroperm; retrieval date: 12.11.2023.

[6] Tape Wound Cores – Design Manual TWC-500. Magnetics Inc. 1998.

[7] Soft Magnetic Materials and Semi-finished Products. PB-PHT001, Vacuumschmelze GmbH & Co. KG, 2002.

[8] Z. Xia, Y. Kang, Q. Wang. Developments in the Production of Grain-Oriented Electrical Steel. *Journal of Magnetism and Magnetic Materials*, Vol. 320(23)23, 2008, pp. 3229–3233.

[9] Gao, Lingyu, Zeng, Lubin, Yang, Jiaxin, Pei, Ruilin. Application of grain-oriented electrical steel used in super-high speed electric machines. *AIP Advances*, Volume 10, 2020, Page 015127, 10.1063/1.5130151.

[10] Vitrovac 6025 Z Tape-wound Cores for Magnetic Amplifier Chokes. of 6025F, Vacuumschmelze Hanau, PK-002.

[11] EMC Products Based on Nanocrystalline VITROPERM. Product Information, Vacuumschmelze GmbH & Co. KG, 2021.

[12] Ferrites and accessories – SIFERRIT Material T38. Data sheet, TDK Electronics AG 2023.

[13] Ferrites and accessories – SIFERRIT Material N87. Data sheet, TDK Electronics AG 2023.

[14] J. Petzold. Advantages of soft magnetic nanocrystalline materials for modern electronic applications. *Journal of Magnetism and Magnetic Materials*, Volumes 242–245, Part 1, 2002, Pages 84–89.

[15] Magnetec GmbH. Datasheet information "Nanoperm® LM Cores for DC biased applications" (www.magnetec.de, Download 29.05.2019).

[16] T. Yamaji, M. Abe, Y. Takada, K. Okada, T. Hiratani. Magnetic properties and workability of 6.5% silicon steel sheet manufactured in continuous CVD siliconizing line. *Journal of Magnetism and Magnetic Materials*, Volume 133, Issues 1–3, 1994, Pages 187–189.

[17] JFE Steel Company. Super Core™ – Magnetic Property Curves, JNEX Core, JNHF Core, Cat.No.F2E-001-01, http://www.jfe-steel.co.jp/en/.

[18] Schneider Electric White Paper. How Amorphous Transformers Enhance Efficiency and Reduce CO_2 Emissions Levels, 2015.

[19] Directive 2009/125/EC of the European Parliament and of the Council of 21 October 2009 Establishing a Framework for the Setting of Ecodesign Requirements for Energy-Related Products.

[20] G. Saage, F. Schnelle: Improving the Power Density of Transformers with Nanocrystalline Cut Cores. Vacuumschmelze Hanau 2021.

[21] Lin, S. Modeling and Design Optimization of Medium Frequency Toroidal Transformers with Foil Winding for Medium Voltage and High Power Applications. Diss., Leibniz-Universität Hannover, 2024.

Chapter 3
Powdered magnetic materials

Lukas Mueller[1]

Most electrical and magnetics engineers are familiar with ferrite, and these materials are covered in great detail in other sections of this book. While ferrites are immensely useful and versatile materials, other magnetic materials have distinct advantages over ferrites in certain applications. For instance, in-line frequency transformers and laminated or ribbon wound magnetic cores offer superior performance compared to ferrites and are used exclusively for this type of application. More information on these types of materials is covered in Chapter 2. Powdered magnetic materials are another distinct magnetic material group that has distinct advantageous compared to ferrite in certain applications. Unfortunately, many engineers are unfamiliar with powdered magnetic materials. This chapter covers the different types of powdered magnetic materials, and their basic performance attributes and highlights their advantages and disadvantages compared to well-known ferrites. Applications that benefit from the use of powdered materials are described as well.

3.1 Powdered material types

Powdered magnetic materials are produced by pulverizing pure magnetic metals or magnetic alloys into a fine powder. The mean particle size of the powders varies from hundreds to a few micrometers [1]. The pulverized magnetic particles are then insulated, mixed with a binder, and compressed under extreme pressures to form a solid magnetic core [1]. The binder acts as a distributed air gap throughout the core, facilitating energy storage and reducing high-frequency eddy current losses in the material. In general, powdered materials can be divided into two broad categories based on their base magnetic material: powdered iron and powdered alloy.

These two categories can be further subdivided by the type of iron or alloy base material used. The choice of the base material will influence the attainable permeability and magnetic characteristics of the material. While the choice of base material has a large impact on the magnetic characteristics of the powdered core, different processing techniques further change the performance of these materials.

[1]Micrometals, Inc., Anaheim, USA.

Table 3.1 Overview of powdered material performance characteristics

Base material	**Iron**	**Carbonyl iron**	**Molypermalloy**	**Permalloy**	**Sendust**	**Silicon–iron**
Material composition	Fe	Fe	Ni–Fe–Mo	Fe–Ni	Fe–Si–Al	Fe–Si
Permeability range	22–100	4–35	14–550	14–160	14–160	14–125
Saturation flux density (T)	0.97–1.89	0.5–1.76	0.75–0.9	1.32–1.54	0.79–0.9	1.41–1.76
DC saturation constant Γ (T^2)	0.04–0.09	0.25–0.49	≈0.08	0.2–0.24	≈0.8	≈0.18
AC saturation constant Γ (T^2)	0.6–1	0.25–0.49	≈0.08	0.2–0.24	≈0.8	≈0.18
Curie temperature (°C)	770	770	460	500	500	700
Thermal conductivity (W/mK)	2.0–4.2	≈1	2.6–6.6	3.8–14	3.7–7.5	2.2–11.9
Relative cost	1.0–2.7	2.0–3.7	8.4	5.8–8.0	1.3–2.6	1.9
		Typical frequency range (MHz)				
EMI filter	DC–100	0.01–3000	–	DC–3	DC–5	DC–30
Power conversion	DC–0.1	1–60	0.01–1	0.01–1	0.01–5	0.01–0.1
High Q inductor	–	0.01–200	0.01–1	–	0.01–1	–
Broadband transformer	–	40–700	–	–	–	–
		Typical core loss (mW/cm^3)				
Reference permeability	60	10	60	60	60	60
60 Hz, 0.5 T	28.8	19.4	≈9.1	10–15.7	6.4–8.6	14.3
50 kHz, 0.1 T	2276	696	165–360	150–650	80–323	424–676
100 kHz, 14 mT	127	19.1	7.3–13	6–28	7–14.3	≈28.1
1 MHz, 4 mT	530	20.5	21–54	13–166	16–50	≈95
		Typical DC bias stability				
Reference permeability	60	10	60	60	60	60
$\%\mu_r$ at 100 Oe	40	98.5	50–55	68–96	44–64	74–84

A brief summary of the broad types and performance attributes of powdered magnetic materials is given in Table 3.1.

Each type of material has its unique operating characteristics that are described in greater detail in the following section.

3.1.1 Powdered iron

3.1.1.1 Iron

Powdered iron materials are made from iron particles. The various methods of producing these particles are beyond the scope of this book; additional information on the production of the base powdered iron can be found in [1]. While all powdered iron cores are composed of iron, there is a multitude of different grades or mixes of powdered iron, each having different magnetic properties like initial permeability, saturation flux density, and core loss.

Powdered iron materials were historically identified by the type of base powder they used [1]. Micrometals introduced a different naming convention in the form of numbered mixes that represent various unique powdered iron materials. While these mixes were originally developed by micrometals, other producers of powdered iron, also label their materials using a mix number. In addition to the mix number, micrometals assigned each mix a unique color code. These color codes allow engineers to quickly identify the type of material used by visually inspecting the magnetic core. Similarly, other powdered iron producers have adopted these color codes for their powdered materials. It is important to note that while mixes from different powdered iron producers have similar specifications, they are not identical. There can be noticeable differences in performance, especially in regard to core loss or the complex permeability of these materials versus frequency. An overview of various powdered iron mixes and their base properties is shown in Table 3.2.

While there is a large variety of powdered iron types, there are a few attributes they all have in common. Powdered iron materials generally feature the highest saturation flux densities of all powdered materials. The initial permeability of

Table 3.2 Overview of powdered iron mixes from micrometals

Mix	Initial permeability	B_{sat} (T)	μ_{diff} at 50 Oe	Core loss (mW/cm^3) 100 kHz, 14 mT	Color code toroid
−18	55	1.78	42	46	Green/red
−19	55	1.82	43	54	Red/green
−26	75	1.85	41	83	Yellow/white
−28[a]	22	1.67	–	–	–
−30	22	1.67	20	129	Green/gray
−33[b]	33	1.71	29	180	–
−34	33	1.71	29	82	Gray/blue
−35	33	1.73	29	119	Yellow/gray
−38	85	1.87	44	103	Gray/black
−40	60	1.84	40	127	Green/yellow
−41	75	1.89	25	–[c]	Green/green
−42[d]	40	–	–	–	Blue/red
−44	90	1.89	40	–[e]	Black/white
−45	100	1.89	43	61	Black/black
−52	75	1.85	46	58	Green/blue

[a]Discontinued, substituted by −30.
[b]Discontinued, substituted by −34.
[c]Low Q material.
[d]Discontinued.
[e]Low Q material.

powdered iron materials is generally also higher than that of powdered carbonyl iron materials. Some grades of powdered alloy are available in higher initial permeabilities than powdered iron.

Powdered iron features a high amplitude permeability, meaning that the effective permeability when a large-signal AC excitation is applied can be up to three times higher than the starting permeability. Powdered iron materials tend to have the highest core losses among powdered materials. The high core losses have mostly displaced powdered iron cores in power conversion inductors, like PFCs or buck inductors, operating above 50 kHz. The higher core losses make powdered material popular in both electromagnetic interference and spike filter inductors. The high core loss damps unwanted resonances, helping reduce instabilities caused by EMI filter resonances and reducing the output impedance of the EMI filter.

Powdered iron materials tend to be one of the lowest-cost magnetic materials available to designers, making them especially useful in cost-sensitive applications.

3.1.1.2 Carbonyl iron

Powdered carbonyl iron forms the base powder for carbonyl iron cores. For more information on the production of carbonyl iron, see [1,2]. Similar to powdered iron, many different variations of this material exist. The materials were traditionally identified by the carbonyl iron type used to make the cores [1]. Similar to powdered iron, micrometals introduce numbered mixes and color codes to identify the various forms of this material. As with the powdered iron, some other manufacturers of carbonyl iron powdered cores have also adopted the different mix designation and color cores. Once again, it is important to highlight that while the base properties, like initial permeability, are similar between micrometals' mixes and mixes from other producers, the materials are rarely identical. Especially at high frequencies, there can be noticeable differences between core loss, complex permeability versus frequency and lifetime versus operating temperature.

An overview of the various carbonyl iron mixes from micrometals is shown in Table 3.3.

Table 3.3 Overview of carbonyl powdered iron mixes from Micrometals

Mix	Initial permeability	B_{sat} (T)	μ_{diff} at 200 Oe	Core loss (mW/cm^3) 1 MHz, 4 mT	Color code toroid
−1	20	1.73	16	20.5	Blue/clear
−2	10	1.48	9.6	9.19	Red/clear
−3	35	1.76	21	20.5	Gray/clear
−4	9	1.48	8.6	20.5	Blue/white
−6	8.5	1.44	8.3	9.08	Yellow/clear
−7	9	1.46	8.7	9.19	White/clear
−8	35	1.46	21	21.7	Yellow/red
−10	6	1.38	6	8.94	Black/clear
−14	14	1.52	12	10.7	Black/red
−15	25	1.75	18	21.7	Red/white
−17	4	1.28	4	8.36	Blue/yellow

As with powdered iron, carbonyl iron powdered materials have a number of attributes in common. These types of material generally feature saturation flux densities in excess of 1.3 T. The initial permeability of these materials is generally the lowest among powdered materials, ranging from 4 to 35μ. Carbonyl iron materials tend to have lower core losses than powdered iron cores. Depending on the material grade or mix, eddy current losses are extremely low. This allows inductor designs with high Q with frequencies up to 200 MHz [3] and effective EMI filter designs into the GHz range. Carbonyl iron cores generally feature a low amplitude permeability, along with small variations in inductance with temperature or frequency. This gives carbonyl iron-based inductors a high level of stability in tuned resonant circuits without having to rely on a lossy air gap in the core.

3.1.2 Powdered alloy

3.1.2.1 Molypermalloy

The base magnetic material consists of roughly 79% nickel, 17% iron, and 4% molybdenum. Molypermalloy powder, often abbreviated to MPP, cores were among the first type of powdered alloy core commercialized [4].

MPP cores feature low hysteresis losses and can be produced with the highest initial permeability of any powdered material, with some manufacturers providing permeabilities up to 550μ. The initial permeability of MPP cores is also insensitive to changes in temperature, varying by less than +/1% between −60 to 100 °C. Micrometals offer temperature-stabilized MPP cores that have an even lower variation in initial permeability with temperature for magnetic sensor applications.

The saturation flux density of MPP cores is around 0.75–0.9 T, depending on the manufacturer and permeability grade. While among the lowest saturation flux densities for powdered materials, MPP cores still have significantly higher saturation flux densities than ferrites. The low hysteresis losses, high achievable initial permeabilities, and reasonable saturation flux density make MPP cores suitable for a variety of inductive components such as flyback transformers, inductors in DC/DC power converters, and high Q resonant inductors operating in the tens to hundreds of kHz range. High-temperature stability makes MPP cores also suitable for highly tuned filter inductors, resonant inductors, or magnetic sensors where a change in inductance would compromise the performance.

The main downside of MPP cores is their high cost caused by being mainly composed of nickel. This makes the use of MPP cores uneconomical in inductive components meant for consumers or other low-cost applications. Today, MPP cores are mainly used in high-performance or high-reliability applications such as military, aerospace, drilling, or other safety-critical applications.

3.1.2.2 Sendust

The Sendust alloy material was first synthesized by Hakaru Masumoto in 1936 in Sendai, Japan, as an alternative to nickel-based magnetic materials, such as Permalloy. The material is composed of iron, aluminum, and silicon. The Sendust material features both low eddy and hysteresis loss. When first introduced, Sendust powdered cores had higher core losses than MPP. Recent advances in processing techniques allowed specialized grades of Sendust materials to achieve similar or even lower losses than MPP. Saturation flux density ranges between 0.8 and 1 T, similar to MPP. Sendust has the unique property that it does not experience magnetostriction, resulting in low to no audible noise from magnetic components using Sendust [4]. Sendust is generally available in permeabilities up to 125μ, while micrometals offers grades of Sendust up to 160μ.

Sendust costs somewhat more than powdered iron while having low core losses and acceptable DC bias performance. This makes Sendust a popular choice for power conversion inductors, such as PFC, buck, or boost converter chokes in consumer applications. Because Sendust costs so much less than MPP while providing similar performance, Sendust has replaced MPP in many consumer applications. However, Sendust's performance is not as stable with temperature as MPP. Therefore, MPP remains the material of choice for high-reliability applications, magnetic sensors, or applications requiring initial permeability exceeding 160μ.

3.1.2.3 Permalloy

Powdered Permalloy cores use an alloy of around 50% nickel and 50% iron. Powdered Permalloy cores have a saturation flux density of 1.4–1.6 T while featuring significantly lower core loss than powdered iron. When Permalloy powdered cores were first introduced, their core loss was noticeably higher than MPP cores. The higher core loss restricted the use of powdered Permalloy cores to high-performance DC filter inductors or lower frequency switching inductors. Recent advances in material processing have resulted in low loss Permalloy grades with similar losses to MPP cores while retaining their high saturation flux density and excellent DC bias stability. Permalloy cores are available in permeabilities up to 160μ.

Permalloy has become a popular choice for both filter and switching inductors because of its low core loss and high DC bias stability, Permalloy cores tend to be the second most expensive option after MPP, because of their high nickel content.

3.1.2.4 Iron–silicon

Iron–silicon cores are composed, as the name implies, of iron and silicon. The silicon content of powdered iron–silicon cores tends to range from 3% to 7%, unlike electrical steel, which has a silicon content of up to only 3.5% [5]. While electrical steel can be grain-oriented to improve magnetic performance, the silicon–iron used in powdered cores is not oriented because the favored magnetization direction could not be maintained during the manufacturing process. The higher silicon content in powdered silicon–iron increases the resistivity of the material, reducing eddy current losses. The additional silicon also decreases the saturation flux density of the material somewhat compared to pure iron or electrical steel; however, saturation flux densities still range from 1.5 to 1.7 T, which is higher than in powdered Permalloy cores. Silicon–iron cores feature a good DC bias stability while generally having higher core losses than all other types of powdered material except standard iron powder. Higher core loss restricts the use of silicon–iron cores to filter inductors or low-frequency switching inductors used in uninterruptible power supplies (UPS) or grid-tied inverters where cost and DC bias stability are more important than core loss. While the core loss is higher than in Sendust, MPP, or Permalloy, silicon–iron cores tend to have a relatively stable inductance behavior with a frequency of up to 30 MHz, making the material an excellent choice for differential mode DC filter inductors.

3.1.2.5 Hybrid materials

Different powdered core manufacturers offer hybrid alloy materials that do not use the traditional raw powder materials described above. The goal of these hybrid materials tends to be to fill the gaps between the available standard materials. For example, Micrometal's Optilloy series consists of a set of three hybrid alloy materials that aim to offer a lower-cost alternative to Permalloy. Micrometals' OC material is a hybrid material that features core losses similar to Permalloy while having a DC bias stability between Sendust and Permalloy, while being significantly lower cost than Permalloy. This fills the gap in material performance between Sendust and Permalloy.

3.2 Powdered materials vs. ferrites

As stated in the Introduction, most electrical engineers are familiar with the performance of ferrites. The disadvantages and advantages of using powdered materials compared to ferrites are summarized below.

3.2.1 Disadvantages

There are a number of attributes where ferrites are superior to powdered materials.

3.2.1.1 Initial permeability

Ferrites are available in relative starting permeabilities up to 15,000μ, whereas powdered materials are only available in starting permeabilities up to 550μ. This makes ferrites vastly superior in applications requiring a high permeability, such as common mode chokes and medium frequency transformers.

3.2.1.2 Linearity

The high permeability of ferrites generally requires them to be gapped to obtain the desired relative permeability in inductor applications. The discrete air gap provides a high level of linearity during operation. Both the amplitude and differential inductance of gapped ferrite cores are virtually constant and equal up to their saturation flux density. Powdered materials, in contrast, have a non-linear inductance with AC and DC excitation. The soft-saturation behavior of powdered cores leads to a gradual decrease in differential permeability with DC excitation. While this effect is beneficial in some applications, it makes the analysis of powdered materials more difficult. Powdered iron materials also feature a strong increase in permeability when operated under a high AC excitation, again leading to highly non-linear operation.

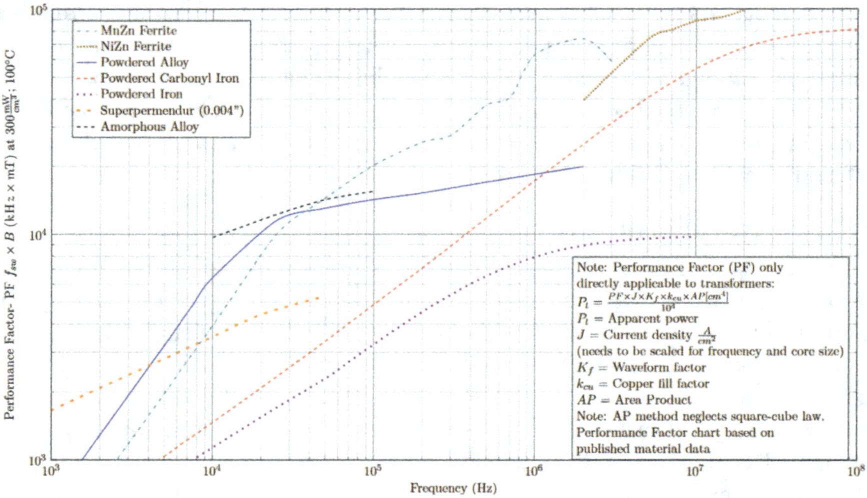

Figure 3.1 Performance factor of various magnetic material types [6]

3.2.1.3 Core loss

Ferrites will generally feature lower core losses than powdered materials, depending on the grade of the material compared. This becomes especially apparent when comparing the peak performance factor of various magnetic material types, as shown in Figure 3.1. This makes ferrites clearly superior in any application where core loss is the only design consideration. Note that the performance factor is only indicative of core loss and not the overall performance of the magnetic component. For instance, Q and core loss are not necessarily linked, as material permeability and AC copper losses are also important design considerations.

3.2.1.4 Thermal aging

Powder iron and powder carbonyl iron cores have organic content and undergo thermal aging. When cores made from these types of powdered materials are exposed to or generate elevated temperatures, a permanent decrease in both inductance and quality factor (Q) will gradually occur. The extent of these changes is highly dependent on time, temperature, core size, frequency, and flux density. It is essential that these properties are considered in any design operating at or above 75 °C. Iron powder cores tolerate temperatures down to −65 °C with no permanent effects.

In high-power applications where core loss is contributing to the total temperature, a decrease in quality factor will translate into an increase in eddy current losses, which will further heat the core and can lead to thermal runaway. Designs where core loss exceeds copper loss should be avoided. Hysteresis losses are unaffected by the thermal aging process. Thermal aging generally does not need to be considered in applications operating at line frequencies.

There are no analytical equations to predict the thermal aging behavior of powder iron and powder carbonyl iron cores. However, some manufacturers provide thermal aging charts based on accelerated lifetime testing. Micrometals provide a thermal aging and lifetime predictor in their free online design tool.

Thermal aging only occurs in powder iron and powder carbonyl iron cores. Powder alloy cores do not experience a significant permanent change in magnetic attributes with prolonged exposure to elevated temperatures.

3.2.2 Advantages

3.2.2.1 Saturation flux density

The saturation flux density of ferrites generally ranges up to 0.45 T, whereas the lowest saturation flux density for powdered materials is around 0.75 T, going all the way to 1.87 T. The higher saturation flux density allows powdered materials to store significantly higher energy per unit volume than ferrites.

3.2.2.2 High-temperature operation

Ferrites tend to have a Currie temperature between 100 and 400 °C, whereas the lowest Curie temperature among powdered materials is around 450 °C. This makes powdered alloy materials much more suitable for extreme temperature environments.

3.2.2.3 Soft-saturation

The gradual drop-off in inductance can be an advantage in certain applications. For power converters operating in continuous conduction mode, the inductance must be sufficiently high enough to maintain a continuous current flow. To this end, it is beneficial to have a higher inductance or permeability at low DC magnetization forces and have a gradual drop in material permeability as the current increases. Soft-saturation also increases the reliability of the inductor, as a temporary overload will only cause a temporary decrease in inductance, whereas a ferrite-based design can be pushed into saturation, losing its inductance entirely. The permeability vs. magnetization force for a variety of powdered materials and a gapped ferrite is shown in Figure 3.2.

3.2.2.4 Temperature stability

Permeability, core loss, and saturation flux density of ferrites can vary greatly in the temperature range commonly used in power electronics. In contrast, because of the higher Curie temperature of powdered materials and the distributed air gap, the magnetic parameters vary much less with temperature. Molypermalloy and Permalloy show very low variation in initial permeability with temperature.

3.2.2.5 Lower AC copper losses

The AC copper losses in the inductor winding are roughly proportional to the spatial average of the squared magnetic field subjected to the winding [7–9]:

$$P_{ac,winding} \propto \langle |B^2| \rangle \tag{3.1}$$

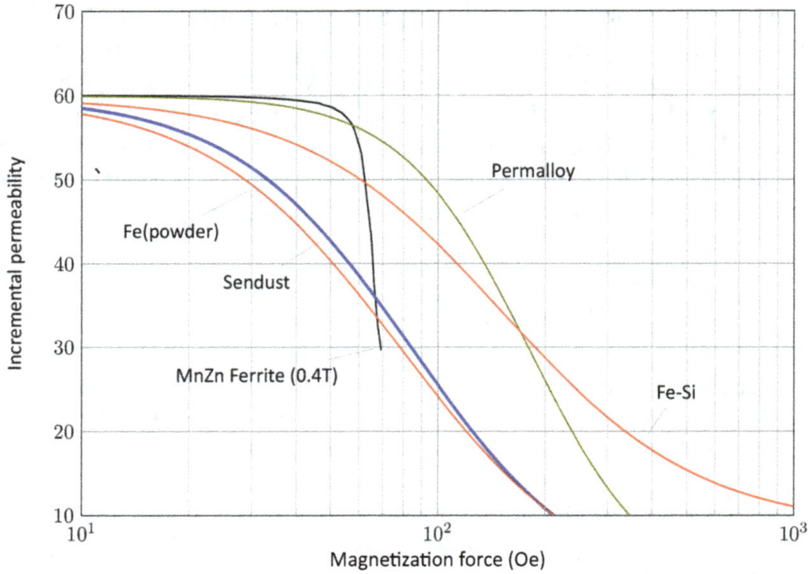

Figure 3.2 Incremental permeability versus magnetization force for a variety of powdered material types and ferrite

Ferrite-based inductors require discrete air gaps, which create strong localized magnetic fields that spill into the winding, greatly increasing the spatial average of the squared magnetic field, and with it, AC copper losses. Adding multiple air gaps into a ferrite-based inductor reduces the magnetic field in the winding; however, this approach results in higher production costs. Additionally, the ferrite pieces between two gaps have no thermal connection to the rest of the core, potentially resulting in a localized hot spot. Powdered cores have a distributed air gap and are available in a large range of permeabilities. The lack of a discrete air gap greatly reduces the spatial average of the squared magnetic field in the winding, resulting in AC copper losses of a fraction of that of a gapped ferrite inductor. A comparison between the spatial average of the squared magnetic field based on the winding fill factor in a gapped ferrite and a powdered carbonyl iron PQ32/30 core is shown in Figure 3.3. The comparison shows up to 5.5 $\times$ lower AC copper losses with a powdered material compared to ferrite with a single air gap.

Additional information on calculating the AC resistance and copper losses of inductors, as well as on how to minimize the AC copper losses in gapped ferrite designs, can be found in Chapter 5.

3.2.2.6 Core loss stability with DC bias

The core loss in ferrites generally increases when operated under DC bias conditions [10,11]. In contrast, powdered cores show only a small variation in core loss

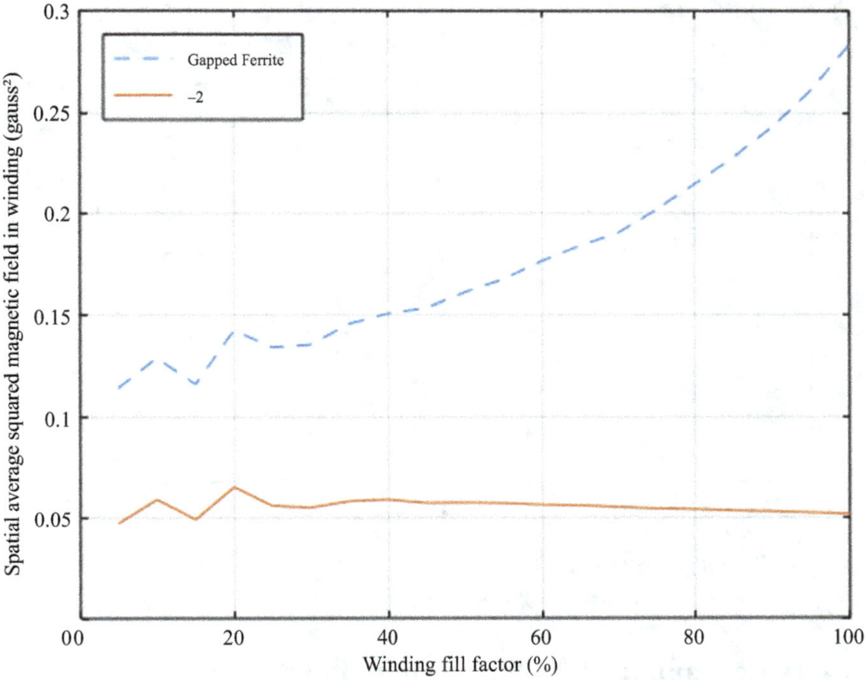

Figure 3.3 Comparison of the spatial average of the square magnetic field in winding window of a PQ32/30 based on winding fill factor

with DC bias operation [11,12]. While ferrite losses still tend to be lower than in powdered cores, the difference in losses becomes smaller when the inductor is operated with DC bias currents.

3.2.2.7 Smooth complex permeability roll-off

Ferrites tend to experience a sharp peak in their real permeability value before the imaginary part of the permeability rises. This is due to the ferromagnetic resonance that occurs in the ferrite. Most powdered materials, with the possible expectation of ultra-low loss Sendust material grades, feature a gradual decrease in real permeability while featuring a gradual increase in imaginary permeability [13]. In addition, imaginary permeability does not significantly exceed the real permeability at any frequency, resulting in an equivalent material quality factor (Q) of 1 at high frequencies. The high imaginary permeability allows the material to feature a high impedance, while providing good dampening, making powdered materials suitable for EMI filter inductors operating over a broad frequency range [13]. The complex permeability of a powdered carbonyl iron with an initial permeability of 35μ is shown in Figure 3.4.

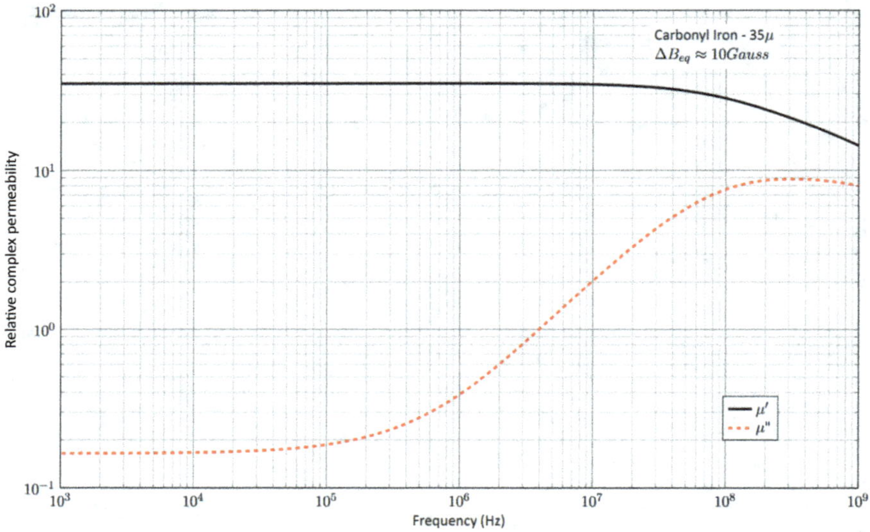

Figure 3.4 Complex permeability behavior of a carbonyl iron material with a starting permeability of 35 μ

3.3 Typical applications for powdered materials

The distinct advantages of powdered magnetic materials over ferrites make them particularly suitable for certain applications. Given the different magnetic characteristics of the various powdered materials, each grade of material will have certain applications they are more suitable for than others. An overview of generally recommended powdered material type by application is shown in Table 3.4. Detailed information for the broad application types and why that particular powdered material is suitable is summarized here.

3.3.1 DC filter inductors

These types of inductors are generally found as lower-frequency filter inductors in DC/DC converters. DC filter inductors need to carry a high DC current while maintaining a high differential permeability; allowing these inductors to provide a high impedance to unwanted noise signals, blocking them. For DC filters operating below 500 kHz, the only material consideration in ferrite or powdered materials is having as high a material saturation constant Γ as possible. The losses in a DC inductor can be related to the core size, saturation constant, and equivalent stored energy using

$$P_{loss} = \frac{4\rho \times E_{eq}^2}{k_{cu} K_g \Gamma} \tag{3.2}$$

with

$$E_{eq} = \frac{1}{2} L_{eq} I_{nom}^2 \tag{3.3}$$

Table 3.4 Typical powdered material used by application

Base material	**Iron**	**Carbonyl iron**	**Molypermalloy**	**Permalloy**	**Sendust**	**Silicon–iron**
Light dimmer chokes	•				•	
Inrush current suppressor	•					
Thyristor/SCR protection	•					
AC line choke	•					•
Differential mode EMI <30 MHz	•				•	•
Differential mode EMI >30 MHz	•			•	•	•
CCM DC inductor <50 kHz	•				•	•
CCM DC inductor 50 kHz to 1 MHz				•	•	•
CCM DC inductor 1 MHz>		•			•	
DCM DC inductor		•			•	
Flyback transformer			•		•	
High Q inductor		•	•		•	
Broadband transformer		•				
Class D inductor		•		•		
UPS/inverter inductor	•					•
Hall effect sensor		•				•
RF current sensor		•				

where K_g is the core geometry constant of the core shape [14,15]. The approximate Γ value of powdered materials is shown in Table 3.1. If the value of Γ for a certain powdered material is unknown, a possible approximation is [14]

$$\Gamma \approx 0.09 \times B_{sat}^2 \tag{3.4}$$

The Γ value of MnZn ferrites is generally between 0.04 and 0.054.

Permalloy, silicon–iron, and hybrid powdered materials have significantly higher Γ values than ferrites, enabling them to feature much lower losses than gapped ferrite designs using the same inductor volume or significantly lower volume assuming equal inductances [16].

As the frequency at which the filter must be effective increases, both the saturation and frequency characteristics of the material need to be considered. Powdered carbonyl iron materials tend to provide a stable real permeability over a wide frequency range, making them suitable for high-frequency filters [13]. Frequency and DC bias are two separate parameters; the performance cannot be expressed using one variable or equation. Material datasheet charts or computer design tools are best used to obtain the optimum filter material for a given magnetization force and frequency.

Table 3.5 Material type providing highest real differential permeability based on magnetization force and frequency

	0 Oe	**50 Oe**	**100 Oe**	**200 Oe**	**500 Oe**	**1000 Oe**
10 kHz	MPP 550μ	Fe–Ni 125μ	Fe–Ni 90μ	Fe–Si 60μ	Fe–Si 26μ	C–Fe 14μ
100 kHz	MPP 550μ	Fe–Ni 125μ	Fe–Ni 90μ	Fe–Si 60μ	Fe–Si 26μ	C–Fe 14μ
500 kHz	MPP 205μ	Fe–Ni 125μ	Fe–Ni 90μ	Fe–Si 60μ	Fe–Si 26μ	C–Fe 14μ
1 MHz	MPP 205μ	Fe–Ni 125μ	Fe–Ni 90μ	Fe–Si 60μ	Fe–Si 26μ	C–Fe 14μ
5 MHz	Sen–160μ	Fe–Si 90μ	Fe–Si 60μ	Fe–Si 40μ	C–Fe 20μ	C–Fe 14μ
10 MHz	Sen–125μ	Fe–Si 90μ	Fe–Si 60μ	Fe–Si 40μ	C–Fe 20μ	C–Fe 14μ
50 MHz	Sen–125μ	Fe 55μ	Fe 55μ	C–Fe 35μ	C–Fe 20μ	C–Fe 14μ
100 MHz	C–Fe 35μ	C–Fe 35μ	C–Fe 35μ	C–Fe 35μ	C–Fe 20μ	C–Fe 14μ
500 MHz	C–Fe 35μ	C–Fe 35μ	C–Fe 35μ	C–Fe 35μ	C–Fe 20μ	C–Fe 10μ

MPP = Molypermalloy, Fe–Ni = Permalloy, Fe–Si = iron–silicon, Fe = powdered iron, and C–Fe = carbonyl iron. For additional information on optimizing the choice of magnetic material for DC inductors, see Chapter 6.

The material choice that maximizes the real differential permeability for a variety of powdered materials is shown in Table 3.5. Using the chart, designers can get an idea of which powdered material type is best suited for a given magnetization force and frequency.

3.3.2 AC line chokes

Powdered materials, especially powdered iron, are well suited for use in 50/60 Hz line filters. Similar to DC inductors, the high saturation flux density of powdered materials allows the design of more compact or efficient inductors compared to ferrites. The high losses of powdered iron materials also dampen undesirable high-frequency resonances in the filter. For more information on the design of AC line chokes using powdered materials, see [17,18].

3.3.3 DC/DC power conversion inductors

These types of inductors are commonly found in buck/boost/PFC or any other type of DC/DC power converter topology. In DC converters, both core loss and energy storage requirements have to be considered. The total losses for a DC inductor with a certain ripple current ratio (F_I) are given by

$$P_{loss} \approx k_{core}\Gamma V_e \left(\frac{F_I}{F_I+1}\right)^2 + \frac{\rho P_{app}^2 d^2 \left(1+\frac{F_I^2}{3}\right)(MLT)}{4k_{cu}W_a A_e^2 \Gamma f_{sw}^2}\left(\frac{1+F_I}{F_I}\right)^2 \tag{3.5}$$

with

$$F_I = \frac{\Delta I}{I_{dc}} \tag{3.6}$$

assuming AC copper losses are low and can therefore be neglected. The equation shows the dependency on both core loss (k_{core}) and energy storage capability (Γ).

The optimum trade-off between the two parameters depends on a variety of factors, such as desired ripple current, frequency, power level, and core size.

The higher energy storage capabilities Γ of powdered materials make them good choices for DC inductors, especially if a low ripple current ratio (F_I) is desired.

Sendust materials will be best for designs where low core loss and low cost are a requirement. Silicon–iron materials will be best suited for inductor designs that need to feature a stable inductance with high currents operating at lower switching frequencies. An example would be inductors found in uninterruptible power supplies or other types of inverters. Permalloy will provide the best performance where both stable inductance and low core loss are crucial, and there are less strict cost constraints. Powdered carbonyl iron, with its low permeability and stable inductance behavior, can find use in inductors for very high-power or high-frequency operation (>1 MHz).

For additional information on the optimum selection of materials for DC power conversion inductors, see Chapter 6.

3.3.4 RF applications

Powdered carbonyl iron materials are best suited for RF applications operating above 1 MHz, as these materials have a reasonably stable inductance behavior to at least 100 MHz. Powdered carbonyl cores can also be used in high Q resonant inductors, where their low initial permeability enables a high Q without having to resort to gapping the core. Carbonyl iron can also be used in broadband transformers and RF current sensors operating in the tens to hundreds of MHz range. An overview of achievable Q values vs. frequency for a variety of inductor designs can be found in [19]. The normalized quality performance factor for select powdered materials on a T106 toroid using a single-layer winding is shown in Figure 3.5.

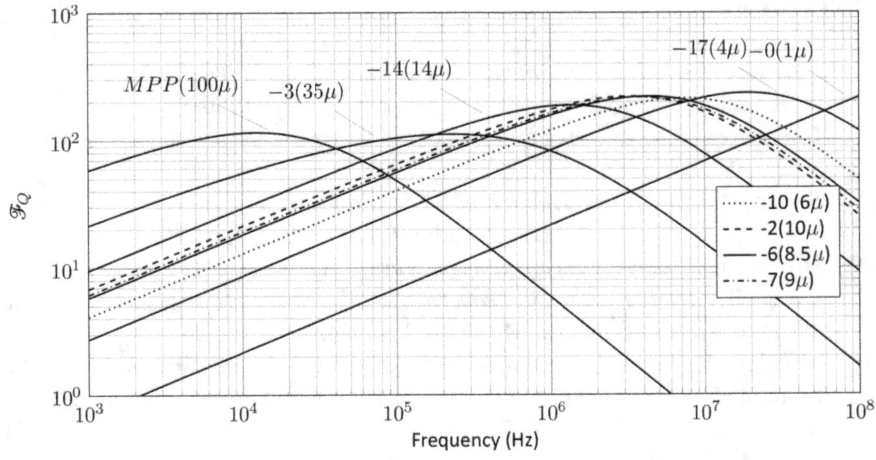

Figure 3.5 Normalized quality performance factor for 100 μ MPP and various carbonyl iron mixes

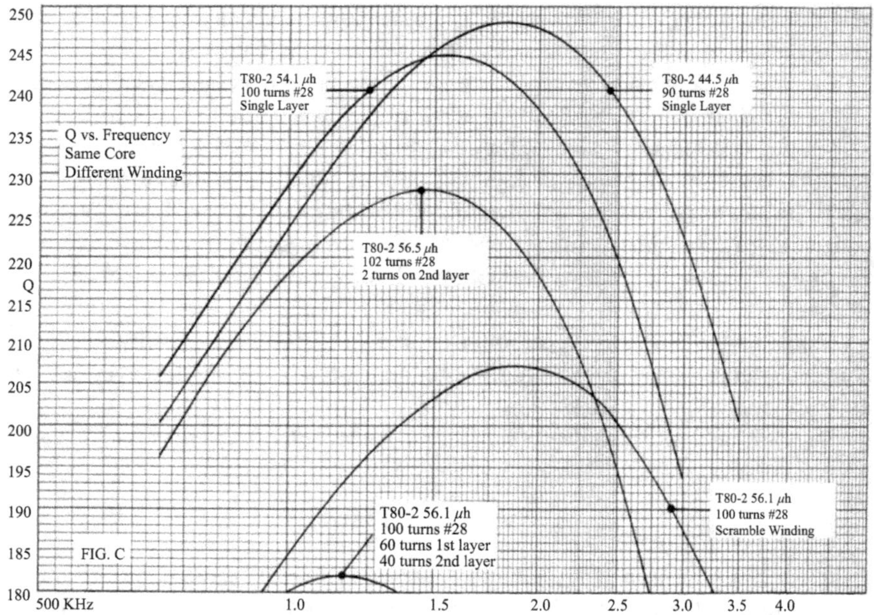

Figure 3.6 Changes in quality factor based on winding construction [19]

Note that the choice of wiring, winding arrangement, core size, and inductance will influence achievable Q drastically. An example demonstrating changes in Q due to the winding arrangement is shown in Figure 3.6.

For additional information on maximizing the Q of inductors, see Chapter 6.

References

[1] Polydoroff WJ. *High-Frequency Magnetic Materials: Their Characteristics and Principal Applications*. New York, NY: Wiley; 1960.

[2] Cox J. *Iron Powder Cores for High Q Inductors.* Micrometals, Inc: Anaheim, CA; 1997.

[3] Silver HW. *The ARRL Handbook for Radio Communication*, 100th Edition. Newington, CT: The American Radio Relay League, Inc.; 2023.

[4] Bozorth RM. *Ferromagnetism*. New York, NY: Wiley; 1993.

[5] Lyman T. *Metals Handbook, Properties and Selection*, Volume 1, 8th Edition. 1961. American Society for Metals Address: Metals Park, USA.

[6] Mueller LK. Inductor Optimization Based on Choice of Different Magnetic Materials. PSMA – Power Magnetics at High Frequency Workshop. 2024.

[7] Sullivan CR and Zhang RY. Simplified design method for litz wire. In: *2014 IEEE Applied Power Electronics Conference and Exposition – APEC 2014*; 2014. p. 2667–2674.

[8] Hu J and Sullivan CR. Optimization of shapes for round-wire high-frequency gapped-inductor windings. In: *Conference Record of 1998 IEEE Industry Applications Conference. Thirty-Third IAS Annual Meeting (Cat. No. 98CH36242)*. vol. 2; 1998. p. 907–912.

[9] Van den Bossche A and Valchev VC. *Inductors and Transformers for Power Electronics*. Boca Raton, FL: CRC Press; 2005.

[10] Niedermeier G and Esguerra M. *Measurement of Power Losses with DC-Bias – The Displacement Factor*; 2000.

[11] Muhlethaler J, Biela J, Kolar JW, *et al.* Core losses under the DC bias condition based on Steinmetz parameters. *IEEE Transactions on Power Electronics*. 2012;27(2):953–963.

[12] Oliver CG. Measurement and modeling of core loss in powder core materials. ETTC Meeting – TTA Annual Meeting. 2012.

[13] Hartmann M, Ertl H, and Kolar JW. EMI filter design for high switching frequency three-phase/level PWM rectifier systems. In: *2010 Twenty-Fifth Annual IEEE Applied Power Electronics Conference and Exposition (APEC)*; 2010. p. 986–993.

[14] McLyman WT. *Transformer and Inductor Design Handbook*. Boca Raton, FL: CRC Press; 2011.

[15] Erickson RW and Maksimovic D. *Fundamentals of Power Electronics*. 2nd Edition. Berlin: Springer; 2001.

[16] Boillat DO, Kolar JW, and Mühlethaler J. Volume minimization of the main DM/CM EMI filter stage of a bidirectional three-phase three-level PWM rectifier system. In: *2013 IEEE Energy Conversion Congress and Exposition;* 2013. p. 2008–2019.

[17] Cox J and Nicol D. *Power Conversion & Line Filter Applications – Issue L*. Anaheim, CA: Micrometals; 2007.

[18] Hurley WG and Wölfle WH. Transformers and Inductors for Power Electronics – Theory, *Design and Applications*. New York, NY: Wiley; 2013.

[19] Cox J, Barbeito J, and Nicol D. Q Curves for Iron Powder Cores. Anaheim, *CA*; 1997.

Chapter 4

Measurement methods for magnetic properties of μm-thick steels used for high-frequency inductor cores

Nguyen Gia Minh Thao[1], Jumpei Tanase[2], Koichiro Nambu[3] and Keisuke Fujisaki[2]

4.1 Introduction

The μm-thick steels fabricated by the rolling process may be promising magnetic materials for inductor cores utilized in various power electronics applications at high frequencies [1]. Besides, the annealing process can be used to noticeably enhance the electromagnetic and mechanical properties of electrical steels and alloys in [2–5], but the characteristics at high frequency of the annealed μm-thick steels have not yet been considered thoroughly. In addition to the hot rolling process, some cold rolling techniques to improve the magnetic properties and hardness of steels were examined [6,7].

Wang *et al.* [8] studied on impacts of the cooling stage in the annealing process on magnetic properties of 6.5 wt.% Si alloy. Furthermore, a curable magnetic composite with low temperature and nanocrystalline block cores was researched and made for power electronic applications as described in [9,10], respectively. The comparison of magnetic properties between the flake and spherical powder cores was studied by Thao *et al.* [11]. However, it is fairly difficult for the mass production of the composite, nanocrystalline and powder cores in [9–11].

As presented in [12–14], some μm-thick iron powders with flaky shapes were fabricated and assessed to have relatively decent magnetic characteristics; nevertheless, the mass production of flaky-shaped powders for commercial purposes is fairly difficult. In addition, an effective technique to measure and evaluate the relative magnetic permeability and conductivity of fairly thin steels with a thickness of 0.1–0.25 mm was introduced in [15]. Besides, Fukuda and Shimotomai [16] researched and assessed electrical steel with a thickness of 0.08 mm. A thin electrical steel with a grain orientation and thickness of 30 μm has also been developed

[1]Shimane University, Department of Mechanical, Electrical and Electronic Engineering, Japan
[2]Toyota Technological Institute, Department of Advanced Science and Technology, Japan
[3]Osaka Sangyo University, Department of Mechanical Engineering, Japan

and examined in [17]. Furthermore, the manufacturing of grain-oriented silicon steels with a thickness of 5–8 μm was shown in [18], but the detailed impacts of high excitation frequencies of up to 1 MHz on the magnetic characteristics of those thin steels have not yet been considered and examined thoroughly.

In recent years, power electronic converters, motor drive inverters and wireless chargers utilized in electric vehicles (EVs) have been designed and operated at high frequencies from hundreds of kilohertz to megahertz for significantly reducing the sizes of magnetic inductor cores and improving the overall efficiency of the whole system. Therefore, magnetic materials suitable and effective for the inductor cores in these cases should be studied and developed.

According to our best knowledge, there are very few works on the detailed magnetic characterization and presentation of effective measurement and calculation techniques of annealed 1-μm-thick electrical steel at a high frequency of up to 1 MHz, which are useful for the design and make of the magnetic inductor cores for high-frequency power electronic applications in commercial EVs. In [19], our two-page digest very briefly showed several measured results of the two 1-μm-thick steels in a virtual conference; nonetheless, details of the improved measurement technique, proposed calculation method and computation results were not presented.

To try substantially reducing the core loss and improving the relative permeability of the 1-μm-thick steel fabricated by the continuous rolling process for power electronics and related applications at high frequencies, we study and implement a proper annealing process for this steel. This chapter presents the magnetic properties of the annealed 1-μm-thick steel compared to that of the non-annealed 1-μm-thick steel of the same material in terms of the core loss density and relative permeability in numerous experiments where the excitation frequency is up to 1 MHz. The composition of the 1-μm-thick steel is 99.63% Fe, 0.003% C, 0.35% Mn, 0.009% P and 0.007% S; the non-annealed steel is made using the continuous rolling process. With careful consideration of the electron backscatter diffraction data and actual durability of the annealed steel samples obtained at various temperatures, the annealing process for the steel is suitably performed at 450 °C in the atmosphere of Ar. Furthermore, an improved measurement method for determining the core loss density and relative magnetic permeability of the non-annealed and annealed steel parts attached to ring cores is shown in detail. Experimental results show significantly better performances on the core loss density and relative permeability of the annealed steel than that of the non-annealed one at high frequencies. Additionally, a calculation method to thoroughly validate the accuracy of the measured results is proposed. Lastly, physics-based insights and explanations for the obtained results are also provided.

In addition, the peak flux density transferring through the investigated steel part in this chapter is fixed as $B_{Fe1,m}$ = 0.3 T at all excitation frequencies of up to 1 MHz. Whereas, the prior works in [13,14] examined the significantly smaller peak flux density, i.e. $B_{Fe,m}$ = 50 mT at a peak frequency of 100 kHz [13] and $B_{Fe,m}$ = 3 mT at a peak frequency of 1 MHz [14]. Because of the significant nonlinearity of the magnetic characteristics, the detailed measurement data and thorough assessment of magnetic characteristics of the designed steel with a pretty large flux density of 0.3 T at high frequencies of up to 1 MHz is helpful for related applications, especially inductor cores used in high-frequency power electronic converters and wireless chargers for EVs.

4.2 Fabrication of non-annealed and annealed steels

Figure 4.1 shows the dimension of the 1-μm-thick steel sheet considered in this chapter. The composition of the 1-μm-thick steel is 99.63% Fe, 0.003% C, 0.35% Mn, 0.009% P and 0.007% S; this steel was made with the support from Maruyoshi Kogyo Co., Ltd., in Japan. Additionally, the electron backscatter diffraction (EBSD) of the considered steel sheets can be measured and obtained using a scanning electron microscope SU6600 from Hitachi, as shown in Figure 4.2. As illustrated in Figure 4.3(a), the non-annealed steel has a structure in which a stretched form is observed, and the (001) and (111) planes are mixed.

In this study, the annealing process is suitably performed at a temperature of 450 °C in the atmosphere of Ar. The temperature was raised from room temperature to 450 °C within 1 h and then held for 1 h; finally, the 1-μm-thick steel was cooled in a furnace. As depicted in Figure 4.3(b), the horizontally elongated tissues were observed. Moreover, the (001) plane is the main crystal orientation. The crystal rotates from the (111) plane to the (001) plane, and the crystal orientation of the (111) plane is slightly reduced as compared to that annealed at 350 °C, as shown in Figure 4.4(a).

According to Figure 4.4(a), when the steel is annealed at a temperature of 350 °C that is a little lower than the typical condition, a horizontally elongated tissue was observed. The (001) plane is the major crystal orientation, but the crystal orientation of the (111) plane similar to the material in Figure 4.3(a) is observed. Whereas, when the steel is annealed at a temperature of 550 °C, as shown in Figure 4.4(b), most of the entire surface becomes the (001) plane, and we see that the recrystallization was partially performed. Moreover, the annealed steel sample, in this case, is easily torn and damaged, which causes significant difficulty in manually making an experimental ring core comprising the steel part as well as measuring its magnetic properties properly. Whereas, once the steel is annealed at a temperature of 850 °C as given in Figure 4.4(c), almost the entire surface is still with the (001) plane; nevertheless, the recrystallization was performed. Despite the annealing in the atmosphere of Ar, it reacts with a small amount of oxygen, and oxidation occurs; this causes difficulty in preparing a good sample for the iron loss

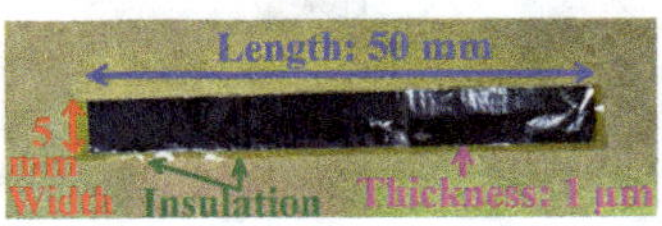

Figure 4.1 Dimension of 1-μm-thick steel considered in this chapter

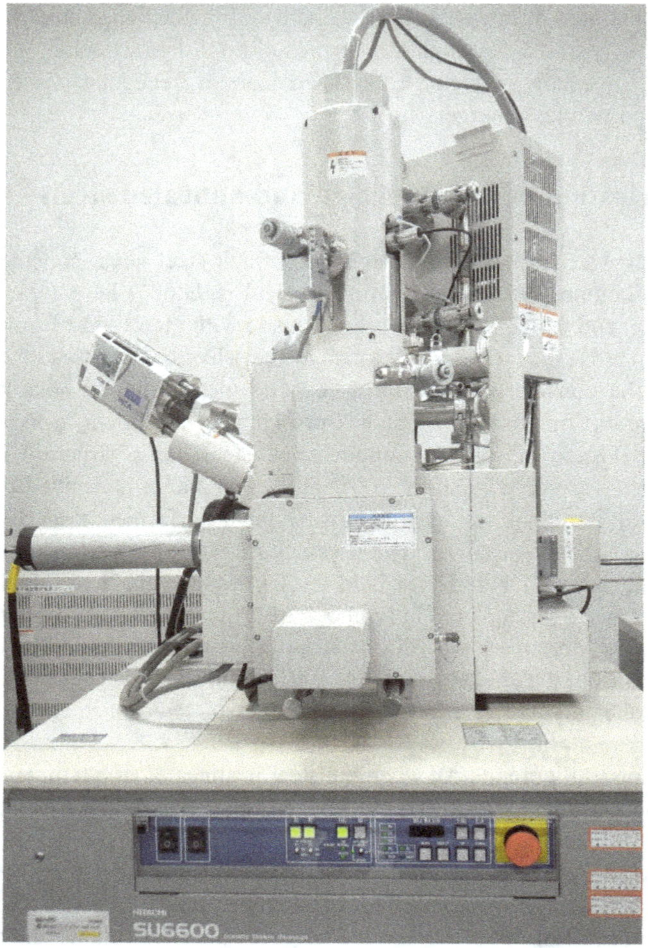

Figure 4.2 Scanning electron microscope SU6600 from Hitachi

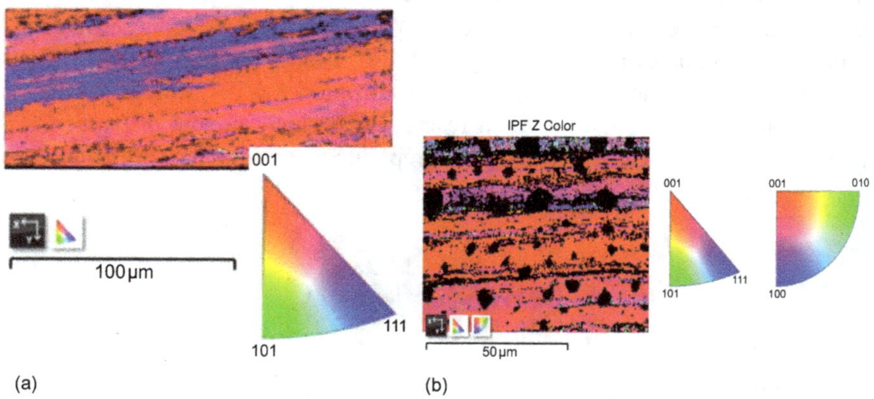

Figure 4.3 EBSD data of two 1-μm-thick steels considered in this research: (a) non-annealed steel and (b) annealed steel at a temperature of 450 °C

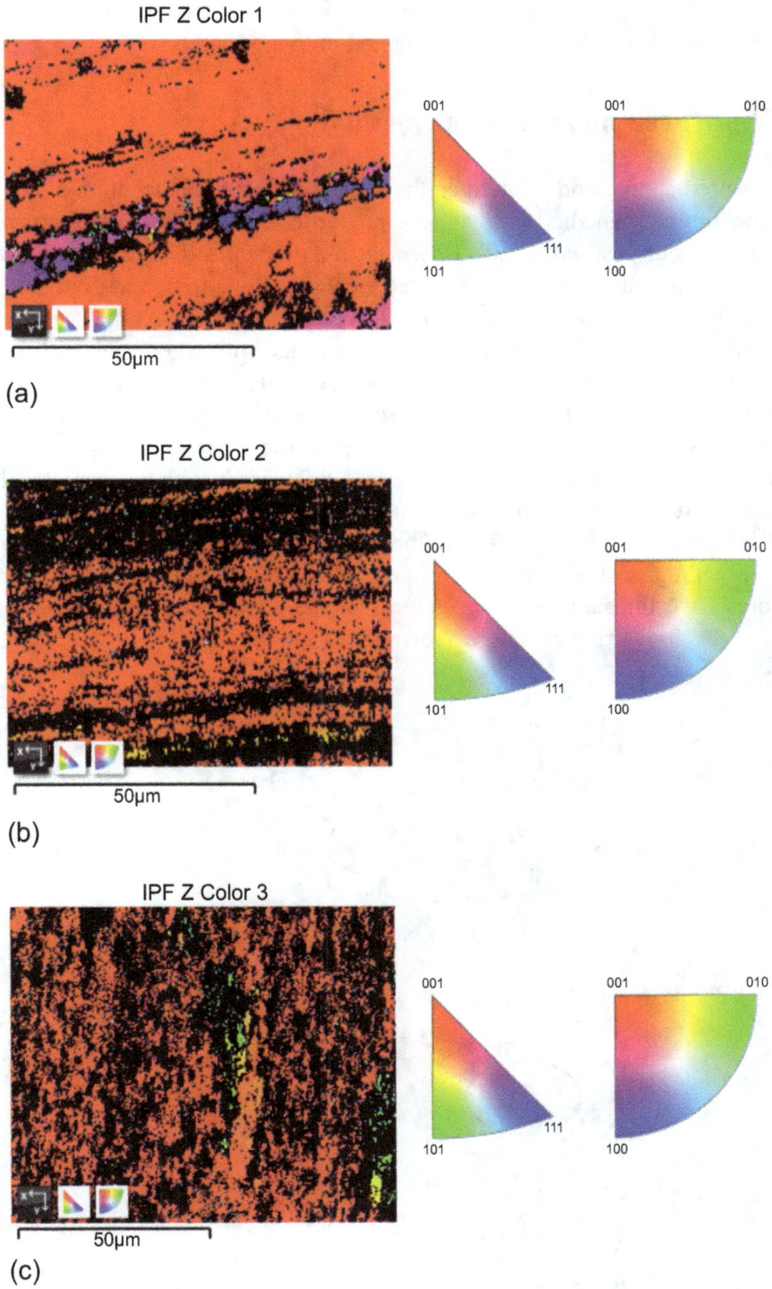

Figure 4.4 EBSD data of 1-μm-thick steel annealed at different temperatures: (a) at a temperature of 350 °C, (b) at a temperature of 550 °C and (c) at a temperature of 850 °C

evaluation. From Figures 4.3 and 4.4, the annealing at a temperature of 450 °C is suitable for the 1-μm-thick steel investigated in this research.

4.3 Improved measurement method

The measurement method is based on the sinusoidal excitation technique. The basis of the measurement method for magnetic properties of 4-μm-thick steel was briefly introduced in our past work [20]. However, the method in [20] cannot ensure the peak flux density through the investigated steel part $B_{Fe,m}$ to be fixed. In detail, the peak magnetic flux density is slightly changed in a small range of $B_{Fe,m}$ = 0.262–0.264 T, but probably variable in a large range with other steels or ring cores under test. Furthermore, our previous two-page conference digest in [19] did not present a calculation method for careful validation of the measured results.

For measurement and evaluation purposes, three separate ring cores are made in this work. As shown in Figure 4.5, the first ring core is with the annealed 1-μm-thick steel part attached, while the second ring core is with the non-annealed steel part attached; the two ring cores have the same dimension. The last one is a plastic ring core without a steel part, which is used for reference purposes. All the three ring cores have the same number of turns of the copper coils. The related parameters of the above ring cores are shown in Tables 4.1–4.3.

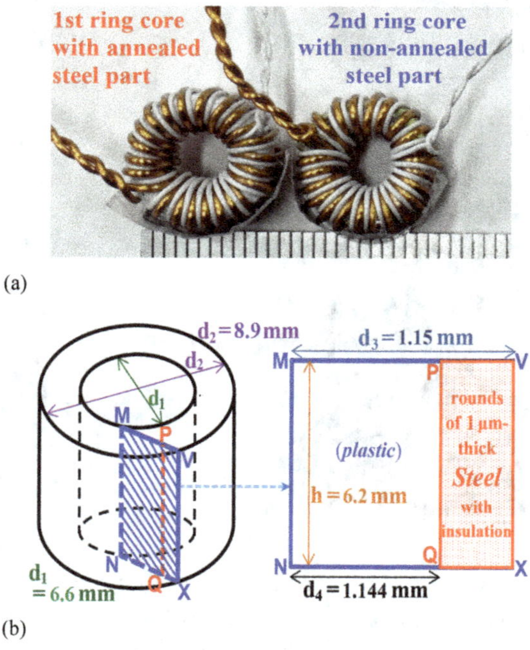

Figure 4.5 Details of ring cores with steel parts considered: (a) first and second ring cores with the same dimension and coil turns, and (b) dimension of ring core with steel part and insulation attached on plastic chassis

Table 4.1 Parameters of first ring core with annealed steel part

Parameter	Symbol	Unit
Peak magnetic flux density	$B_{\text{Ring1,m}}$	T
Magnetic retentivity	$B_{\text{Ring1,c}}$	T
Magnetic field intensity	$H_{\text{Ring1,m}}$	A/m
Magnetic coercivity	$H_{\text{Ring1,c}}$	A/m
Cross-sectional area	S_{Ring1}	m^2
Magnetic flux of first ring core	Φ_{Ring1}	Wb
Peak flux density through annealed steel part	$B_{\text{Fe,Ring1,m}}$	T
Relative permeability of annealed steel part	$\mu_{\text{Fe,Ring1,r}}$	
Magnetic length of annealed steel part	$l_{\text{Fe,Ring1}}$	m

Table 4.2 Parameters of second ring core with non-annealed steel part

Parameter	Symbol	Unit
Peak magnetic flux density	$B_{\text{Ring2,m}}$	T
Magnetic retentivity	$B_{\text{Ring2,c}}$	T
Magnetic field intensity	$H_{\text{Ring2,m}}$	A/m
Magnetic coercivity	$H_{\text{Ring2,c}}$	A/m
Cross-sectional area	S_{Ring2}	m^2
Magnetic flux of second ring core	Φ_{Ring2}	Wb
Peak flux density through non-annealed steel part	$B_{\text{Fe,Ring2,m}}$	T
Relative permeability of non-annealed steel part	$\mu_{\text{Fe,Ring2,r}}$	
Magnetic length of non-annealed steel part	$l_{\text{Fe,Ring2}}$	m

Table 4.3 Parameters of third ring core without steel part

Parameter	Symbol	Unit
Peak magnetic flux density	$B_{\text{Ring3,m}}$	T
Magnetic retentivity	$B_{\text{Ring3,c}}$	T
Magnetic field intensity	$H_{\text{Ring3,m}}$	A/m
Magnetic coercivity	$H_{\text{Ring3,c}}$	A/m
Cross-sectional area	S_{Ring3}	m^2
Magnetic flux of third ring core without steel part	Φ_{Ring3}	Wb

As described in Figure 4.5(b), the cross section of the first or second ring core with a steel part is illustrated by the frame MVXN. In fact, the area of the overall frame MVXN is the summation of areas of the subframes MPQN and PVXQ, i.e. $S_{\text{MVXN}} = S_{\text{MPQN}} + S_{\text{PVXQ}}$. In the first ring core including the annealed steel part, the material of the chassis area indicated by the frame MPQN is plastic, and the other section shown by the frame PVXQ is the annealed steel part together with the insulation layer, i.e. the 5-film Kapton 20EN polyimide film [19,21]. As expressed in Figure 4.1, the insulation with a thickness of 5 μm which is yellow [21] is suitably affixed below the steel part. Conversely, in the second ring core comprising the non-annealed steel part, the material of the chassis area indicated by the frame MPQN is still plastic, but the other section shown by the frame PVXQ is the non-annealed steel part with the same insulation.

Moreover, the third ring core without a steel part, which has a similar number of turns of the two copper coils to those of the first two ring cores with steel parts as depicted in Figure 4.5(a), is utilized to measure the peak magnetic flux of the non-steel ring core $\Phi_{Ring3,m}$ under the same peak field strength with the first two ring cores, i.e. $H_{Ring3,m} = H_{Ring2,m} = H_{Ring1,m}$.

Whereas, in the third ring core as presented in the right part of Figure 4.5(b), the material of the chassis area indicated by the frame MPQN is also plastic; however, the other area covered by the frame PVXQ only consists of the insulation in use without a steel part.

The experimental measurement system for the three ring cores is operated with the sinusoidal excitation method. In detail, the *B–H* analyzer SY-8219 (Iwatsu) [22] is directly connected to a linear amplifier HSA 4014-IW (NF Corporation) to determine the core loss density P_{cm} and relative magnetic permeability μ_r of the considered ring cores [11,20]. According to the information of the manufacturer, the excitation frequency limit of the *B–H* analyzer is $f = 1$ MHz.

First, the primary input current I_1 and the secondary open-circuit voltage V_2 of a ring core are measured to compute the magnetic field strength *H* and flux density *B*, as shown in (4.1) and (4.2), respectively. After *H* and *B* are determined, the core loss density P_{cm} is calculated by (4.3). For each test, we have conducted five independent measurement times to thoroughly verify the repeatability and accuracy of the measured findings

$$B = \frac{1}{N_2 S} \int V_2 dt \tag{4.1}$$

$$H = \frac{N_1 I_1}{l} \tag{4.2}$$

$$P_{cm} = \frac{f_0}{\rho W_m} \int H dB \tag{4.3}$$

where $N_1 = N_2 = 23$ are the number of the primary and secondary coil turns of each ring core, respectively, $l = 0.05$ m is the magnetic path length, $\rho = 7874$ kg/m^3 is the steel density, W_m is the weight of the steel part attached in the ring core and *S* is the ring cross-sectional area. In detail, $S_{Ring1} \approx S_{Ring2}$ as shown in Figure 4.5(a); $S_{Ring3} \approx S_{Ring1} - S_{Fe,Ring1}$ where $S_{Fe,Ring1}$ is the area of the annealed steel part attached in the first ring core. Moreover, $S_{Fe,Ring1} \approx S_{Fe,Ring2}$, where $S_{Fe,Ring2}$ is the area of the non-annealed steel part attached in the second ring core.

In this work, the main principle of the introduced measurement method is to appropriately adjust the peak magnetic field strength of the first or second ring core for forcing the peak magnetic flux density through the steel part attached to the first or second ring core to be fixed at 0.3 T, respectively. The measurement method has five key steps as below:

Step 1: The first ring core comprising the annealed 1-μm-thick steel part is suitably connected to the *B–H* analyzer SY-8219 from Iwatsu for measurement and evaluation. In settings of the *B–H* analyzer, the frequency of the sinusoidal excitation waveform is varied from f = 10, 50, 100, 200, 300 kHz and up to 1 MHz. In each test of this step, the peak magnetic field strength $H_{\mathrm{Ring1,m}}$ of the first ring core is appropriately set to force the magnetic flux density of the steel part $B_{\mathrm{Fe,Ring1,m}}$ to be fixed at 0.3 T at all the excitation frequencies (Figure 4.6).

Step 2: After Step 1 is completed, the first ring core is removed from the *B–H* analyzer, and the third ring core without a steel part is connected to the *B–H* analyzer SY-8219 for measuring the peak magnetic flux through the non-steel area $\Phi_{\mathrm{air}(i),\mathrm{m}}$ in the first ring core, as given by (4.4). In this step, the peak magnetic field strength $H_{\mathrm{Ring3,m}}$ of the third ring core should be chosen to be equal to $H_{\mathrm{Ring1,m}}$ of the first ring core performed in Step 1 at each excitation frequency.

Step 3: After Steps 1 and 2 are completed, from the six known parameters $B_{\mathrm{Ring1,m}}$, $H_{\mathrm{Ring1,m}}$, S_{Ring1}, $B_{\mathrm{Ring3,m}}$, $H_{\mathrm{Ring3,m}}$ and S_{Ring3}, where $H_{\mathrm{Ring1,m}} = H_{\mathrm{Ring3,m}}$, the peak flux density through the annealed steel part $B_{\mathrm{Fe,Ring1,m}}$ and its relative magnetic permeability $\mu_{\mathrm{Fe,Ring1,r}}$ in the first ring core at each excitation frequency are computed by (4.5) and (4.6), respectively, where the index is i = 1. We note that $H_{\mathrm{Ring1,m}}$ is appropriately set in Step 1 to force $B_{\mathrm{Fe,Ring1,m}}$ to be fixed at 0.3 T at all the excitation frequencies (Figure 4.6).

Step 4: After the measurement and evaluation of the first ring core are completed, we repeat Step 1 with the second ring core including the non-annealed steel part (instead of the first ring core) and then repeat Step 2 with the third ring core without a steel part where $H_{\mathrm{Ring3,m}} = H_{\mathrm{Ring2,m}}$.

Step 5: Lastly, the peak flux density through the non-annealed steel part $B_{\mathrm{Fe,Ring2,m}}$ and its relative magnetic permeability $\mu_{\mathrm{Fe,Ring2,r}}$ in the second ring core can be estimated by (4.5) and (4.6), respectively, where the index is i = 2. We also note that

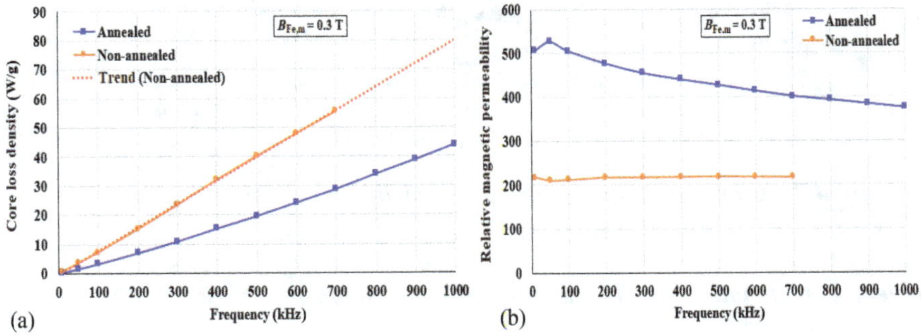

Figure 4.6 Measured results of core loss density and relative permeability, $B_{Fe,m}$ = 0.3 T: (a) core loss density P_{cm} and (b) relative permeability $\mu_{Fe,r}$

$H_{\text{Ring2,m}}$ is suitably chosen in Step 4 to make $B_{\text{Fe,Ring2,m}}$ to be fixed at 0.3 T at all the excitation frequencies (Figure 4.6).

$$\Phi_{\text{Ring}(i),\text{m}} = \Phi_{\text{Fe,Ring}(i),\text{m}} + \Phi_{\text{air}(i),\text{m}} \approx \Phi_{\text{Fe,Ring}(i),\text{m}} + \Phi_{\text{Ring3,m}} \tag{4.4}$$

where the index is $i = 1$ for the first ring core examined in Steps 1–3 or $i = 2$ for the second ring core considered in Steps 4 and 5. Furthermore, $\Phi_{\text{Fe,Ring}(i),\text{m}}$, $\Phi_{\text{Ring}(i),\text{m}}$ and $\Phi_{\text{Ring3,m}}$ are the peak magnetic flux values of the steel part attached in the first or second ring core, the first or second ring core comprising the steel part and the third ring core without a steel part, respectively,

$$B_{\text{Fe,Ring}(i),\text{m}} \approx \frac{\Phi_{\text{Ring}(i),\text{m}} - \Phi_{\text{Ring3,m}}}{S_{\text{Fe,Ring}(i)}} = \frac{B_{\text{Ring}(i),\text{m}} S_{\text{Ring}(i)} - B_{\text{Ring3,m}} S_{\text{Ring3}}}{S_{\text{Fe,Ring}(i)}} \tag{4.5}$$

$$\mu_{\text{Fe,Ring}(i),\text{r}} = \frac{B_{\text{Fe,Ring}(i),\text{m}}}{\mu_0 H_{\text{Ring}(i),\text{m}}} \tag{4.6}$$

where $\Phi_{\text{air}(i),\text{m}} \approx \Phi_{\text{Ring3,m}}$, $S_{\text{Ring3}} \approx S_{\text{Ring}(i)} - S_{\text{Fe,Ring}(i)}$ and $\mu_0 = 4\pi \times 10^{-7}$ H/m.

4.4 Measured results in experiments

The measured results of P_{cm} and $\mu_{\text{Fe,r}}$ are shown in Figures 4.6 and 4.7, where $B_{\text{Fe,m}}$ is fixed at 0.3 T, and the excitation frequency is changed from f = 10, 50, 100, 200 kHz and up to 1 MHz. The validation and discussion of the results obtained in the tests are presented in Sections 4.5 and 4.6, respectively.

Additionally, Figure 4.7 shows the characteristics of core loss density and frequency P_{cm}/f, and Table 4.4 depicts the estimated coefficients for the hysteresis

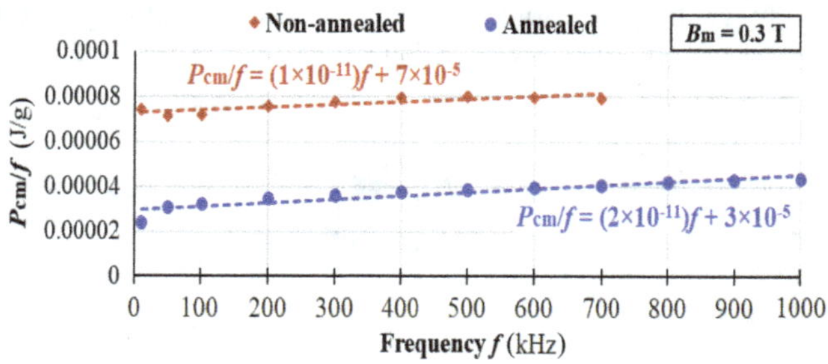

Figure 4.7 Characteristics of core loss density and frequency, $B_{Fe,m}$ = 0.3 T

Table 4.4 Estimated coefficients of core loss components in Figure 4.7

Core loss components	Non-annealed steel	Annealed steel
Hysteresis coefficient	7×10^{-5}	3×10^{-5}
Eddy current coefficient	1×10^{-11}	2×10^{-11}

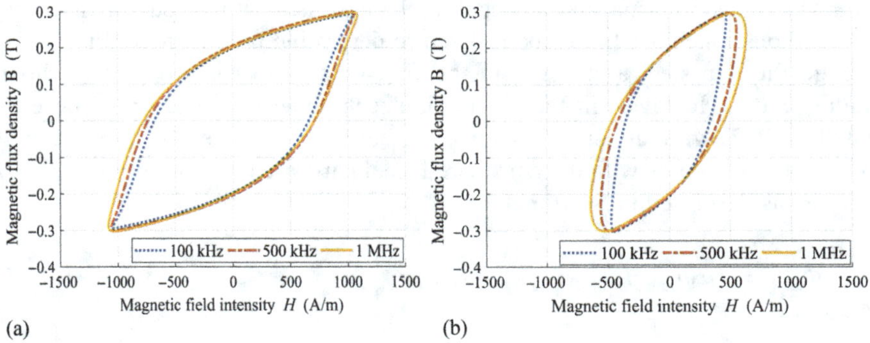

Figure 4.8 Obtained B–H curves of two 1-μm-thick steels, $B_{Fe,m}$ = 0.3 T: (a) non-annealed steel and (b) annealed steel at a temperature of 450 °C

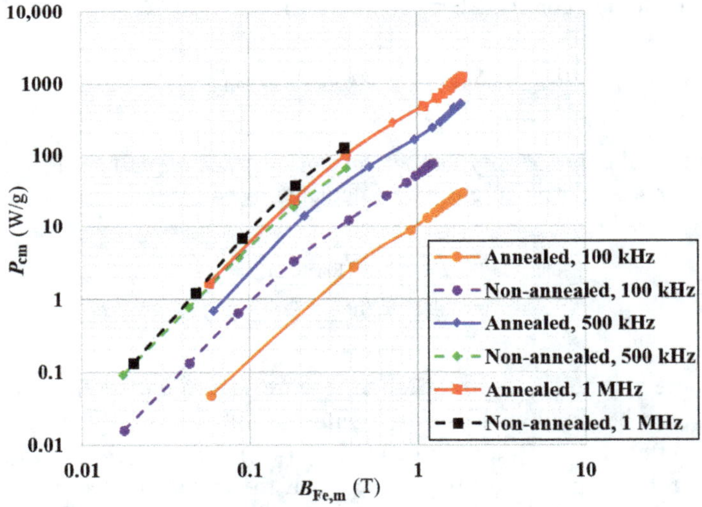

Figure 4.9 Measured P_{cm} of two steels under various $B_{Fe,m}$ and f in additional tests

and eddy current loss components in the core losses of the two tested steels. Furthermore, the obtained *B–H* curves of the two steels are described and compared in Figure 4.8. In addition, to thoroughly check and confirm the advantages of the annealed steel, the measured P_{cm} of the two steels under various $B_{Fe,m}$ and f are presented and compared in Figure 4.9.

4.5 Proposed calculation technique for validation

To carefully validate the correctness and reliability of the measurement method presented in Section 4.3, a calculation technique without using the third ring core is introduced and implemented. In detail, in the measurement method, the third ring core without a steel part must be utilized to determine the magnetic flux $\Phi_{\text{air}(i)}$ through the non-steel area $S_{\text{air}(i)}$ in the first or second ring core as given in Figure 4.5(b). However, in the calculation technique, $\Phi_{\text{air}(i)}$ can now be estimated by (4.7) without using the third ring core. This means that only the first and second ring cores with the considered steel parts are measured with the *B*–*H* analyzer as described in Section 4.3.

$$\Phi_{\text{air}(i)} = \left(\mu_0 \mu_{\text{r,air}} H_{\text{Ring}(i)} S_{\text{air}(i)}\right) \times k_{\text{coil}} \tag{4.7}$$

where $S_{\text{air}(i)} = S_{\text{Ring}(i)} - S_{\text{Fe,Ring}(i)}$, $\mu_{\text{r,air}} \approx 1$, and the index $i = 1$ or 2 for the first or second ring core, respectively. Additionally, k_{coil} in (4.7) is the filling factor of the coil and often has a value in the range of 0%–100%; in this chapter, for simplicity, we can set $k_{\text{coil}} = 100\%$. From (4.4) and (4.7), the peak flux through the investigated steel part in the ring core is determined in (4.8)

$$\Phi_{\text{Fe,Ring}(i),\text{m}} \approx B_{\text{Ring}(i),\text{m}} S_{\text{Ring}(i)} - \mu_0 H_{\text{Ring}(i),\text{m}} S_{\text{air}(i)} \tag{4.8}$$

After that, the flux density through the considered steel part $B_{\text{Fe,Ring}(i)}$ attached in the first or second ring core can be calculated by (4.9)

$$B_{\text{Fe,Ring}(i),\text{m}} \approx \frac{B_{\text{Ring}(i),\text{m}} S_{\text{Ring}(i)} - \mu_0 H_{\text{Ring}(i),\text{m}} S_{\text{air}(i)}}{S_{\text{Fe,Ring}(i)}} \tag{4.9}$$

We note that $H_{\text{Ring}(i),\text{m}}$ of each ring core is properly adjusted to force $B_{\text{Fe,Ring}(i),\text{m}}$ to be fixed at the desired value of 0.3 T at all the excitation frequencies. After $B_{\text{Fe,Ring}(i),\text{m}}$ is calculated by (4.9) where $S_{\text{air}(i)} = S_{\text{Ring}(i)} - S_{\text{Fe,Ring}(i)}$, the relative permeability $\mu_{\text{Fe,Ring}(i),\text{r}}$ of the annealed steel part included in the first ring core (where $i = 1$) or the non-annealed steel part comprised in the second ring core (where $i = 2$) can be computed by (4.6).

As depicted in Figure 4.10 (with different H_{m}) and Figure 4.11 (with various B_{m} and *f*), the results obtained from the improved measurement method in Sections 4.3 and 4.4 are nearly similar to that from the proposed calculation technique in many experiments.

4.6 Discussion on obtained results

As depicted in Figure 4.6(a), at the frequency $f = 700$ kHz, the core loss density values P_{cm} of the non-annealed and annealed 1-μm-thick steels are 55.41 and 28.71 W/g, respectively; furthermore, at $f = 1$ MHz, P_{cm} of the annealed steel is 43.95 W/g, while the trended core loss of the non-annealed steel is around 80 W/g

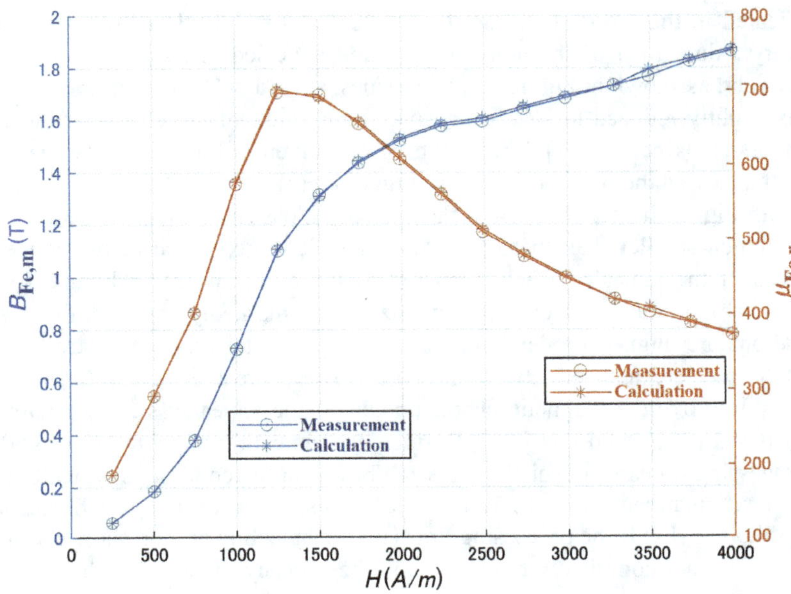

Figure 4.10 *Comparisons of obtained $B_{Fe,m}$ and $\mu_{Fe,r}$ of annealed steel under various H_m between measurement and calculation methods, f = 1 MHz*

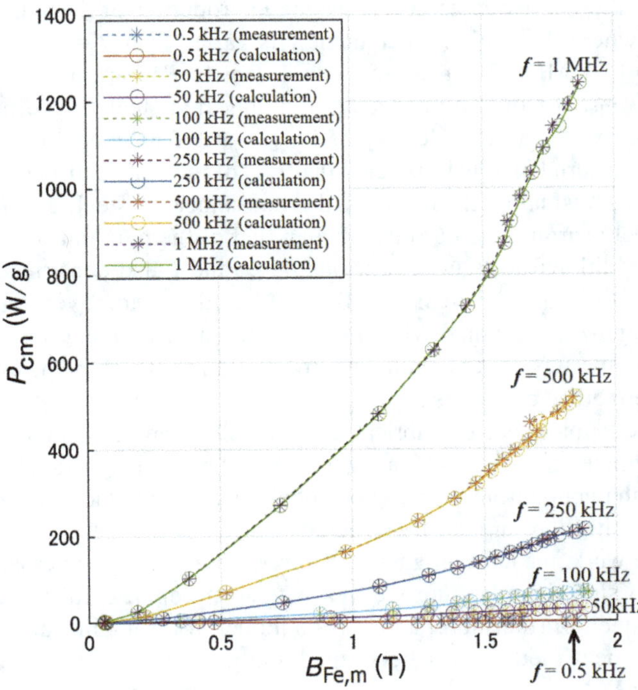

Figure 4.11 *Comparisons of obtained P_{cm} of annealed steel under various B_m and f between measurement and calculation methods*

as indicated in the red dashed line. In addition, at $f = 700$ kHz, the relative permeability values $\mu_{Fe,r}$ of the non-annealed and annealed steels are 216 and 401, respectively, as shown in Figure 4.6(b). Besides, at $f = 1$ MHz, $\mu_{Fe,r}$ of the annealed steel is slightly reduced to 375, which may be because of the effects of the eddy current, as explained in [11]. The results of P_{cm} given in Figure 4.9 also have the same tendency as the ones described in Figure 4.6(a).

From Figure 4.7 and Table 4.4, the hysteresis loss component of the annealed steel is much smaller than that of the non-annealed one, which helps reduce the core loss of the annealed steel compared to the other shown in Figure 4.9. In Figure 4.3(b), although the crystal grain size is not markedly changed, the crystal orientations are aligned properly; the strain in the crystal grains and stress in the annealed steel are also effectively removed. These are the key factors to substantially improve the magnetic characteristics of the annealed steel compared to that of the non-annealed one [3,4]. Besides, the annealing temperature of 450 °C and processing times were suitably selected based on the conditions typically used in the heat treatment field [2,3] and actual responses of the tested steels in Figures 4.3 and 4.4. If the processing times of the annealing are too short, the strain in the steel is not completely removed, and the efficacy in reducing the core loss may be decreased.

4.7 Conclusion

This chapter thoroughly assessed the magnetic properties of the annealed 1-μm-thick steel, where the excitation frequency is varied from $f = 10$, 50, 100, 200 kHz and up to 1 MHz. First, the core loss density P_{cm} and relative magnetic permeability $\mu_{Fe,r}$ of the non-annealed and annealed steels made of the same material at a fixed peak flux density of $B_{Fe,m} = 0.3$ T were examined and compared. The measured results illustrated that P_{cm} of the annealed 1-μm-thick steel is around 48.2% smaller than that of the non-annealed steel, and $\mu_{Fe,r}$ of the annealed steel is around 185.6% larger than that of the non-annealed one. After that, many additional tests under various $B_{Fe,m}$ and f also confirmed the much lower P_{cm} of the annealed steel than that of the other. Moreover, the results of $B_{Fe,m}$, $\mu_{Fe,r}$ and P_{cm} obtained with the presented measurement and calculation methods in numerous experiments are almost similar, which helps evaluate and verify the reliability of the results.

Thus, the 1-μm-thick steel fabricated by the rolling process and then processed in the appropriate annealing, which has noticeably better magnetic characteristics than that of the non-annealed one, is potentially used for inductor cores in power electronics applications at high frequencies.

In future work, the annealing temperature will be suitably increased to 700 °C or 800 °C to significantly improve the magnetic properties of the considered 1-μm-thick steel. Furthermore, an analysis model of the 1-μm-thick steel using the finite element method will be studied and developed to validate the measured findings.

References

[1] M. Komatsubara, K. Sadahiro, O. Kondo, T. Takamiya, and A. Honda, "Newly developed electrical steel for high-frequency use," *Journal of Magnetism & Magnetic Materials*, vol. 242–245, pp. 212–215, 2002.

[2] M. C. Guimaraes, F. L. Alcantara, R. U. C. Moreira, N. E. F. Lima, and J. G. Ank, "Improvement of magnetic properties of a reversible hot rolled grain-oriented electrical steel," *IEEE Transactions on Magnetics*, vol. 52, no. 5, p. 2002704, 2016.

[3] D. Li, L. Zhang, G. Li, Z. Lu, and S. Zhou, "Reducing the core loss of amorphous cores for distribution transformers," *Progress in Natural Science: Materials International*, vol. 22, no. 3, pp. 244–249, 2012.

[4] E. J. Hilinski and G. H. Johnston, "Annealing of electrical steel," *Proc. of 2014 4th International Electric Drives Production Conference*, pp. 1–8, Nuremberg, Germany, 30 September–1 October 2014.

[5] D. L. Rodrigues, T. S. P. Nishikawa, A. A. Almeida, F. J. G. Landgraf, and R. V. Martin, "The effect of recovery annealing on the magnetic and mechanical properties of nonoriented electrical steels," *IEEE Transactions on Magnetics*, vol. 50, no. 4, p. 2002204, 2014.

[6] H. Kikuchi, F. Ito, T. Murakami, and K. Takekawa, "Relationship between magnetic properties and hardness and its effect on recovery and recrystallization in cold-rolled steel," *IEEE Transactions on Magnetics*, vol. 51, no. 11, p. 2004804, 2015.

[7] S. Steentjes, N. Leuning, J. Dierdorf, *et al.*, "Effect of the interdependence of cold rolling strategies and subsequent punching on magnetic properties of NO steel sheets," *IEEE Transactions on Magnetics*, vol. 52, no. 5, p. 2001504, 2016.

[8] X. Wang, H. Liu, H. Li, and Z. Liu, "Effect of cooling rate on order degree of 6.5 wt.% Si electrical steel after annealing treatment," *IEEE Transactions on Magnetics*, vol. 51, no. 11, p. 2005704, 2015.

[9] Y. Yan, W. Sun, S. Gao, T. Ge, K. D. T. Ngo, and G.-Q. Lu, "Design, fabrication, and characterization of a low-temperature curable magnetic composite for power electronics integration," *IEEE Transactions on Magnetics*, vol. 54, no. 1, p. 2800106, 2018.

[10] M. Soinski, J. Leszczynski, C. Swieboda, and M. Kwiecien, "Nanocrystalline block cores for high-frequency chokes," *IEEE Transactions on Magnetics*, vol. 52, no. 11, p. 2801904, 2014.

[11] N. G. M. Thao, K. Fujisaki, L. Ton That, and S. Motozuka, "Magnetic comparison between experimental flake powder and spherical powder for inductor cores at high frequency," *IEEE Transactions on Magnetics*, vol. 57, no. 2, p. 8400407, 2021.

[12] M. Hara, M. Namba, S. Tajima, M. Tani, T. Hattori, and Y. Kaneko, "Properties of iron core fabricated from flaky-shaped and annealed pure iron powder," *Materials Transactions*, vol. 59, no. 12, pp. 1920–1927, 2018.

[13] N. Nakamura, Y. Ozaki, and K. Higuchi, "Flaky-shaped iron powder "KIP MG150D" for soft magnetic applications," *Kawasaki Steel Technical Report*, no. 47, pp. 60–64, 2002.
[14] T. Tosaka, I. Nagano, S. Yagitani, and Y. Yoshimura, "Determining the relative permeability and conductivity of thin materials," *IEEE Transactions on Electromagnetic Compatibility*, vol. 57, no. 2, pp. 352–360, 2005.
[15] S. Ueno, M. Enokizono, Y. Mori, and K. Yamazaki, "Vector magnetic characteristics of ultra-thin electrical steel sheet for development of high-efficiency high-speed motor," *IEEE Transactions on Magnetics*, vol. 53, no. 11, p. 6300604, 2017.
[16] Y. Fukuda and M. Shimotomai, "Magnetic properties of thin sheets (50 μm) of Nd-(FeCoNiTi)-B sintered magnets," *IEEE Transactions on Magnetics*, vol. 9, no. 5, pp. 40–46, 1994.
[17] M. Nakano, K. Ishiyama, and K.L. Arai, "Production of ultra thin grain oriented silicon steel sheets," *IEEE Transactions on Magnetics*, vol. 31, no. 6, pp. 3886–3888, 1995.
[18] R. Ramprasad, P. Zurcher, M. Petras, and M. Miller, "Magnetic properties of metallic ferromagnetic nanoparticle composites", *Journal of Applied Physics*, vol. 96, no. 1, pp. 519–529, 2004.
[19] N. G. M. Thao, J. Tanase, K. Nambu, F. Keisuke, and E. Tsuchida, "Experimental investigation and comparison of magnetic properties at high frequency between non-annealed and annealed 1 μm-thick steels," *Digest Book of the 2021 IEEE International Magnetics Conference*, digest JA-11, pp. 1199–1200, Lyon, France (virtual), 26–30 April 2021.
[20] N. G. M. Thao, K. Fujisaki, H. Mamiya, S. Kuroda, Y. Mitarai, and N. Motohashi, "Magnetic characterization of 4 μm-thick steel made by continuous rolling process for power electronic applications in high frequency," *Proc. of the 23rd International Conference on Electrical Machines and Systems*, pp. 1031–1034, Hamamatsu, Japan, 24–27 November 2020.
[21] DuPont™ Kapton® 20EN polyimide film, [Online] accessed on 18 April 2024: www.dupont.com/electronics-industrial/kapton-polyimide-film.html
[22] B-H Analyzer SY-8218/SY-8219, Iwatsu Electric Co., Ltd., Japan.

Chapter 5
Winding types in high-frequency power electronics applications

Alexander Stadler[1] *and Lukas Reißenweber*[1]

The materials and structure of a winding significantly influences the properties of inductive components. Even a single winding leads to losses and parasitic capacitive coupling to the core and the environment. The properties of a winding can be separated into the magnetic and parasitic capacitive couplings, the losses, and the thermal behavior. An improvement of one property often causes a deterioration of the others. Therefore, the optimum design of a winding always entails a compromise of the aforementioned properties.

The properties can be influenced by numerous variables, including the type of conductor used, such as solid conductors made of round wire, flat wire, or foil. Additionally, more specialized conductors, such as high-frequency (HF) litz wire or waveguides, and their materials. The configuration of one or more windings, along with their placement relative to the core, introduces additional degrees of freedom that must be accounted for. Similarly, the inclusion of insulating foils - such as those used to enhance resistance against electrical fields - or chambers within the coil former to reduce capacitance, further influence the design. Finally, factors such as active or passive thermal management and the application-specific voltage and current characteristics are also critical considerations.

Initially, an overview regarding inductive (magnetic) and capacitive (electric) coupling including the occurring loss mechanisms, is given. Based on this, the characteristics of different windings are examined using the example of an inductor for a 1.6 kW buck converter, which transforms an input voltage of 240 V to an output voltage of 180 V at a switching frequency of 100 kHz. The focus lies on the winding losses of different windings with HF litz wire, foil, or flat wire and their parasitic properties.

5.1 Inductive and capacitive coupling

Inductive and capacitive couplings are dependent on each other. Without a magnetic core, a good magnetic coupling results in general also in a high parasitic

[1]Coburg University of Applied Sciences and Arts, Department for Electrical Engineering and Computer Sciences, Germany

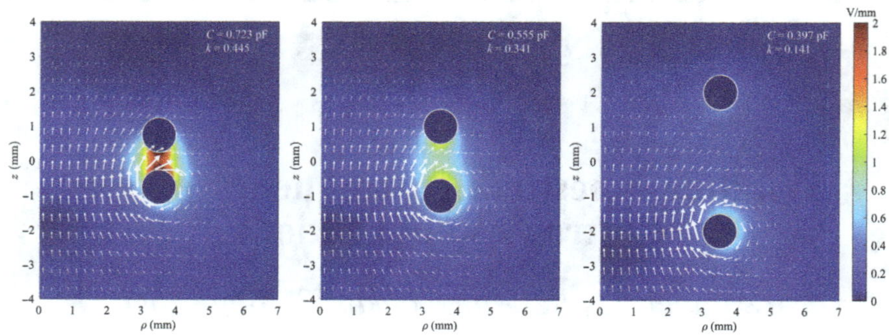

Figure 5.1 Magnitude of the electrical field strength and arrows proportional to the magnetic flux density of two single conductors with different spacing. For the electrical field strength, the lower turn has U = 1 V, and the upper turn has U = 0 V. For the arrows proportional to the magnetic flux density, the lower turn has I = 1 A, and the upper turn has I = 0 A.

coupling capacitance and vice versa. Figure 5.1 shows two single conductors in a two-dimensional axial symmetrical arrangement with different spacing in the z-direction. The diameter of the cross-section of the conductors is 1 mm, and the inner diameter of the turns is 10 mm. The resulting electric field strength is shown, with the lower and upper conductors having potentials of 1 V and 0 V, respectively. The sizes of the arrows are proportional to the magnetic flux density outside the conductor, with the lower and upper conductors carrying currents of 1 A and 0 A, respectively. The calculated values of the capacitance and the coupling factor of the arrangement is denoted in each sub-figure. The reduction in magnetic coupling with increasing spacing between the windings is evident from the decreasing amplitude of the arrows, which are proportional to the flux density and located within the upper conductor loop. The decrease in capacitance is shown by the reduction of the electrical field strength with increasing distance. This shows that a good magnetic coupling simultaneously leads to an increase in the usually undesirable parasitic capacitance. However, parasitic capacitances can be used in specific applications, e.g., as resonant elements in DC/DC power converters (DC direct current) [1].

The inductance L of an arrangement is calculated from the stored magnetic energy W_m. This can be calculated from the volume integral of the magnetic energy density w_m and from the magnetic field strength $\vec{H}$ and the magnetic flux density $\vec{B}$ [2,3]:

$$W_m = \iiint_{V_\infty} w_m \, dV = \frac{1}{2} \iiint_{V_\infty} \vec{H} \cdot \vec{B} \, dV = \frac{1}{2} L I^2 \rightarrow L = \frac{2 W_m}{I^2} \tag{5.1}$$

The inductance of a winding with N turns is made up of the respective self-inductances L_{ii} of the individual conductors and their mutual inductances L_{ij}:

$$L = \sum_{i=1}^{N} L_{ii} + \sum_{i=1}^{N} \sum_{j=1, j \neq i}^{N} L_{ij}. \tag{5.2}$$

The self-inductance L_{ii} results from the outer L_o and the inner self-inductance L_i. The outer one is calculated by the magnetic flux Φ_i passing through the turn related to the current I causing it

$$L_\text{o} = \frac{\Phi_i}{I} \quad \text{with } \Phi_i = \iint_{A_i} \vec{B} \cdot \text{d}\vec{A}_i, \tag{5.3}$$

where the magnetic flux results from the area integral of the magnetic flux density that penetrates the inner surface $\vec{A}_i$ of the turn.

The inner inductance is determined from the magnetic energy $W_\text{m,i}$ in relation to the square of the causing current. For a round conductor, it can be shown that this is proportional to the length l of the conductor loop and a factor including the magnetic field constant μ_0 [2,3]

$$L_\text{i} = \frac{2}{I^2} W_\text{m,i} = \frac{\mu_0 l}{8\pi}. \tag{5.4}$$

The self-inductance is therefore given as follows:

$$L_{ii} = L_\text{o} + L_\text{i} = \frac{\iint_{A_i} \vec{B} \cdot \text{d}\vec{A}_i}{I} + \frac{\mu_0 l}{8\pi}. \tag{5.5}$$

In the example in Figure 5.1, the self-inductance of the turns is always the same at all distances. The mutual inductance L_{ij}, conversely, changes. It results from the proportion of the magnetic flux Φ_{ij} that is caused by the current I_j in the other winding. Where the magnetic flux can be calculated from the area integral of the magnetic flux density $\vec{B}_j$:

$$L_{ij} = \frac{\Phi_{ij}}{I_j} \quad \text{with } \Phi_{ij} = \iint_{A_i} \vec{B}_j \cdot \text{d}\vec{A}_i. \tag{5.6}$$

The coupling factor k_{ij} describes the relation between the mutual inductance and the self-inductance:

$$k_{ij} = \frac{L_{ij}}{L_{ii}} = \frac{\Phi_{ij}}{\Phi_{ii}}. \tag{5.7}$$

Similar to the calculation of the inductance from the magnetic energy, the capacitance C of an arrangement can be calculated by the electric energy W_e. This can be calculated from the volume integral of the electric energy density w_e and from the electric field strength $\vec{E}$ and the electric flux density $\vec{D}$ [1–3]:

$$W_\text{e} = \iiint_{V_\infty} w_\text{e} \, \text{d}V = \frac{1}{2} \iiint_{V_\infty} \vec{E} \cdot \vec{D} \, \text{d}V = \frac{1}{2} C U^2 \rightarrow C = \frac{2 W_\text{e}}{U^2}. \tag{5.8}$$

5.2 Inductive and capacitive coupling of a multi-layer air coil

Using the example of a multi-layer air coil, we will now examine the influence of a layer structure with regard to the magnetic and capacitive coupling. Each coil has up to $N = 60$ turns with a diameter of the round copper wire of $d_{cu} = 2$ mm. The thickness of the insulation is 50 μm and has a relative permittivity (real part) of $\varepsilon_r = 3$. The winding is arranged in an orthogonal way. This means that the turns of the layers lie on top of each other and are not shifted in the spaces between the neighboring turns, as for an orthocyclic winding. A comparison of orthogonal and orthocyclic windings with regard to their capacitance is presented in [1] and will be taken up again in the example in Section 5.8. The inner diameter of the winding is $d_{cf} = 15.9$ mm. The layer widths (number of turns per layer) of the arrangements are $N_{layer} = \{5, 10, 15\}$. A diagram of conventional windings with different layer widths is shown in Figure 5.2(a). A diagram of the winding progress for a conventional winding is depicted in Figure 5.2(b).

5.2.1 Inductive coupling

Figure 5.3(a) depicts the resulting total inductance calculated by evaluating (5.1). Each arrangement is successively expanded, one turn at a time. It can be seen that a wider layer leads to a lower overall inductance. This can be explained by the fact that, on average, there are larger distances between the turns, which leads to a lower magnetic coupling. This reduces the mutual inductances in (5.2). In addition, the mean value of the self-inductances is lower, as the mean diameter of the windings is smaller here. This influence is small compared to the lower coupling. The mean coupling is shown in Figure 5.3(b). The highest coupling is achieved for two directly adjacent turns. As the number of turns increases, average coupling decreases. From the point of a full layer, the drop in the average coupling is reduced as the turns above couple well with the turns below. Nevertheless, the average coupling continues to decrease as the average distance increases.

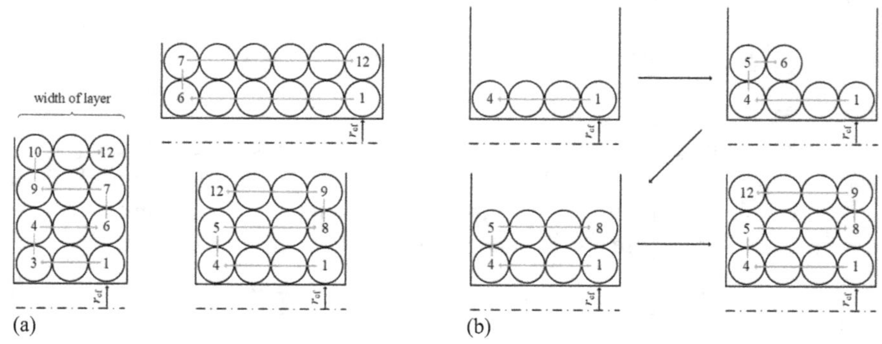

Figure 5.2 (a) Diagram of windings with the same number of turns and different layer width and (b) principle of winding up a conventional winding

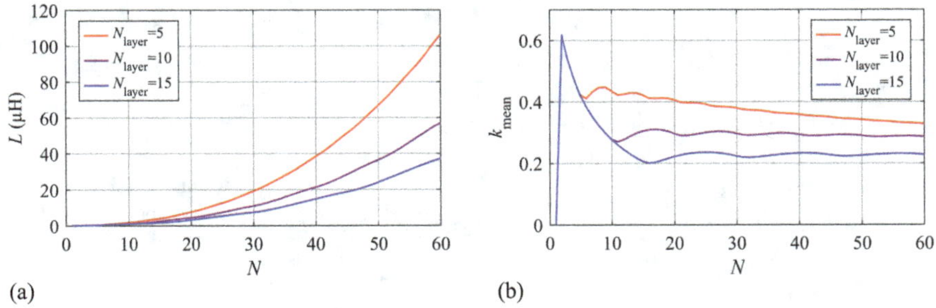

Figure 5.3 Windings with different numbers of turns per layer over the number of turns: (a) overall inductance and (b) mean coupling coefficient

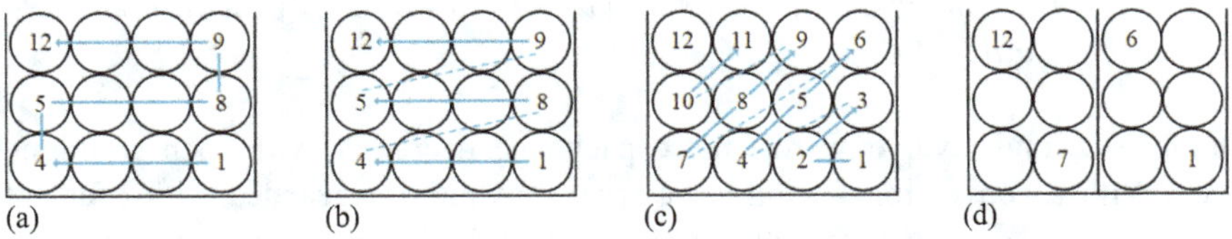

Figure 5.4 Diagram of (a) conventional, (b) flyback, (c) progressive, and (d) chamber winding with decreasing capacitance from left to right

The self-inductances and mutual inductances can be significantly increased by using a magnetic core. The considerations made can be fundamentally transferred to this. However, influences such as the fringing fields of possible air gaps or eddy currents in the core have to be taken into account.

5.2.2 Capacitive coupling

As already mentioned, a high magnetic coupling also leads to a high capacitive coupling. This can be strongly influenced by appropriate winding arrangements. Figure 5.4 depicts different winding arrangements. The capacitance decreases from left to right. In the following, the results of a conventional winding (Figure 5.4(a)), a flyback winding (Figure 5.4(b)), and a conventional winding with two chambers (Figure 5.4(d)) are compared. In addition, the capacitance curves of foil windings with widths comparable to the layer width of the round wire windings are shown. The thickness of the foil was chosen so that the cross-section remained the same as for the round wire. The thickness of the insulation is 50 μm. Figure 5.5 shows the results from the evaluation of (5.8) for different winding arrangements with different layer widths and numbers of turns.

With a single-layer winding, a very low capacitance is achieved. In contrast, a winding with a full second layer will lead to a very high capacitance. Overall, large layer width increases the capacitance and a high number of layers reduces it. In single-layer windings, a local maximum is reached with the second turn, as the highest and lowest potentials are located directly next to each other. From the

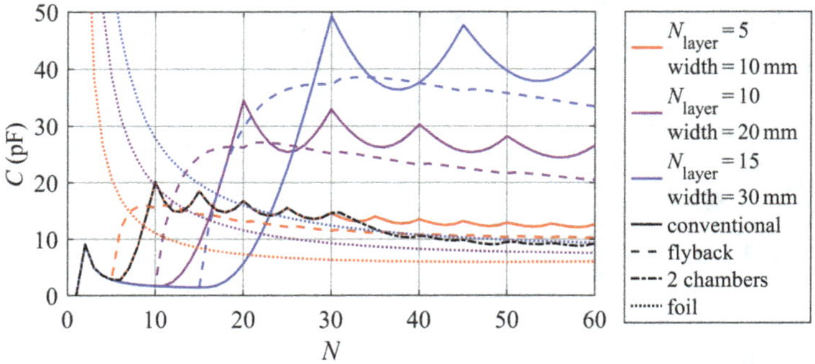

Figure 5.5 Capacitance over the number of turns for conventional and flyback windings with different numbers of turns per layer N_{layer} and for a two-chamber winding as well as for foil windings with different layer widths

complete second layer onwards, the capacitance decreases with each additional layer. With a conventional winding, a local maximum is reached with each full layer. The capacitance is initially reduced when the layer is filled and increases again from the half to the full layer. With a flyback winding, the capacitance initially increases as the next layer is filled. However, a completely filled layer always has a lower capacitance than a conventional winding. Over a large part of the incomplete second layer and in the middle section of the incomplete third layer, the capacitance of the flyback winding is higher. Subsequently, it is always lower than the conventional winding. In principle, an advantage in terms of capacitance can often be achieved with flyback winding. However, in order to implement the arrangement, the conductor must be reset from the end of the respective full position back to the beginning over all turns. This reduces the filling factor and the local magnetic coupling. The additional portion of the conductor for the return contributes to the losses due to its additional length and the fact that it is in the field between the layers. In practice, a good option for reducing the capacitance is to divide the winding into chambers. This reduces the layer width and voltage in each chamber by the number of chambers N_{ch}. Since the voltage is squared in the energy (5.8), the electric energy in the chamber is reduced by a factor of N_{ch}^3 compared to a non-chambered winding. When the capacitances of all chambers are added together, meaning that the result of one chamber is multiplied by N_{ch}, the total capacitance remains lower by the factor N_{ch}^2. However, the capacitance between the chambers is not taken into account. The factor is therefore not fully achieved in reality [2,3] and is in the range from N_{ch} to N_{ch}^2. In this example, the winding is divided into two chambers, which is often sufficient in practice. Apart from the volume of the rib (width = 0.8 mm), the dimension of the chamber winding is comparable to the winding with N_{layer} = 10. Until the first chamber is completely filled, the capacitance of the winding behaves like the conventional winding with a layer width of N_{layer} = 5. From the beginning of the second chamber onwards, the capacitance is reduced further.

With foil windings, the highest capacitance is also achieved with the second layer, respectively, the second winding. Due to the large layer width, very high capacitance values occur here. The capacitance decreases rapidly with each additional layer turn due to the series connection. For windings with foils, low numbers of turns are critical in terms of capacitance. The wider the foil, the higher the capacitance.

A core also has a significant influence here. It depends on whether the core is grounded ($U_{\text{core}} = 0$ V) or floating ($Q_{\text{core}} = 0$ As). The influence of the core on capacitance is discussed in Section 5.8.

5.3 Winding losses

In addition to the effects on the inductive and capacitive coupling, the winding layout and type of conductor influence the occuring losses. Here, we only provide a brief overview. A detailed explanation can be found in [3,4]. In essence, the winding losses can be divided into the ohmic root mean square (RMS), the skin, and the proximity losses:

$$P_{\text{win}} = P_{\text{RMS}} + P_{\text{skin}} + P_{\text{prox}}. \tag{5.9}$$

The losses depend on the effective value of the current of all harmonics (FFT) of any current waveform, the conductor cross-section, and the geometrical structure of the winding. Also, the material and the temperature have an influence. The RMS losses depend on the geometry of the conductor (area A_{cs} and length l) as well as on the temperature-dependent electrical conductivity of the conductor material $\sigma(T)$, the RMS value of the current and the DC resistance R_0

$$P_{\text{RMS}} = \frac{1}{\sigma(T)} \frac{l}{A_{\text{cs}}} I_{\text{RMS}}^2 = R_0 I_{\text{RMS}}^2. \tag{5.10}$$

The electrical conductivity of the conductor material (mostly copper or aluminum) reduces as the temperature increases. The relationship can be described with [5]

$$\sigma(T) = \sigma(T_0) \cdot \frac{1}{1 + \alpha_{T_0} \cdot (T - T_0)}. \tag{5.11}$$

Here, $\sigma(T_0)$ stands for the electrical conductivity of the material at the temperature T_0 [$\sigma_{\text{Cu, 20°C}} \approx 5.8 \cdot 10^7\,\text{A/(Vm)}$, $\sigma_{\text{Al,20°C}} \approx 3.7 \cdot 10^7\,\text{A/(Vm)}$] and α_{T_0} for the (linear) resistance-temperature coefficient ($\alpha_{\text{Cu,20°C}} \approx 3.93 \cdot 10^{-3}\,\text{K}^{-1}$, $\alpha_{\text{Al,20°C}} \approx 4.0 \cdot 10^{-3}\,\text{K}^{-1}$). The skin losses depend on the penetration depth δ, the equivalent conductor layer thickness, which depends on the angular frequency $\omega = 2\pi f$ and the permeability μ:

$$\delta = \sqrt{\frac{2}{\omega\sigma(T)\mu}}. \tag{5.12}$$

The relationship between the increased losses due to the skin effect and the RMS losses is defined by the skin factor F_s, which can be calculated for round wires with the radius r_w using the modified zero-order I_0 and first-order I_1 Bessel functions and the skin constant α:

$$F_s = \frac{R}{R_0} = \frac{1}{2}\mathrm{Re}\left\{\alpha r_w \frac{I_0(\alpha r_w)}{I_1(\alpha r_w)}\right\} \text{ with } \alpha = \frac{1+j}{\delta}. \tag{5.13}$$

As f approaches 0, the equation converges to the value of the RMS losses. The RMS and skin losses can be calculated by

$$P_{RMS+skin} = R_0 I_{RMS}^2 F_s = P_{RMS} + P_{skin} = R_0 I_{RMS}^2 + R_0 I_{RMS}^2 (F_s - 1). \tag{5.14}$$

Figure 5.6(a) depicts the alternate current (AC) resistance of different wire geometries with the same cross-section and inner radius , this is referred to as the DC resistance. The inner radius is r_{cf} = 10 mm, and the cross-section is

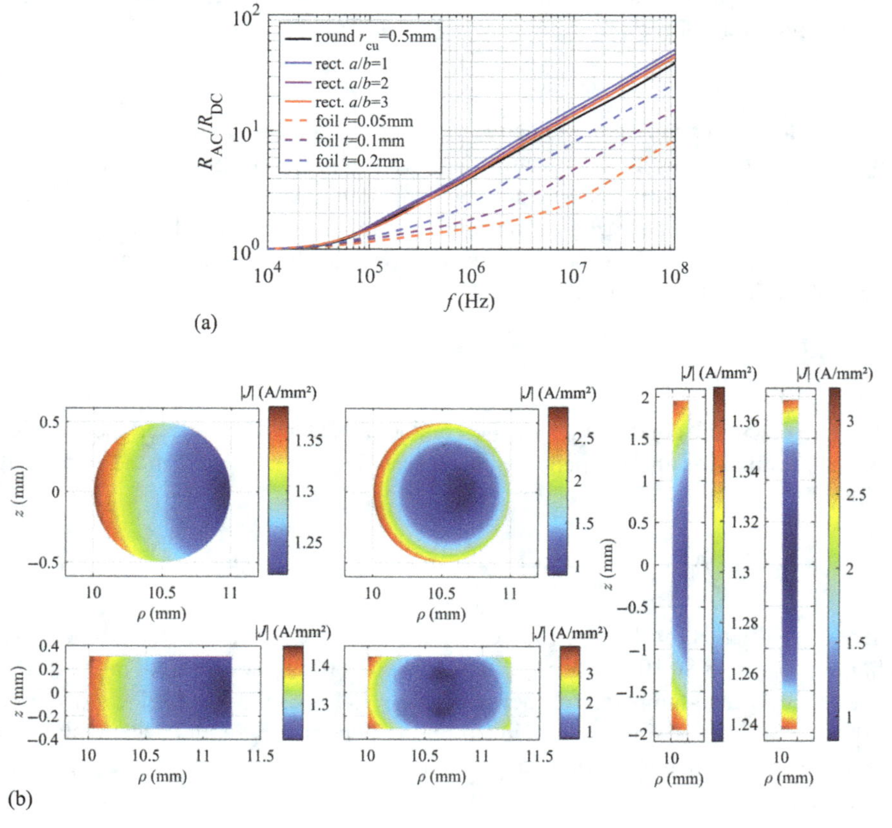

Figure 5.6 (a) R_{AC}/R_{DC} resistance of different conductor geometries and (b) distribution of the current density in the conductor cross-section of round, rectangular and foil. Left at f = 10 kHz and right at f = 100 kHz.

$A_{cs} = 0.7854\ \text{mm}^2$. Therefore, the radius of the round copper wire is $r_{cu} = 0.5$ mm. The ratio of the sides a/b is given for rectangular conductors. Here, $a/b = 1$ corresponds to a square cross-section. The other cross-sections with $a/b > 1$ correspond to flat wire edgewise windings, as shown in Figure 5.6(b). For the windings with foil, the thickness was varied with the width adjusted accordingly.

For all conductors, the resistance increases with increasing frequency. This is due to the current density distribution. As shown in Figure 5.6(b) for selected conductors, the current density increases with increasing frequency at the inner edge. More with rectangular cross-sections than with round wires. The foil behaves best here. However, the advantage of the foil in terms of skin effect is often overcome at higher frequencies by increased proximity losses in multi-layer windings due to fringing fields of air gaps in cores [6] and by the layer proximity effect.

When a conductor is exposed to a time-varying external magnetic field, eddy currents are induced, causing proximity losses. These magnetic fields can, for example, be caused by neighboring turns, windings, or an air gap. For a homogeneous external field $\widehat{H}_{\text{ext}}$, the time average of the proximity losses in a round conductor can be calculated by means of proximity factor D_s. Which, like the skin factor from (5.13), depends on the radius of the round wire, the skin constant from (5.13) and the modified zero-order and first-order Bessel functions:

$$P_{\text{prox}} = \frac{l}{\sigma(T)}\widehat{H}^2_{\text{ext}}D_s \text{ with } D_s = 2\pi\text{Re}\left\{\alpha r_w \frac{I_1(\alpha r_w)}{I_0(\alpha r_w)}\right\}. \tag{5.15}$$

Alternatively, the calculation can also be carried out using a model with a complex permeability (round [3,7], rectangular [8]). Figure 5.7(a) shows the resistance resulting from the RMS and skin losses as well as the proximity losses for a winding with 30 turns and different layer widths. As the circumference of the outer layers increases, the RMS and the skin resistance of the arrangement with the smallest layer width but the highest number of layers is the highest.

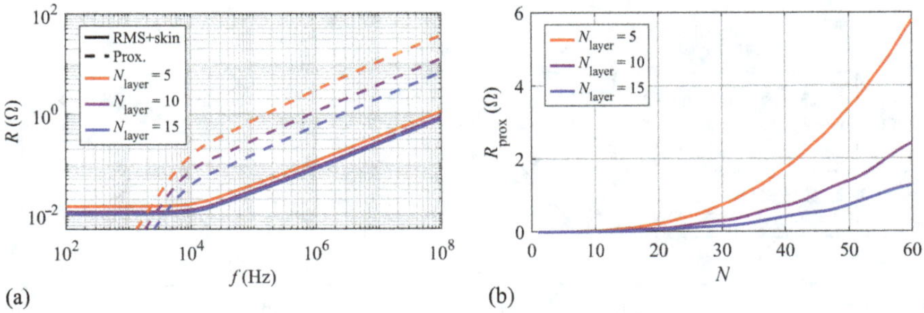

Figure 5.7 (a) Resistance due to RMS and skin losses and due to proximity losses of windings with N = 30 turns and different layer widths. (b) Resistance due to proximity with different numbers of turns N = {1 … 60} and different layer width at f = 100 kHz.

However, the number of layers has a significant influence on the proximity losses due to the superposition of the fields. A high number of layers leads to a good magnetic coupling with a relatively low parasitic capacitance but also to high proximity losses. Figure 5.7(b) shows the resistance due to the proximity losses with different numbers of turns $N = \{1 \ldots 60\}$ and different layer widths at $f = 100$ kHz. Here, it can be seen that a high number of layers leads to high proximity losses.

In the medium frequency range, the losses can be reduced by HF litz wire. By constructing the conductor from thinner strands insulated from each other, the skin effect is reduced on the one hand and the proximity effect by external fields like an air gap on the other. The small diameters ensure an approximately homogeneous current distribution in the strands up to higher frequencies, which reduces the skin effect. To counteract the current displacement in the bundle, the strands are twisted in such a way that each strand interchanges position with the other strands over a short distance. The smaller cross-sectional areas yield eddy currents cannot expand as much, which reduces the proximity losses. However, each strand also leads to a magnetic field that causes inner proximity losses in the remaining strands of the wire. Litz wire with a high number of strands in particular exhibit considerable inner proximity losses even at lower frequencies. In addition, the structure leads to a lower filling factor and reduced heat conduction. Figure 5.8(a) shows the resistance due to the RMS and skin losses of two litz wires with 30 turns and different layer widths. One litz wire has 245 strands with a diameter of 100 μm, and the other litz wire has 60 strands with a diameter of 200 μm. The dimensions of the litz wire are comparable to the round wire shown above. Due to the lower fill factor of around 60%, the litz wires have a higher RMS resistance compared to the round wire in Figure 5.7(a). In this case, a high number of layers increases the average diameter and, thus, the resulting RMS and skin resistance. The superposition of the fields also leads to higher proximity losses. The litz wire with a smaller strand diameter rises later than the coarser strand. Figure 5.8(b) shows the proportion due to proximity losses. These are significantly reduced compared to the round wire. Up to approx. 1 MHz, the windings with the fine strand have lower proximity losses. Above this, the inner proximity effect dominates, and the litz wire with the higher number of strands has higher losses.

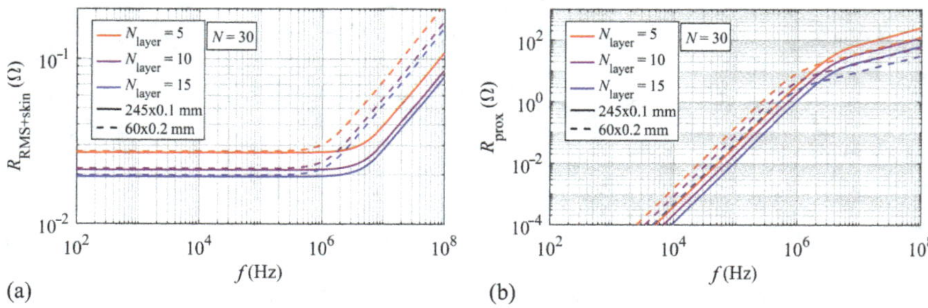

Figure 5.8 Resistance of litz wire windings with N = 30 turns and different layer widths due to (a) RMS+skin losses and (b) proximity losses

The RMS and skin losses of a litz wire are calculated in the same way as for a round wire [3,4]:

$$P_{\mathrm{RMS+skin,litz}} = R_0 I^2_{\mathrm{RMS}} F_{\mathrm{s,strand}} \tag{5.16}$$

The losses differ from the round wire by the skin factor $F_{\mathrm{s,strand}}$. For this, the radius of the strand must be used in (5.13) instead of the radius of the solid wire. As mentioned before, a time-varying external magnetic field induces eddy currents and thus proximity losses. This includes wires connected in parallel, such as the strands of the litz wire, which is then known as internal proximity loss. The proximity losses of litz wire can be calculated by

$$P_{\mathrm{prox,litz}} = \frac{l}{\sigma(T)} N \widehat{H}^2_{\mathrm{ext}} D_{\mathrm{s,litz}}. \tag{5.17}$$

For the proximity factor, the radius of the strand is used in (5.15) instead of the radius of the solid wire. One source of error in the method is the averaging of the field strength, which is otherwise inhomogeneous in the litz wire [9]. Real litz wires often do not behave ideally. Rossmanith *et al.* [4] introduce quality parameters for this purpose. However, the litz wire must be measured in advance. Derivation for the calculation of losses of HF litz wires can be found in [3,4,10]. A practicable alternative for calculating the proximity losses of litz wire is to implement a complex permeability in the area of the conductor [11,12]. Meeker [13] presents an improved complex permeability model for litz wire that is valid for all strand diameters, strand spacings, and frequencies. This approach is used in Section 5.4.

Figure 5.9 compares the resistances of foil windings of different widths with the curves from the litz wire with 60×200 μm stands. Figure 5.9(a) shows the resistance due to the RMS and skin losses. The thickness of the foil was adjusted so that the DC resistance matches the corresponding layer width of the litz wire. Figure 5.9(b) shows the proportion due to proximity losses. A smaller number of layers also leads to lower RMS and skin losses as well as proximity losses here. Compared to the litz wire, the foil has an advantage in terms of skin losses as long

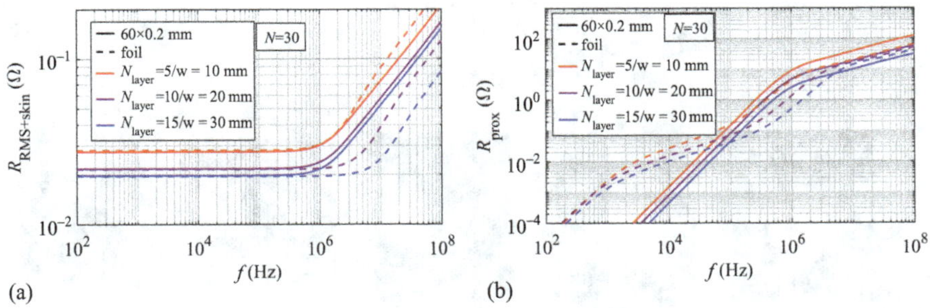

Figure 5.9 Resistance of litz wire and foil windings with N = 30 turns and different layer width or width due to (a) RMS+skin losses and (b) proximity losses

as the layer thickness is low. In terms of proximity losses, the foil is higher than the litz wire in the lower frequency range and then lower in the middle range. However, this statement only applies as long as there is no core with an air gap. In the case of a core with an air gap, significant eddy currents can occur in the foil due to the fringing field. This will become clear in Section 5.4. Approaches for the analytical calculation of the losses of foil windings with certain simplifications are shown in [3,14].

5.4 Design of an inductor for a buck converter

In this section, we develop a power inductor for a buck converter. We look at the losses occurring in different conductor types [HF litz wire, foil, and flat wire (rect)]. The influence of a single (SG) or distributed air gap (DG) on the winding losses is analyzed. In addition, the capacitance of non-cast and cast components and their effects are examined. Selected inductors were realized and compared with each other in terms of their losses during operation using thermography. The parasitic capacitance is evaluated by means of small signal measurement. For the design, we use a simplified practical approach that allows a design to be implemented quickly and compare this to a more precise method. In Figure 5.10, prototypes with (a) foil winding, (b) litz wire, and (c) cast litz wire are shown. Beside them, (d) shows the used core with one central air gap and (e) the core with a DG.

The converter is to supply a load R_{load} during operation with a constant power of $P = 530$ W from an input voltage of $U_0 = 240$ V. The capacitor C is big enough to stabilize the output voltage $U_c = 60$ V. The duty cycle is $\delta = U_c/U_0 = 0.25$. The converter operates at the limit between continuous and discontinuous modes. This means $i_L = 0$ A at the beginning and end of the switching period. The switching frequency is $f = 100$ kHz. The ideal equivalent circuit diagram of the converter is illustrated in Figure 5.11(a). The current waveform normalized to one period is shown in Figure 5.11(b).

The peak-to-peak value of the current ripple Δi_L results from the continuous direct current $I_R = P/U_c$, which in the case of the ideal (loss less) converter results from the ratio of the power to the output voltage:

$$\Delta i_L = 2 \cdot I_R = 2 \cdot \frac{P}{U_C} = 17.67\,\text{A}. \tag{5.18}$$

(a) (b) (c) (d) (e)

Figure 5.10 Inductor prototypes with (a) foil-winding, (b) litz wire, cast litz wire, and used cores with (c) central single air gap (SG) and (d) distributed air gap (DG)

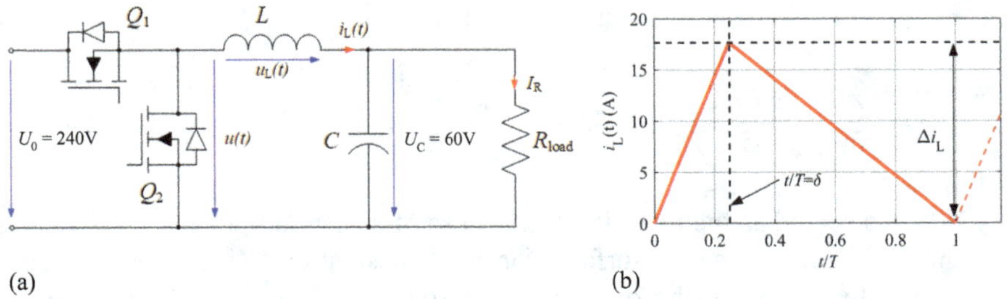

Figure 5.11 *(a) ECM of synchronous buck and (b) current waveform normalized to one period*

This allows the required inductance L to be calculated

$$L = \frac{(U_0 - U_C) \cdot \delta \cdot T}{\Delta i_L} = 25.54\ \mu\text{H}. \tag{5.19}$$

Further specifications for the design are as follows:

- The inductor is mounted on a circuit board together with other components. Heat is dissipated by natural convection at a maximum temperature of 50 °C above the circuit board.
- Heat class F design (155 °C hot-spot temperature).
- Limitation of the overall height to 40 mm.
- With respect to possible additional losses in the semiconductors and electromagnetic compatibility (EMC), the design should be as low-capacitance as possible. Stray (electro)magnetic fields should also be minimized with regard to operational safety.
- All voltages in the system are below 1000 V.
- Vacuum-impregnated or (partially) cast version.

In the next step, we estimate what power loss we can allow for the component under natural convection. For this purpose, McLyman [15] provides an approximation formula for compact structures:

$$\frac{\Delta T}{\text{K}} \approx 450 \cdot \left(\frac{P/\text{W}}{A/\text{cm}^2}\right)^{0.826} \tag{5.20}$$

For the maximum hot-spot temperature of 155 °C, the (average) surface temperature of the inductor should not be higher than 120 °C (estimated). The relative temperature increase compared to the surroundings is therefore $\Delta T = 120\ °\text{C}–50\ °\text{C} = 70\ \text{K}$. We chose a PQ32/30 core because it is within the required height (<40 mm from specification) with a height of 30.35 mm. In addition, the design results in low stray fields, as the winding is mainly enclosed by the core. The area to be used for calculating the permissible losses can be estimated using the surface of a cuboid of the main dimensions of the core. For numerous core variants, values are given in [15].

The following total power loss can be permitted:

$$\frac{P}{\mathrm{W}} = \frac{A}{\mathrm{cm}^2} \cdot \left(\frac{\Delta T/\mathrm{K}}{450}\right)^{\frac{1}{0.826}} \approx 4.94\ \mathrm{W}. \tag{5.21}$$

N95 is selected as the material. In the case of a cast component, the winding and the core use the same surface for heat dissipation. Therefore, the core and copper losses should be distributed approximately equally. (The design is power loss limited and not saturation limited.) Now, we determine which amplitude of the flux density $\widehat{B}_{\mathrm{ripple}}$ must be obtained in order to get a power loss of approximately 4.94 W/2 = 2.47 W. For this purpose, we use the information in the core data sheet. With a flux density of 200 mT at a core temperature of 120 °C and a frequency of 100 kHz, the losses are specified as 7.56 W. As a first approximation, core losses behave in the square of flux density in the vicinity of measured values. The approximation via sine values, although we have a triangular waveform of magnetic flux density, is also an approximation that is made in order to achieve a quick design. Methods such as the principle of equivalent frequency [3,16] lead to more exact results here. The pre-magnetization by the mean value of the current is considered at this point with a factor of $a_{\mathrm{bias}} = 1.5$:

$$2.47\mathrm{W} \overset{!}{=} a_{\mathrm{bias}} \cdot 7.56\mathrm{W} \cdot \left(\frac{\widehat{B}_{\mathrm{ripple}}}{200\mathrm{mT}}\right)^2 \rightarrow \widehat{B}_{\mathrm{ripple}} \approx 93.37\,\mathrm{mT} \tag{5.22}$$

This can now be used to determine the number of turns

$$\widehat{B}_{\mathrm{ripple}} \geqq \frac{L \cdot \widehat{i}_{\mathrm{ripple}}}{N \cdot A_{\mathrm{e}}} \rightarrow N \geqq \frac{L \cdot \widehat{i}_{\mathrm{ripple}}}{\widehat{B}_{\mathrm{ripple}}} \cdot A_{\mathrm{e}} \approx 15.69 \tag{5.23}$$

The next higher whole value $N = 16$ is selected. In this case, the selected integer number of turns is very close to the calculated number. The recalculation of the maximum flux density with (5.23) and the expected loss with (5.22) do not deviate from the previous values.

5.4.1 *Simplified loss estimation*

For the design of the winding, the standard coil former is used. In case of HF ripple current, a design with a litz wire is typically a good choice. However, foil can also be a good solution. This is examined after the design with litz wire. To achieve a good compromise between the cost and benefit of litz wire, a ratio of AC to DC (or RMS) losses of $F_{\mathrm{r}} = 1.6$ should be allowed. Care must be taken to ensure that the additional AC losses, e.g., in the area of the air gap, are not concentrated on a few turns in order to avoid local overheating. For the copper losses remain

$P_{win} = 4.94–2.47\ W = 2.47\ W$. Therefore, the ohmic DC resistance can be calculated by

$$P_{win} = R_{AC} \cdot I_{RMS}^2 = F_r \cdot R_{DC} \cdot I_{RMS}^2 = F_r \cdot R_{DC} \cdot \left[\underbrace{\left(\frac{\Delta i_L}{2}\right)^2}_{\text{DCvalue}} + \underbrace{\left(\frac{\Delta i_L}{2 \cdot \sqrt{3}}\right)^2}_{\text{RMS of ripple}} \right] \tag{5.24}$$

$$\rightarrow R_{DC} \approx \frac{P_{win}}{0.5 \cdot \Delta i_L^2} \approx 15.8\,\text{m}\Omega \tag{5.25}$$

Due to the high ripple proportion of the current waveform, considerable proximity losses are to be expected in the region close to the air gap. This entails the risk of local overheating. Litz wire is produced from a strand diameter of 20 μm. However, these are very expensive. In most applications, litz wires with a strand diameter of 50 μm are cost-justifiable if no more cost-effective solution, such as a DG, is available to reduce proximity power losses. We first try the less expensive litz wire with a strand diameter of $d = 71$ μm $[d/(2\delta) \approx 0.14]$

$$R_{DC} = \frac{N \cdot l_N}{\sigma_{Cu,145^\circ C} \cdot N_s \cdot \pi \cdot \left(\frac{d}{2}\right)^2} \rightarrow N_s = \frac{N \cdot l_N}{\sigma_{Cu,145^\circ C} \cdot R_{DC} \cdot \pi \cdot \left(\frac{d}{2}\right)^2} = 466.67 \tag{5.26}$$

N is the number of turns, l_N is the average length of the turns und N_s the number of strands. The 145 °C copper temperature is assumed because, as expected, it is slightly higher in the poorly coolable litz wire winding than on the surface of the inductor (assumption 120 °C). A safety margin of 10 °C to 155 °C of thermal class F would be just acceptable in practice. A number of $N_s = 480$ is finally selected, as this type is currently available from stock in the example. The delivery times for customer-specific litz wire can be considerable. A litz wire with a strand diameter of 100 μm and 245 strands (calculated 235.25) and another with a strand diameter of 200 μm and 60 strands (calculated 58.81) are selected as an alternative. Finally, the length of the air gap is still missing to obtain the desired inductance. At this point, no calculation is presented. Approximations for this calculation can be found in [15,17]. The deviations increase with the length of the air gap. In the following, we consider two configurations of cores with air gap. The first is a central air gap with a length of 2.8 mm. The second is a DG (three equal gaps) with an overall length of 2.2 mm. The difference in the overall length of the air gaps is related to the fringing fields of the air gaps. Large air gaps have a smaller magnetic resistance in relation to their length than small air gaps. The fringing field and the distribution of the magnetic flux density at $I_{DC} = 17.67$ A are shown in Figure 5.12 (a) for the central air gap and in Figure 5.12(c) for the DG.

Next, we examine the winding losses. The calculations are based on a copper temperature of 145 °C as a worst-case scenario. First, we want to compare the three different litz wires with each other. Litz wire with fine strands has a lower filling

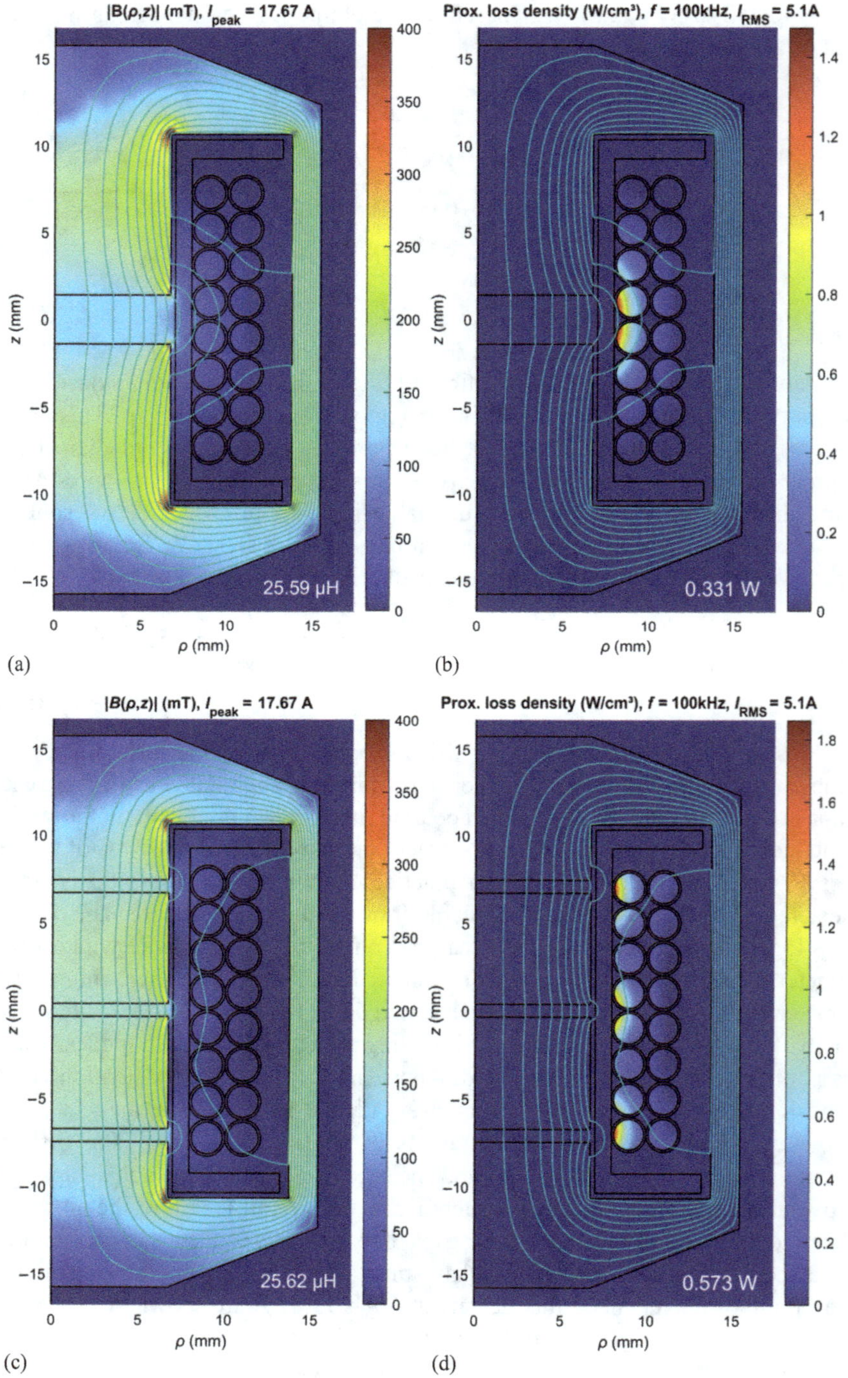

Figure 5.12 *(a) Magnetic flux density with SG, (b) proximity loss density with 245×100 μm litz wire and SG, (c) magnetic flux density with three air gaps and (d) proximity loss density with 60×200 μm litz wire and three air gaps*

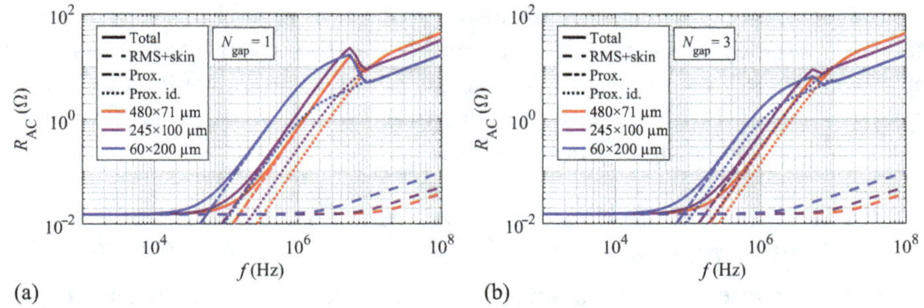

Figure 5.13 AC resistance of different litz wire with (a) one central air gap and (b) three DG

factor than those with coarse strands. We use 0.56 for the 71 μm strand, 0.57 for the 100 μm strand, and 0.59 for the 200 μm strand. Figure 5.13(a) shows the calculated AC resistances (solid) for the three litz wire types, the proportions due to the RMS and skin losses (dash), and those due to the total proximity losses (dash-dot). The proximity losses (Prox id.) that would occur with an ideally realized homogeneously DG (dot) are also shown.

The difference between the total proximity losses and those with the ideal air gap is the proportion that occurs due to the fringing field of the air gap. It can be seen that for all strands the skin losses only become relevant above 1 MHz. The proportion due to proximity losses constitutes the majority of the resistance. The bend in the curves of the resistance due to total losses or the total proximity losses between 2 MHz and 10 MHz is caused by the decrease in the real part of the permeability. When the value falls to 1 the core behaves like air. The fringing field of the air gap disappears and the curves of the proximity losses, including the fringing field effect, converge to the ideal proximity loss curves without the fringing field. The coarser the strand grain size, the higher the proximity losses. The fact that the resistance of the finer strands is higher from 10 MHz is related to the inner proximity effect caused by the high number of strands. With the simplified assumption regarding the current composition from the DC component and the RMS of the ripple from (5.24), for the finest strand there is a loss of $P_{v,litz} = 1.76$ W ($P_{v,RMS+skin} = 1.59$ W, $P_{v,prox} = 0.17$ W) and for the middle $P_{v,litz} = 1.90$ W ($P_{v,RMS+skin} = 1.57$ W, $P_{v,prox} = 0.33$ W). The proximity losses of the medium strand are twice as high as those of the fine strand. Both are below the permissible loss. Figure 5.12(b) shows the distribution of the loss density by proximity effect in the middle litz wire with a single air gap in the core. The losses occur particularly in the area of the fringing field. The losses of the coarse strand $P_{v,litz} = 2.91$ W ($P_{v,RMS+skin} = 1.59$ W, $P_{v,prox} = 1.32$ W) are too high due to the high proportion of proximity losses. A DG can be used to reduce the proximity losses caused by the fringing field of the air gap. The proximity loss density of the coarse litz wire with a DG is shown in Figure 5.12(d). This allows the losses of the coarse litz wire to be reduced to an acceptable value $P_{v,litz} = 2.16$ W ($P_{v,RMS+skin} = 1.59$ W, $P_{v,prox} = 0.57$ W).

The reduced power loss density now also extends over several turns. In practice, the lower costs for the coarser litz wire must be weighed against the higher costs for a DG. The influence of the DG on the AC resistance is shown for all litz wire types in Figure 5.13(b). It can be seen that the proportion of resistance due to proximity loss can be significantly reduced.

5.4.2 More precise loss calculation using Fast Fourier Transformation (FFT)

For a more precise loss calculation, the respective frequency components of the current signal must be considered. To do this, the current signal is split into its frequency components using FFT. Figure 5.14(a) shows the current signal from Figure 5.11(b) and two others with different duty cycles. The RMS value of all signals calculated using the simplified approach from (5.24) is identical, which also applies for the calculated losses. When the signals are transformed into the frequency spectrum, they exhibit a different distribution of the RMS value components, as in Figure 5.14(b). It can be seen that higher values are obtained, particularly with a duty cycle of 0.9 at frequencies between 300 kHz to 600 kHz. With a duty cycle of 0.5, the value at 200 kHz has almost completely disappeared and the values at higher frequencies are reduced.

To calculate the winding losses, the squared effective values are multiplied by the respective resistance value. For the inductor with coarse litz wire and SG, the losses at the respective frequencies for a duty cycle of 0.25 are shown in Figure 5.15(a). In addition to the previously considered components, the 200 kHz component also contributes significantly to the losses. Figure 5.15(b) shows the cumulative sum of the losses for different duty cycles. The highest losses are reached at a duty cycle of 0.9 related to the higher RMS values in the 300–600 kHz range [Figure 5.14(b)].

Figure 5.16 shows the comparison of the calculated losses of the middle litz wire with a SG. The first bar graphs show the calculations with the simplified method and next to it the evaluation with FFT at a duty cycle of 0.25. Slightly higher proximity losses are calculated with the more precise method. This is due to

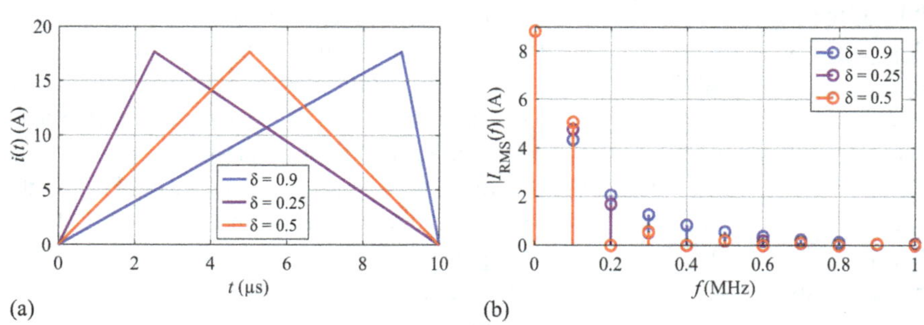

Figure 5.14 (a) Current waveform with different duty cycles and (b) spectrum of effective currents for different duty cycles

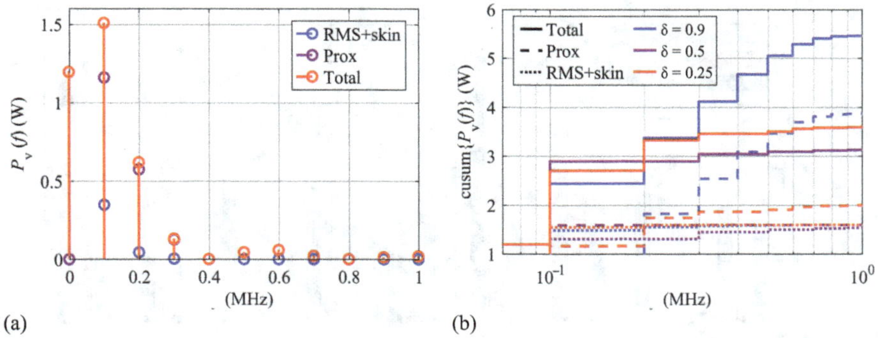

Figure 5.15 (a) Loss components over frequency of inductor with SG and coarse litz wire at a duty cycle of 0.25 and (b) cumulative losses over frequency of the same inductor for different duty cycles

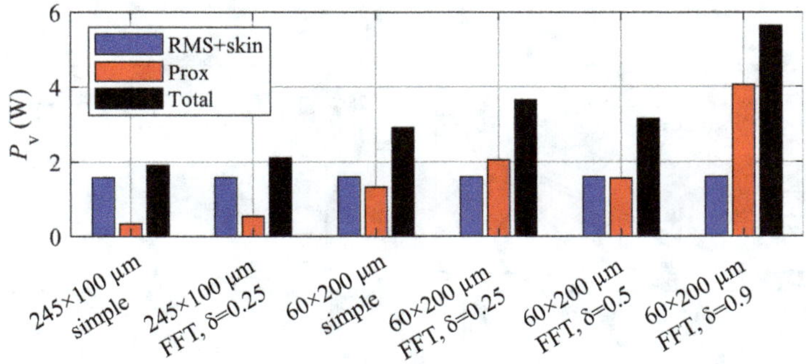

Figure 5.16 Comparison of the results of the simplified loss calculation with the more precise calculation by FFT using the example of the inductor with one air gap and different strands

the relatively flat shape of the resistance curve in the range of the significant amplitudes. The calculated losses on the coarse winding with an air gap at a duty cycle of 0.25 are also shown. Here, a greater difference in the proximity and thus the total losses between the simple calculation and the precise calculation is observed. The increase in resistance is also significantly higher for this inductor in the relevant range. The comparison with the calculations at cycles of 0.5 and 0.9 is also illustrated. The losses at 0.9 in particular are significantly higher than those estimated using the simple method. The losses calculated with the precise method for the duty cycle of 0.25 are shown in Figure 5.17.

Instead of a litz wire, a cost-effective copper foil with a thickness of 0.1 mm could also be a solution. The large area of the foil in the area of the fringing field of the air gap leads to very high proximity losses. To reduce the effects of the fringing field, we position the foil further away from the air gap. Therefore, we increase the

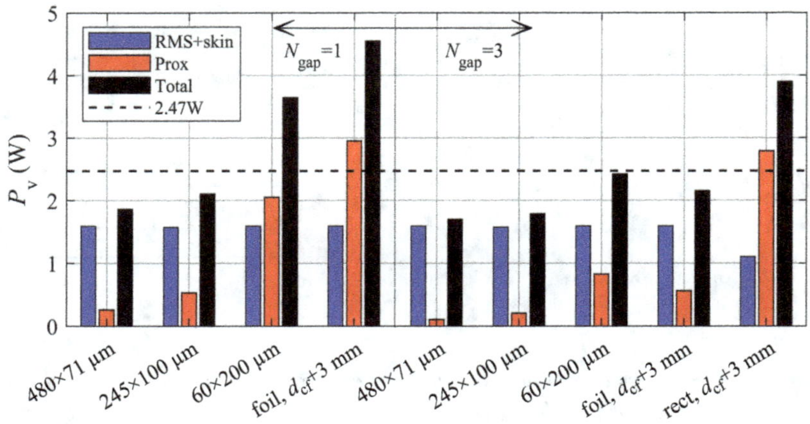

Figure 5.17 Comparison of losses of all investigated windings with one and with three air gaps at a duty cycle of 0.25

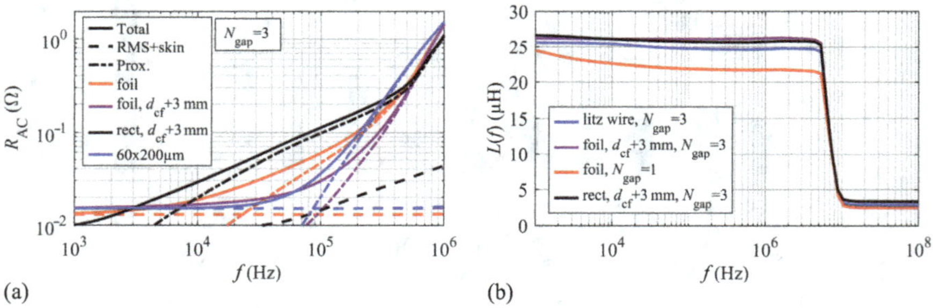

Figure 5.18 (a) AC resistance of foil, flat wire, and litz wire winding with DG, (b) inductance over frequency with DG for litz wire, foil, d_{cf} +3 mm and rect, d_{cf} +3 mm and of foil with SG

diameter of the coil former d_{cf} by additional 3 mm (d_{cf} + 3 mm). The resulting increase in RMS and skin losses is compensated by the significantly higher reduction in proximity losses. However, with a central air gap, the losses $P_{v,foil}$ = 4.54 W ($P_{v,RMS+skin}$ = 1.59 W, $P_{v,prox}$ = 2.95 W) cannot be reduced sufficiently. Only with a DG are the losses below the requirement $P_{v,foil,d_{cf}+3mm}$ = 2.15 W ($P_{v,RMS+skin}$ = 1.59 W, $P_{v,prox}$ = 0.56 W). Figure 5.18(a) shows the AC resistance curves of the foil windings with the original coil former (foil) and the coil former with additional 3 mm in diameter (foil, d_{cf} + 3 mm). This clearly shows the reduction of resistance due to proximity losses in the range of the switching frequency. Up from the switching frequency, the resistance is even lower than those of the coarse litz wire. We also aim to examine how a single-layer edge-wise wound flat wire behaves. The winding is also moved away from the air gap by the same distance as the foil and has a DG. The calculated AC resistance of the flat wire design is also depicted in Figure 5.18(a). The RMS losses are reduced by using a larger cross-section. However, the proximity losses clearly exceed the other designs.

In Figure 5.17, the estimated losses of all investigated inductor designs are depicted. The losses of the windings with SG are shown on the left, and the losses with the DG on the right. The total losses and the proportions by RMS and skin as well as by proximity are shown. All winding types have comparable RMS and skin losses. However, as shown in this section, skin losses play a subordinate role. It can be seen that the proximity losses can be significantly reduced by a DG. With SG, the fine and medium-litz wire are below the permissible losses, where as the coarse strand and the copper foil exceed these. The losses of the flat wire are not shown here, as these are much too high. With the DG, the proximity losses resulting from the air gap are reduced so far that the coarse stranded wire and the foil can also be taken into consideration. The flat wire is still too high.

In addition, Figure 5.18(b) shows the inductance curves as a function of frequency. The values of the foil and the flat wire with an additional 3 mm in diameter of the coil former are slightly higher than the variant with litz wire. Also shown is the curve of the foil with the original coil former and a central air gap. The fields of the occurring eddy currents counteract the primary one. This leads to a strong reduction in inductance, even at low frequencies.

5.4.3 Analysis of losses by thermography

To validate the estimated losses, selected prototypes with different types of litz wire and foil with cores with SG and DG were measured during the stationary operation of the step-down converter using thermography. The accuracy of the thermographic images was verified using temperature sensors. Figure 5.19 shows the stationary temperature distribution of the cast inductors. The maximum temperatures occurring correspond to the calculated losses shown in Figure 5.17. At an ambient temperature of 50 °C, all prototypes would be below the permissible surface temperature of 120 °C (245×100 μm SG = 95.2 °C, 245×100 μm DG = 88.1 °C, 60×200 μm DG = 103.3 °C, foil d_{cf} + 3 mm DG = 99.2 °C).

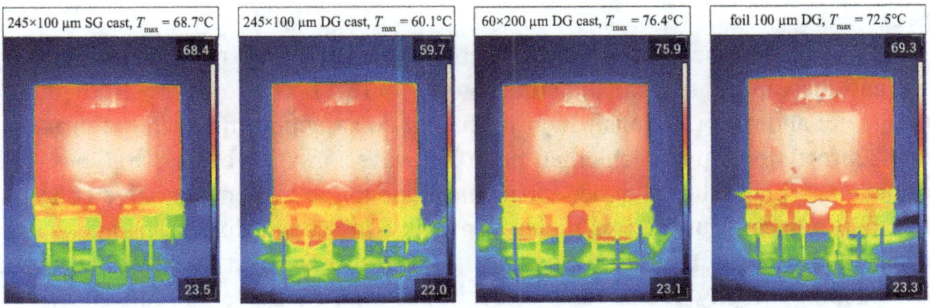

Figure 5.19 Stationary temperature distribution of cast prototypes of (a) litz wire 245×100 μm with SG and with DG, (b) litz wire 245×100 μm, (c) litz wire 60×200 μm and (d) foil d_{cf} + 3 mm

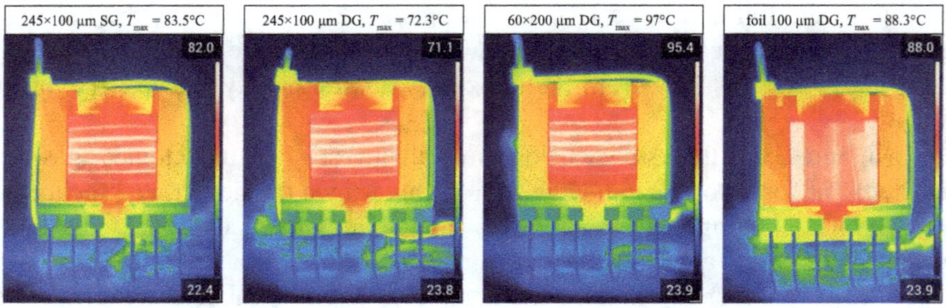

Figure 5.20 Stationary temperature distribution of non-cast prototypes of (a) litz wire 245×100 μm with SG and with DG, (b) litz wire 245×100 μm, (c) litz wire 60×200 μm and (d) foil d_{cf} + 3 mm

In the case of non-cast inductors, it is to be expected that the temperature of the winding is higher relative to the core, as a shared surface is no longer used for heat dissipation. The surface area of the core is significantly larger than that of the winding. Figure 5.20 compares the non-cast prototypes. As with the cast prototypes, the increase in temperature is in line with the calculations. At an ambient temperature of 50 °C, the inductor with 60×200 μm strands would exceed the maximum surface temperature of 120 °C with 123.1 °C. All others (245×100 μm SG = 111.1 °C, 245×100 μm DG = 98.5 °C, foil d_{cf} + 3 mm DG = 114.4 °C) comply with the specification.

5.4.4 Capacitance of the prototypes

After having observed the winding losses, the following section looks at the effect of the different windings regarding the parasitic capacitance. We evaluate the influence of a grounded core compared to a floating core and the influence of the casting. We start with the litz wire. In order to achieve high accuracy in the calculation (whether analytical or with FEM), precise knowledge of the geometry and the material data is required. Figure 5.21 depicts the cross-section of the litz wire 60×200 μm. The dimensions are simplified to circles for the simulation. The insulation around the stranded wire is a woven structure, here a mixture of polymer (typically $\varepsilon_{\mathrm{r}} \approx \{3 \ldots 3.3\}$) and air ($\varepsilon_{\mathrm{r}} = 1$). The value $\varepsilon_{\mathrm{r}} = 2.5$ is used for the simulations. In addition, the exact knowledge of the geometry is also relevant for enabling an exact calculation of the losses.

Figure 5.22 depicts the voltage distribution in different arrangements with litz wire and the capacitance calculated according to (5.8). The turns are arranged (a) orthocyclic and (b) orthogonal. In any winding construction in which the turns are wound directly on top of each other, there is always a mixture of the two types. The orthocyclic winding has a higher capacitance due to the higher capacitive coupling to shifted adjacent turns (also higher proximity losses and lower leakage inductance). In the case of the prototypes, most of the turns are next to each other as the high packing density means that there is not enough space to shift the turns. Figure 5.22(c) depicts the voltage distribution in a flyback winding and (d) in a conventional two-chamber winding. The core is floating ($Q = 0$).

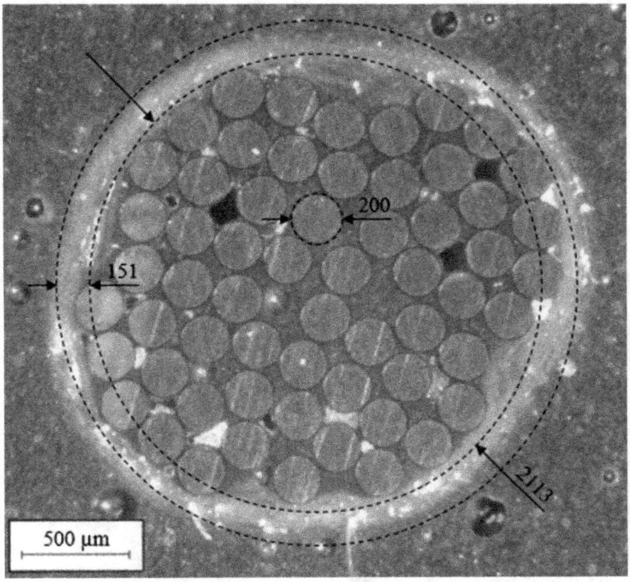

Figure 5.21 Cross-section of litz wire with 60×200 µm and woven polymer structure as insulation

Figure 5.23 shows the non-cast (a) conventional and (b) the two-chamber winding with a grounded core. In both cases, the capacitance increases significantly compared to the floating cases in Figure 5.22. Furthermore, (c) shows the foil winding and (d) the winding with flat wire with a floating core and non-cast. The flat wire winding has a relatively low capacitance due to the single-layer structure despite the small spacing. The additional 3 mm in the diameter of the coil former can also be clearly seen in the illustrations.

Figure 5.24 gives an overview of the calculated capacitance values. In general, grounding the core and casting leads to a higher capacitance. Casting results in a greater increase in capacitance compared to grounding the core.

The realized prototypes were measured using an impedance analyzer. The potential of the cores was floating. Figure 5.25(a) depicts the magnitude and (b) the argument of the complex impedance of selected prototypes. The parasitic parallel capacitance was calculated from the inductance and the first resonance frequency. The casting of the component with the litz wire leads to a doubling of the capacitance. The non-cast flyback winding has a lower capacitance than the non-cast conventional winding. The results of the litz wire prototypes agree with the results of the simulations. The capacitance of the cast foil winding is slightly lower. This is due to a slightly smaller layer width in the prototype and the non-ideal winding.

The effect of higher capacitance on the current waveform in the application is shown in Figure 5.26. In (a) two periods of the current and (b) the section at the moment of switching of the non-cast inductor with litz wire and the cast foil winding

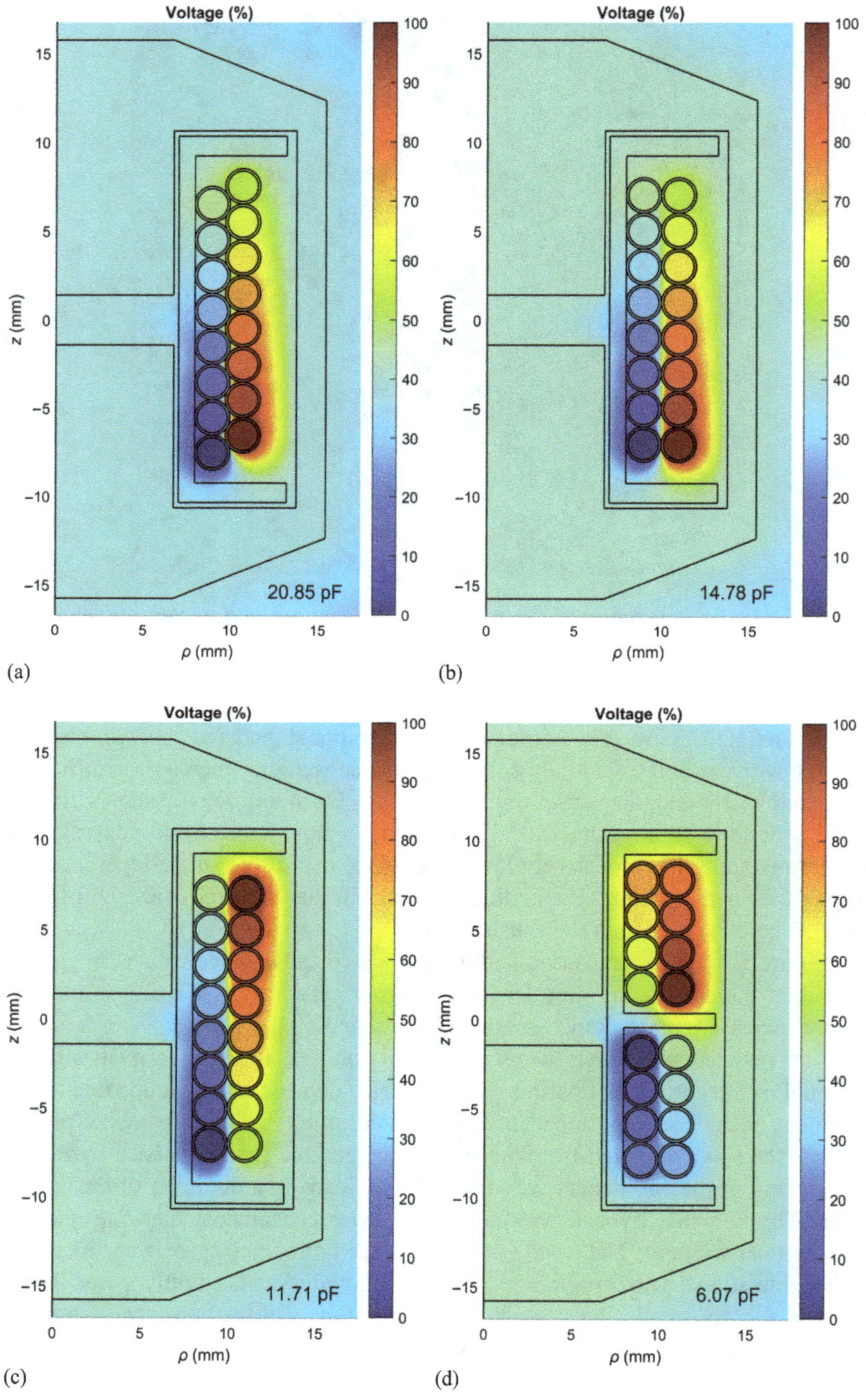

Figure 5.22 Capacitance with the floating core of conventional non-cast (a) orthocyclic, (b) orthogonal and of (c) flyback and (d) two-chamber winding

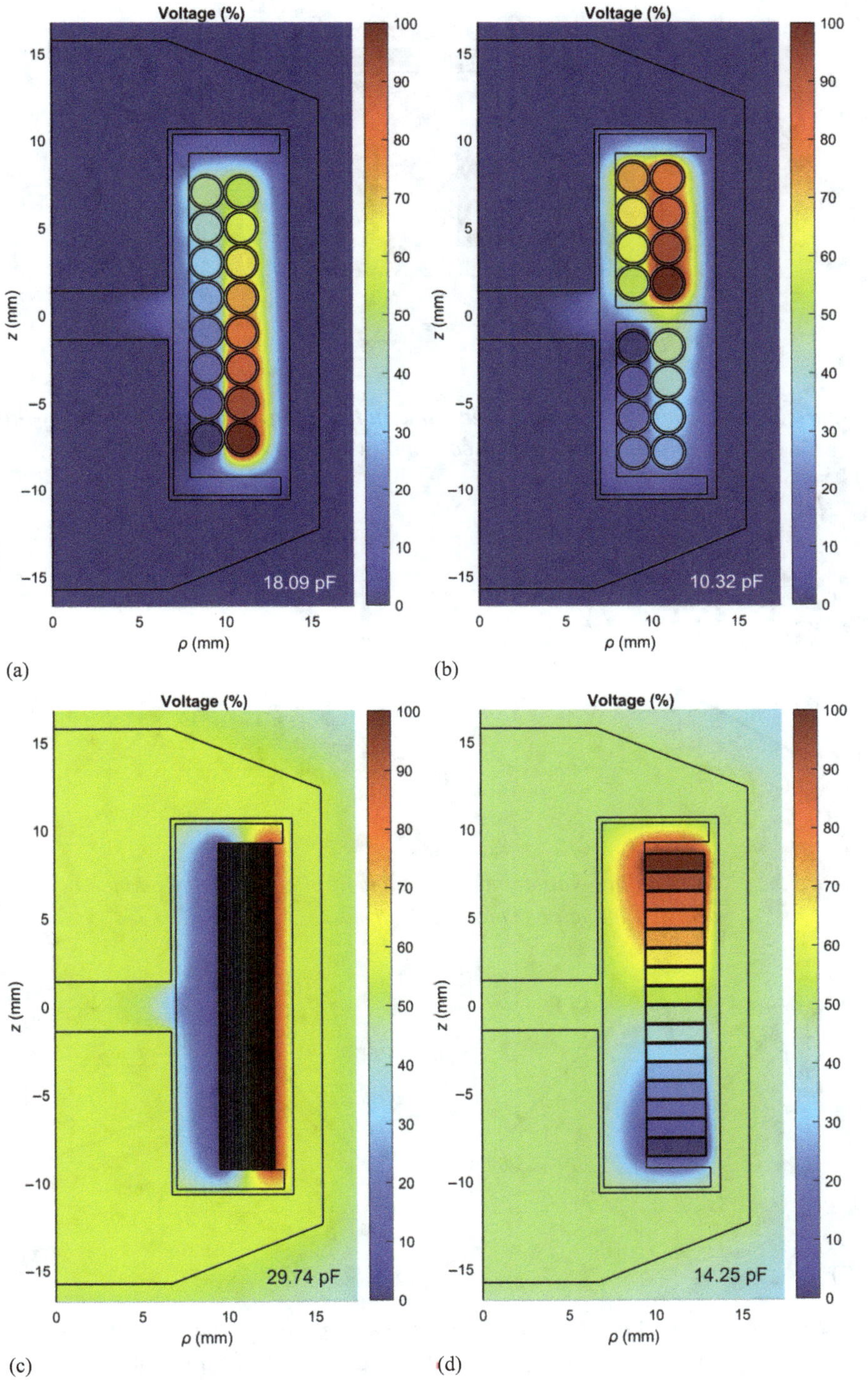

Figure 5.23 Capacitance of non-cast inductors with a grounded core of (a) orthogonal conventional, (b) two-chamber winding and with floating core, (c) foil, and (d) flat wire winding

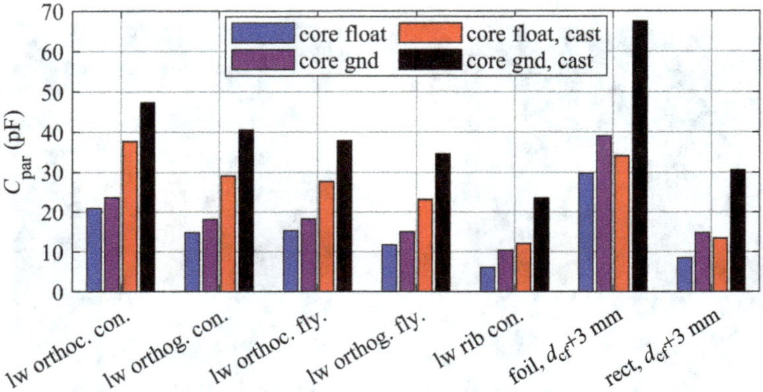

Figure 5.24 Comparison of parasitic capacitance of investigated inductors with floating and grounded core and with and without cast

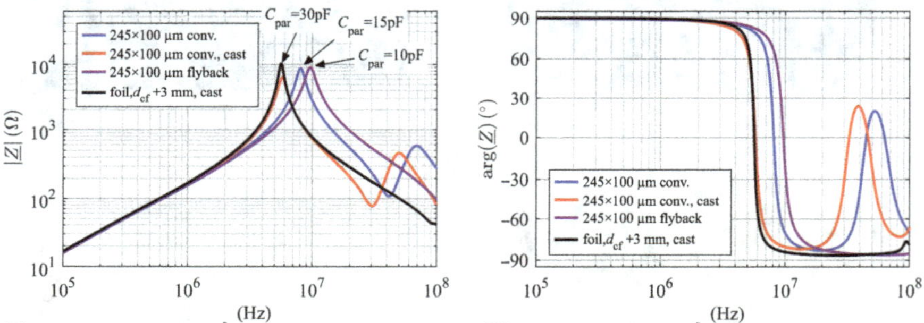

(a) (b)

Figure 5.25 (a) Absolute value and (b) argument of complex impedance of selected inductor prototypes with DG

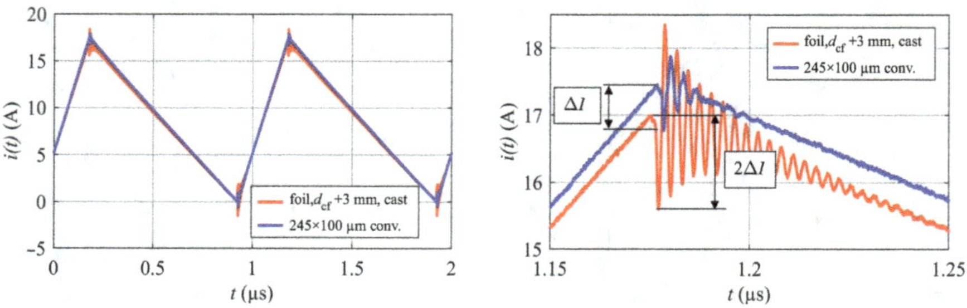

(a) (b)

Figure 5.26 (a) Two periods of the current and (b) the section at the moment of switching of the non-cast inductor with litz wire and the cast foil winding with DG is depicted. This shows that, according to $I = C\frac{du}{dt}$, the foil winding (30 pF) with twice as high capacitance as the litz wire winding (15 pF) also has twice as high a current drop, accompanied by further oscillations. This increases semiconductor losses and negatively impacts EMC performance. The distribution of capacitance in a winding is playing an increasingly important role, especially with the ever-steeper voltage slopes in SiC and GaN applications. Non-linearities can occur [18,19], which can result in higher stress to the insulation or partial discharges. This can lead to a reduced lifetime of the component. Models can be used for the analysis in which, for example, each winding is modeled by an RLC equivalent circuit diagram and these are coupled magnetically and electrically [20–22]. The effects of the design of an inductor on the signal characteristics can thus be analyzed in advance.

References

[1] Biela J, and Kolar JW. Using transformer parasitics for resonant converters - a review of the calculation of the stray capacitance of transformers. *IEEE Transactions on Industry Applications*. 2008;44(1):223–233.

[2] Zacharias P. *Magnetic Components: Basics and Applications*. Springer Fachmedien Wiesbaden; 2022.

[3] Albach M. *Induktivitäten in der Leistungselektronik*. Springer Fachmedien Wiesbaden; 2017.

[4] Rossmanith H, Doebroenti M, Albach M, and D. Exner Measurement and characterization of high frequency losses in nonideal litz wires. *IEEE Transactions on Power Electronics*. 2011;26(11):3386–3394.

[5] Lindner H, Brauer H, and Lehmann C. *Taschenbuch der Elektrotechnik und Elektronik*. Carl Hanser Verlag; 2018.

[6] Stadler A, and Gulden C. Copper losses of litz-wire windings due to an air gap. In: *2013 15th European Conference on Power Electronics and Applications (EPE)*; 2013. p. 1–7.

[7] Nan X, and Sullivan CR. An improved calculation of proximity-effect loss in high-frequency windings of round conductors. In: *IEEE 34th Annual Conference on Power Electronics Specialist*, Acapulco, Mexico, *2003. PESC '03*, vol. 2; 2003. p. 853–860.

[8] Moreau O, Popiel L, and Pages JL. Proximity losses computation with a 2D complex permeability modelling. *IEEE Transactions on Magnetics*. 1998;34 (5):3616–3619.

[9] Roßkopf A, Bär E, Joffe C, and Bonse C. Calculation of power losses in litz wire systems by coupling FEM and PEEC method. *IEEE Transactions on Power Electronics*. 2016;31(9):6442–6449.

[10] Ferreira JA. Analytical computation of AC resistance of round and rectangular litz wire windings. *IEE Proc B Electr Power Appl UK*. 1992;139(1):21.

[11] Stadler A, Huber R, Stolzke T, and Gulden C. Analytical calculation of copper losses in litz-wire windings of gapped inductors. *IEEE Transactions on Magnetics*. 2014;50(2):81–84.

[12] Nan X, and Sullivan CR. An equivalent complex permeability model for litz-wire windings. *IEEE Transactions on Industry Applications*, 2009;45 (2):854–860.

[13] Meeker DC. An improved continuum skin and proximity effect model for hexagonally packed wires. *J. Comput. Appl. Math.*; 2012;236(18): 4635–4644.

[14] Stadler A, Albach M, and Macary F. The minimization of copper losses in core-less inductors: application to foil- and PCB-based planar windings. In: *2005 European Conference on Power Electronics and Applications*, Dresden, Germany; 2005. p. 7.

[15] McLyman CWT. *Transformer and Inductor Design Handbook*. 3rd ed. No. 121 in Electrical and Computer Engineering. Marcel Dekker; 2004.

[16] Duerbaum T, and Manfred A. Core losses in transformers with an arbitrary shape of the magnetizing current. In: *6th European Conference on Power Electronics and Applications (EPE'95)*; 1995.

[17] Snelling EC. *Soft Ferrites – Properties and Applications*. Iliffe Books Ltd; 1969.

[18] Florkowski M, Furgal J, and Pajak P. Analysis of fast transient voltage distributions in transformer windings under different insulation conditions. *IEEE Transactions on Dielectrics and Electrical Insulation*. 2012;19(6):1991–1998.

[19] Cremasco A, Rothmund D, Curti M, and Lomonova EA. Voltage distribution in the windings of medium-frequency transformers operated with wide bandgap devices. *IEEE Journal of Emerging and Selected Topics in Power Electronics*. 2022;10(4):3587–3602.

[20] Reissenweber L, Stadler A. Extension of a calculation model based on the Pi line theory for transient voltage distribution in inductors by complex permittivity and frequency-depending complex permeability. In: *2019 21st European Conference on Power Electronics and Applications (EPE '19 ECCE Europe)*; 2019. p. P.1–P.8.

[21] Reissenweber L, Stadler A. Modeling of a power transformer including higher order resonances. In: *2020 22nd European Conference on Power Electronics and Applications (EPE'20 ECCE Europe)*; 2020. p. P.1–P.9.

[22] Cremasco A, Curti M, Pourkeivannour S, and Lomonova EA. Medium-frequency transformers for fast rise time PWM voltages: modelling and design considerations. In: *2021 Sixteenth International Conference on Ecological Vehicles and Renewable Energies (EVER)*; 2021. p. 1–6.

Chapter 6

Magnetic material selection for power inductors and transformers

Alyssa Brown[1], Elaine Ng[1] and Alex Hanson[1]

6.1 Background

How should magnetic materials be evaluated for power electronics applications? For a design targeting a specific operating frequency, one might choose a material for its relative permeability $\mu_r = \mu/\mu_0$, the saturation flux density B_{sat}*, volumetric core loss $P_{v,core}$, or any number of other parameters. *All else equal*, it is almost always better to have higher permeability, higher saturation flux density, and lower core loss. However, "all else" is rarely equal. Consider choosing among the following in Table 6.1.

Table 6.1 Which of the following materials is "best?"

Material	μ_r	$\hat{B}$ (mT)	B_{sat} (mT)
1	1000	200	700
2	100	300	400
3	500	100	800

Saturation and permeability have already been defined; $\hat{B}$ is defined as the maximum ac flux density that can be imposed on the material at a given frequency without violating thermal limits (essentially a metric for core loss). It may be unclear which material in Table 6.1 is best. Material 1 has the highest permeability, Material 2 has the lowest core loss, and Material 3 has the highest saturation flux density.

It is tempting to say that the choice of the best material depends on the application. This observation is true, but is usually made to evade the question. Let us meet the question head-on: Even if we had a specific application in mind, could we then clearly tell which material is the best choice?

[1]Chandra Family Department of Electrical and Computer Engineering, University of Texas at Austin, USA

*For the purposes of this chapter, B_{sat} can be the actual saturation flux density of the material or can be a saturation-based limit that includes design margin, e.g. $0.75B_{sat}$. This distinction is particularly relevant for powdered iron cores with relatively smooth saturation characteristics (Chapter 3).

To address this question, we need to identify categories of applications with different demands and an easily calculable figure of merit for each category. This chapter begins by presenting procedures for selecting materials at a particular frequency by deciding if the intended application is limited by saturation or core loss. Designs that fall into these cases are simply constrained by B_{sat} (if saturation is limited) and $\widehat{B}$ (if core loss is limited); permeability is negligible as long as it is sufficiently high. We then derive a calculable metric for how much permeability is necessary for inductors and transformers, and we find that the permeabilities of most magnetic materials are more than sufficient.

At extremely high frequencies (above tens of MHz), core loss may become so high that it is better to forego the magnetic core and rely on air-core inductors. The boundary where this should occur has historically been hazy, leaving applications operating at 1–100 MHz without proper guidance. This chapter discusses recent research about a more rigorous threshold for deciding when to use a magnetic-core inductor over an air-core inductor.

Finally, we conclude with a discussion of the performance factor $\mathcal{F}$ as a useful metric for comparing one or more materials at different frequencies. The performance factor thus provides quantitative guidance about the power density gains that one might achieve by increasing operating frequency.

6.2 AC-dominated cases

Let us start with the material selection task for the category of sinusoidal ac inductors. These are single-winding components with purely ac sinusoidal current and voltage excitations. Resonant inductors such as those in resonant dc–dc converters or radio frequency (rf) matching networks would fall into this category.

Our goal is to design the smallest possible inductor for a given power level or, in other words, to process the most power possible for a given inductor geometry. We evaluate the power handling of the inductor by focusing on its VA rating, $\text{VA} = V_{max}I_{max}$. The first component, V_{max}, is the maximum voltage amplitude that can be applied to the inductor while staying within material limits on B field; it is calculated through Faraday's law as $V_{max} = N\omega B_{max}A_c$, where A_c is the core cross-sectional area and N is the number of turns on the inductor. The other component, I_{max}, is the maximum current amplitude that can flow in the inductor for a given current density limit J_{max} and is given by $I_{max} = J_{max}\frac{A_w}{N}$, where A_w is the window area. Then, the VA rating of the inductor is given by

$$\text{VA} = \omega B_{max}J_{max}A_cA_w \tag{6.1}$$

Let us pause to appreciate (6.1). It tells us the maximum power that can be processed by an inductor of a given geometry or the minimum size that can be achieved for a given power level. The result is general across different inductors for different applications, as the design and application details of the inductor (the target inductance L, the number of turns N, and the gap length l_g) do not appear in the equation.

If we focus our attention on evaluating the core material alone, clearly B_{max} is our metric of choice. The more B field the material can sustain, the smaller the inductor can be for a given power level, or the more power the inductor can process for a given size.

To experienced designers, this conclusion will be sensible, even obvious. Still, consider that (6.1) makes no mention of permeability; we therefore tentatively conclude that permeability does not matter. We will quantify this statement in Section 6.7 and find that it is largely true in practice. Of the hypothetical materials listed in Table 6.1, Material 1 has the highest permeability, but (6.1) tells us to avoid being tempted by that high permeability and instead focus on B_{max}.

It remains for us to decide what B_{max} is—in particular, whether it is set by saturation or by core loss. Whether the core will saturate depends on whether its peak instantaneous B field B_{pk} is greater or less than the saturation B field of the material B_{sat}. By contrast, core loss depends not on B_{pk} but rather on B_{ac}, the amplitude of the ac component of the flux (dc B fields do not directly cause core loss). The limit on B_{ac} that determines if the core will overheat was defined earlier as $\widehat{B}$.

Either saturation or core loss may limit the design of the inductor. One can imagine a material whose core loss is so low that the B field (sinusoidal in this case) would be allowed to swing all the way from $+B_{sat}$ to $-B_{sat}$ without overheating; one can also imagine a material whose core loss is so large that the B field would only be allowed to swing from $+\widehat{B}$ to $-\widehat{B}$, where excitation at an amplitude $\widehat{B} < B_{sat}$ causes the inductor to reach thermal limits before saturation.

We therefore arrive at a method to evaluate a core material for a sinusoidal ac inductor application in which miniaturization is the primary goal:

1. Unless permeability is close to 1, ignore it.
2. Identify the B field amplitude that results in the core reaching a loss density limit (chosen, perhaps, between 100 and 500 mW/cm^3), called $\widehat{B}$, using core loss data from the datasheet. Further identify B_{sat} from the datasheet.
3. Since this analysis is for an ac inductor, the peak B field B_{pk} and the ac amplitude of the B field B_{ac} are the same. Therefore, the core will be limited by the smaller of B_{sat} or $\widehat{B}$. If B_{sat} is smaller, we call the material *saturation limited* in this design scenario; if $\widehat{B}$ is smaller, we call the material *core loss limited*. In other words, $B_{max} = \min(B_{sat}, \widehat{B})$.
4. Whichever material has the greatest B_{max} wins

From the list of hypothetical materials listed above, every material is core loss limited—it is only possible to impose an ac B field of $\widehat{B}$, which means that it is impossible to reach B_{sat} using ac excitations with these materials at this frequency. This was not guaranteed to be the case for every material, but we will find later that core loss is in fact likely to limit ac-dominated designs. Material 2 is the best choice for this ac-dominated application, and the subsequent design of the inductor should be based on core loss rather than saturation.

6.3 DC-dominated cases

Let us expand the list of cases that we can handle by considering dc-dominated or choke inductors. These are cases in which the inductance is chosen to be so large that the ac current ripple is small, and the inductor current and B field are mostly dc. In the limit of a true dc choke inductor, there is no core loss, and the selection of core material prioritizes keeping the (large) peak B field below saturation rather than keeping the (small) ac B field within core loss limits.

The most commonly used metric to quantify the available performance of a magnetic core in this application is the core geometric constant K_g, which is derived by identifying the target inductance, the peak allowable instantaneous B field $B_{pk} = B_{sat}$, and the maximum allowable winding resistance R. These constraints are combined with the wire resistivity ρ and the wire packing factor K_u to yield

$$K_g = \frac{A_c^2 A_w}{l_w} \geq \frac{\rho L^2 I_{pk}^2}{B_{sat}^2 R K_u} \tag{6.2}$$

where K_g is only a function of the core cross-sectional area A_c, the window area A_w, and the mean length of a turn l_w. Using core geometries with larger K_g, one can construct inductors with lower loss (R) for a given energy storage ($\propto LI_{pk}^2$) or a higher energy storage for a given loss. Regarding material selection, B_{sat} is the relevant metric here.

These cases can also be examined in a manner similar to the ac inductor case derived above. Assume an inductor has dc current flowing through it and, in steady state, a sinusoidal voltage v applied to it. This results in a sinusoidal current superimposed on the dc current. We begin with expressing the magnetic flux through the inductor as $NA_cB_{sat} = I_{pk}L$, where L must be $\int vdt/\Delta i_{pkpk}$, yielding

$$\frac{\left(\int vdt\right) I_{pk}}{\Delta i_{pkpk}} = NA_cB_{sat} \tag{6.3}$$

where Δi_{pkpk} is the peak-to-peak inductor current.

Let us perform a small change of variables by noting that $I_{pk} = I_{dc} + I_{ac} = I_{dc} + \Delta i_{pkpk}/2$ and defining I_{ac}/I_{dc} as the ripple ratio $\mathcal{R}$. Incorporating this observation and the fact that the integral of a half-sinusoid yields $2/\omega$ times the amplitude gives us

$$\frac{V_{pk}}{\omega}\left(\frac{1+\mathcal{R}}{\mathcal{R}}\right) = NA_cB_{sat} \tag{6.4}$$

We therefore have a relationship between the size of the applied voltage and the maximum allowable B field. A limitation on current can be set as $I_{rms} = \sqrt{I_{dc}^2 + I_{ac}^2} = J_{max,rms}A_w/N$ in a manner similar to the ac case above. The power handling capability can then be expressed as

$$V_{pk}I_{rms} = 2\pi f B_{sat} J_{max,rms} A_w A_c \frac{\mathcal{R}}{1+\mathcal{R}} \tag{6.5}$$

According to this metric, to maximize power handling capability, we again choose a core material with the highest B_{sat}.

We can calculate a slightly different metric, $V_{pk}I_{dc}$ as follows:

$$I_{rms} = \sqrt{I_{dc}^2 + I_{ac}^2} = I_{dc}\sqrt{1+\mathcal{R}^2} \tag{6.6}$$

$$V_{pk}I_{dc} = 2\pi f B_{sat} J_{max,rms} A_w A_c \frac{\mathcal{R}}{(1+\mathcal{R})\sqrt{1+\mathcal{R}^2}} \tag{6.7}$$

The interpretation of this equation becomes even clearer if we imagine the inductor to be in a buck converter with a PWM voltage and a triangular ac current. In this case, the integral of voltage when the switch is off is $\int vdt = V_{out}(1-D)T$. Propagating this through the derivation yields

$$P = V_{out}I_{dc} = \frac{2}{(1-D)} f\, B_{sat} J_{max,rms} A_w A_c \frac{\mathcal{R}}{(1+\mathcal{R})\sqrt{1+\mathcal{R}^2}} \tag{6.8}$$

where P is the power processed by the converter. This specific case makes it very clear that a material with higher B_{sat} would permit greater power for the same geometry, or a smaller geometry for a given power level.

Yet another interesting way to examine the inductor is through its peak energy storage requirement, $\frac{1}{2}LI_{pk}^2$, where $L = (2V_{pk}/\omega)/(\Delta i_{pkpk})$ for sinusoidal voltage excitations. Using the same relationships above between the peak-to-peak current, ac current, dc current, and ripple ratio, we arrive at

$$\frac{1}{2}LI_{pk}^2 = \frac{1}{2}\frac{V_{pk}}{\omega}I_{dc}\frac{(1+\mathcal{R})^2}{\mathcal{R}} \tag{6.9}$$

where we have solved $V_{pk}I_{dc}$ previously:

$$\frac{1}{2}LI_{pk}^2 = \frac{1}{2}\frac{1+\mathcal{R}}{\sqrt{1+\mathcal{R}^2}} B_{sat} J_{max,rms} A_w A_c \tag{6.10}$$

It is clear from all of the above expressions that the power processing capability (whichever way we quantify it) is proportional to $B_{sat}A_wA_c$, which means that a higher B_{sat} material can process more power for a given size or can yield a smaller inductor for a given power rating. The popular K_g metric offers similar guidance.

We therefore arrive at a method to evaluate a core material for a dc-dominated case in which miniaturization is the primary goal:

1. Unless the permeability is close to 1, ignore it.
2. In this case, B_{sat} is the parameter of interest and can be found from the datasheet.
3. Whichever material has the greatest B_{sat} wins.

Which of the hypothetical materials listed at the beginning of the chapter would be best for such an application? Only B_{sat} matters here, so we would clearly choose Material 3. Even though Material 3 has modest permeability and the worst core loss characteristic, it is clear that neither the permeability nor the core loss determine the minimum size of this dc-dominated inductor.

6.4 In between AC- and DC-dominated cases

In dc-dominated cases, we can be sure that the design will be limited by the need to keep the high peak B field under saturation and will not be limited by the need to control core loss, which arises from the small ac B field. When the ripple ratio is substantial, however, we may not have such confidence. However, we can say the following:

- If the design is limited by saturation, then we should choose the material with the highest B_{sat}. There will be core loss, but if our desire is to process the most power for a given volume, then the core loss will not play a major role.
- If the design is limited by core loss, then we should choose the material with the highest $\widehat{B}$.

The question, then, is how we can know which case we should consider. In the ac-dominated case, this became a question of whether B_{sat} or $\widehat{B}$ was smaller. When dc and ac components are present, we must examine the ripple ratio, $\mathcal{R} = I_{ac}/I_{dc} = I_{ac}/(I_{pk} - I_{ac})$, which is usually set by the circuit design. Rearranging yields

$$\frac{I_{pk}}{I_{ac}} = 1 + \frac{1}{\mathcal{R}} = \frac{1+\mathcal{R}}{\mathcal{R}} \tag{6.11}$$

The peak current determines the peak B field, which we use to determine whether saturation will occur; the ac current determines the ac B field, which we use to calculate core loss. Therefore, the question is whether a given material in a given application has lower B_{sat}/I_{pk} or lower $\widehat{B}/I_{ac}$. If the former is lower, then the peak B field will reach saturation while core loss is still relatively small. If the latter is lower, then the ac B field will reach core loss limits while the peak B field still falls short of saturation. We can quantify the question as whether

$$\widehat{B} > B_{sat}\frac{\mathcal{R}}{1+\mathcal{R}} \tag{6.12}$$

is true or not [1]. If it is, then the design will be saturation limited; if not, then the design will be core loss limited.

This yields the following, more general material selection procedure:

- Unless permeability is close to 1, ignore it.
- Identify whether the design is limited by core loss or saturation by asking whether $\widehat{B} > B_{sat}\frac{\mathcal{R}}{1+\mathcal{R}}$.
- If the design is core loss limited for a given material, $B_{max} = \widehat{B}$.
- If the design is saturation limited for a given material, $B_{max} = B_{sat}\frac{\mathcal{R}}{1+\mathcal{R}}$.
- Whichever material has the greatest B_{max} wins.

Consider an example in which $\mathcal{R} = 0.5$ and therefore $\frac{\mathcal{R}}{1+\mathcal{R}} = 1/3$ and let us re-examine the materials in Table 6.1. For Material 1, $\widehat{B}$ is less than $B_{sat}/3$, which means that the design will be core loss limited for this material despite the application's substantial dc current. The same applies for Material 3. For Material 2, $\widehat{B}$ is greater than $B_{sat}/3$, indicating that Material 2 will be saturation limited.

The material with the greatest B_{max} is Material 1, whose $\widehat{B}$ is greater than the $B_{sat}/3$ of Material 2 and the $\widehat{B}$ of Material 3. Interestingly, Material 1 had neither the best core loss nor the best saturation characteristic among the three materials, yet it would be the preferred choice for this application.

Finally, note that Material 1 had the best permeability among the three materials, but that isn't why it was the best choice.

6.5 Is saturation or core loss the likely limit?

By surveying available materials, we can begin to get a sense of when to expect saturation or core loss to limit a design. In the AC inductor case, Figure 6.1 (curve labeled "pure ac") shows the envelope of ferrites with the highest $\widehat{B}$ (equal to $\widehat{B}\frac{1+\mathcal{R}}{\mathcal{R}}$ for $\mathcal{R} \to \infty$) at frequencies ranging from 10 kHz to 100 MHz compared with typical ranges for B_{sat}. These are color coded to distinguish between MnZn materials (blue) and NiZn materials (red). We see that $\widehat{B}$ becomes less than B_{sat} at frequencies above a few tens of kHz. In the hundreds of kHz and beyond, we expect ac inductors to be core loss limited.

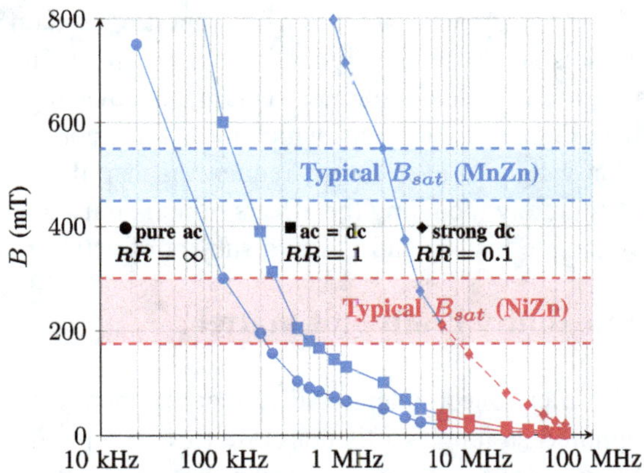

Figure 6.1 Curves of $\widehat{B}(1 + \mathcal{R})/\mathcal{R}$ for existing materials at different ripple ratios. Blue markers and regions represent MnZn materials; pink represents NiZn materials. When a curve crosses B_{sat} (denoted by the shaded regions), core loss determines the design instead of saturation, which occurs as low as 40 kHz for purely ac applications. By 10 MHz, only the most strictly dc component designs would be limited by saturation (such as EMI filter inductors).

On the opposite end of the spectrum, we cannot plot a curve for truly dc-dominated cases ($\mathcal{R} = 0$), which would be infinite for all frequencies and show B_{sat} to be the dominant parameter. Nevertheless, we can plot a strongly dc case ($\mathcal{R} = 0.1$) and the intermediate case ($\mathcal{R} = 1$, typical of critical or boundary conduction mode or BCM). We see that BCM designs are likely to be core loss limited beyond 100–200 kHz. More strongly dc cases may be saturation limited into the MHz range; nevertheless, past a few MHz, core loss is so intense that it dominates over saturation concerns even with a 10% ripple ratio.

Recall that Figure 6.1 shows the envelope of the best available $\widehat{B}$ among ferrites at a given frequency. The conclusions of this section are therefore not a guarantee that a design will be core loss limited or saturation limited if the available core material choices do not include those with the lowest loss. The saturation flux density ranges given are also approximate. Nevertheless, it is a useful exercise to reflect on classes of possible applications and to build accurate intuition about the likely limits of those designs.

6.6 What about non-sinusoidal waveforms?

The data in Figure 6.1 is taken for sinusoidal excitations, and our prior derivations have been based on primarily sinusoidal voltage excitations, yielding sinusoidal currents (plus dc current). Many power electronics applications do not conform to this paradigm and often use PWM voltage excitations and triangular-plus-dc currents.

Non-sinusoidal excitations are well known to cause additional core loss compared to sinusoidal excitations [2]. In addition, while ac flux density swings are causally responsible for core loss, it is nevertheless well known that the same ac flux density swing will incur more core loss when superimposed on a dc bias [3–5]. Therefore, core loss is liable to be higher in many real-life power electronics applications than Figure 6.1 would suggest. This means that $\widehat{B}(1 + \mathcal{R})/\mathcal{R}$ would tend to be lower than Figure 6.1 suggests, which means that core loss is more likely to constrain inductor designs at even lower frequencies than those suggested in the previous section. Accurately modeling the effects of non-sinusoidal waveforms and of dc bias has a venerable history and is still a subject of active research.

6.7 Does permeability really not matter?

We have repeatedly mentioned, intentionally, that the permeability of the core material is to be ignored so long as it is not close to 1. We previously argued this from the fact that permeability does not appear in the VA equations above, namely (6.1), (6.2), (6.5), (6.7), (6.8), and (6.10). Still, this is not satisfying and remains counter-intuitive. Why should permeability not matter, and how close to $\mu_r = 1$ does a material need to be before we should worry?

Permeability matters, up to a point, in the same way that the conductivity of wire matters up to a point. Once the resistance of a wire is low enough to be negligible relative to the other impedances in an electric circuit, obtaining a higher conductivity wire has a diminished impact on circuit operation. Similarly, once the permeability of a core material is high enough, additional permeability causes little further change in the magnetic behavior of a component. The electric wire and magnetic core are both important elements in their respective systems, but they cease to determine system performance once they are good enough.

Inductors, specifically, tend to have gaps in the magnetic path, which dominate the total reluctance of that path. Importantly, nothing else about the component design would change if the core permeability decreased and the gap were also shrunk to compensate for the increased core reluctance. The number of turns, peak B field, ac B field, saturation B field, and core loss would all stay the same. Thus, the performance of the component (power handling per unit volume) would be completely unchanged by a modest decrease in core permeability.

We can begin to quantify our logic above by noting that the maximum B field that a material can handle, B_{max}, is given by

$$B_{max} = \frac{LI}{NA_c} \tag{6.13}$$

where B_{max} could arise from the peak current $I = I_{pk}$ causing saturation $B_{max} = B_{sat}$, or it could arise from the ac current I_{ac} causing core loss to reach a limit $B_{max} = \widehat{B}$.

The inductance of a gapped structure is given by

$$L = \frac{N^2}{\frac{l_g}{\mu_0 A_c} + \frac{l_c}{\mu_0 \mu_r A_c}} \tag{6.14}$$

Here, we can verify that a decrease in core permeability can be offset by decreasing the gap length l_g to maintain constant L without also changing B_{max} or N which determine losses. Further, if we started with an optimized design, we can conclude that this is the *best* way to compensate for a reduced-permeability core, as any change to N or A_c would increase losses or volume. Therefore, decreasing core permeability has no impact at all on the performance of the inductor.

Nevertheless, all good things must come to an end. If the permeability were decreased to the point where the required gap disappeared, then the target B_{max} and L could no longer be maintained simultaneously. The point when $l_g = 0$ yields a critical relative permeability for inductors, $\mu_{r,critL}$:

$$\begin{aligned}\mu_{r,critL} &= \frac{Ll_c}{\mu_0 A_c N^2} \\ &= \frac{Ll_c}{\mu_0 A_c \frac{L^2 I^2}{A_c^2 B_{max}^2}} \\ &= \frac{A_c l_c \frac{1}{2}\frac{1}{\mu_0} B_{max}^2}{\frac{1}{2} L I^2}\end{aligned} \tag{6.15}$$

This analysis can apply to ac, dc, or in-between inductors, with saturation or with core loss as the limiting physical factor. The derivation does not change, except that the limit on B may be expressed more specifically. Therefore,

$$\mu_{r,critL} = \begin{cases} \dfrac{A_c l_c \frac{1}{2}\frac{1}{\mu_0} B_{sat}^2}{\frac{1}{2} L I_{pk}^2} & \text{if saturation limited} \\[2ex] \dfrac{A_c l_c \frac{1}{2}\frac{1}{\mu_0} \widehat{B}^2}{\frac{1}{2} L I_{ac}^2} & \text{if loss limited} \end{cases} \tag{6.16}$$

When $B_{max} = B_{pk}$ (either because the design is saturation limited or in ac-dominated cases in which $I_{pk} = I_{ac}$), we can interpret the denominator as the peak energy storage required of the component,

$$\mu_{r,critL} = \frac{A_c l_c \frac{1}{2}\frac{1}{\mu_0} B_{max}^2}{E_{store}}, \tag{6.17}$$

while the numerator is the maximum energy that could be stored in the core. Equation (6.17) suggests a useful interpretation: So long as the core has sufficient permeability that the core itself cannot store the required energy, a gap will be used and further increases in permeability are no longer beneficial.

When the application is ac-dominated, we can achieve an even more useful expression by recognizing that the quality factor of the inductor is $Q \triangleq 2\pi E_{store}/E_{loss}$, where E_{store} is the peak stored energy and E_{loss} is the lost energy over a cycle. We make the further approximation that about half of the lost energy in an ac loss-limited inductor is due to the core, while the other half is due to the windings. In other words, $E_{loss} \approx 2P_{core}/f = 2P_{v,core}$

$$\mu_{r,critL} = \frac{2\pi A_c l_c \frac{1}{2}\frac{1}{\mu_0} B_{max}^2}{Q\, E_{loss}} \tag{6.18}$$

$$= \frac{\pi \frac{1}{\mu_0} f\, B_{max}^2}{2Q\, P_{v,core}} \tag{6.19}$$

This last equation is particularly useful because ac inductors are typically built within certain common ranges of quality factor (let us say 3–300) and core loss density (say, 100–500 mW/cm^3). For a given frequency and given material $\widehat{B}$, the minimum required permeability for optimized use in an inductor or energy-storage transformer can be determined. The envelope of the lowest-loss (highest $\widehat{B}$) commercially available ferrite materials is used to plot $\mu_{r,critL}$ in Figure 6.2 for $P_{v,core} = 500$ mW/cm^3 and various values of quality factor. It can be seen that permeabilities above 500 are largely not useful except in the lowest-frequency,

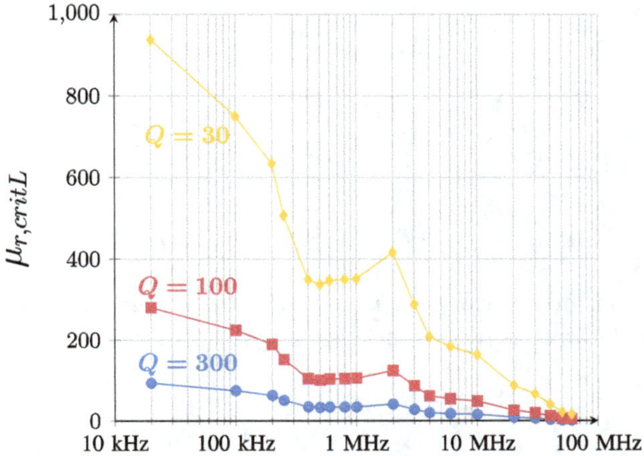

Figure 6.2 Critical relative permeability for ac inductors at different target quality factors using the envelope of best $\widehat{B}$ for existing ferrites. For reasonable values of Q, the required permeability can be well under 100 in the MHz region.

lowest-efficiency cases. In the MHz regime, permeabilities of less than 100 may be sufficient for higher-efficiency applications.

Many core materials with advantageous high-frequency properties such as low loss [6–9] and low permittivity [10–13] also tend to have permeabilities $\mu_r < 150$. This discussion is particularly important for evaluating these materials where relatively low permeabilities might otherwise discourage the use of a magnetic core at all.

6.8 Permeability in transformers

The prior argument focused heavily on the gap as the buffer which could be shrunk if core permeability were decreased. Therefore, one might imagine that the effect of permeability on conventional ungapped transformers would be quite different. Nevertheless, we will find that very high permeability is likewise of little value to transformers. Before quantitatively analyzing the transformer case, let us first seek an intuitive understanding of the role of permeability. This is exemplified in Figure 6.3, where a voltage source is used to deliver power to a resistive load through a transformer whose leakage inductances are neglected and whose magnetizing inductance is intended to be large, in part by virtue of a high-permeability core.

Finite permeability causes the magnetizing inductance to be finite, which is undesirable for this type of transformer. Still, if we could decrease the permeability of the core, independent of everything else, we would find very little difference so long as $\omega L_{mag} \gg R_{load}$. Only when ωL_{mag} approaches R_{load} would the magnetizing

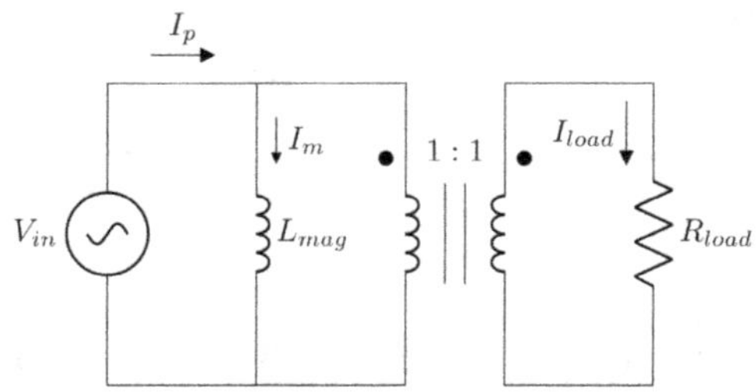

Figure 6.3 Transformer with a turns ratio of 1:1 and finite magnetizing inductance driving a load R_{load}

current become comparable to the primary-referred load current and the overall performance of the transformer would be noticeably worse.

We thus reach a similar conclusion as we did for inductors: additional permeability beyond a certain point is not to be valued highly. The difference is that the impact of permeability on non-energy-storage transformers has diminishing returns (permeability must be *substantially* greater than some threshold, after which increasing permeability provides *little* benefit), whereas we found that permeability in inductors and energy-storage transformers has an absolute threshold (any permeability above $\mu_{r,critL}$ provides *no* benefit).

Let us quantify the threshold permeability for transformers $\mu_{r,critX}$ by identifying it with the condition that $\omega L_{mag} = R_{load}$. Each of the parameters involved (ω, L_{mag}, R_{load}) is obviously application-dependent, so we must cast the problem in a more generalizable way. We begin by considering a transformer with a 1:1 turns ratio and finite magnetizing inductance driving a resistor R_{load} on the secondary, as shown in Figure 6.3. The power delivered to the load on the secondary side for given voltage, and current amplitudes are given by

$$P_{load} = \frac{1}{2} V_{load} I_{load} = \frac{1}{2} V_{in} I_{load} = \frac{1}{2} \omega N A_c B_{max} I_{load} \tag{6.20}$$

where V_{in} was replaced by the maximum voltage amplitude that can be applied to the transformer to yield B_{max}, which could be set by saturation or core loss. The maximum current amplitude allowed to flow on the primary side is defined as

$$I_p = \frac{\widehat{J} A_w/2}{N} \tag{6.21}$$

where $\widehat{J}$ is the maximum allowable current density amplitude in the winding due to loss constraints. This relates to the required load current amplitude through

$$I_{load} = \sqrt{I_p^2 - I_m^2} \tag{6.22}$$

due to the orthogonality of the load current I_{load} and the magnetizing current I_m. The deliverable power to the load is then given by

$$\begin{aligned} P_{load} &= \frac{1}{2} V_{load} I_{load} \\ &= \frac{1}{2} \omega N A_c B_{max} \sqrt{\left(\frac{\widehat{J} A_w}{2N}\right)^2 - \left(\frac{V_{in} l_c}{\omega N^2 \mu_0 \mu_r A_c}\right)^2} \\ &= \frac{1}{4} \omega A_c A_w B_{max} \widehat{J} \sqrt{1 - \left(\frac{2 B_{max} l_c}{\mu_0 \mu_r \widehat{J} A_w}\right)^2} \end{aligned} \tag{6.23}$$

Let us assume now that the transformer will be core loss limited, which we have seen is likely for ac-dominated cases at even modest frequencies. The allowable current density can be related to the core loss density by assuming that a well-designed component will have copper loss approximately equal to the core loss, yielding

$$P_{v,core} A_c l_c \approx \frac{1}{2} \widehat{J}^2 \rho A_w l_w \tag{6.24}$$

which produces

$$P_{load} \approx \frac{1}{4} \omega A_c A_w \widehat{B} \widehat{J} \sqrt{1 - \frac{2 \widehat{B}^2 \rho}{\mu_0^2 \mu_r^2 P_{v,core}} \frac{l_c l_w}{A_w A_c}} \tag{6.25}$$

where we have replaced B_{max} with $\widehat{B}$ specifically for core loss limited designs. Let us define the critical permeability for transformers as

$$\mu_{r,critX} = \frac{\widehat{B}_{max}}{\mu_0} \sqrt{\frac{2\rho}{P_{v,core}} \frac{l_c l_w}{A_c A_w}} \tag{6.26}$$

From Figure 6.4, we see that increasing μ_r much beyond $\mu_{r,critX}$ eventually saturates the power deliverable to the load P_{load} for a given current density $\widehat{J}$ and core loss density$P_{v,core}$, since P_{load} is proportional to $\sqrt{1 - \mu_{r,critX}/\mu_r}$. As long as

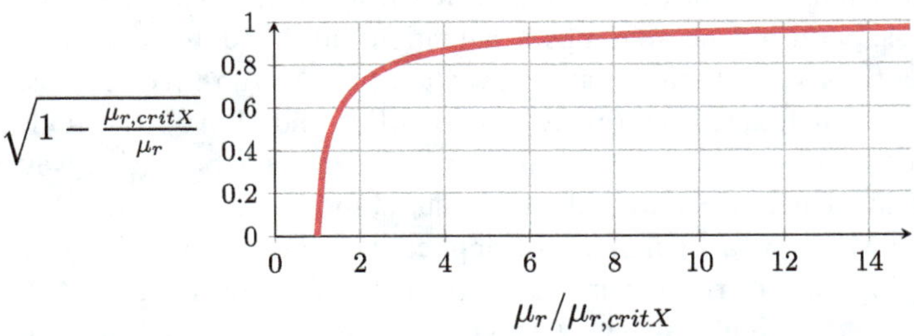

Figure 6.4 Function $\sqrt{1 - \mu_{r,critX}/\mu_r}$ is substantially saturated when $\mu_r > 5\mu_{r,critX}$

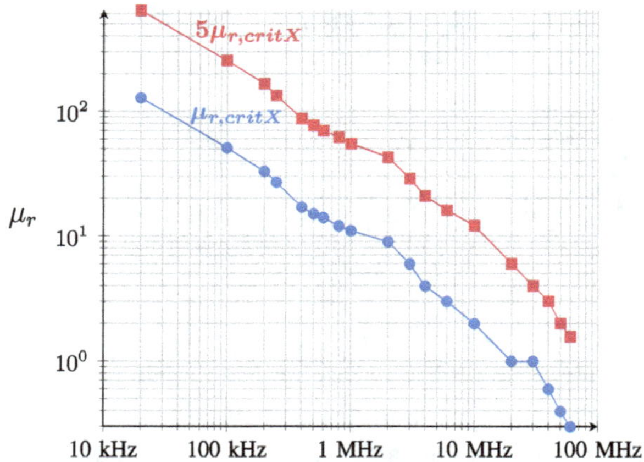

Figure 6.5 Critical relative permeability for transformers using the envelope of the best $\widehat{B}$ for existing ferrites, $P_{v,core} = 500\ mW/cm^3$, $\rho = 1.72 \times 10^{-8}\,\Omega m$, and geometric dimensions for an RM7 core as an example

the relative permeability is much greater than the critical permeability for transformers, additional permeability has little impact on the maximum power that can be delivered to a load. From Figure 6.4, we might quantify the $\gg$ symbol by saying that the relative permeability must be $5\times$ the critical permeability in order to be close enough to the maximum power handling.

To get a rapid numerical sense of $5\mu_{r,critX}$, consider the resistivity of copper ($\rho = 1.72 \times 10^{-8}\,\Omega\text{m}$), a core loss density of 500 mW/cm^3, and the dimensions of an RM7 core. Using numerical data for $\widehat{B}$, we plot $\mu_{r,critX}$ and $5\mu_{r,critX}$ in Figure 6.5. Here, we see again that relatively modest permeabilities satisfy the demands of $5\mu_{r,critX}$, with permeabilities less than 100 being sufficient above just a few hundred kilohertz.

6.9 Transition to air core magnetics

At ordinary power conversion frequencies, magnetic cores are used because they reduce the number of turns required to reach a given inductance. This reduces copper loss, but using a core imposes saturation limits and induces core loss. Still, this tradeoff is almost universally advantageous for sub-MHz applications. At greatly elevated frequencies (many tens of MHz and above), the tradeoff is no longer worthwhile, and coreless or air-core magnetics are preferred. Nevertheless, it is not immediately clear what the best path is for operating frequencies between 1 and 100 MHz. A large number of applications fall into this frequency range, including high-power rf matching networks for semiconductor and other material processing [8,14], high-frequency class D amplifiers for EMI filtering [15] or envelope tracking [16,17], power supplies on chip [9,18,19], applications operating at industrial-scientific-medical bands of 6.78/13.56/27.12 MHz, and others.

We saw in the previous section that materials do not offer increasing performance for ever-increasing permeability; instead, there is a limit above which additional permeability offers little benefit. We also saw that this limit decreases with frequency. Therefore, we might recycle our previous results in order to coarsely predict at what frequencies the required permeabilities $\mu_{r,critL}$ and $5\mu_{r,critX}$ reach 1, indicating that the core itself is only imposing additional restrictions without enabling any benefits. The data is re-plotted in Figure 6.6, where we find that the trend of both $\mu_{r,critL}$ and $5\mu_{r,critX}$ will cross 1 somewhere in the 80–100 MHz range. This initial clue suggests that almost all of the applications discussed above can benefit from the use of a magnetic core.

More rigorous air-core boundary definitions are derived in [20]. Here, we will focus on the derivation of the frequency at which the total power loss of an optimized magnetic-core inductor is equal to that of an optimized air-core inductor.

Consider an inductor with a toroidal geometry, as depicted in Figure 6.7. We begin the derivation by assuming there is an equal distribution of total loss between core and copper loss in magnetic-core structures (approximately true for optimized designs in which neither core nor copper loss should dominate). Also assume a high-frequency application where the maximum peak ac flux density $\widehat{B}$ is limited by a maximum core loss density, i.e., $P_{v,core}$ which, when multiplied by the volume of the inductor, gives an expected power loss due to the core, i.e., $P_{core,mag}$

$$\begin{aligned} P_{diss} &= P_{cu,mag} + P_{core,mag} \\ &= 2P_{cu,mag} = 2A_c l_c P_{v,core} \end{aligned} \tag{6.27}$$

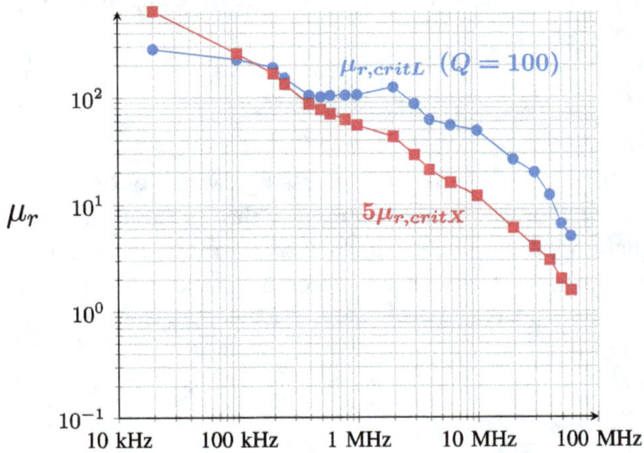

Figure 6.6 Critical permeability for inductors with $Q = 100$ and five times the critical permeability for transformers using the best $\widehat{B}$ data for ferrites

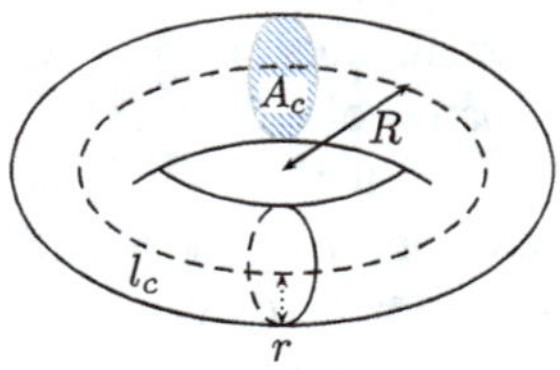

Figure 6.7 Labeled geometry of a toroid with a circular cross-section, showing its cross-sectional radius, r, and its radius of revolution, R

where $P_{cu,mag}$ and $P_{core,mag}$ are the total copper and core losses in a magnetic-core inductor (as opposed to the copper loss in an air-core inductor, $P_{cu,air}$). The current is restricted by the maximum copper loss from (6.27)

$$P_{cu,mag} = \frac{1}{2} I_{max}^2 R = I_{max}^2 \frac{1}{2} \frac{2\pi r N}{\delta\sigma\left(\frac{l_c}{N}\right)} \tag{6.28}$$

where σ is the conductance of the wire and δ is the skin depth (at the frequencies of interest, current conduction is almost certain to be skin depth limited). Since we are considering a toroid with a circular cross-section as depicted in Figure 6.7, with cross-sectional radius r and radius of revolution R, the cross-sectional area is $A_c = \pi r^2$ and mean path length is $l_c = 2\pi R$. The maximum allowed current is therefore

$$I_{max} = \sqrt{\frac{\delta\sigma l_c P_{cu,mag}}{\pi r N^2}} = \sqrt{\frac{\delta\sigma A_c l_c^2 P_{v,core}}{\pi r N^2}} \tag{6.29}$$

The ac B field is restricted to $\widehat{B}$ and thus voltage is restricted to $V_{max} = N A_c \widehat{B} \omega$. Thus, the maximum power processing capability of the magnetic-core inductor is

$$P_{mag} = \frac{1}{2} I_{max} V_{max} = \frac{1}{2} A_c l_c \widehat{B} \omega \sqrt{\frac{\delta\sigma A_c P_{v,core}}{\pi r}} \tag{6.30}$$

We would like to find the power processing capability of an air-core inductor with the same total power dissipation P_{diss}; the air-core structure is therefore allowed a higher copper loss due to its lack of core loss. Recalling the assumption that core and copper loss are balanced for magnetic-core components, we can conclude that, for an air-core component, $P_{diss} = P_{cu,air} = 2P_{cu,mag}$, and thus the maximum current is

$$I_{max} = \sqrt{\frac{2\delta\sigma A_c l_c^2 P_{v,core}}{\pi r N^2}} \tag{6.31}$$

The maximum voltage for the air-core structure is related to the maximum current ($V_{max} = L I_{max} \omega$); therefore, the maximum power processing capability of the air-core structure is

$$\begin{aligned} P_{air} &= \frac{1}{2}L\omega I_{max}^2 = \frac{1}{2}\frac{N^2 A_c \mu_0 \omega}{l_c}\frac{2\delta\sigma A_c l_c^2 P_{v,core}}{\pi r N^2} \\ &= \frac{1}{\pi r}A_c^2 l_c \delta\sigma\mu_0\omega P_{v,core} \end{aligned} \tag{6.32}$$

Using (6.30) and (6.32), we can derive the ratio of power processing capability under the constant-total-loss constraint as

$$\begin{aligned} \frac{P_{mag}}{P_{air}} &= \frac{\pi r\widehat{B}}{2A_c\delta\sigma\mu_0 P_{v,core}}\sqrt{\frac{\delta\sigma A_c P_{v,core}}{\pi r}} \\ &= \frac{\widehat{B}}{2\mu_0\sqrt{r\delta\sigma P_{v,core}}} \\ &= \frac{\widehat{B}f^{1/4}\pi^{1/4}}{2\mu_0^{3/4}\sigma^{1/4}r^{1/2}P_{v,core}^{1/2}} \end{aligned} \tag{6.33}$$

When (6.33) is equal to 1, both the air-core and magnetic-core inductors have the same maximum power rating given the same limitations on their total loss (P_{diss}). Equation (6.33) can be re-expressed as a $\widehat{B}$ threshold

$$1 \leq \frac{\widehat{B}f^{1/4}\pi^{1/4}}{2\mu_0^{3/4}\sigma^{1/4}r^{1/2}P_{v,core}^{1/2}} \tag{6.34}$$

$$\widehat{B} \geq \frac{2\mu_0^{3/4}\sigma^{1/4}r^{1/2}P_{v,core}^{1/2}}{f^{1/4}\pi^{1/4}} = 2\sqrt{rP_{v,core}\sqrt{\frac{\mu_0^3\sigma}{\pi f}}} \tag{6.35}$$

If this inequality is true, then the magnetic-core inductor is preferred given a constant total loss budget and the objective of minimizing volume.

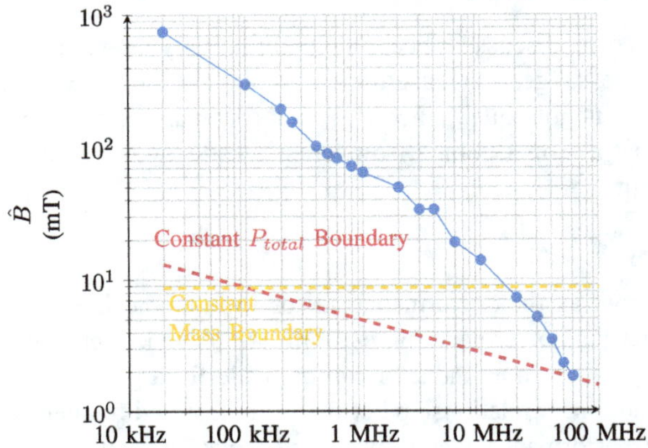

Figure 6.8 Survey of maximum $\widehat{B}$ for magnetic materials from industry and literature data points. The boundary from (6.35) is plotted in red and calculated using a $P_{v,core}$ = 200 mW/cm^3 and a cross-sectional radius r = 5 mm. The constant mass boundary from [20] is plotted as well.

The boundary is plotted on the same graph as an envelope of maximum $\widehat{B}$ for surveyed materials at different frequencies in Figure 6.8. With the total loss held constant, we see that the materials surveyed would outperform an air-core counterpart up to about 60 MHz. Many designs within the 5–50 MHz range use air-core structures, but the boundary given by (6.35) suggests that higher efficiencies can be achieved using magnetic cores with the same effective volume. This is similar to the conclusion we obtained by asking when $\mu_{r,critL}$ and $5\mu_{r,critX}$ cross 1.

In industries where mass is the main concern (biomedical, aerospace, etc.), we can also derive a similar boundary for air-core vs. magnetic core structures while keeping mass constant. The full derivation is in [20] and its final result is plotted in Figure 6.8 as well using typical values of $\rho_{core} = 5$ g/cm^3; $\widehat{J} = 500$ A/cm^2; $r = 5$ mm. We see that magnetic-core components can offer superior performance and mass up to nearly 20 MHz.

6.10 Beyond a single frequency: the performance factor

So far, choosing a material at a particular frequency has mostly reduced to choosing the material with the highest B_{sat} or $\widehat{B}$ and sufficiently high permeability given by $\mu_{r,critL}$ or $\mu_{r,critX}$. However, the techniques presented in the previous sections offer no guidance in choosing a material without a particular frequency in mind or for applications that require a wide range of frequencies. How should a magnetic material be evaluated against itself and against other materials at different frequencies? At high frequencies, the component's energy storage requirement decreases; nevertheless, core loss increases, which reduces the amount of energy a material can store. High-frequency materials also tend to have lower B_{sat}, and therefore both $\widehat{B}$ and B_{sat} tend to impose tighter limits at high frequencies.

To evaluate a material over a range of frequencies, we can begin by referring to the maximum power handling capability of ac inductors

$$\text{VA} = 2\pi f B_{max} J_{max} A_c A_w \tag{6.36}$$

where the tradeoffs discussed become quantitatively apparent. Increasing frequency f seems to suggest increased VA capability of an inductor, *all else equal.* However, all else is not equal, since B_{max} is a (decreasing) function of frequency. We conclude that a proper comparison across frequency requires us to look at the combined quantity

$$\mathcal{F} = fB_{max} \tag{6.37}$$

where $\mathcal{F}$ is known as the performance factor [3,21]. The larger the performance factor, the greater the power handling capability of an ac inductor of a given size or, conversely, the smaller an ac inductor can be made for a given power requirement. Because of the nonlinear relationship between core loss and frequency, it is logical to expect that the performance factor for a given material will increase at low frequencies, peak at some optimal frequency for that material, and then decline at higher frequencies. Using a material at frequencies below its peak $\mathcal{F}$ leaves promised gains from increased frequency on the table; using a material beyond its peak $\mathcal{F}$ pays a greater core loss price than a reduced energy storage requirement is worth.

The performance factor is usually cast in terms of core loss $\mathcal{F} = f\widehat{B}$. It is possible for B_{max} to be set by saturation instead of by core loss (at very low frequencies), in which case $\mathcal{F} = fB_{sat}$ would be more appropriate. By leaving $\mathcal{F}$ in terms of B_{max}, we are able to consider both cases.

Variations of the performance factor appear in expressions for non-ac inductors as well. For example, consider again the case of a saturation-limited buck inductor, where we evaluated the power handling capability of the inductor as

$$V_{pk}I_{dc} = 2\pi fB_{sat}J_{max,rms}A_wA_c\frac{\mathcal{R}}{(1+\mathcal{R})\sqrt{1+\mathcal{R}^2}} \tag{6.38}$$

In this case, while ac and dc components are both present, the power handling of the inductor is still given by the product of frequency and some maximum B field. The same is true regardless if we consider $V_{pk}I_{rms}$, $V_{pk}I_{dc}$, or $V_{out}I_{dc}$.

Interestingly, when we evaluate the buck inductor through its maximum allowable energy storage, $\frac{1}{2}LI_{pk}^2$, there is no explicit dependence on frequency. This makes sense in dimensional terms. Multiplying (6.10) through by frequency yields power-like quantities, which once again contain the term fB_{max}.

The explicit dependence on frequency also does not appear in the expression for the core geometric constant K_g in (6.2). This is an illusion, however, as the required inductance L depends inversely on frequency. The required core geometric constant K_g is therefore inversely proportional to $(fB_{max})^2$.

The performance factor for core loss limited designs is easily calculated if $\widehat{B}$ is already known, producing a plot such as Figure 6.9. We can see that the capability of existing core materials to process more power in a smaller size can continue to increase with frequency up to several tens of MHz. The field of power electronics has been increasing operating frequency for decades with the logic that doing so

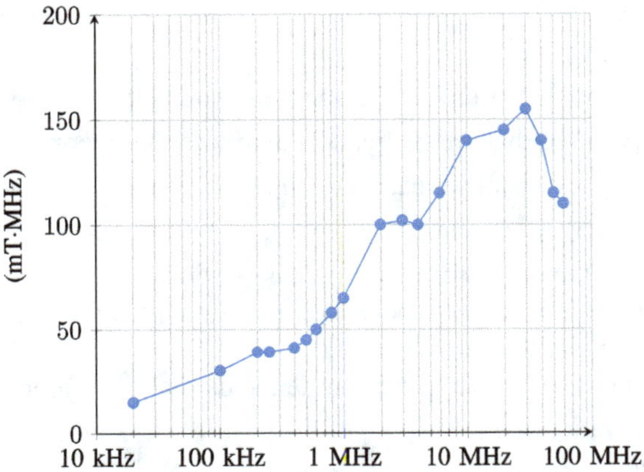

Figure 6.9 Survey of standard performance factor based on $\widehat{B}$

will increase power density, with state-of-the-art frequencies in the hundreds of kHz or perhaps 1 MHz for most applications. We see from the performance factor that this logic is expected to hold for at least one more order of magnitude increase in frequency.

The expected improvement is not linear, however. The standard performance factor only predicts perhaps a factor of 5 improvement from a factor of 100 increase in frequency (from 100 kHz to 10 MHz). Moreover, other frequency-dependent loss mechanisms not reflected in the standard performance factor further erode the expected gains of high-frequency operation. Modified performance factors [3] predict even more modest gains in power density, peaking near 10 MHz with existing materials.

Nomenclature

A_c	Core cross-sectional area
l_c	Path length of magnetic flux through the core
A_w	Window area
l_g	Gap length
μ_r	Dimensionless relative permeability, equal to the permeability divided by the permeability of free space, $\mu_r = \mu/\mu_0$
$\widehat{B}$	Maximum flux density amplitude that a material can sustain before reaching a predetermined loss limit
B_{sat}	Maximum instantaneous flux density that a material can sustain before its dB/dH is unacceptably reduced. Note that this definition of B_{sat} may be substantially lower than the flux density at which dB/dH approaches μ_0
B_{max}	Limit on the flux density that can be applied to a material; B_{max} is used when it is yet undetermined whether saturation or core loss will limit a component's design
J_{max}	Limit on the instantaneous current density that can be applied a component's winding
$J_{max,rms}$	Limit on the rms current density that can be applied to a component's winding
l_w	Mean length of a turn
I_{pk}	Peak current that an application demands that a component sustain
V_{pk}	Peak voltage that an application demands that a component sustain
R	Maximum resistance that an application demands that a component have
ρ	Resistivity of conductors
K_u	Packing factor; the fraction of the window area that can reasonably be filled with copper as opposed to insulation, manufacturing gaps, and gaps between circular conductors
L	Inductance that an application demands that a component have
$\mathcal{R}$	Ripple ratio, defined as half of the peak-to-peak current divided by the dc current
$P_{core,v}$	Core loss per unit volume

P_{core}	Total core loss
P_{cu}	Total copper loss
$\mu_{r,critL}$	Critical permeability for inductors; the permeability above which additional permeability provides no theoretical advantage
$\mu_{r,critX}$	Critical permeability for transformers; the permeability above which additional permeability provides much less advantage
$\mathcal{F}_w$	Performance factor, defined as $f^w B_{max}$. The original performance factor usually omits the subscript, $\mathcal{F}_1 = \mathcal{F} = fB_{max}$
ρ	Resistivity of the winding material

References

[1] A. J. Hanson, "Opportunities in magnetic materials for high-frequency power conversion," *MRS Communications*, vol. 12, pp. 521–530, 2022.

[2] K. Venkatachalam, C. Sullivan, T. Abdallah, and H. Tacca, "Accurate prediction of ferrite core loss with nonsinusoidal waveforms using only Steinmetz parameters," in *2002 IEEE Workshop on Computers in Power Electronics*, 2002. pp. 36–41.

[3] A. J. Hanson, J. A. Belk, S. Lim, C. R. Sullivan, and D. J. Perreault, "Measurements and performance factor comparisons of magnetic materials at high frequency," *IEEE Transactions on Power Electronics*, vol. 31, no. 11, pp. 7909–7925, 2016.

[4] J. Muhlethaler, J. Biela, J. W. Kolar, and A. Ecklebe, "Core losses under the dc bias condition based on Steinmetz parameters," *IEEE Transactions on Power Electronics*, vol. 27, no. 2, pp. 953–963, 2012.

[5] M. S. Rylko, B. J. Lyons, J. G. Hayes, and M. G. Egan, "Revised magnetics performance factors and experimental comparison of high-flux materials for high-current dc–dc inductors," *IEEE Transactions on Power Electronics*, vol. 26, no. 8, pp. 2112–2126, 2011.

[6] R. S. Yang, A. J. Hanson, B. A. Reese, C. R. Sullivan, and D. J. Perreault, "A low-loss inductor structure and design guidelines for high-frequency applications," *IEEE Transactions on Power Electronics*, vol. 34, no. 10, pp. 9993–10005, 2019.

[7] R. S. Yang, A. J. Hanson, C. R. Sullivan, and D. J. Perreault, "Design flexibility of a modular low-loss high-frequency inductor structure," *IEEE Transactions on Power Electronics*, vol. 36, no. 11, pp. 13013–13024, 2021.

[8] R. S. Bayliss, R. S. Yang, A. J. Hanson, C. R. Sullivan, and D. J. Perreault, "Design, implementation, and evaluation of high-efficiency high-power radio-frequency inductors," in *2021 IEEE Applied Power Electronics Conference and Exposition (APEC)*, 2021, pp. 881–888.

[9] D. V. Harburg, A. J. Hanson, J. Qiu, B. A. Reese, J. D. Ranson, D. M. Otten, C. G. Levey, and C. R. Sullivan, "Microfabricated racetrack inductors with thin-film magnetic cores for on-chip power conversion," *IEEE Journal of*

Emerging and Selected Topics in Power Electronics, vol. 6, no. 3, pp. 1280–1294, 2018.

[10] H. Zhao, S. Luan, Z. Shen, *et al.*, "Rethinking basic assumptions for modeling parasitic capacitance in inductors," *IEEE Transactions on Power Electronics*, vol. 37, no. 7, pp. 8281–8289, 2022.

[11] Z. Shen, W. Chen, H. Zhao, *et al.*, "Core energy capacitance of nizn inductors," *IEEE Transactions on Power Electronics*, vol. 38, no. 4, pp. 4235–4240, 2023.

[12] M. Kacki, M. S. Rylko, J. G. Hayes, and C. R. Sullivan, "Magnetic material selection for EMI filters," in *2017 IEEE Energy Conversion Congress and Exposition (ECCE)*, 2017, pp. 2350–2356.

[13] M. Kacki, M. S. Rylko, J. G. Hayes, and C. R. Sullivan, "Measurement methods for high-frequency characterizations of permeability, permittivity, and core loss of MN-ZN ferrite cores," *IEEE Transactions on Power Electronics*, vol. 37, no. 12, pp. 15152–15162, 2022.

[14] R. S. Bayliss, D. J. Perreault, C. R. Sullivan, R. S. Yang, and A. J. Hanson, "Self-shielded high frequency inductor," May 3, 1988, US Patent 4,741,207. [Online]. Available: http://www.google.it/patents/US4741207=0pt

[15] D. T. Nguyen, E. Macias, and A. J. Hanson, "Active emi filter with switch-mode amplifier for high efficiency," in *2022 IEEE Applied Power Electronics Conference and Exposition (APEC)*, 2022, pp. 443–450.

[16] L. R. Kahn, "Single-sideband transmission by envelope elimination and restoration," *Proceedings of the IRE*, vol. 40, no. 7, pp. 803–806, 1952.

[17] P. Asbeck and Z. Popovic, "Et comes of age: Envelope tracking for higher-efficiency power amplifiers," *IEEE Microwave Magazine*, vol. 17, no. 3, pp. 16–25, 2016.

[18] J. Qiu, A. J. Hanson, and C. R. Sullivan, "Design of toroidal inductors with multiple parallel foil windings," in *2013 IEEE 14th Workshop on Control and Modeling for Power Electronics (COMPEL)*, 2013, pp. 1–6.

[19] M. Araghchini, J. Chen, V. Doan-Nguyen, *et al.*, "A technology overview of the powerchip development program," *IEEE Transactions on Power Electronics*, vol. 28, no. 9, pp. 4182–4201, 2013.

[20] M. Solomentsev and A. J. Hanson, "At what frequencies should air-core magnetics be used?" *IEEE Transactions on Power Electronics*, vol. 38, no. 3, pp. 3546–3558, 2022.

[21] S. A. Mulder, "Loss formulas for power ferrites and their use in transformer design," *Philips Components*, p. 16. Eindhoven: Philips, 1994.

Part II

Modelling

Chapter 7

Modeling of inductive components for simulation in power electronic applications

Peter Zacharias[1] *and Alex Hanson*[2]

Modeling physical relationships permits design, optimization, and simulation. It can also make processes and states that cannot be perceived by the human senses accessible for further investigation. Simplifications are made in order to keep the complexity of a model within reasonable limits, which always results in limits to the applicability of the respective model.

In principle, models go farthest when they can be physically justified. However, purely data-driven models are also justifiable when the depth of physics or the inherent complexity of the problem exceeds human capacity. Both types of models are often used to describe magnetic materials and inductive components. This chapter provides an overview of the most important approaches to modeling the relationships between magnetic variables, electric variables, and geometry.

7.1 Linear approaches

The advantages of linear relationships are obvious: the dependencies are easy to describe, the solutions are scalable through linear algebra and superposition, and the calculations are very simple. When systems are nonlinear, linear approximations (small-signal models) can describe small responses to small changes in the influencing variables. The linear combination of nonlinear models, such as those resulting from the application of Kirchhoff's rules to magnetic and electrical circuit diagrams, can also produce some astonishing and useful modeling approaches.

The modeling of inductive components is based on Maxwell's equations, which we briefly review here

$$\text{Ampère's circuital law } \oint_l \vec{H} \cdot d\vec{l} = \sum_\nu I_\nu = \mathbb{F} \tag{7.1}$$

[1]Department of Power Electronics, University of Kassel, Germany
[2]Chandra Department of Electrical and Computer Engineering, University of Texas at Austin, USA

The integral of the H field around a closed loop l is equal to the total current that penetrates the loop. More accurately, the closed loop integral of the scalar product of magnetic field strength H along a path l is equal to the signed sum of the currents I_ν enclosed by l. These (equivalent) integrals and summations are also known as the magnetomotive force (MMF or F). The currents I_ν can come from actual current or, less commonly, the Maxwell displacement current density

$$\text{MMF} = \sum I_\nu = \int_A \left(\vec{J}_c + \vec{J}_d\right) \cdot d\vec{A} = \int_A \left(\sigma\vec{E} + \varepsilon\frac{\partial\vec{E}}{\partial t}\right) \cdot d\vec{A} \tag{7.2}$$

with J_c being the current density of moving charge carriers, J_d being displacement current density, E being the electric field, σ being the conductivity, σ being the permittivity

$$\text{Faraday's law of induction } \frac{d}{dt}\int_A \vec{B} \cdot d\vec{A} = \oint_l \vec{E} \cdot d\vec{l} = V \tag{7.3}$$

The integral of the E field around a closed loop l is equal to (the derivative of) the magnetic flux that penetrates the loop. More accurately, the change over time of the integral of the scalar product of the magnetic flux density $\vec{B}$ and the element $d\vec{A}$ over an area A bordered by l is equal to the line integral of the scalar product $\vec{E} \cdot d\vec{l}$ along the closed path l. These (equivalent) integrals are also known as the electromotive force (EMF), which has units of volts V.

$$\text{Gauss's law} \oint_S \vec{D} \cdot d\vec{S} = \oint_S \varepsilon\vec{E} \cdot d\vec{S} = \int_V \rho \cdot dV = Q \tag{7.4}$$

The total electric displacement flux that penetrates a closed surface S is equal to the charge enclosed by the surface. More accurately, the integral of the scalar product of the displacement current density D and the element $d\vec{S}$ of a surface S of a volume V is equal to the volume integral over the charge density ρ or equal to that of the charge Q enclosed in the volume V

$$\text{Gauss's law for magnetism} \oint_S \vec{B} \cdot d\vec{S} = 0 \tag{7.5}$$

The total magnetic flux that penetrates a closed surface S is zero. More accurately, the integral of the scalar product of magnetic flux density $\vec{B}$ and the element $d\vec{S}$ of a surface S of a volume equals zero. By analogy to Gauss's law for electricity, this law implies that there are no "magnetic charges" but only multipoles.

Ampere's law and Gauss's law can be combined to yield

$$\frac{\partial\rho}{\partial t} + \nabla\vec{J} = 0 \tag{7.6}$$

also known as charge conservation. Conservation of charge can therefore be thought of as a consequence of Maxwell's equations rather than a separate law.

These equations also exist in differential form. They form the basis for numerical field calculations using FEM programs. These are dealt with in a special chapter of this book.

In electrical circuits, Maxwell's equations are simplified in the quasistatic limit to yield $\int E \cdot dl = 0$ [Kirchhoff's voltage law (KVL) that the sum of voltage drops around a loop must be zero, $\sum V = 0$] and $\int J \cdot dS = 0$ [Kirchhoff's current law (KCL) that the total current entering/leaving any closed box must be zero, $\sum I = 0$]. KVL and KCL give circuit engineers a powerful tool to graphically draw the electrical behavior of a physical structure (a circuit diagram) as well as simple mathematical rules for solving for the voltages and currents in the circuit.

Faraday's law and charge conservation gave rise to KVL and KCL and the concept of a "circuit"; we observe that Ampere's law and Gauss's law for magnetism have a similar structure. Therefore, we are motivated to leverage the theoretical analysis that applies to electric circuits, with all of the powerful intuition that many electrical engineers have for circuits and circuit diagrams, and create a "magnetic circuit" to easily describe the behavior of magnetic fields. This gives rise to a magnetic KVL (that the sum of the magnetomotive forces around a loop is zero, $\sum F = 0$) and a magnetic KCL (that the sum of the magnetic flux entering/leaving any closed box must be zero, $\sum \phi = 0$).

Electrical circuit elements relate the voltage across a component to the current through that component. Similarly, magnetic circuit elements relate the MMF across a component to the magnetic flux through the component. The simplest such component has MMF proportional to flux, $\mathbb{F} = \mathbb{R} \cdot \phi$. In electric circuits, such a component R is called a resistor; in magnetic circuits, this component is called a reluctance. To make a difference, the suffix m is used: $\mathbb{R}_m$.

Figure 7.1 shows these relationships diagrammatically, emphasizing the magnetomotive force (magnetic voltage drop) and the flux (magnetic current) for a single branch element.

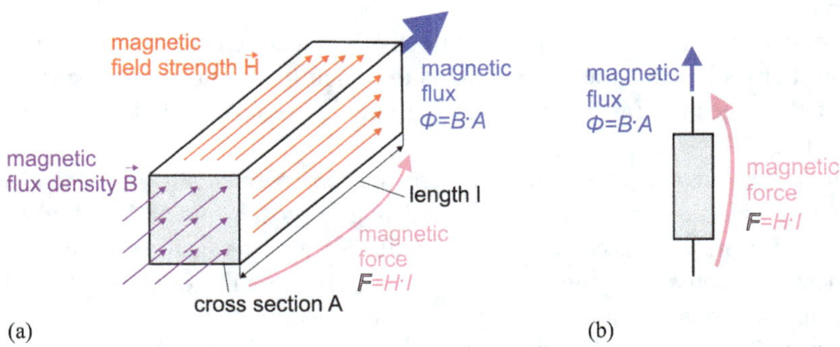

Figure 7.1 A single magnetic branch element with labeled field parameters: (a) sketch with magnetic field parameters and (b) schematic diagram of magnetic reluctance

In general, the magnetic flux can be a nonlinear function of MMF

$$\Phi = B \cdot A = f(H \cdot l) = f(\mathbb{F}) \tag{7.7}$$

Nevertheless, small signal modeling can be applied to predict small changes in flux in response to small changes in MMF (or vice versa). In memoryless components, this linearization ensures a $\mathbb{F} = \mathbb{R}\cdot\phi$ relationship

$$\mathbb{F} = \mathbb{R}_m \cdot \Phi \tag{7.8}$$

Note that the term "magnetic flux" (Φ) semantically denotes something flowing. But nothing is flowing/moving here! Nevertheless, the structures to be described are schematically similar to electrical circuits and can be treated with the same mathematical methods; hence, the vocabulary of magnetic circuits borrows some of the language of electric circuits. The term "magnetic reluctance" is similarly made in analogy to electric resistance. We can expand the analogy to include permeability μ, the proportionality constant between H and B in the linear part of the BH curve. Permeability in magnetic materials plays the same role as conductivity in conductive materials. This is reflected in the geometric equation for the magnetic reluctance $\mathbb{R}_\mathrm{m}$:

$$\mathbb{R}_m = \frac{l}{\mu A} = \frac{l}{\mu_r \mu_0 A} \text{ the physical unit } [\mathbb{R}_m] = \frac{A}{Vs} = H^{-1} \tag{7.9}$$

The permeability is divided into two factors μ_0 and μ_r, for a simpler interpretation. μ_0 is the permeability of the vacuum, while the relative permeability μ_r represents how many times more (or less) permeable a material is than vacuum. Magnetic materials with nonzero dipole moments usually have $\mu_r > 1$; $\mu_r < 1$ (diamagnetism) is uncommon, with the most diamagnetic material (Bismuth) having $\mu_r = 0.9998$. Permeability of $\mu_r \approx 1$ is characteristic of a paramagnetic material, while $\mu_r \gg 1$ is characteristic of ferromagnetic materials. Although there are further subdivisions into ferrimagnetism and antiferromagnetism, only the technically interesting ferromagnetic materials are considered below for the core materials.

Conducting materials can still effectively block changing magnetic flux through Lenz's law and the eddy currents that such flux induces in the conductor. Perfect flux rejection (and rejection at dc) is only possible with superconducting materials.

Note that air and even vacuum have nonzero permeability (magnetic conductivity). This does not change the modeling approach, but it does affect our intuition. We may be accustomed to thinking of air or vacuum as having zero electrical conductivity (a good approximation), but we will not always be able to make such an approximation for magnetic components.

Figure 7.2(a) shows the typical nonlinear relationship between B and H (and hence between MMF and flux) in ferromagnetic materials. The behavior of free space is represented by $B_0 = \mu_0 H$ (green). In a ferromagnetic material, the disordered magnetic dipoles in the initial state become more and more aligned as the

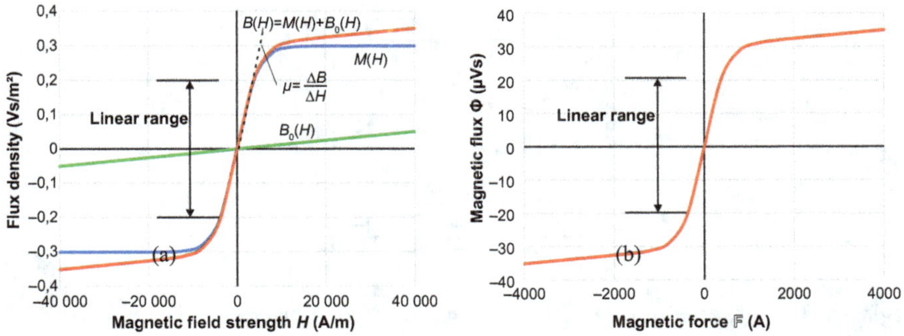

Figure 7.2 Example of characteristic curves of the magnetic properties: (a) related to the differential quantities magnetic field strength and flux density and (b) related to the dependence of the flux Φ *on the magnetic force* $\mathbb{F}$ *by scaling the axes of (a) with the length l and the cross-sectional area A*

magnetic field strength increases, thereby increasing the magnetic flux density. The additional effect of the material is characterized by the magnetization $M(H)$. When the dipoles are all aligned, there is no further increase in magnetization. This is referred to as saturation and can be seen in the $M(H)$ curve becoming flat or the $B(H)$ curve returning to a slope of μ_0.

A typical ferromagnetic material is therefore approximately linear for a wide range of B fields, justifying the use of reluctances to describe magnetic circuit components. Linearity is maintained by avoiding operation with B fields that would enter the saturation region.

In magnetic circuits, complex networks of components (whether linear or nonlinear) can be easily represented by interconnecting them, just like electric circuits. Applying electrical network theory can be validly applied, and the presence of nonlinear components simply limits the analysis tools that may be used, as is true for electric circuits with nonlinear components.

We have represented magnetic materials using the concept of reluctance. To form working magnetic circuits, a driving force for the magnetic flux is still missing. The driver of magnetic flux is, of course, electric current. Therefore, a magnetic circuit with a winding for the electric current will serve as an example of the modeling of magnetomotive force sources (Figure 7.3).

According to (7.3), the driving MMF $\mathbb{F} = I{\cdot}N$ is inserted into the circuit diagram. If Kirchhoff's rules are applied to this mesh in the circuit, the result is

$$\begin{aligned} I \cdot N &= (\mathbb{R}_{m1} + \mathbb{R}_{m2} + \mathbb{R}_{m3} + \mathbb{R}_{m4}) \cdot \Phi \\ \Phi &= \frac{I \cdot N}{\mathbb{R}_{m1} + \mathbb{R}_{m2} + \mathbb{R}_{m3} + \mathbb{R}_{m4}} \end{aligned} \tag{7.10}$$

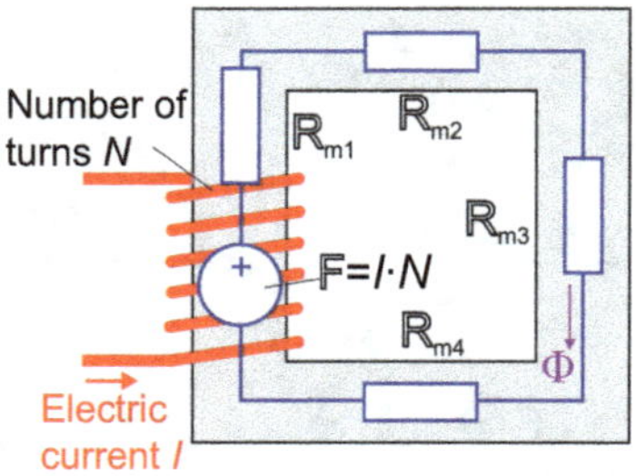

Figure 7.3 *Representation of a magnetic circuit as a principle sketch for the structure and as a circuit diagram*

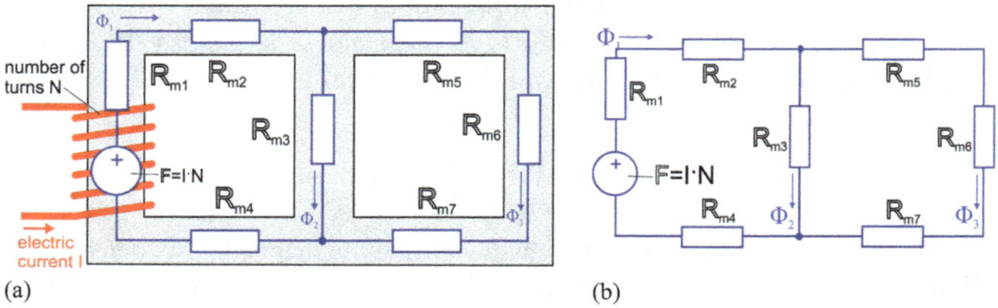

Figure 7.4 *Winding with branched magnetic core: (a) assembly sketch and (b) schematic magnetic circuit*

If one has a branched circuit, as shown in Figure 7.4(b), Kirchhoff's laws can be applied to it and obtain the following equations for the three magnetic fluxes $\mathbf{\Phi}_1$, $\mathbf{\Phi}_2$, and $\mathbf{\Phi}_3$:

$$\begin{aligned} &\mathbf{\Phi}_1 - \mathbf{\Phi}_2 - \mathbf{\Phi}_3 = 0 \\ &(\mathbb{R}_{m1} + \mathbb{R}_{m2} + \mathbb{R}_{m4})\mathbf{\Phi}_1 + \mathbb{R}_{m3}\mathbf{\Phi}_2 = I \cdot N \\ &-\mathbb{R}_{m3}\mathbf{\Phi}_2 + (\mathbb{R}_{m5} + \mathbb{R}_{m6} + \mathbb{R}_{m7})\mathbf{\Phi}_3 = 0 \end{aligned} \tag{7.11}$$

The procedure shown enables the mapping and calculation of magnetic relationships, with only a loose connection to electric circuits [through the current I in Figure 7.4(b)]. It may be desirable to derive the equivalent electrical behavior of such a device (i.e., the relationship between the winding voltage and the winding current).

In the following, a formalism is shown how an electrical model can be generated by transforming a magnetic model. We will show the modeling method but will not go into detail here as to why it works; we instead refer the reader to [5,15,16].

The method to generate the equivalent electric circuit from a magnetic circuit begins by taking the dual circuit from the magnetic equivalent circuit. A general formalism from network theory, known as graphical circuit inversion, is suitable for this purpose. This is shown in Figure 7.5. In the example, a network of different bipoles (black) forms four meshes [the outer circuit is referred to as an (extra) mesh] and four nodes $K_{1\ldots4}$. If you want to generate the associated dual

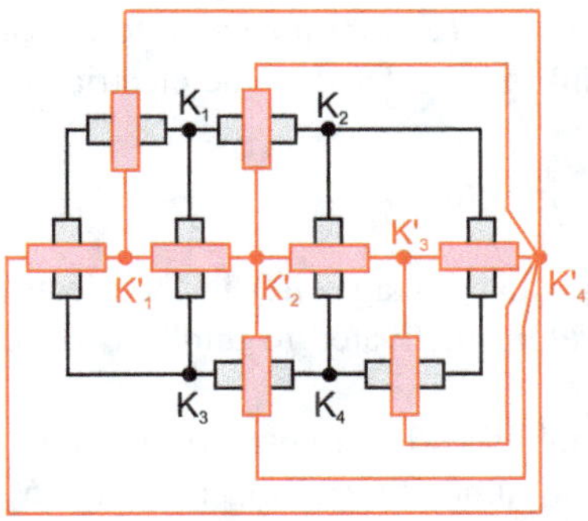

Figure 7.5 Schematic of the construction of a dual circuit (red) from a given circuit arrangement (black)

Table 7.1 Dual elements of magnetic and electrical circuits

Magnetic circuitry	**Electric circuitry**
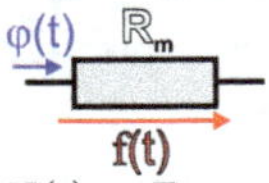 $\mathbb{F}(t) = \mathbb{R}_m \cdot \varphi(t)$ Magnetic resistance R_m $[\mathbb{R}_m]$ = A/Vs = 1/H	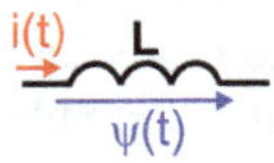 $\psi(t) = L \cdot i(t)$ Inductance L $[L] = H$ (Henry)=Vs/A
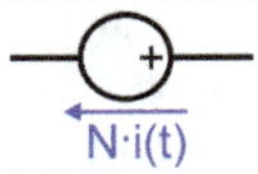 MMF $F = I{\cdot}N$	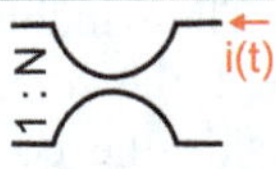 Ideal transformer $I = F/N$

structure, you enter a node inside each mesh (red). The branches of the dual structure are obtained by interconnecting the new nodes with components that pass through the original magnetic components. Thus, the red circuit topology is the dual topology to the black one, where the black components are magnetic circuit elements and the red components are their duals.

In order to carry out these operations, the branches and their dual equivalents must be known. In this chapter, two elements are sufficient to describe the magnetic world: magnetic reluctance and magnetomotive force sources.

Magnetic reluctance and inductance represent the dual components of the magnetic and electrical worlds. This can be plausibly justified by considering a loop of magnetic material with a single conducting turn wrapped around it. The MMF is related to the flux through

$$\mathbb{F}(t) = \mathbb{R}_m \cdot \varphi(t) = \frac{l}{\mu_r \mu_0 A} \cdot \phi(t) \tag{7.12}$$

for the magnetic reluctance $\mathbb{R}_m$ ($\mathbb{F}(t)$) – magnetic force, $\phi(t)$ magnetic flux).

Similarly, the flux linkage is related to the electric current by

$$\psi(t) = L \cdot i(t) = \frac{\mu_r \mu_0 A}{l} \cdot i(t) \tag{7.13}$$

where ψ is the magnetic flux linkage, *i*(*t*) is the electric current, and L is the inductance). We are therefore motivated to think of inductance and reluctance as duals of one another.

The MMF source establishes the connection between the magnetic and electrical worlds through the winding of the number of turns *N*. The dual component of an electrical circuit for this purpose is an ideal transformer [5,14]. The ideal transformer representing the electrical coupling to the magnetic world has an *N*:1 turns ratio, such that electrical current (amperes) is scaled into MMF (ampere-turns).

At the same time, the electrical circuit and the magnetic circuit are isolated from one another. The connection of the magnetic circuit to the surrounding electrical circuit is established via the transformer (or multiple transformers for multi-winding components). With a sufficiently detailed magnetic circuit, it is therefore possible to represent the magnetic component and its coupling to an electric circuit using a circuit simulator, for example.

An example of the application of reluctance modeling and transformation to an equivalent electric circuit is given in Figure 7.6 for a two-winding transformer. The two winding are wound on one magnetic leg, an arrangement referred to as close coupling. The windings build up an MMF with current, which drives the magnetic flux through the core. Since the permeability of the air is not zero, flux lines also close outside the core. These stray flux components are taken into account in Figure 7.6 by the magnetic reluctances $\mathbb{R}_{\mathrm{ms1}}$ and $\mathbb{R}_{\mathrm{ms2}}$. The basic structure of such an arrangement is shown in Figure 7.6(a) together with the associated magnetic equivalent circuit diagram. Figure 7.6(b) visualizes the conversion into an electrical

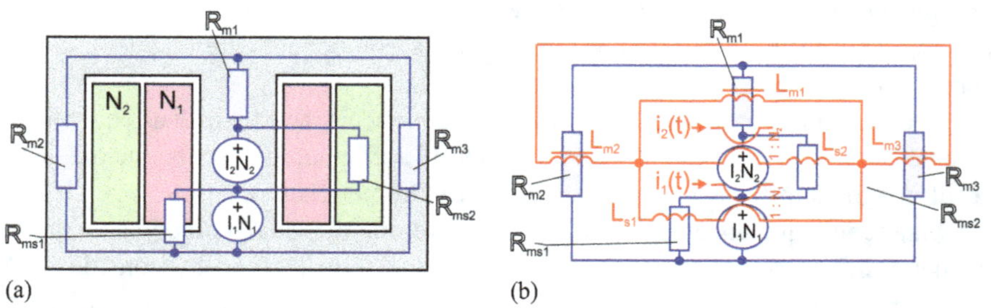

Figure 7.6 Transformer with close coupling of primary and secondary winding: (a) Sketch of the mechanical arrangement with magnetic equivalent circuit diagram and (b) conversion into an electrical equivalent circuit diagram

equivalent circuit diagram according to the rules for the graphical inversion of a circuit shown above.

If the elements of the electrical circuit are rearranged slightly without changing the function, the electrical equivalent circuit diagram in Figure 7.7 is obtained. It reflects the familiar T-equivalent circuit diagram of a transformer. For better differentiation, flux components with and without a ferromagnetic core are labeled in both the magnetic and the electrical equivalent circuit diagram. This circuit can be used to calculate the electrical behavior in a circuit. In particular, the size of the inductances for mapping the core can be derived from the design formula for inductance (14). The inductances of the core sections are generally much larger than those of the leakage sections. The latter are largely linear because, as in the example shown, the stray fields are built up in air or other non-magnetic material.

For a simulation, this structural analysis can also be used for nonlinear core sections. For this purpose, the $I(\psi)$ characteristics of the corresponding sections are stored in the simulation program. The leakage inductances can be approximately determined experimentally from test windings. At higher frequencies, however, the field displacement in the winding material plays an important role, and more precise results are obtained from FEM calculations for the magnetic field.

In the previous example, the proportion of the magnetic field that is not conducted via the core plays a role. This proportion increases the greater the distance between the windings. Figure 7.8 shows such a structure. The windings are

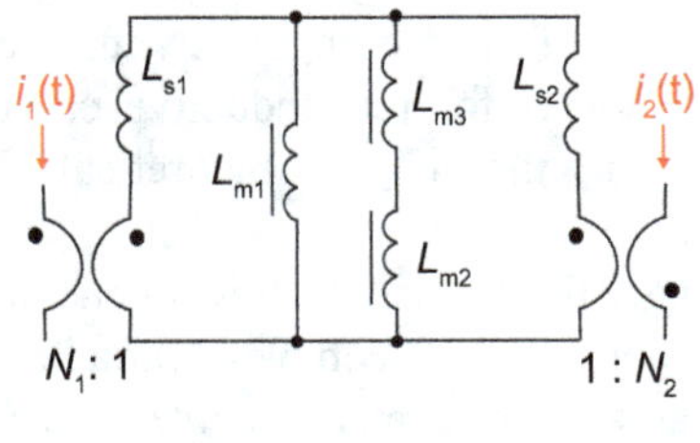

Figure 7.7 "T" equivalent circuit diagram of a transformer derived from Figure 7.6 (winding starts for windings wound clockwise are marked with a dot)

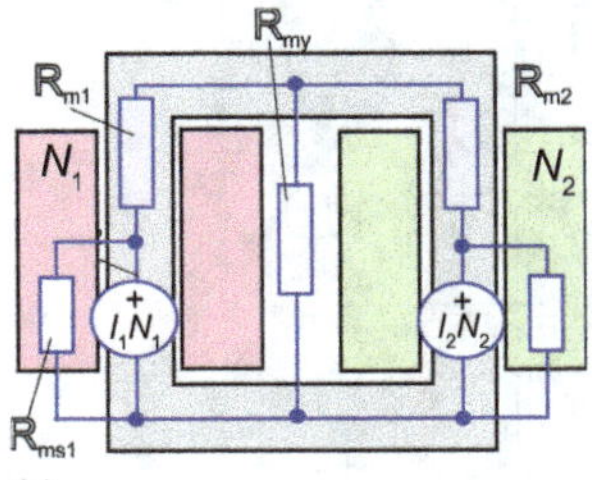

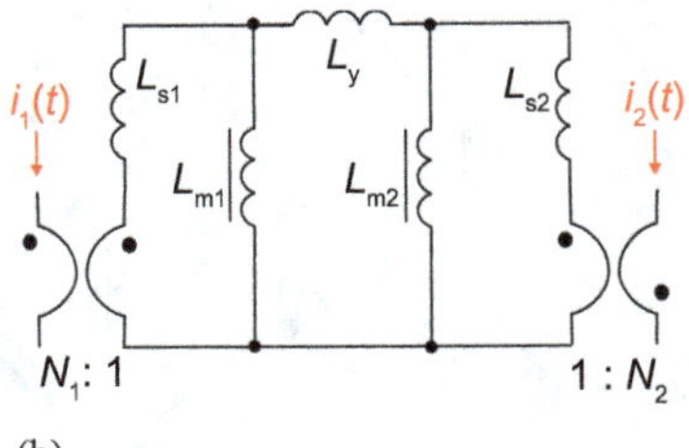

Figure 7.8 Two windings with weak coupling: (a) basic arrangement of the windings on the core and magnetic equivalent circuit diagram, (b) electrical equivalent circuit diagram created after reshaping (winding starts for windings wound clockwise are marked with a dot)

arranged on two separate legs. There is an upper and a lower yoke between the two legs. These are the beginning and end of magnetic field lines that propagate in space with the permeability μ_0. In Figure 7.8(a), this part of the magnetic field is taken into account with the magnetic reluctance for its propagation $\mathbb{R}_{my}$. The non-linked flux components of the windings are recorded as in Figure 7.6(a). If the rules of the graphical inversion of circuit topologies from Figure 7.8(a) are applied, the electrical equivalent circuit diagram shown in Figure 7.8(b) is obtained. As before, there are also flux components of the winding that do not close with the core. However, the inductance components L_{s1} and L_{s2} to be assigned here are typically smaller than the yoke leakage inductance L_y by a factor of 10–20. The two examples in Figures 7.6 and 7.8 show coupled windings and their modeling with lossless inductive components.

7.2 Consideration of losses in inductive components

In the previous section, only the inductive part of individual components was considered. However, what all inductive components have in common is that they cause losses [1,3]. These can be divided into winding losses and core losses. While winding losses are generally linear (though perhaps frequency dependent when the skin and proximity effects must be considered), core loss is, in general, nonlinear. When dealing with nonlinear core loss, the aim is to model the behavior of the components as adequately as possible [4]. This enables circuit simulators to better and better model the behavior of these components in an electrical circuit.

For the calculation of losses for real inductive components, please refer to Chapter 9. In this section, only the different theoretical approaches and their justifications are explained.

In the small signal range, ferrite manufacturers generally specify the frequency dependence of the permeability as a complex variable for the core materials. Figure 7.9 shows such a diagram for the material N87. The permeability is given here

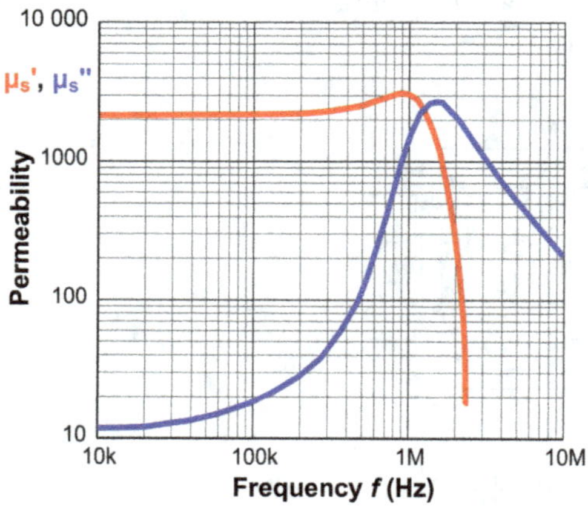

Figure 7.9 Components of the complex permeability of ferrite N87 (measured on R34 cores at flux densities of <0.25 mT, source: TDK/EPCOS [7])

in the components for the imaginary part and the real part. The real part provides the inductive component, while the imaginary part provides the loss component for small signal modulation. The following definition equation is used for this:

$$\begin{aligned} \underline{Z}_L &= j\omega\underline{L} = j\omega\underline{\mu}\frac{A}{l} = j\omega\left(\mu_s' - j\mu_s''\right)\mu_0\frac{A}{l} \\ \underline{Z}_L &= j\omega\mu_s'\mu_0\frac{A}{l} + \mu_s''\mu_0\omega\frac{A}{l} = j\omega L_s + \mathbb{R}_s \end{aligned} \tag{7.14}$$

Equation (7.15) shows that the total impedance of an inductor can be thought of as an equivalent series connection of two components. The inductive component is calculated as usual; the core loss component connected in series is then calculated using (7.16)

$$\mathbb{R}_s = \mu_s''\mu_0\omega\frac{A}{l} \tag{7.15}$$

The core loss component is therefore frequency dependent in this representation. Recall that winding losses, not considered here, may also be frequency dependent.

The definition of complex permeability is an extension of the magnetic equivalent circuit diagram. It reflects a series circuit in the "magnetic world." According to the rules of duality in the previous section, a series circuit becomes a parallel circuit when transitioning to the "electrical world." Figure 7.10 compares the loss components of an R34 toroid for N87 and N27 interpreted both as electrical series resistance and as parallel resistance

$$\frac{1}{\mathbb{R}_p} + \frac{1}{j\omega L_p} = \frac{1}{\mathbb{R}_s + j\omega Ls} = \frac{\mathbb{R}_s - j\omega Ls}{\mathbb{R}_s^2 + (\omega Ls)^2},$$

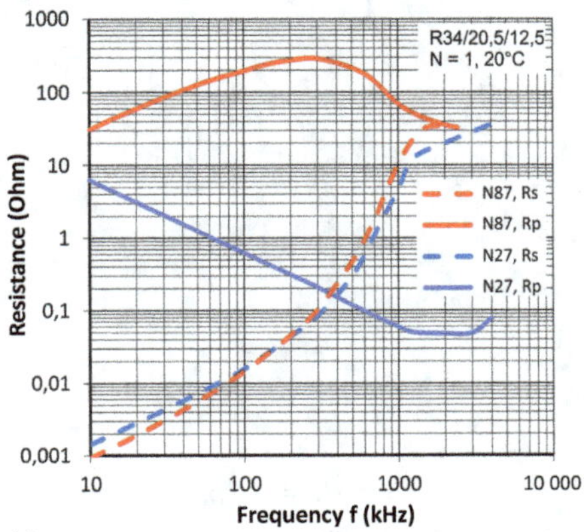

(a)

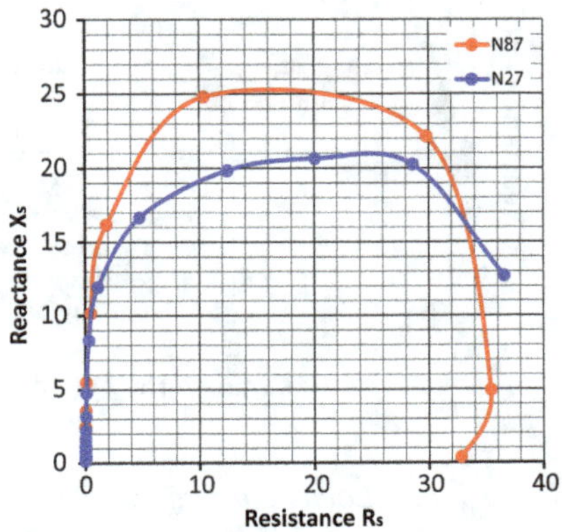

(b)

Figure 7.10 Results from small signal measurements: (a) components of the complex permeability of ferrites N27 and N87 and (b) calculated Z*-locus curves for R34 cores from N27 and N87 based on [7]* i.e.

$$\frac{1}{\mathbb{R}_p} = \frac{\mathbb{R}_s}{\mathbb{R}_s^2 + (\omega Ls)^2} \tag{7.16}$$

$$\mathbb{R}_p = \frac{\mathbb{R}_s^2 + (\omega Ls)^2}{\mathbb{R}_s} = \mathbb{R}_s + \frac{(\omega Ls)^2}{\mathbb{R}_s} \tag{7.17}$$

$$\mathbb{R}_p = \mu_s''\mu_0\omega\frac{A}{l} + \frac{\left(\mu_s'\mu_0\omega\frac{A}{l}\right)^2}{\mu_s''\mu_0\omega\frac{A}{l}} = \left(\mu_s'' + \frac{\left(\mu_s'\right)^2}{\mu_s''}\right)\mu_0\omega\frac{A}{l} \tag{7.18}$$

Proceed in the same way to determine L_p. It is therefore easy to determine the elements of the parallel set circuit (Figure 7.11). The impedance locus curve can be easily constructed with this information (Figure 7.10(b)). For a simple lossy inductance, a curve with a constant increase in the imaginary part is to be expected. However, the shape shown in Figure 7.10(b) is reminiscent of the section of a circle that intersects the real axis. A parallel connection of RLC would result in a circle that is symmetrically intersected by the real axis. This is not the case here. However, the shape is closer to the locus curve shown.

This model simulation is justified for MnZn ferrites, as in the present case, by the fact that they have very high permeability. The permeability of NiZn ferrites is several orders of magnitude lower [7]. No electric polarization occurs in metallic magnetic materials. If the permeability is assumed to be $\mu_r = \mu_i = 2200$, this results

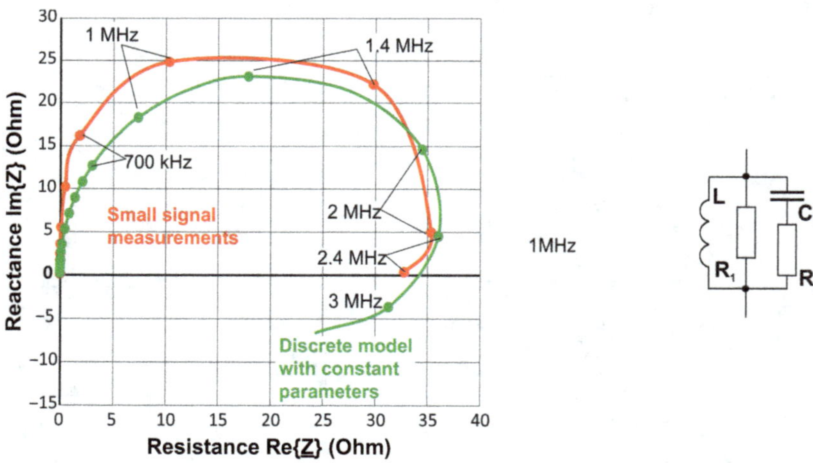

(a) (b)

Figure 7.11 Locus curve of an R34 × 12.5 × 20.5 core (N = 1) based on the small-signal parameters for the complex permeability compared to a locus curve (a) for interconnection of linear, constant elements according to Figure 7.10(b) (μ_r = 2200 → L = 2.78 μH, C = 1.72 nF, R_1= 60 Ohm, R_2 = 20 Ohm) based on [7] in an inductance for the R34 core of $L = 2.78$ µH. At approx. 2.3 MHz, the positive part of the complex permeability in Figure 7.9 disappears. This leads approximately to a capacitance of

$$C \approx \frac{1}{(2\pi \cdot 2.3\ \text{MHz})^2 2.78\ \mu\text{H}} = 1.72\ \text{nF} \tag{7.19}$$

If the capacitance of an R34 toroidal core is calculated using the formula of a cylindrical section of a magnetic section interspersed with an alternating magnetic field (Figure 7.13), the result is

$$C_p = \varepsilon_r \varepsilon_0 \frac{l_e}{4\pi} \Rightarrow \varepsilon_r = \frac{4\pi C_p}{\varepsilon_0 l_e} = \frac{4\pi \cdot 1.72\ \text{nF}}{\varepsilon_0 \cdot 82.06\ \text{mm}_\text{e}} = 29,800 \tag{7.20}$$

However, in the upper-frequency range of the calculated locus curve, the frequency distribution differs considerably from the measured values. Such a model is hardly suitable for quantitative analyses in the upper-frequency range. Qualitatively, the frequency behavior is in good agreement.

If, for example, an operating frequency of 100 kHz is assumed, a series resistance of 0.0148 Ohms is calculated for the R34 core from the measurement data for the complex permeability. The reactance is also $j1.71$ Ohms for these small-signal measurements. Equation (7.18) can be used to calculate an equivalent parallel resistance:

$$R_p = \frac{R_s^2 + (\omega Ls)^2}{R_s} = \frac{(0,0148\ \text{Ohm})^2 + (1.71\ \text{Ohm})^2}{0,0148\ \text{Ohm}} \tag{7.21}$$

$$R_p = 197.6\ \text{Ohm}$$

When using this information, the question arises as to whether anything can be deduced from it for the large-signal behavior. Figure 7.12(a) shows a diagram from the data sheet of N87 cores, also measured on R34 toroidal cores. For a temperature of 25 °C and a load with $B_{\text{max}} = 200$ mT at 100 kHz, the reading is approx. 620 kW/m^3 = 620 mW/cm^3. The measurement conditions correspond to the R34 core

$$\begin{aligned} \hat{B} \cdot A_e &= \frac{\sqrt{2} \cdot U}{2\pi f} \\ U &= 2\pi f \hat{B} \frac{\cdot A_e}{\sqrt{2}} = \sqrt{2}\pi \cdot 100\ \text{kHz} \cdot 0.2T \cdot 0.0000826\ \text{m}^2 = 7.34\ \text{V} \end{aligned} \tag{7.22}$$

$$P = \frac{P}{V} \cdot V_e = 620\ \text{mW/cm}^3 \cdot 6.78\ \text{cm}^3 = 4.2\ \text{W} \tag{7.23}$$

$$R_p = \frac{U^2}{P} = \frac{(7.34\ \text{V})^2}{4.2\ \text{W}} = 12.83\ \text{Ohm} \tag{7.24}$$

If one looks at Figure 7.12(b), it becomes clear that the loss resistance assumed in parallel must be voltage-dependent. This means that as the amplitude of the voltage or flux density increases, R_p decreases (or, equivalently, G_p increases). The application of

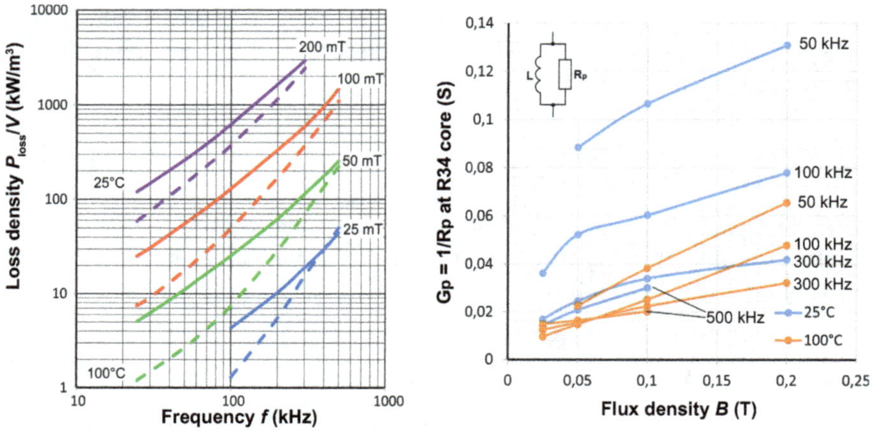

(a) (b)

Figure 7.12 Loss density for N87 ferrite, measured at R34 toroid cores (a) according to N87 datasheet (TDK/EPCOS) and calculated loss conductance for different frequencies (b)

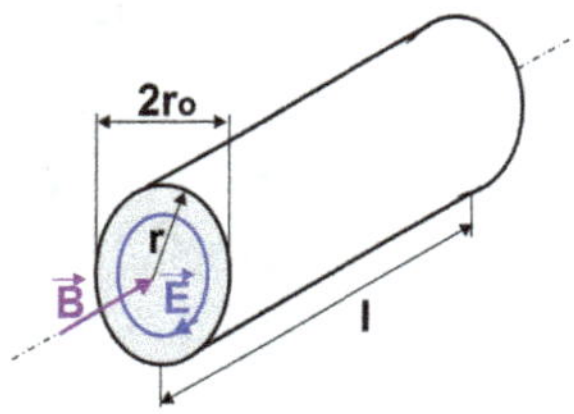

Figure 7.13 Cylindrical magnetic conductor with magnetic and electric field quantities

the complex notation is based on harmonic excitation and constant coefficients of the elements involved. It is only symbolic here, but not for quantitatively exact calculations. Figure 7.13 is used to estimate the influence of the geometry.

If the law of induction is applied together with a complex conductivity σ, the result for the conductance $Y_p = 1/Z_p$

$$\underline{Y}_p = \frac{l\underline{\sigma}}{4\pi} \tag{7.25}$$

$$P_{loss} = U^2 \cdot \mathrm{Re}\left\{\underline{Y}_p\right\} = \frac{\hat{U}^2}{2} \cdot \frac{l \cdot \mathrm{Re}\{\underline{\sigma}\}}{4\pi} \tag{7.26}$$

With these simplifying assumptions, the radius of the magnetic branch in the linear modulation range has no influence on the size of the losses, only the length l. If the total losses of the core are considered using the Steinmetz formula, the cross-sectional area has an influence, as the following estimate for 1 winding shows:

$$P_{loss} = \left(k \cdot f^a \hat{B}^b\right) \cdot l \cdot \pi r_0^2 = k \cdot f^a \cdot \left(\frac{\hat{U}}{\omega}\right)^b \cdot l \cdot \left(\pi r_0^2\right)^{1-b} \quad (7.27)$$

This means that the core shape also has an influence on the actual losses. Other variables also have an influence, such as temperature, premagnetization, shape of the AC voltage, speed of remagnetization [3].

Many of these influences act via a nonlinear relationship. The nonlinear character of the electrical properties of ferrite N87 can be easily determined by measurements. An arrangement with a largely homogeneous field as shown in Figure 7.14(a) can be used to determine basic dependencies. It allows both DC and AC measurements [2]. The measurements that led to the results of the DC behavior were carried out on a 2.75 mm thick ferrite plate. This means that at a maximum DC voltage of 5 V, the electric field strength was 1818 V/m. In a temperature range from 20.5 °C to 100 °C, a very strong change in DC resistance can be observed. At the same time, the dependence on the voltage or the electric field strength is significant.

If the capacitance is measured as a function of the frequency on an arrangement according to Figure 7.14a and the permittivity ε_r is determined from this, Figure 7.15, for example, is obtained. Please note that the permittivity of MnZn ferrite is extremely high; therefore, any air gap at all between the ferrite and the conducting plate

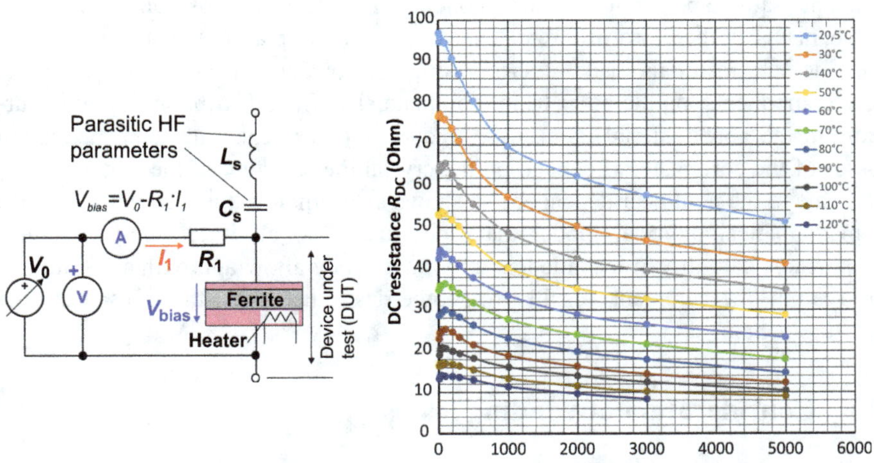

(a) (b)

Figure 7.14 Experimental setup to determine the electrical and dielectric properties of ferrite: (a) L_s = 296 nH, C_s = 1.01 nF, R_{DC} = 49.5 Ohms, $\rho_{DC.20\,°C}$ = 10.77 Ohm·m and (b) DC characteristics of an N87 ferrite plate calculated from the measured data [2]

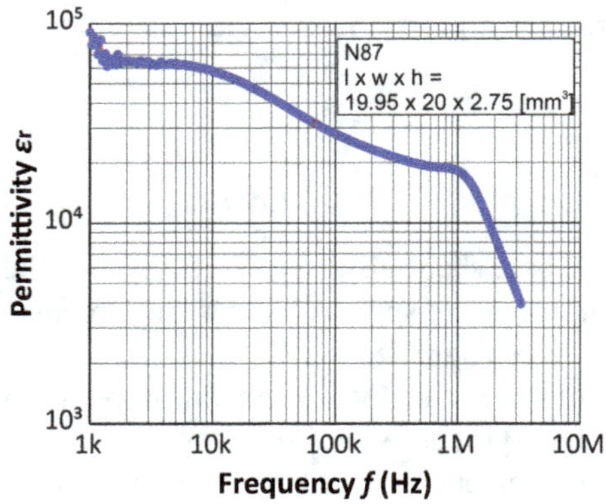

Figure 7.15 Dependence of relative permittivity $\varepsilon_r(f)$ from small-signal measurements at 20 °C [2]

will seriously affect the measurements. For the measurements on the N87 ferrite plate, the contact surfaces were silver-plated to avoid air inclusions. MnZn ferrites are among the most frequently used in power electronics. The example discussed here primarily addresses the complexity of physical relationships.

The high values of the permittivity are typical for MnZn ferrites and show that the dielectric AC currents make a contribution to the behavior of MnZn cores that must be taken into account. The permittivities of NiZn ferrites are several orders of magnitude lower than those of MnZn ferrites [6].

The same applies to the DC conductivity. The composition of the loss mechanisms is therefore different. Dielectric currents play practically no role in metallic core materials made of sheets, strips or powder. The eddy currents in the core components, on the other hand, dominate. The field distribution of eddy currents in the core material is strongly frequency-dependent. The remagnetization losses of the core material depend on the crystalline structure of the material.

Due to the nonlinear dependencies shown as examples, it is difficult to specify a mathematical–physical model that does justice to all influencing variables. This makes data-driven loss models important. They allow approximate solutions by scaling measurement results to geometries other than those used in the measurements.

7.3 Consideration of winding losses

In the previous sections, the content focused primarily on the presentation of the properties of the magnetic core. In many real applications (such as transformers and resonant inductors), however, the losses in the core are similar to the losses in the winding for well-designed components. Figure 7.16 serves to visualize the problem. In the model shown, the symbols for describing the properties of the core material, which were discussed in the previous sections, can be seen on the right.

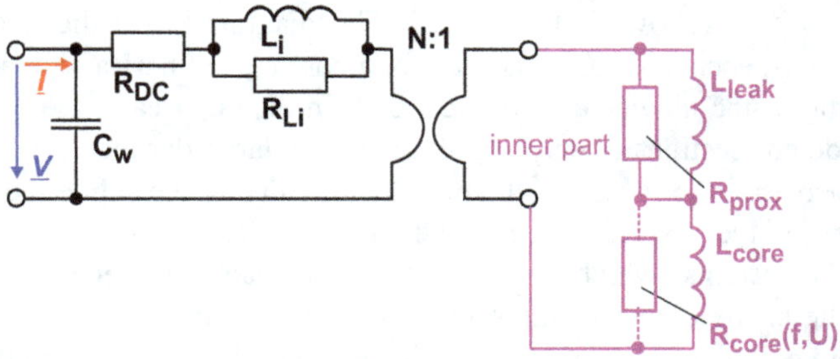

Figure 7.16 Extension of the model of an inductive component to the winding

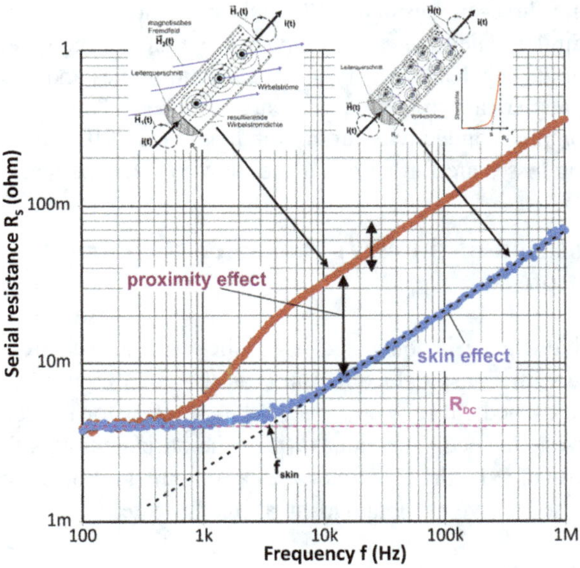

Figure 7.17 Measured serial resistance at the same conductor as single winding (blue) or 15 windings (Ø45 mm × 65 mm, (3 mm × 4 mm) = 12 mm², Cu) (brown) [8].

On the left-hand side, R_{DC} represents the DC winding losses. In addition, there are frequency-dependent losses due to the skin effect. The parallel connection of the internal inductance L_i and R_{skin} only approximates this effect. Actually, an LR chain conductor should be used here for the dependence of the losses $\sim f^{0.5}$. The frequency-dependent increase in the conductor resistance results from an increasing displacement of the electric flow field by the inner magnetic field of the conductor.

The proximity effect is a similar effect. It is caused by induced eddy currents in the winding material as a result of the magnetic field of neighboring conductors. The higher the frequency, the higher the induced voltages in the material. However, the lower the penetration depth into the material. This also results in a dependence of the losses $\sim f^{0.5}$, as shown in Figure 7.17. This diagram shows the measurements of the frequency-dependent AC resistance for a coreless cylindrical coil and a single turn consisting of the unwound cylindrical coil. In the latter case, the mutual influence of conductor sections is only slight due to the large distance. The frequency-dependent increase in resistance is therefore mainly due to the skin effect.

In a single-layer cylindrical coil, windings are always close together. This leads to eddy currents, which are caused by the magnetic field of neighboring windings. The shape of the winding cross-section also plays a role here. With a short solenoid coil, the losses increase as the magnetic field strength in the winding increases. This is characteristic of the proximity effect. Its effect starts at lower frequencies than the skin effect. The effects of both effects add up largely independently of each other. In principle, the loss elements of the winding can be determined using three-dimensional FEM field calculations, e.g., with ANSYS or COMSOL or similar. The greater the number of windings, the greater the number of meshes required. This leads to very long calculation times and potentially also to relatively large numerical errors with very large numbers. This can be remedied by introducing a complex conductivity for the winding space [9,10]. In Chapters 5 and 9, core and winding losses are further explained, how to calculate them analytically.

7.4 Composition of the Ψ(*I*) characteristic curves in simulation

If a nonlinear inductance is connected in parallel with a linear resistor, the magnetization curve with sinusoidal excitation is as shown in Figure 7.18. Without the loss resistor, the dashed curve is obtained. When the resistor is added, a trajectory of a closed curve is formed. The values $i(t)$ of the horizontal axis follow the time, while on the vertical axis, the flux linkage as a time integral of the voltage follows the time. A closed curve is obtained at a steady state. The area of the curve

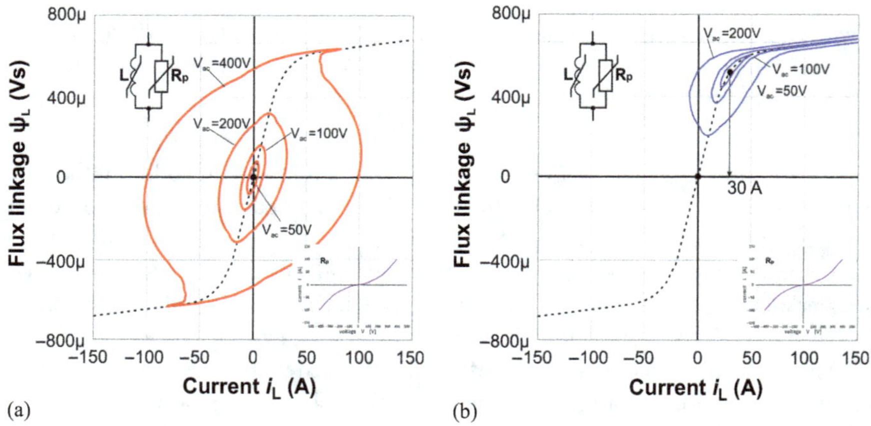

Figure 7.18 Calculated Ψ(I) loop at various levels without (a) and (b) with pre-magnetization of 30 A.

corresponds to the losses during one period. This would be the first approach to simulating a circuit with the core losses if R_p is experimentally adjusted to the operating point. If the amplitude of the sinusoidal voltage is chosen to be smaller, the flux linkage at the inductance also becomes smaller. The enclosed area of the $\Psi(I)$ loop gets smaller too. However, if a magnetization is added (in the simulation, the inductor was designed as a controllable gyrator [5], Chapter 5), the shape of the $\Psi(I)$ loop changes. If the model only contains the two elements shown, the losses do not change. Since this partly contradicts the observations, a dependence on the bias magnetization of I_{bias} can also be assumed. However, as explained above, this resistance is actually dependent on both voltage and temperature.

The oldest of these models goes back to Charles Proteus Steinmetz (1865–1923) [17]. It maps the measured data for the losses per volume to the frequency and the flux density at sinusoidal voltage. This numerical model was later extended to include a temperature term for the data

$$\frac{P_{core}}{V_{core}} = k \cdot f^a \cdot B^a \left(c_{t0} + c_{t1} T + c_{t2} T^2\right) \tag{7.28}$$

Further improvements were achieved by adding additional physical correlations in the treatment of the measurement results [11–20]. Since the losses ultimately result from very different causes, this limits the general validity of data-driven models. However, significant improvements can be achieved in combination with FEM calculations (see Chapter 9). The results can also be used to parameterize the electrical equivalent circuits described above.

References

[1] Rodriguez-Sotelo, D., Rodriguez-Licea, M. A., Araujo-Vargas, I., Prado-Olivarez, J., Barranco-Gutiérrez, A. I., and Perez-Pinal, F. J. Power Losses Models for Magnetic Cores: A Review. *Micromachines (Basel)*. 2022; 13(3): 418. doi: 10.3390/mi13030418

[2] Zacharias, P. *Model Approach for Simulation of the Large-Signal Performance of Inductive Components Using a N87 Core*. EPE/ECCE 2023, Aalborg, Denmark

[3] Ishikura Y., Imaoka J., Noah M., and Yamamoto M. *Core loss evaluation in powder cores: A comparative comparison between electrical and calorimetric methods; Proceedings of the International Power Electronics Conference (IPEC-Niigata 2018-ECCE Asia)*; Niigata, Japan. 20–24 May 2018; pp. 1087–1094.

[4] Jiandong D., Yang L., and Hao L. *Research on ferromagnetic components J-A model—A review; Proceedings of the International Conference on Power System Technology (POWERCON)*; Guangzhou, China. 6–8 November 2018; pp. 3288–3294.

[5] Zacharias, P. Magnetic Components: Basics and applications. *Springer Nature*, ISBN 9783658372071

[6] Ranga Mohan, G., Ravinder, D., Ramana Reddy A.V., and Boyanov, B.S. Dielectric properties of polycrystalline mixed nickel–zinc ferrites. Elsevier, Materials Letters, 40 (1999), p. 39–45

[7] TDK/EPCOS Data sheet: Ferrites and accessories, SIFERRIT material N87, February 2023.

[8] Zacharias, P. What Challenges are Near-Ideal Switches to Inductive Components in Power Electronics?, *ETG Conference "Power electronics components and their application." 20–21 Juni 2023, ETG-Fb. 171*, Bad Nauheim, Germany.

[9] Nan, X., and Sullivan, C. R. An equivalent complex permeability model for Litz-wire windings. *40th IEEE Industry Applications Society Annual Meeting*, 2005, pp. 2229–2235.

[10] Sullivan, C. R. Power Magnetics Design and Measurement of Power Magnetics. ARPA-E Magnetics Workshop, 2011.

[11] Mühlethaler, J., Biela, J., Kolar, J. W., and Ecklebe, A. Improved Core-Loss Calculation for Magnetic Components Employed in Power Electronic Systems. *IEEE Transactions on Power Electronics*, Vol. 27, No. 2, 2012

[12] Jieli Li; Abdallah, T.; and Sullivan, C.R. "Improved calculation of core loss with nonsinusoidal waveforms," *Conference Record of the 2001 IEEE Industry Applications Conference, 2001. Thirty-Sixth IAS Annual Meeting*, vol. 4, pp. 2203–2210.

[13] Albach, M.; Durbaum, T.; and Brockmeyer, A. "Calculating core losses in transformers for arbitrary magnetizing currents a comparison of different approaches," *27th Annual IEEE Power Electronics Specialists Conference, 1996. PESC '96 Record.*, vol. 2, no., pp. 1463–1468.

[14] Tellegen, B. D. H.: The Gyrator, a New Electric Network Element. *Philips Research Reports*, (3), 81–101, April 1948.

[15] Bloom, Gordon Ed and Rudolf P. Severns. "The generalized use of integrated magnetics and zero-ripple techniques in switchmode power converters." *1984 IEEE Power Electronics Specialists Conference (1984)*: 15–33.

[16] G. W. Ludwig and S. -A. El-Hamamsy, "Coupled inductance and reluctance models of magnetic components," in *IEEE Transactions on Power Electronics*, vol. 6, no. 2, pp. 240–250, April 1991, doi: 10.1109/63.76810

[17] C. P. Steinmetz (1892): "On the law of hysteresis" in *Proceedings of the IEEE*, vol. 72, no. 2, pp. 197–221, 1984, doi: 10.1109/PROC.1984.12842

[18] K. Venkatachalam, C. R. Sullivan, T. Abdallah, *et al.*, "Accurate prediction of ferrite core loss with nonsinusoidal waveforms using only Steinmetz parameters," in 2002 IEEE Workshop on Computers in Power Electronics, 2002. *Mayaguez*, Puerto Rico: IEEE, 2002, pp. 36–41.

[19] A. Van Den Bossche, V. Valchev, and G. Georgiev, "Measurement and loss model of ferrites with non-sinusoidal waveforms," in 2004 *IEEE 35th Annual Power Electronics Specialists Conference (IEEE Cat. No.04CH37551).* Aachen, Germany: IEEE, 2004, pp. 4814–4818

[20] S. Barg, K. Ammous, H. Mejbri, *et al.*, "An Improved Empirical Formulation for Magnetic Core Losses Estimation Under Nonsinusoidal Induction," *IEEE Transactions on Power Electronics*, vol. 32, no. 3, pp. 2146–2154, 2017.

Chapter 8
Measuring magnetic core loss

George Slama[1]

8.1 Introduction

We have created our own problem. The constant quest for lower loss magnetic material, operating at higher and higher frequencies, coupled with the demand to increase the efficiency of magnetic devices (transformers and inductors), has made it increasingly more difficult to measure those same losses. This is compounded by the market's desire to make electronic devices smaller and with more features. Each new feature ultimately places additional demands on power. Are the core loss measurements taken on standardized R10–R30 toroidal cores still relevant to moulded core inductors that are only 5 × 5 × 1 mm?

The fact that it has become challenging to measure core loss is a testimony to the success of achieving the goal of reducing the losses. Traditional core loss testing has been done by the core manufacturers themselves. Because cores can be used in so many different ways, under varied conditions, the testing has only been done using sinusoidal waveforms under a few standardized conditions. For example, IEC 60401-3:2015 sets out datasheet guidelines for soft ferrite cores. Two major problems have arisen. Today, most power supplies are using some form of pulse width modulated dc voltage, resulting in triangle current waveforms. If this is occurring at or near 50 per cent duty, the sine wave data would be in fact overstated since it has been demonstrated that core loss for square waves is about 15 per cent less than the equivalent frequency sine wave. However, to achieve a high ratio voltage step down in buck regulators, the duty cycle needs to be very short. This results in the opposite, very high core losses. The first problem is we do not have core loss data for the types of waveforms we are using to drive cores. This often results in magnetic devices running hotter than expected and subsequent redesigns.

The second problem has several parts. The first is miniaturization coupled with high frequency has resulted in very small inductors with small inductance. Traditional core loss testing has been done on medium-sized toroids in the range of R10-R30 [IEC 62044-2 with outer diameter (OD) to inner diameter (ID) ratio no

[1]Würth Elektronik, USA

greater than 1.4 per IEC 62044-3]. As the operating frequency is now in the low megahertz range, these toroids are physically large in relation to the wavelength of the frequencies used, bringing the possibility of incomplete flux penetration (skin effect) and dimensional resonance. An additional complication is that most miniature inductors have moulded cores. There is no discrete core (like an E core), but the winding is surrounded by the core material, which has been pressed into place to its final form. With no discrete core to test, a representative core needs to be fabricated that is dimensionally large enough to measure yet small enough not to be affected by the high-frequency effects. If the measured values are too small, the instruments will be imprecise. Finally, the electrification of transportation and the development of solid-state transformers for the power distribution grid take us to the opposite side of the scale, where the standard test toroid is much smaller than the final magnetic core. These are the measurement challenges.

Related to this is the desire to design magnetics more efficiently – with fewer design-prototype-test iterations. This entails the ability to model the device accurately. The accuracy of the models depends on good empirical data because until recently, it has not been possible to model the core loss behavior from fundamental physical principles though research is being done in this area. As noted earlier, the current data is not in the form desired. Artificial intelligence has been proposed to help fill this gap, but it needs large amounts of high-quality data to train. The old computer adage, garbage in = garbage out, applies here. In this chapter, we look at what it takes to get high-quality data.

8.2 What is core loss?

There is abundant literature to explain the physics of core loss. Here, we are interested in its measurement. Simply put, core loss is the power absorbed by the core under the stated conditions of applied time varying magnetic field and dissipated as heat. The unit is watts. In magnetics, we speak of specific core loss, which is simply the core loss absorbed by a unit of mass or volume. Mass is used with steel or ribbons, and volume is used with ferrites and powdered cores.

When an alternating voltage is applied to a magnetic device, the power measured is referred to as the apparent power (*S*, units VA), which is composed of two elements. The reactive power (*Q*, units VARs) is ideally returned to the system, and the active (a.k.a. real, true) power (*P*, units Watts) is the power consumed or lost. The phase angle between the active power and the apparent power is theta

$$P = I_{rms} \cdot V_{rms} \cdot \cos\theta \qquad (8.1)$$

Theta θ is the angle between true power (watts, resistive) and apparent power (VA). This is the phase angle between the current and voltage waveforms. Positive is inductive, and negative is capacitive. The $\cos\theta$ is power factor. Power factor is the ratio of true power (*P*) to apparent power (*S*) (Figure 8.1).

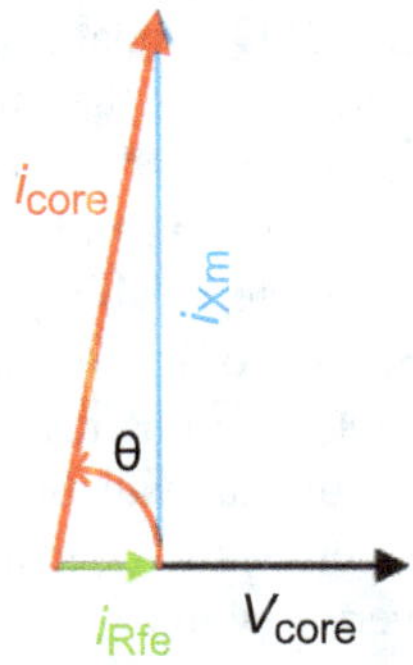

Figure 8.1 *The measured current, i_{core}, is the vector sum of the reactive current, i_{xm}, and the real loss current, i_{rfe}, with θ being the angle between it and the voltage*

8.3 Initial conditions

Before discussing testing a core, consideration needs to be given to the core's initial conditions. Because cores can have a remanent flux density, that is after being excited, there is a residual flux, meaning it does not return to zero naturally – each time a core is tested, it may be starting at different residual flux level. For consistent measurement, the core needs to be reset. Cores made from ferrites can be heated to beyond they curie temperatures for 30 min, then cooled down. All cores can be reset by applying a large alternating flux and deducting it gradually to zero. The flux can be inducted in the windings or it can be an external field. IEC 60367-1, section 6 Magnetic conditioning, is an excellent reference [1].

Another consideration is the size of the core. For powdered cores and ferrites, according to IEC 62044-2, the core is ideally a ring or toroid with a continuous magnetic path. The 'standardized' size according to IEC 63300 is 26.9 mm outside diameter, 14.7 mm inner diameter with 11.2 mm height. Commonly used sizes range from R10 to R30, where the ratio between the outer diameter and inner diameter is less than two. Cores consisting of two or more parts need to be assembled and held with a defined and constant clamping force. The standard for laminated cores is the 25 cm Epstein frame using strips 30 × 304.8 mm arranged in a square pattern with overlapping corners.

8.4 Source waveforms for testing

8.4.1 Sine

Sinusoidal signals from a waveform generator are typically amplified by a wideband RF amplifier. One must be careful to minimize distortion and should measure it. A distorted sine wave can appear surprisingly clean on an oscilloscope just as a clearly distorted wave may measure surprisingly clean. Measurement is the only way to know for sure. A forthcoming IEC standard specifies less than 1 per cent total harmonic distortion. The biggest challenge is the impedance mismatch between the output of the amplifier, typically 50 ohms, and the device under test, usually only a few ohms. A simple step-down transformer can be used to improve the mismatch. This can be a suitable medium to high permeability core with enough primary turns to have an exciting current of 10 per cent or less of the load current. The square of the turns ratio reflects the impedance. A 3:1 turns ratio gives a 50:5.6 ohm impedance transfer. Note that this reduces the voltage available to drive the flux density, which may become a limiting factor. Numerous high quality RF amplifiers are designed to be stable, working properly into very low impedances loads.

8.4.2 Rectangular

A half bridge, full bridge or a more sophisticated circuit can be used to produce rectangular voltage waveforms. This may be a simple rectangular waveform with a duty cycle, or they may be trapezoidal waveforms with dead time between equally positive and negative pulses or pulses, each with different duty cycles and voltages within the cycle. Here, a switcher is preferred over an arbitrary waveform generator and wide-band amplifier. To provide fast slew rates for the vertical edges the amplifier would need to work to very high frequencies (MHz). It will tend to have artifacts such as overshoot and ringing that considerably distort the waveform. IEC 63300 paragraph 4.2.3 has strict requirements for rise time (<1 per cent), overshoot (<5 per cent), overshoot duration (<10 per cent), droop (<2 per cent) and fall time (<1 per cent, all of the period).

8.4.3 Arbitrary

Waveforms that are neither sine nor rectangular are usually application-specific and can be generated by any suitable means that reproduces them correctly. Arbitrary wave generators can be programmed to reproduce almost any shape. The output power needs to be increased by a wide-band power amplifier.

8.4.4 DC bias

All the previously mentioned waveforms may be used in applications with DC bias, and, therefore, this needs to be incorporated into the core loss testing. It has been shown that DC bias does affect the core loss. DC bias can be provided through a third winding on the test core or injected into the drive signal path where the dc source is isolated from the ac source by a suitable inductor, and the ac source is isolated from the dc source by a suitable capacitor. When using half bridge drives, different voltages on each half can be used to introduce a bias. However, in order to set up a known dc flux in the core, it is necessary to know the magnetization curve of the material (a separate measurement) and, therefore, the dc magnetic field strength required [2].

A method known as the ‘triple pulse test’ [3] is a variation of the half-bridge with separate programmable power supplies on each half. Under suitable control, the positive and negative pulses can be at different amplitudes and durations. Using this method, the first pulse is large enough to move the core flux to some maximum value. This is followed by the second pulse of a lower magnitude and opposite polarity. Subsequent pulses are of the same value in both directions. This results in a net dc bias and ac ripple on the core under test. The final pulse brings the core back to zero flux. The middle set of pulses only runs for a few cycles, enough to stabilize and capture the waveform for the measurement. This way, core heating is minimized during the test.

8.5 Methods to measure core loss

There are three general methods that are used to measure core loss. The 'AC method' is any circuit where voltage, current and phase are measured at the device, and power is calculated. The precision can vary widely and is very dependent on accurate phase measurement, which becomes increasingly more difficult as frequency increases. There are numerous variations of this method, each trying to address the phase accuracy, usually with success but at the expense of introducing their own complications. This tends to make them more application-specific.

The 'DC method' is any circuit where the system input and output power are measured (typically dc power), and the core loss is the difference after accounting for system losses (switches, diodes, etc.). Here, the challenge is to accurately determine the system losses for each of the test conditions, which is not a simple task considering multiple test conditions.

The third way is the 'calorimetric' method. This is considered the most accurate and the most time-consuming, requiring its own careful setup. There are three primary ways to measure when using a liquid medium: single vessel, reference and double vessel with reference (Figure 8.2). The methods are outlined in the IEC 63300 standard, and the details will not be repeated here. The challenge with the single vessel method is that the thermal constant of the medium must be known beforehand. This is not the case with the other two methods which use a separate reference heater as a comparison. For larger cores with greater power losses, large chambers and double-walled chambers are used where the thermal medium is usually air [4–6]. The details are in the literature and will not be covered here. Since the calorimetric method is considered the most accurate, it makes an excellent method to create a reference standard for verifying results from the electrical methods, at least for a limited number of conditions.

Error analysis requires understanding the tolerances of thermal measuring devices. Many common probes, like thermocouples and thermocouple readers, have 1–2 per cent tolerances each making the user susceptible to digital delusion – reading and using more digits than are actually significant. Thermocouples can be sensitive to placement. One technique used is to connect several thermocouples located at different spots together for an average reading. Infrared thermometers and thermal imaging cameras have become very popular but have the same 1–2 per cent tolerances and are sensitive to the emissivity of the surface being measured.

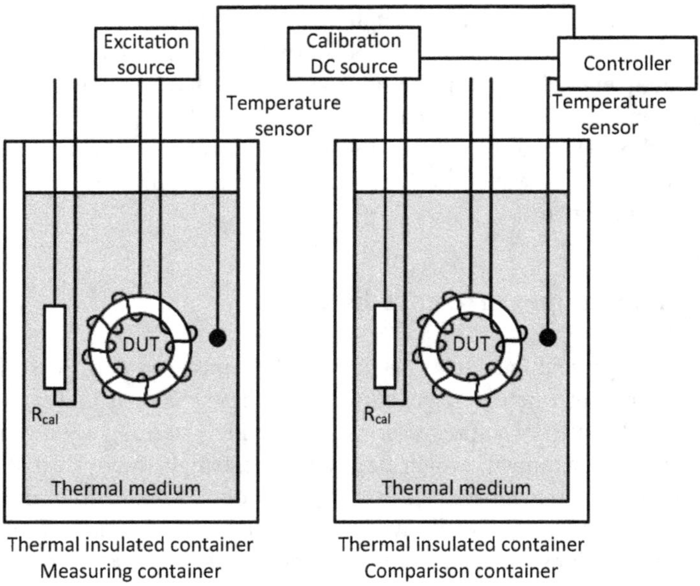

Figure 8.2 Calorimetric method using a reference vessel and liquid fluids

8.6 Circuits

8.6.1 Epstein frame

In the low-frequency world of laminated cores fabricated from thin sheet material (electrical steel), the 25 cm Epstein frame continues to be the preferred choice. IEC standard 60404-2:2008 Magnetic materials – Part 2: Methods of measurement of the magnetic properties of electrical steel sheet and strip by means of an Epstein frame defines the procedure. The 'frame' consists of a square made by laying long thin strips of electrical steel, overlapping at the corners to a fixed height. Each of the four legs passes through an exciting coil and measuring coil, both with 175 turns, connected in series. The exciting coil generates the flux and the measuring coil is used for measurement to negate the effects of winding losses. A wattmeter is typically the measuring instrument.

8.6.2 Classic circuit

The classic two-winding core loss measuring circuit using the ac method is shown in Figure 8.3. It accommodates virtually any type of source, from an oscillator and power amplifier for sinusoidal waveforms or a switching circuit to produce rectangular waveforms. The source is connected to one winding through a current sensor. The current sensor may be a resistive shunt, a coaxial shunt, a current probe or a current transformer. The second winding, typically of an equal number of turns, is used to measure the applied voltage and determine the induced flux

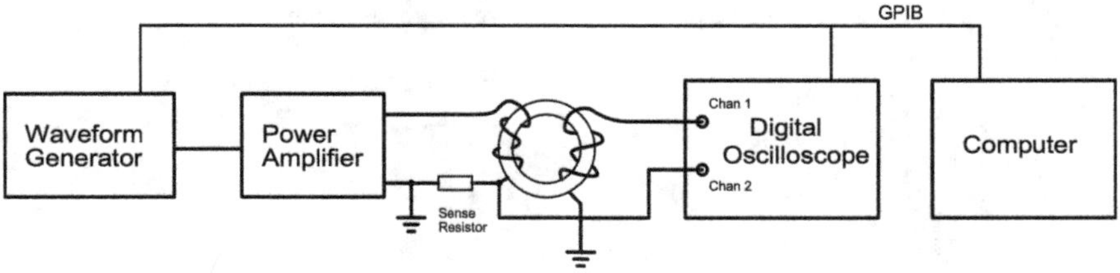

Figure 8.3 Classic circuit with current sense resistor using a digital oscilloscope under computer control to acquire data

density. In this way, the resistive losses of the driven winding are excluded from the measurement. This also assumes near-perfect coupling between the two windings. They are often wound together (bifilar), but at high frequencies, the interwinding capacitance needs to be considered. Low permeability powder cores, which are distributed gapped cores, are often tested with a single winding because the coupling to a second winding is poor. This necessitates estimating the winding loss to subtract it from the measured loss.

The average core loss power is the voltage times the current integrated over one cycle.

The circuit is sensitive to phase error. As the angle θ approaches 90°, the cos θ approaches zero, resulting in higher measurement error. Low loss, low permeability cores have angles very close to 90. The error can easily exceed 100 per cent

$$Error = \frac{\cos(\theta + \phi) - \cos(\theta)}{\cos(\theta)} 100 \tag{8.2}$$

where θ is the phase angle and ϕ is the error. Figure 8.4 illustrates the high percentage of error possible from very small errors in phase as the angle approaches 90°.

Nevertheless, with careful setup and characterization of the measuring devices (probes) and circuits, it is possible to get good measurements into the low 10s of MHz with sine waves.

The classic circuit can also be used with a single winding to measure inductors. In this case, the power loss of the winding needs to be removed from the measurement. The challenge is not dc power loss (I^2R) but the ac power loss resulting from skin and proximity effect of high-frequency currents in the winding. It is worth noting that the ac resistance is not a simple matter of measuring the small signal impedance of an equivalent air core (no core) to remove core loss. A magnetic field around a coil is produced by the current flowing through it. The greater the current, the stronger the magnetic field and its influence on the windings. The small current from an impedance analyzer will in no way simulate the actual magnetic field and its effects on the winding impedance. Additionally, the magnetic core around the winding shapes the magnetic field around those same windings and generally concentrates it. Therefore, the major influences on ac impedance are missing in a small signal measurement and cannot provide the desired information. Nevertheless, it is often the best available information.

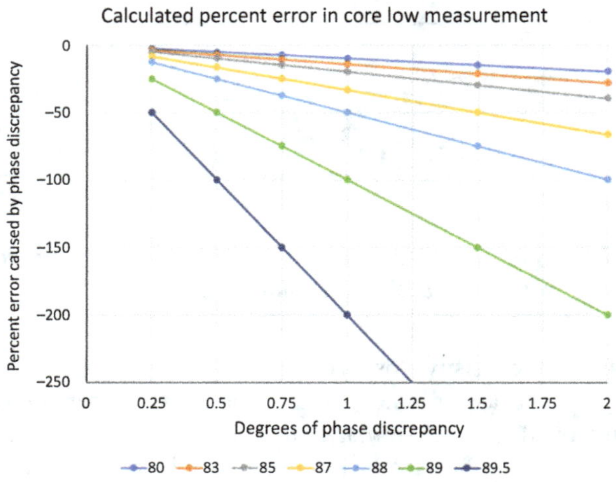

Figure 8.4 Calculated per cent error in core loss measurement with the phase angle close to 90°

The classic circuit can easily accommodate additional dc biasing through a third winding or by appropriate injection into the circuit.

8.6.3 Resonant circuits

Many interesting resonant circuits have been proposed to get around the classic circuit's short coming of phase sensitivity. The resonant method uses a capacitor connected in series or in parallel with the test inductor to cancel the phase angle of the inductor under test, resulting in, under ideal conditions, only the core loss being measured with voltage and current in phase. This is simple enough for a single-point measurement if one does not require the test at an exact frequency. This means that only sinusoidal waveforms can be tested. It has also been shown that inductive cancellation can be used for arbitrary waveforms, but this appears to be more an academic exercise than a practical method [7]. Almost all references promoting this method choose to ignore the capacitors' contribution to the losses, stating that if high-quality capacitors (i.e. mica) are used, the contribution is minimal. The proof offered is that the losses measured appear to match the manufacturer's data, which, coincidentally, are measured without using resonance.

The first trade-off is frequency. The capacitor must be a specific value to match the inductor under test in order to resonate at the desired frequency. Capacitors come in discrete values, which are not likely to be the exact value needed. This entails either series and paralleling several capacitors to achieve the desired value or the possibility of a variable capacitor for tuning. Core loss does not change abruptly, so using a frequency near the desired frequency would be sufficient for most uses. It merely grates against the engineer's ego that he cannot make it precisely as he wants it. The frequency shift is also complicated by the fact that magnetic core material's permeability, hence, its inductance will vary with temperature and amplitude of applied flux density for the same frequency. This makes it tedious to maintain a fixed frequency under varying conditions by constantly adjusting the capacitance.

The second concern is the capacitor's own parasitics are measured with the inductors. The capacitor parasitics, like an inductor, vary with the application conditions (voltage, temperature). It is difficult to determine accurately the ESR and ESL of a capacitor to the point where many who suggest the method, simply exhort the user to use high quality devices and ignore the losses. In other words, leave them as part of the core loss. Whether this is satisfactory depends on the situation.

A third concern is that the method does not lend itself well to automated testing. Today, the desire and need are for large volumes of data under a wide range of operating conditions for use with computer design programs and materials characterization using artificial intelligence. This can only be reasonably accomplished with automated testing. The resonant methods appear to be too specialized for general application in automated testing. Additionally, it is restricted to sinusoidal waveforms only, whereas many power circuits use rectangular or trapezoidal voltage waveforms with triangular current waveforms. Though the method can be useful in specific applications, for general testing over a wide range of materials, frequencies, duty cycles, flux density levels and temperatures, it is not practical.

A further technique using resonant circuits is partial cancellation. This method addresses some of the previously mentioned concerns with exact resonance. Here, the goal is not an exact match but to shift the phase away from the sensitive area of 80–90° to an area less sensitive 30–60°. There are two schools of thought surrounding this method. The first is not to be concerned about the actual phase because even though the resulting calculation will be off, it will be considerably better than without the partial cancellation [7]. The second idea is to decompose the waveforms using fast Fourier transform (FFT) into their harmonics and then make corrections and calculations at the harmonic level and sum together [8].

8.6.4 Other circuits

Calorimetric methods only require driving the test coil by any suitable means because the measurement is based on temperature and equivalent dc power. The challenge is accurate temperature measurement. It needs to be pointed out that the most common temperature measurement devices such as thermocouples or thermal cameras, typically have tolerances of ±2 °C. This makes them less precise than the multiple digits often shown after the decimal point. The winding losses are included in the measurement and need to be estimated and subtracted.

Circuits for the dc method are often a simplified version of the application or a carefully built tester. The challenge is to quantify all the losses in the system for each test condition. They cannot all be considered to be linear.

8.6.5 Circuit fixtures

Many a paper includes a photograph of the test setup strewn with wires and cabling in what appears to be a random arrangement. This might work for a single proof of concept, but it is less than ideal. To make consistent, repeatable measurements, especially as the frequency increases and the core loss decreases, it is wise to use a fixture. This can be a printed circuit board that holds everything in a fixed position. This facilitates the connection of probes and provides a knowable calibration plane. Small cores or inductors can be directly mounted, whereas larger cores with leads can be fixed to terminals. The few commercial systems available for core loss measurement all use fixtures or terminals for connecting the device under test.

8.7 Measurement equipment

8.7.1 Power analyzer (digital wattmeter)

Today, the analog wattmeter of old has been replaced by the digital power analyzer, a convenient specialized instrument designed and calibrated to measure power under a wide range of frequencies and with multiple phases. The current state-of-the-art measurement limit is 10 MHz. The meters have floating inputs, high-bit analog-to-digital (A/D) converters (16–18 bits) and are optimized to have very low phase delay between measurement channels. They are especially useful for motor drives, and the recent emphasis on electric vehicles, which operate between 5 and 20 kHz range, makes them an ideal tool. They come with impressive accuracy in the low-frequency range 50–1000 Hz, and very good accuracy to 100 kHz, but beyond 500 kHz, most instruments suffer from significant phase delay, which is expressed as per cent error per kilohertz. A close examination of the specifications shows the initial impressive figures of fractional percentage can quickly degrade to 10 per cent error on phase alone at 1 MHz. See Figure 8.5 for a typical trend line for meters rated to 1 MHz.

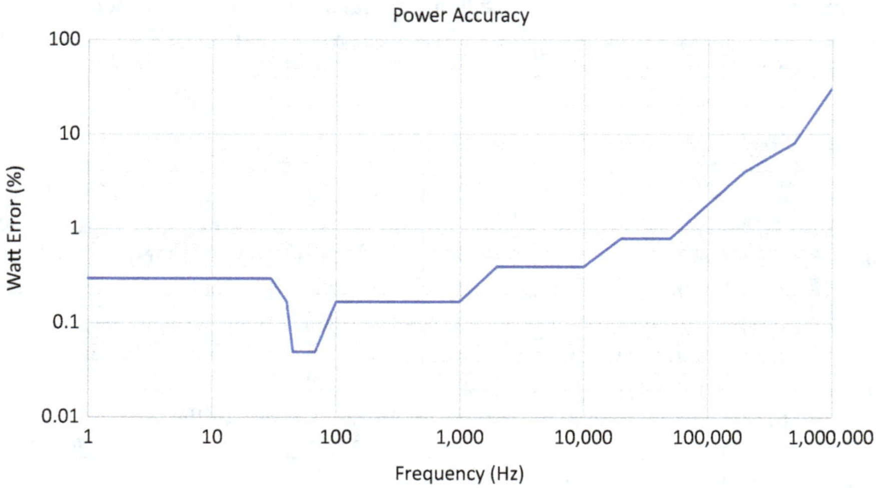

Figure 8.5 Typical wattmeter power error vs. frequency specification. Note that the logarithmic vertical scale compresses the range.

Even if it is poor, when the manufacturer documents it, it at least provides a known reference for measurement. Not surprisingly, power analyzers come with 4 mm connection jacks and are used with long leads revealing the fact they are not really intended for high-frequency measurements.

The advantages of the power analyzer for core loss measurements are that generally no special voltage or current probes are required. For special cases, matched high-voltage probes and high-current sensors are available (Figure 8.6). The current sensor is built into the instrument, calibrated and compensated accordingly. The floating inputs simplify connections to the circuit. The meters have very small full-scale ranges like 5 mA and 2 V to make sensitive power measurements. Like most modern digital instruments, the displays can show multiple parameters like voltage, current, power, phase, the waveforms of each, the harmonics through FFT and can output all the data through an interface to a computer. Used within their capabilities, they are extremely convenient.

Some models may require the user to make an adjustment to the readings to account for the internal current shunt resistance, which gets larger as the sensitivity increases and the input impedance of the voltage meter. Because the floating inputs allow freedom as to where the device is connected in the test circuit, the correction applied varies. The correction value may be small but can increase quickly at low power and with frequency. For the most accurate results, the shunt loss needs to be accounted for based on its position in the circuit and the current range used for the reading (Figure 8.7).

Figure 8.6 The back of a typical multi-phase power analyzer (Norma 6000) with 4 mm voltage input connections and a triaxial current shunt module that contains a ROM chip with calibration/correction data. Simply plug into the instrument, connect and start using it.

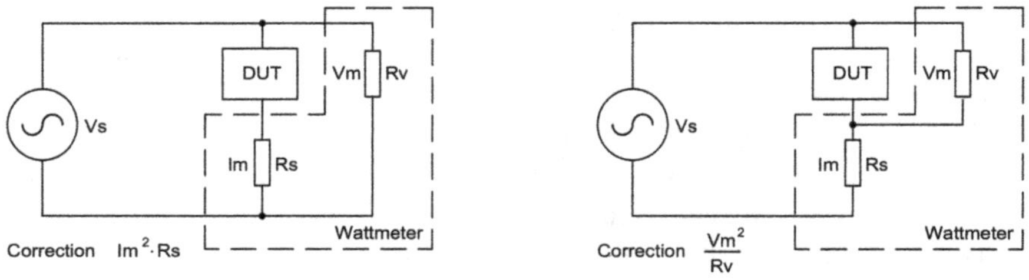

Figure 8.7 Correction for the impact of the measuring instrument in the circuit. Wattmeters have floating inputs, so they measure current on the high or low side and voltage across only the device under test (DUT) or both the DUT and current sense, each requiring a different compensation adjustment.

8.7.2 BH analyzers

BH analyzers fall into this same category of power analyzer with respect to measuring core loss. They are complete systems optimized for the task of measuring magnetic devices rather than power in general. BH analyzers are specialized for measuring numerous core characteristics but are somewhat limited in drive levels and waveforms depending on external amplifiers for additional power. Accessories allow for some expansion of capabilities such as dc bias and thermal chambers to measure across various temperatures, both hot and cold. Their published accuracy limitations are similar to power analyzers with the same state-of-the-art limit of 10 MHz. One important benefit is they use fixtures for consistent measurement setup with short leads helping to mitigate inconsistencies at high frequencies. The BH analyzer interface provides for entering the test core's physical dimensions, which enables it to calculate setups and resulting magnetic parameters for the user directly, greatly simplifying the process of measurement. It can display all the common magnetic $(B_m, B_r, H_m, H_c, \mu_a, P_c, \theta)$. The instrument can be programmed to do sweeps of B or H making data collection semi-automatic. Similarly, all data can be downloaded through interfaces to a computer.

8.7.3 Other systems

Whereas the BH Analyzer is a single instrument incorporating everything, there are turnkey systems on the market that are made up of individual off-the-shelf instruments connected together and programmed by proprietary software. These systems are either oscilloscope-based or use discrete A/D converter modules for data acquisition. They include power supplies, a waveform generator, an amplifier and a connection interface. Although they take up more space, they can be customized, and individual instruments can be upgraded over time. They save the time and expense of setting up, debugging and calibrating the hardware and creating the software to operate it. It is possible to calibrate the individual instruments, but the system calibration depends on the manufacturer.

Manufacturers of both types of digital systems have a strong advantage. They can characterize the voltage and current sensors for amplitude, frequency and phase error, then store the data internally to use as a correction factor when doing measurements. This increases apparent accuracy, simplifying the task for the user.

8.7.4 Digital oscilloscope

The digital oscilloscope has become the Swiss army knife of measuring instruments on any power electronics test bench. This is one of the most common instruments used today to measure power. Digitization of analog signals combined with ample numeric processing power allows the oscilloscope to mimic numerous other instruments by virtue of simply applying mathematics to the voltage and current measurements. Most oscilloscope manufacturers offer software upgrades to integrate power measurements into their instruments. Transducers such as resistive shunts, coaxial shunts, hall-effect, or current transformers convert current into voltage. By measuring voltage on one channel and current on another, multiplying the two and integrating over time gives the user an instant power analyzer. Phase error between channels can be compensated by built-in channel deskew functions and corresponding fixtures. This easily works at low frequencies, but as frequencies increase, phase error in the current transducer itself, the frequency response of the voltage probe, and the acquisition rate are some of the problems that start to degrade the results.

What are the differences between the oscilloscope and the true power analyzer? First, oscilloscopes up until very recently have employed only 8-bit A/D converters whereas power analyzers going back more than two decades were using 16-bit A/D converters. Today 18-bit is common for power analyzers and 10–12 bit for new oscilloscopes. This can be a somewhat misleading comparison, and one must consider the context. For a digital oscilloscope to mimic an analog oscilloscope, it must have a very high sampling rate, well in excess of its frequency bandwidth, to capture fast transient signals. They are also designed to measure much higher frequencies and arbitrary waveforms with fast edges. The trade-off has been fewer conversion bits for faster performance. This is partially compensated for by the front-end amplifier which provides for a wide range of input voltage ranges that is scaled up or down to make maximum use of the available bits. Because power analyzers have traditionally been designed primarily for low-frequency sinusoidal waveforms, they have had the advantage of higher bit converters operating at slower speeds. It can be shown that 8 bits achieve about 0.03 per cent error for rms voltages, and 10 bits achieve better than 0.01 per cent accuracy [9]. Adding more bits has diminishing returns.

Examining the specifications of power analyzers shows that sampling rates are hardly higher than the Nyquist frequency for the highest rated frequency and, in some cases, less, where they depend on averaging to fill in the gaps. Oscilloscopes, instead, are always sampling beyond the Nyquist frequency. With oscilloscopes, the sampling rate will be a determining factor in the measurable phase shift. One must also appreciate the use of interpolation algorithms used to connect the dots (the samples).

A big advantage of power analyzers over oscilloscopes is that they have floating inputs. This allows connecting the voltage and current probes in the most convenient way without regard to potentially shorting to ground as you would with an oscilloscope, which is always ground referenced. Differential probes are available for oscilloscopes, but the differential amplifier adds phase delay and linearity error, which must be compensated. Differential probes often have high input ratios, reducing the input voltage, which may need to be further amplified by the oscilloscope. This is all an additional error. Power analyzers have very sensitive front ends, often with 5 mA current and 2 V voltage full-scale ranges. For measuring small amounts of core loss, this can be essential.

Oscilloscopes, conversely, are designed with increased bandwidth to measure higher frequencies, requiring ever faster conversion rates, hence the tendency to use fewer bits to allow a faster conversion rate. Today, 10- and 12-bit scopes are becoming mainstream with high sampling rates of starting 2.5–5.0 GSa/s (giga samples per second). Nevertheless, at 1 MHz, one nanosecond delay between channels represents a 0.36 per cent phase error. At 1 MHz, an uncompensated one-meter cable represents a 10 ns delay or 3.6 per cent. Previously, Figure 8.4 shows how only a 0.25° phase discrepancy could cause a 50 per cent error.

The phase error is not limited to the phase delay between channels. This can be removed in most oscilloscopes by the deskew feature. A typical deskew procedure uses a fixture that connects both the voltage and current probe to a fast-rising rectangular waveform. The timing on one channel is adjusted to match the other channel. Two potential problems exist. First, this calibration happens at only one frequency, so it does not account for phase changes in the probes or front-end circuit at different frequencies. The second is that the adjustment itself may not be fine enough, lacking resolution when the adjustment increments are too large. Unlike analog controls, digital controls have discrete steps that are limited by the sampling rate. Table 8.1 illustrates the difficulties of getting a precise match when testing with frequencies above 1 MHz.

The challenge with phase matching has another dimension. The phase shift of the probes over a range of measurement frequencies changes. This is more prevalent with the current sensors. Resistive shunts and current transformers introduce

Table 8.1 Examples of phase shifts from small sampling rates and mismatches at high frequencies

Problem	1 MHz	5 MHz	10 MHz
Sample rate 5 GSa/s	0.072	0.36	0.72
Sample rate 2.5 GSa/s	0.144	0.72	1.44
Probe mismatch – 1 ns	0.36	1.8	3.6
Current sense parasitic inductance – 1 nH	0.18	0.9	1.8

a small inductance, which causes a delay. The solution is to characterize the sensor's phase using an impedance analyzer or VNA and later compensate for it. Core loss measurements are usually taken at many different frequencies. Therefore, the probes, particularly the current transducer (shunt or transformer) need to be characterized over the frequency of intended use.

8.7.5 *Resistive shunt*

The simplest current transducer is the resistive shunt. The problems are parasitic inductance from internal meander, spiraling or winding, a low sense voltage and heat dissipation. The most common solution is to use several resistors in parallel with a practical limit being 8–10 surface mount parts. A larger number introduces parasitic inductance because the interconnections become too long. These can easily be integrated into a pcb test fixture where half the number is mounted on each side of the pcb, reducing connection lengths. At high frequencies, the devices to be measured are small, and the sensor should be correspondingly small. For larger shunts, at lower frequencies, axial resistors can be paralleled. Carbon composition types have less parasitic inductance and are suitable for 1 MHz and lower. Note that VNA inputs are 50 ohms which matches many current transformers, but many split-core current probes expect to be used with 1 MΩ inputs.

According to IEC 62044-3, the parasitic inductance of the sampling resistor shunt can be calculated by the following equation:

$$L \le \frac{R}{2 \cdot \pi \cdot f} \sqrt{2 \cdot \delta_a} \tag{8.3}$$

where L is the parasitic inductance of the sampling resistor (H), R is the resistance of the sampling resistor (Ω), δ_a is the allowable relative error of the voltage drop resulting from inductance typically given as 0.1 per cent and f = frequency of the test (Hz) (Figure 8.8).

8.7.6 *Coaxial Shunt*

A coaxial shunt is a resistive shunt constructed in such a way as to minimize the effect of parasitic inductance on the measurement. In principle, it is like a coaxial cable where the inducted magnetic field from the current following through the

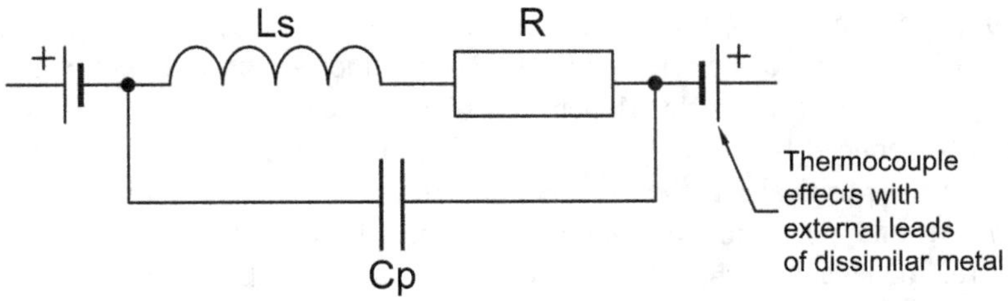

Figure 8.8 Equivalent circuit of a shunt resistor [10]

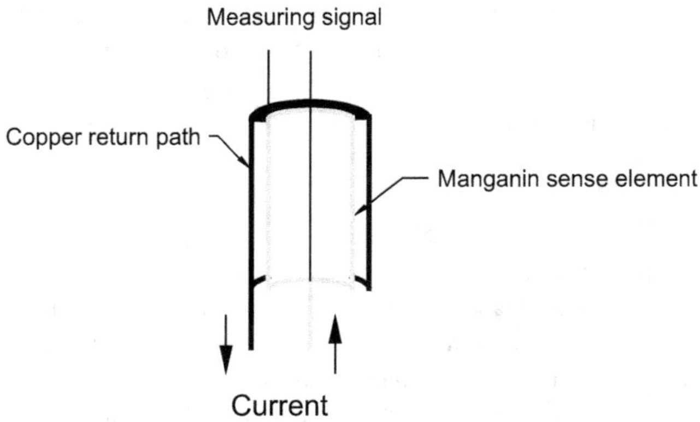

Figure 8.9 Cross section of a coaxial shunt. The inner element is resistive, and the outer return is copper [4,11]

inner conductor is cancelled by the return current in the outer shield (Figure 8.9). In this way, the magnetic field is contained between the elements and does not affect (induce a voltage) in the sense leads, one of which passes through the center where the field is minimal. The inner element is made from a resistive alloy like Manganin that has a constant resistance over a wide temperature range. The shunts come in a wide variety of sizes, shapes and terminations. One objection when measuring small devices is their size, which tends to be large. As shunts, there is no galvanic isolation. With extremely low inductance they can be used to very high frequencies.

8.7.7 Current transformer (or probe)

Current transformers offer galvanic isolation from the measured circuit allowing positioning where it is most convenient rather than in the ground return path as is required for resistive shunts. Their turns ratio can be used to attenuate large currents if necessary. Their output voltage is much higher than a shunt, which increases the signal-to-noise ratio for better readings. Finally, they do not heat up. Whether the device is split core clamp type or a closed-core toroid, the most important consideration is bandwidth. Bandwidth is always specified as the −3 dB point in a frequency vs. output plot. The −3 dB point is where the signal is attenuated to 70 per cent of its reference value, or in other words, a 30 per cent error! For the amplitude to be within 1 per cent of the reference value, the bandwidth needs to be reduced by at least one decade from the −3 dB points. However, the phase is just as critical. At the −3 dB point in amplitude, the phase will be 45° shifted! At one per cent amplitude error, the phase will still be 6° shifted. Therefore, to have less than 1 per cent phase shift, the frequency needs to be reduced by a factor of 60 from the −3dB points. As an example, a 200 MHz rated current transformer would only be useful to 3.3 MHz (Figure 8.10).

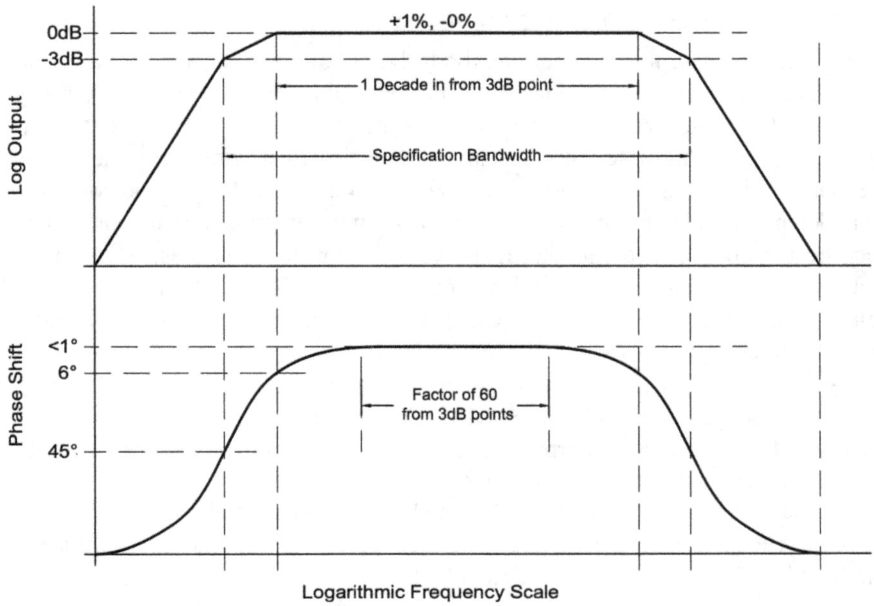

Figure 8.10 Frequency and phase response of a current transformer. You must be well away from the −3 dB points to have reasonable accuracy. From [12] www.pearsonelectronics.com.

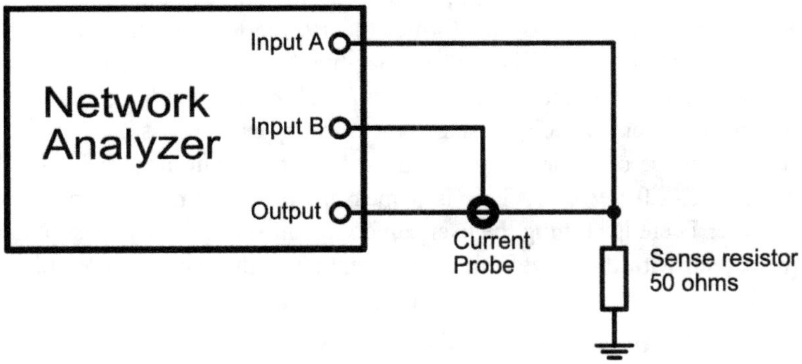

Figure 8.11 Method to measure phase response of a current probe by comparing to flat frequency response RF-rated resistor [13]

The solution is to characterize the phase shift over the frequencies of interest and make corrections during calculations. This requires measuring against an ideal resistor. High-frequency 50 ohm resistors are available with very low parasitic inductance, which can be used as a reference. Bear in mind that the analyzer will have its own tolerance of phase measurements that need to be verified and accounted for (Figure 8.11).

8.7.8 Oscilloscope reading error

Voltage error in oscilloscopes affects both the voltage and current reading channels. The measurement error consists of probe attenuation error, reading error (dc linearity), frequency response error, signal range error and ENOB (effective number of bits) error. All these errors can add up to a significant figure per channel and are then multiplied to obtain the power. When looking at each error and combining them, keep in mind that these are the specification limits. Most instruments operate very near nominal, meaning the actual error is considerably less. Some manufacturers publish charts with typical error and specification limits. The actual values can only be determined by testing with a calibrator. The error calculation gives an outer limit.

Since an oscilloscope input channel can only receive a small voltage, the measured voltages need to be attenuated. The first attenuator is the probe itself, usually at 10:1 with a typical error of 1–2 per cent for passive probes. Active probes are often better at 0.5 per cent (Table 8.2).

Probes have a frequency response rated to −3 dB point like the oscilloscope. At 70 per cent of the true value, this is a massive error at both ends of the frequency bandwidth. When the harmonics of rectangular waveforms are considered, the rated bandwidth needs to be substantially larger. Using a 1 MHz switching frequency as the baseline with 51 harmonics plus 10× for good frequency response puts the requirement out to 500 MHz. Realistically, harmonics this far out have little effect on the result. As with current transformers, a decade from the 3 dB points is the minimum. Consideration should also be given to probe capacitance. At frequencies above 1 MHz, probe capacitance can easily load the source being measured. In a two-winding test, according to transformer theory, the load on the secondary will be transferred to the primary. The probe capacitive loading allows a current to flow in the sense winding, which is transferred to the driving winding, adding to the measured drive current, which is used to measure the power loss.

The next source of error is the DC linearity of the input amplifier, usually on the order of 1.5–2.0 per cent. This is a measure of how linear the oscilloscope amplifiers are. There is nothing the user can do to adjust to this. The best practice is to fill the screen with the waveform using as much of the range as possible.

Table 8.2 Typical active and passive probe specifications

Probe type	**Active differential**	**Passive**
Example	Tek P6246	Tek P109B
Bandwidth	400 MHz	100 MHz
Passband ripple	± 0.25 dB	−3 dB at 100 MHz
Attenuation	1×, 10×	10×
Linearity	2 per cent	1.3 per cent
Harmonic distortion	1.5 per cent	N/A
Input capacitance	1 pF	13.3 pF
Loading @ 1 MHz	$159k\Omega$	$12k\Omega$

The measuring range is a source of error. Most instruments have a range tolerance based on full scale, which means that the error increases at the lower end of the scale simply because the error is a larger portion of the measurement. Below 10 per cent of full scale, the error is potentially high, and the A/D will effectively lose 4-bit resolution [9]. The signal should be at least 80 per cent full scale to use the most bits on the A/D converter.

Finally, there is ENOB, an effective number of bits, which is a measure of the dynamic range of an A/D converter. It measures the real-world performance, which is always less than the ideal bit number. The resolution is determined by the number of bits, but the hardware limits the actual vertical resolution with noise and distortion (non-linearities). A typical system will operate at least 2–3 bits lower than the A/D bit resolution. It is shown in [9] that at 8 bits ENOB, the RMS error is less than 0.03 per cent, meaning that the number of bits is not a significant contributor to the error. But this would take a 12-bit A/D to achieve.

An example calculation: Probe attenuation ratio error, P_{attn}: 2 per cent DC gain accuracy, DC_{ga}: 1.5 per cent probe frequency response, P_{freq}: 0.25 dB = 3 per cent signal range, S_{rng}: 80 per cent ENOB: 8 bits (of 12 bits), offset error, OS_{err}: 0

$$Error = (P_{attn} + DC_{ga} + P_{freq} + \frac{S_{rng}}{(2^{ENOB})}) + OS_{err} \tag{8.4}$$

$$Error = (0.02 + 0.015 + 0.03 + 0.8/2^8 + 0) = 6.8\% \tag{8.5}$$

8.7.9 Total error

In the final analysis, the total power loss error will contain the amplitude error of the voltage and current measurements plus the phase error between the two signals

$$\frac{\Delta P_l}{P_l} = \frac{\Delta U_{rms}}{U_{rms}} + \frac{\Delta I_{rms}}{i_{rms}} + \frac{\Delta \cos\,\theta}{\cos\,\theta} \tag{8.6}$$

As noted earlier, the phase error is the most critical and requires the most attention during setup, calibration and testing. The time delay can also be caused by the signal path – that is the length of the cables and/or by the response time of the circuitry. The deskew function in an oscilloscope can remove a large portion of this. On some oscilloscopes, the deskew function uses fixed increments, which may not be fine enough at high frequencies. The sampling rate of the instrument will limit what is possible.

The time delay of cable lengths can be calculated by

$$\tau = \frac{L_{cable}\sqrt{\varepsilon}}{C} \tag{8.7}$$

where L_{cable} is the length of a cable in meters, ε is the permittivity of the dielectric material in the cable and C is the velocity of light.

This translates into a phase delay of

$$\Delta\theta = -360^{\circ} \cdot f \cdot \tau \tag{8.8}$$

Short, equal-length cables become more important as the test frequency increases into the MHz range.

It bears repeating that this is the worst-case error, and the actual error, assuming the instruments measure near nominal, will be considerably less. If the total calculated error is within ±20% the nominal error will likely be less than half and should be considered adequate. It will be more than sufficient for design purposes.

8.8 Calculation methods

After acquiring measurement data, it needs to be processed. The first step is often to average it over a number of cycles to smooth it. Then, any calibration or characterization data should be applied to correct it for sensor parasitics. Finally, the power can be determined, and the result divided by the core effective volume (used for powdered and ferrite cores) or by core mass (used for steels and ribbons) to put into standard form

$$P = I_{rms} \cdot V_{rms} \cdot \cos\theta \tag{8.9}$$

One method proposed in [8] is to compensate for the phase discrepancy by shifting the current waveform horizontally (time axis) to align with a reference phase angle, which is determined from an impedance sweep on the component under test and FFT analysis. This method is considered more accurate than a fixed deskew because it considers the frequency response over a large spectrum.

Cross power is another method of post-processing data [14]. The method converts the time axis waveforms into the frequency spectrum using FFT and then executes the calculation with no phase difference. Compensation for the amplitude and phase of the detection circuit is then applied. This captures the time-axis waveforms with little error by returning the frequency spectrum having the error compensation to the time-axis waveforms.

8.9 Temperature

During the measurement of core loss, the core itself experiences a temperature rise. At high frequency and high excitation, the core may heat very quickly, skewing the results. Therefore, the excitation time needs to be minimal, and readings need to be taken quickly. Testing under computer control is the most practical method of accomplishing this consistently in a few seconds. The source is preset and activated, allowed to settle, then the readings are taken including monitoring the temperature of the core and shut down. This only takes a few seconds. Smaller cores can be immersed in mineral oil to help disperse the heat.

The second use of mineral oil is the opposite, to help heat the core evenly to an elevated test temperature. Cores are rarely operated at room temperature therefore testing should be performed at or near the expected operating temperature or over a range to characterize the material. Powdered cores, steel based laminations, or ribbon and amorphous/nanocrystalline cores all have relatively stable core loss over temperature. However, typically ferrite core losses vary widely with temperature, having high losses at cold, then decreasing as the temperature approaches 80–100 °C and then rising again.

Anytime air is used to adjust the temperature, there are several points to keep in mind. First, and often not realized, is that the cooling coils in a thermal chamber, unlike the heating elements, which operate with pulse width modulation, only operate at a fixed temperature difference from the ambient. Cycling the cooling on and off and stirring the air produces an average temperature. Most cooling systems also need a small delay time before restarting. Therefore, the device under test in a thermal chamber needs to be shielded from the direct airflow from the cooling coils and be in turbulent air. It takes a considerable time to stabilize at the target temperature. If the core losses are high at cold temperatures, such as with ferrites, the interval between each test needs to be long. Unlike metals, ferrites are very poor conductors of heat. Using a medium like oil is not possible because it will congeal. The outside surface of the core may cool quickly, but the inside takes time.

8.10 Summary

Measuring core loss accurately between 20 Hz and 20 MHz is possible. Various instrument manufacturers have made is relatively easy below 500 kHz. Beyond that, all instruments degrade quickly. However, if proper attention is given to the limitations of the equipment and careful compensation is made for the voltage, current and especially the phase angle, measurements into the MHz range can be made with reasonable and definable accuracy.

References

[1] IEC 60367-1:1982. Core for Inductors and Transformers for Telecommunications Part 1: Measuring Methods. International Electrotechnical Commission; 1982.

[2] Baguley CA, Madawala UK, and Carsten B. The Influence of Temperature and Core Geometry on Ferrite Core Losses under DC Bias Conditions. International Symposium of Power Electronics; 2008.

[3] Wang J, Yuan X, and Rasekh N. Triple Pulse Test (TPT) for Characterizing Power Loss in Magnetic Components in Analogous to Double Pulse Test (DPT) for Power Electronic Devices. IECON 2020. *IEEE Industrial Electronics Society*; 2020.

[4] Zacharias P. *Magnetic Components: Basics and Applications*. Berlin: Springer; 2022.

[5] Van den Bossche A and Valchev VC. *Inductors and Transformers for Power Electronics*. Boca Raton, FL: Taylor & Francis; 2005.

[6] Xiao C, Chen G, and Odendaal W. Overview of Power Loss Measurement Techniques in Power Electronics Systems. vol. 43. *IEEE Transactions on Industry Applications*; 2007.

[7] Hou D, Mu M, Lee FC, *et al.* New High-Frequency Core Loss Measurement Method with Partial Cancellation Concept. vol. 32. *IEEE Transactions on Power Electronics*; 2017.

[8] Rasekh N, Wang J, and Yuan X. A New Method for Offline Compensation of Phase Discrepancy in Measuring the Core Loss with Rectangular Voltage. vol. 2. *IEEE Industrial Electronics Society*; 2021.

[9] Analogue to Digital Converter (ADC) – Bit Count Superstition. Newtons4th Ltd.; 2020. D000123.

[10] Sato T and Sakaki Y. 100 kHz–10 MHz Iron Loss Measuring System. vol. MAG-33. IEEE Transactions on Magnetics; 1987.

[11] Okamure S and Okabe T. A New Device for Measurements of Pulses or High Frequency Currents. vol. IM-23. *IEEE Transactions on Instrumentation and Measure*; 1974.

[12] Frequency Response Phase Shift; 2024 [cited 2/2/2024]. Available from: https://pearsonelectronics.com/pdf/Frequency

[13] Thottuvelil VJ, Wilson TG, and Owen HAJ. High Frequency Measurement Techniques for Magnetic Cores. vol. 5. *IEEE Transactions on Power Electronics*; 1990.

[14] Nagahama R. AC Core Loss Measurement on High Phase Angle Material Core Loss; 2024. Available from: https://www.psma.com/sites/default/files/uploads/files/ACcore loss measurement on high phase angle material (Ryu Nagahama, Iwatsu Electric Company).pdf [cited 2024-01-31].

Chapter 9

Software for inductive components

Alfonso Martínez[1]

Historically, the application of software in magnetics for power electronics has been something personal: Each engineer created their own in the language they already knew, which in 99% of the cases has been a spreadsheet.

While this solution allowed a quick start-up of the design process due to the ease of implementation, it also resulted in unintended dead-ends, as often the set of operations in spreadsheet software was not suitable for the advanced operations required for electromagnetic modeling.

This produced a curious effect which, to the best of the author's knowledge, is something unique for power magnetics: Development in the state of art for design of inductive components has never really reached most engineers in their day-to-day work, even after half a century.

There were some exceptions to this rule; some engineers who were sufficiently versed in programming produced software aimed at designing inductors and transformers with modern techniques. Examples of these include Intusoft Magnetics Designer, Colonel Wm. T. McLyman's Titan software, and Rafael Asensi's PExprt, which was bought by Ansoft and later Ansys.

However, the reality is that most magnetics in the last 30 years have been designed using Steinmetz's equation (from 1896) and Ohm's law (1827), with the most advanced ones using Dowell's equation (1966). While being correct, these methods are not thoroughly developed for high switching frequencies and the electromagnetic effects derived from non-sinusoidal waveforms.

In this chapter, we first review the basic equations for each of the different aspects interesting for the design of magnetic components and then add and compare new approaches.

9.1 Data structure

Before getting into modeling magnetics with software, we need to describe the magnetics with software. This means having in memory all the data needed for the

[1]Würth Elektronik, Germany

calculation from both the excitation waveform and the magnetic component, with the amount of detail necessary highly dependent on the model.

For example, if we want to use the Steinmetz equation, we would need the peak of the magnetic flux density in the core and its frequency, while if we were using the iGSE (a method with a higher accuracy for non-sinusoidal waveforms that we will look at in the coming pages), we would need a sampled period of the waveform.

As we want to cover many models in this chapter, we need a data structure that is able to store all the information about our magnetic component and its excitation regardless of the model used.

In order to describe our magnetics in such a way, we would need the following information:

- Dimensions of the core, according to IEC 60401-2, are commonly found in any manufacturer's datasheet. From this information, we can extract
  - Effective volume, needed to calculate the total core losses.
  - Effective area, needed to calculate the magnetic flux density.
  - Effective length, needed to calculate the reluctance of the core.
  - Dimensions of the winding window, needed to calculate the reluctance of the core, and the leakage inductance.
- Dimensions of the core legs, needed to calculate the reluctance of the core, the inductance, and winding losses due to the fringing effect.
- Steinmetz coefficients of the core material, or in its absence, the volumetric losses graph provided by the manufacturer, needed to calculate the volumetric core losses.
- Initial permeability of the core material, needed to calculate the reluctance of the core.
- Number of turns for each of the windings in our coil, needed to calculate the magnetic field and winding losses.
- Number of parallels for each of the windings in our coil, needed to calculate the winding losses.
- Wire information for each of the windings in our coil:
  - Conducting diameter, if the wire is round, or height and width if the wire is rectangular or foil, needed to calculate the winding losses.
  - Thickness of the insulating coating, needed to calculate the leakage inductance.
  - Number of conductors present in the wire, typically for Litz wire, needed to calculate the winding losses and leakage inductance.
- Coordinates of each of the turns inside the winding window, referred to as a fixed origin, needed to calculate the winding losses and the leakage inductance.

Although it looks like a lot of information, especially the information about the coil for complex transformers, it can be greatly automated with the proper software.

There are already existing data structures designed specifically for the purpose of holding the information of a magnetic structure, such as the Open-Source Magnetic Agnostic Structure (MAS) [1], which is the one that will be used in the rest of this chapter.

In the following sections, we will assume that we already have a MAS file with the information of the inputs and the magnetic component loaded and readily accessible by the program.

All the models described here will use information that is either commonly known to the engineer, as the number of turns, or that is provided by the manufacturer of the part, such as the permeability of the magnetic core.

9.2 Magnetizing inductance

The magnetizing inductance (L_{mag}) represents the amount of magnetic energy that is linked with the magnetic core, and in the case of transformers, the amount of energy that is coupled with the secondary.

It is usually modeled as

$$L_{mag} = \frac{N^2}{\mathfrak{R}} \tag{9.1}$$

where N is the number of turns and $\mathfrak{R}$ is the reluctance of the core, which represents the resistance of the core to the magnetic flux.

As the number of turns is usually a known parameter of the design, the difficulty of predicting the magnetizing inductance resides in the reluctance, which depends mainly on the geometry of the magnetic core.

The total reluctance of the core depends on the reluctance of the core, and the reluctance of any air gaps that may exist along the path of the magnetic flux.

The reluctance of the core is calculated by

$$\mathfrak{R}_{core} = \frac{l_{eff}}{\mu A_{eff}} \tag{9.2}$$

where l_{eff} and A_{eff} are the effective length and effective area of the core, respectively, and is the absolute permeability of the core material, calculated as the product of the magnetic constant, a physical constant equal to $1.25663706212 \times 10^{-6}$, and the relative permeability of the core, for which the initial permeability is commonly used, as is a parameter provided by the manufacturer of the magnetic core.

The reluctance of the air gap is more complex to model, especially due to an effect called fringing, a deformation of the magnetic flux lines when they cross from a high permeability material (such as the magnetic core) to a low permeability one (such as the air gap). This deformation causes an increase in the volume occupied by magnetic flux, and therefore, a decrease in the theoretical reluctance calculated by the classical equation (9.2) that we would have if the magnetic flux lines were crossing the air gap straight.

To account for this decrease, we use the term fringing factor, which reflects how much the volume occupied by the flux lines in the air gap has increased.

Several approaches exist in the literature [2–6] to calculate this air-gap reluctance/fringing factor, where Partridge's model (described in [6]) is the most used in the past, and Zhang *et al.* [2] is the most advanced with the best accuracy.

The code implementation for the Partridge model is shown in the snippet below, where we directly apply the formula from [6] to calculate the fringing factor and apply that factor to the classical expression of the gap reluctance (9.2).

```
double get_gap_reluctance(double length, double area, double
distanceClosestNormalSurface) {
    double reluctance;
    double fringingFactor = 1;
    double vacuumPermeability = 1.25663706212e-6;
    // where length is gap length
    // area is the core cross sectional area
    // distanceClosestNormalSurface is the distance to the nearest
surface parallel to its cross section, typically the top or bottom plate,
or another gap, in the case of distributed gaps.

    if (length > 0) {
        fringingFactor = 1 + length / sqrt(area) * log(2 * 2 *
distanceClosestNormalSurface / length);
    }

    return length / (vacuumPermeability * area * fringingFactor);
};
```

In the code for Zhang *et al.*, we first calculate the perimeter of the gap, as it will mark the amount of fringing flux we have in our gap. With it then calculate the reluctance due to the fringing flux and combine it in parallel with the classical one (9.2), called internal in this model. Finally, we calculate the fringing factor from the calculated reluctance, inversely, as in the previous method.

```
enum class ColumnShape : int { RECTANGULAR, ROUND };

double get_gap_reluctance(double length, double area, ColumnShape
gapShape, double distanceClosestNormalSurface, double gapSectionWidth,
double gapSectionDepth) {
    double vacuumPermeability = 1.25663706212e-6;
    double perimeter = 0;
```

```
    auto reluctanceInternal = gapLength / (vacuumPermeability * area);
    double reluctanceFringing = 0;
    double fringingFactor = 1;
    // use gapSectionWidth for diameter if gapShape is round

    if (gapShape == ColumnShape::ROUND) {
        perimeter = std::numbers::pi * gapSectionWidth;
    }
    else {
        perimeter = gapSectionWidth * 2 + gapSectionDepth * 2;
    }

    if (length > 0) {
        reluctanceFringing = std::numbers::pi / (vacuumPermeability *
perimeter * log((2 * distanceClosestNormalSurface + length) / length));
    }
    double reluctance = 1. / (1. / reluctanceInternal + 1. /
reluctanceFringing);

    if (length > 0) {
        fringingFactor = length / (vacuumPermeability * area *
reluctance);
    }
    return reluctance;
};
```

Table 9.1 Reluctance error for different cores and gaps. Difference between measured value and calculated value.

Models	**PQ 28/20 grinded (%)**	**E42 21/20 spacer (%)**	**ETD 59 spacer (%)**	**E 55/28/21 spacer (%)**
No fringing flux	25.48	9.829	9.36	66.76
Zhang	17.87	5.86	5.11	19.10
Mühlethaler	23.14	7.301	9.052	8.231
Partridge	17.77	5.324	4.944	22.66
Stenglein	17.5	5.992	5.468	29.48

In order to evaluate which model is better, a verification was built with measurement data taken from different papers, and all the models were evaluated against it (Table 9.1).

It can be appreciated how the error is reduced in all cases by taking into account the fringing flux, in most cases substantially, which is translated into an increased accuracy in predicting the magnetizing inductance.

9.3 Core losses

Core losses are one of the most important parameters to predict in a magnetic device, as the magnetic core is the part where the temperature has the largest impact: All the magnetic parameters depend on the temperature of the material, including permeability and core losses. The core losses produce heat, and the heating can increase the core losses, which can produce larger core losses and more heating.

To avoid an infinite loop of positive feedback that would make many magnetics unusable, some materials such as ferrite are designed in a way that they have minimum core losses at a temperature between 60 °C and 100 °C.

In order to have a working inductive component, we need to make sure that the heating due to the losses minus the cooling of our system stays in a balance near the temperature with the minimum core losses for our material.

Having a good model that predicts the core losses of our magnetic component can save us a lot of iterations in the laboratory.

Traditionally, magnetic engineers have used the Steinmetz equation [7] to predict the losses, as most manufacturers provide coefficients for that model. This equation consists of a curve fit of several measurement points using the following equation:

$$P_v = kf^{\alpha}B^{\beta} \tag{9.3}$$

where f is the switching frequency of the magnetic field, B is the amplitude of the magnetic flux density, P_v is the loss per volume and k, α, and β are the Steinmetz coefficients, provided by the manufacturer for each material.

The problem with this model is that it was developed for sinusoidal waveforms for the magnetic field, which is not the common case today. Many other models were developed to take into account different waveforms, some of them based on independent measurements (which are not presented here, as they are only valid for a small subset of materials), some of them for the material of some specific manufacturer [14,15], and some of them were developed with the coefficients already widely available [8–13].

In this section, we will focus on the original Steinmetz equation [7] and the most accurate model, iGSE [8].

The implementation of the Steinmetz model is quite straightforward. We obtain the alternating current (AC) peak of the waveform, the Steinmetz coefficients from the core, and apply the equation from [7] to obtain the volumetric losses, which are finally multiplied by the effective volume to obtain the total losses.

```
double get_core_losses(double magneticFluxDensityAcPeak, double
frequency, double effectiveVolume, double temperature,
std::map<std::string, double> steinmetzCoefficients) {
    double frequency = excitation.get_frequency();

    double k = steinmetzCoefficients["k"];
    double alpha = steinmetzCoefficients["alpha"];
    double beta = steinmetzCoefficients["beta"];
```

```
    double ct0 = steinmetzCoefficients["temperatureCoefficients"]["ct0"];
    double ct1 = steinmetzCoefficients["temperatureCoefficients"]["ct1"];
    double ct2 = steinmetzCoefficients["temperatureCoefficients"]["ct2"];

    double volumetricLosses = k * pow(frequency, alpha) *
pow(magneticFluxDensityAcPeak, beta);
    volumetricLosses *= (ct2 * pow(temperature, 2) - ct1 * temperature +
ct0);

    return volumetricLosses * effectiveVolume;
};
```

The iGSE method is based on the assumption that the losses in the core are due to two different physical mechanisms: one that depends on the hysteresis loop (including possible subloops inside), which depends on the shape of the waveform (this part will be represented in the integral we will do later); and one that depends on the eddy currents induced inside the grains of the core material, which depends on the frequency of the switching and the maximum amplitude of the voltage that they subjected to, which depends on the maximum amplitude of the magnetic flux density (this is the part that will be treated outside the integral)

$$P_v = \frac{1}{T}\int_0^T k_i \left|\frac{dB}{dt}\right|^{\alpha} (\Delta B)^{\beta-\alpha} dt \tag{9.4}$$

where

$$k_i = \frac{k}{(2\pi)^{\alpha-1}\int_0^{2\pi} |\cos\theta|^{\alpha} 2^{\beta-\alpha} d\theta} \tag{9.5}$$

In comparison with vanilla Steinmetz, the implementation for iGSE is more complex, as it involves two functions. In the first one, we obtain the k_i coefficient by considering each of the steps of the waveform in a subloop and applying the formula from [8], integrating numerically using a for loop.

Once we obtain the k_i coefficient, together with the rest of the parameters from the manufacturer, we use them to calculate the total losses in the second function.

In this second function, we repeat the procedure of considering every step of the waveform as a subloop and calculate the variation of magnetic flux density in between them. We will assume that they are uniformly distributed and use the frequency to calculate the time difference between one step and the other.

This technique of considering waveform steps as subloops might look more time-consuming than the method proposed by the author, which is detecting and processing only a few subloops. This might be true for special cases, but for the general case, the effort to identify and obtain these subloops from the steps would be computationally equal to just using the steps as subloops.

To finish the model, we multiply the integration result by the part that does not depend on the waveform shape, and with the effective volume, we obtain the final losses.

```
double calculate_ki(std::map<std::string, double> steinmetzCoefficients,
int numberPointsSamplesWaveforms=512) {
    double k = steinmetzCoefficients["k"];
    double alpha = steinmetzCoefficients["alpha"];
    double beta = steinmetzCoefficients["beta"];
    std::vector<double> theta_vect(numberPointsSamplesWaveforms);
    std::generate(theta_vect.begin(), theta_vect.end(), [n = 0,
&numberPointsSamplesWaveforms]() mutable {
        return n++ * 2 * std::numbers::pi / numberPointsSamplesWaveforms;
    });
    double theta_int = 0;
    for (auto& theta : theta_vect) {
        theta_int += pow(fabs(cos(theta)), alpha) * pow(2, beta - alpha)
* 2 * std::numbers::pi /
                        numberPointsSamplesWaveforms;
    }
    double ki = k / (pow(2 * std::numbers::pi, alpha - 1) * theta_int);
    return ki;
}

double get_core_losses(std::vector<double> magneticFluxDensityWaveform,
double frequency, double effectiveVolume, double temperature,
std::map<std::string, double> steinmetzCoefficients) {

    double magneticFluxDensityWaveformMaximum =
*std::max_element(magneticFluxDensityWaveform.begin(),
magneticFluxDensityWaveform.end());
    double magneticFluxDensityWaveformMinimum =
*std::min_element(magneticFluxDensityWaveform.begin(),
magneticFluxDensityWaveform.end());

    double magneticFluxDensityPeakToPeak =
magneticFluxDensityWaveformMaximum - magneticFluxDensityWaveformMinimum;

    OpenMagnetics::SteinmetzCoreLossesMethodRangeDatum steinmetzDatum;
    if (is_steinmetz_datum_loaded()) {
        steinmetzDatum = get_steinmetz_datum();
    }
    else {
        steinmetzDatum =
get_steinmetz_coefficients(core.get_functional_description().get_material
(), frequency);
    }
```

```
    double k = get_ki(steinmetzCoefficients);
    double alpha = steinmetzCoefficients["alpha"];
    double beta = steinmetzCoefficients["beta"];
    double ct0 = steinmetzCoefficients["temperatureCoefficients"]["ct0"];
    double ct1 = steinmetzCoefficients["temperatureCoefficients"]["ct1"];
    double ct2 = steinmetzCoefficients["temperatureCoefficients"]["ct2"];

    double volumetricLossesSum = 0;
    double timeDifference = 1 / frequency / numberPointsSamplesWaveforms;

    for (size_t i = 0; i < magneticFluxDensityWaveform.size() - 1; ++i) {
        volumetricLossesSum += pow(fabs((magneticFluxDensityWaveform[i +
1] - magneticFluxDensityWaveform[i]) / timeDifference), alpha) *
timeDifference;
    }

    double volumetricLosses = ki * pow(magneticFluxDensityPeakToPeak,
beta - alpha) * frequency * volumetricLossesSum;
    volumetricLosses *= (ct2 * pow(temperature, 2) - ct1 * temperature +
ct0);

    return volumetricLosses * effectiveVolume;
}
```

Table 9.2 Core loss error for different cores and gaps

Models	3C90 (%)	3C94 (%)	3F4 (%)	N27 (%)	N30 (%)	N49 (%)	N87 (%)
Steinmetz	58.9	39.3	58.5	35.2	25.2	34.4	37.5
iGSE	54.9	31.2	58.5	29.1	8.5	33.1	31.8
Barg	75.9	45.2	49.5	36.9	35.5	36.2	38.0
Roshen	47.3	35.5	NA	33.0	NA	45.4	61.9
Albach	54.8	30.9	58.0	29.2	8.4	32.4	31.9

In order to evaluate which model is better, a verification was built with measurement data taken from the MagNet project, and all the models were evaluated against it (Table 9.2).

It can be observed how the base error for core losses is much higher than for other models, which can be explained by the complexity of modeling the different effects that happen in a ferromagnetic material under nonlinear excitation.

9.4 Skin effect losses

There is a concept that has been going on around for years: Increasing the switching frequency of the converters will decrease the losses in the magnetic components, allowing smaller sizes; however, while increasing the frequency reduces the magnetic flux density swing in the magnetic core, and therefore, the core losses, it is not necessarily true that the total losses will decrease.

The reason for the losses increasing at a higher frequency is the eddy currents happening in the turns of the component. These eddy currents are commonly divided into two types: skin effect losses, which are provoked by the current circulating inside the wire where the losses are induced, and proximity effect, provoked by external magnetic fields, usually produced by other nearby turns or magnetic core air gaps.

In this section, we will elaborate more on the skin effect and how to model the loss due to it.

When an AC current circulates inside a conductor, it produces an internal radial-induced magnetic field, which in turn induces currents that oppose the main AC current in the center of the wire and reinforces it in the ring area closer to the surface.

The total effect is translated into a reduction of the conducting cross-sectional area of the wire which is proportional to the frequency of the current, which increases the effective ohmic resistance and the loss in the wire. It is equivalent to having a wire that is reduced in area as the frequency goes higher.

To avoid this problem, smaller wire diameters bundled together, but insulated from each other by enamel coating, are used. These wires suffer less from this effect but they are more expensive, and they have less conducting area to insulation area ratio than thicker wires. Even without all these negative aspects, they might not decrease the winding losses, as we will see in the next section.

The modeling of skin effect losses is something that has been researched for decades [16–21]. Most authors start working with Maxwell equations and develop their models by adding simplifications that are valid under certain conditions (that are enough for them) in order to end up with an analytical equation that can be handled by the computation method of the time.

An example of such is the classical method of AC resistance by Dowell, who created a simple equation that is valid for a transformer as long as the conducting of the windings is balanced over the switching period and there is no gap, which at his time it was a valid condition, but highly decreases the accuracy of its model for modern magnetics like flybacks, or gapped transformers for LLC converters.

As computation power increases, modern models have less necessity for simplifications and become more general, which is the case for the two models that we will implement here: Albach's [16] and Kutkut's [19], which use complex functions for solving Maxwell equations in round and rectangular conductors, respectively.

The following calculations are shown for one individual turn, and they must be repeated for each of the turns in our coil.

For round and Litz wire, we use Albach's model, which calculates the skin effect losses as a proportion of the DC losses

$$P_{round} = I_{rms}^2 R_{DC} F_{skin} \tag{9.6}$$

where

$$F_{skin} = \frac{1}{2}\mathrm{Re}\left\{ \alpha r_s \left[\frac{I_0(\alpha r_s)}{I_1(\alpha r_s)} + \frac{N(N-1)r_s}{r_{outer}} \frac{I_1(\alpha r_s)}{I_0(\alpha r_s)} \right] \right\}, \quad \alpha = \frac{1+j}{\delta} \tag{9.7}$$

with r_s being the radius of the individual strand and in case the wire is made out of several conductors (e.g., Litz wire), r_{outer} being the outer radius of the wire, N being the number of conductors or strands, δ being the skin depth, and I_0 and I_1 being the modified Bessel functions of first kind of order 0 and 1, respectively, whose implementation we have included in the code below.

This proportion is called the skin factor, which is always greater than 1 for AC current. Albach calculates this factor with two terms: one that depends on the skin depth, valid for all wires, and an extra term that depends on the number of strands, which reflects the effect of the losses in the magnetic field strength from surrounding strands.

```
double vacuumPermeability = 1.25663706212e-6;
double copperPermeability = 0.999994;
double copperResistivity = 0.00000001678;
double copperTemperatureCoefficient = 0.004041;

std::complex<double> modified_bessel_first_kind(double order,
std::complex<double> z) {
    std::complex<double> sum = 0;
    std::complex<double> inc = 0;
    std::complex<double> aux = 0.25 * pow(z, 2);
    size_t limitK = 1000;
    for (size_t k = 0; k < limitK; ++k)
    {
        double divider = tgammaf(k + 1) * tgammaf(order + k + 1);
        if (std::isinf(divider)) {
            break;
        }
        inc = pow(aux, k) / divider;
        sum += inc;
        if (std::abs(inc) < std::abs(sum) * 0.0001){
            break;
        }
    }

    auto bessel = sum * pow(0.5 * z, order);
    return bessel;
}

double calculate_skin_depth(double frequency, double temperature) {
    double resistivity = copperResistivity * (1 +
copperTemperatureCoefficient * (temperature - 20));
    double skinDepth = sqrt(resistivity / (std::numbers::pi * frequency *
vacuumPermeability * copperPermeability));
    return skinDepth;
};
```

```
double calculate_skin_factor_round_wire(double conductingDiameter, double
outerDiameter, int numberConductors, double frequency, double
temperature) {

    double skinDepth = calculate_skin_depth(frequency, temperature);
    double wireRadius = conductingDiameter / 2;
    double wireOuterRadius = outerDiameter / 2;

    std::complex<double> alpha(1, 1);
    alpha *= wireRadius / skinDepth;
    double factor = 0.5 * (alpha * (modified_bessel_first_kind(0, alpha)
/ modified_bessel_first_kind(1, alpha) + numberConductors *
(numberConductors - 1) * pow(wireRadius, 2) / pow(wireOuterRadius, 2) *
modified_bessel_first_kind(1, alpha) / modified_bessel_first_kind(0,
alpha))).real();

    return factor;
}

// Use strand conducting and outer diameter in the case of Litz
double calculate_skin_losses_round_wire(double conductingDiameter, double
outerDiameter, double dcResistance, double currentRms, double frequency,
double temperature) {
    double skinFactor = calculate_skin_factor(conductingDiameter,
outerDiameter, frequency, temperature);
    double dcLossTurn = dcResistance * pow(currentRms, 2);
    double turnLosses = dcLossTurn * (skinFactor - 1);
    return turnLosses;
}
```

For rectangular and foil wires, we use Kutkut's method. This author bases his analysis on the fact that the current density in a conductor tends to adopt a circular/elliptical shape, which produces extra losses in rectangular conductors, especially in the corners

$$F_{skin} = \left[1 + \left(\frac{f}{f_l}\right)^{\alpha} + \left(\frac{f}{f_h}\right)^{\beta}\right]^{\frac{1}{\gamma}}, \quad \gamma \simeq 11, \beta = \frac{\gamma}{2} = 5.5, \alpha = 2 \tag{9.8}$$

where

$$f_l = \frac{3.22}{8\sigma\mu_0 b' a'} \tag{9.9}$$

$$f_h = \frac{\pi^2}{4\sigma\mu_0 \mathrm{a}'^2} \cdot K^{-2}\left(\sqrt{1 - \frac{\mathrm{a}'^2}{\mathrm{b}'^2}}\right) \tag{9.10}$$

where a and b are the thinnest dimension of the rectangular conductor and the thicker, respectively, f is the frequency, and σ is the conductivity of copper. γ, α, and β are obtained by taking the limits $f \rightarrow 0$ and $f \rightarrow \infty$ in F_{skin}. More information can be found in the source paper [19].

Kutkut [19] calculates the elliptical semi-major and semi-minor axis of the equivalent ellipsis to the rectangular shape and uses it together with some parameters extracted from simulations to calculate the AC resistance factor.

The code for this method, including an implementation of the complete elliptic integral, is

```
double vacuumPermeability = 1.25663706212e-6;
double copperResistivity = 0.00000001678;
double copperTemperatureCoefficient = 0.004041;

double comp_ellint_1(double x) {
    double k;
    double m;
    double a;
    double g;
    double a_old;
    double g_old;
    double two_n;
    double M_PI_2 = 1.57079632679;

    if ( x == 0.0 ) { return M_PI_2; }
    k = fabs(x);
    m = k * k;
    if ( m == 1.0 ) { return std::numeric_limits<double>::quiet_NaN(); }

    a = 1.0;
    g = sqrt(1.0 - m);
    two_n = 1.0;
    for (int i=0;i<100;i++) {
        g_old = g;
        a_old = a;
        a = 0.5 * (g_old + a_old);
        g = g_old * a_old;
        two_n += two_n;
        if ( fabs(a_old - g_old) <= (a_old * DBL_EPSILON) ) break;
        g = sqrt(g);
    }
    return std::numbers::pi / 2 / a;
}

double calculate_skin_losses_rectangular_wire(double conductingWidth,
double conductingHeight, double dcResistance, double currentRms, double
frequency, double temperature) {
    double bPrima = std::max(conductingWidth, conductingHeight) / 2;
    double aPrima = std::min(conductingWidth, conductingHeight) / 2;
```

```
    double resistivity = copperResistivity * (1 +
copperTemperatureCoefficient * (temperature - 20));

    double fl = 3.22 * resistivity / (8 * vacuumPermeability * bPrima *
aPrima);
    double fh = pow(std::numbers::pi, 2) * resistivity / (4 *
vacuumPermeability * pow(aPrima, 2)) * pow(comp_ellint_1(sqrt(1 -
pow(aPrima, 2) / pow(bPrima, 2))), -2);
    double gamma = 11;
    double beta = 5.5;
    double alpha = 2;

    double acResistanceFactor = pow(1 + pow(frequency / fl, alpha) +

pow(frequency / fh, beta), 1 / gamma);
    double dcLossTurn = dcResistance * pow(currentRms, 2);
    double turnLosses = (acResistanceFactor - 1) * dcLossTurn;
    return turnLosses;
}
```

9.4.1 Magnetic field strength and proximity effect losses

9.4.1.1 Magnetic field strength

We discussed how skin effect losses are caused by the magnetic field induced by the current flowing through the same wire where they are produced, and in this section, we discuss the losses caused by the external fields, which are commonly, and in some cases wrongly, referred to as proximity effect losses.

However, before calculating the proximity losses, unless we are in a really simple inductor or non-interleaved transformer, we need to calculate the magnetic field strength produced by turns and air gaps in the magnetic core, as this provide us with the external field that surrounds each turn, and therefore the field that will induce the eddy current and generate losses.

In magnetic components, these external fields have two main sources. The first one is the current that circulates through nearby neighboring turns, which induce magnetic field strength in a plane perpendicular to the circulation of the current. When this field crosses a conducting material, like the rest of the turns in a winding, it induces eddy currents in the material, which provoke losses and temperature rise (Figure 9.1). The strength of this field does not depend so much on the distance to other turns but on the two-dimensional (2D) magnetomotive force that is produced between the rest of the turns and the turn where we are evaluating, which is the reason why some people choose not to use the term "proximity."

The second source is the magnetic field that circulates through the magnetic core, induced by our turns. As we discussed in the magnetizing inductance section, when the magnetic flux in the core changes medium, as happens when it crosses an air gap, its flux lines are distorted and expanded, taking more volume than is occupied by the air gap (Figure 9.2).

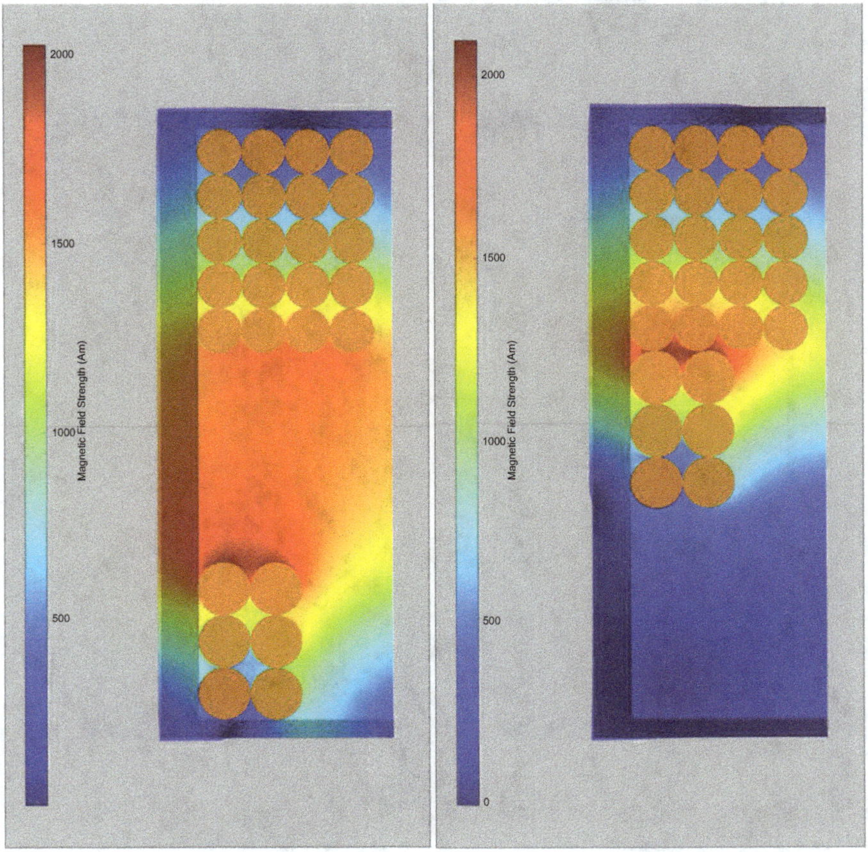

Figure 9.1 *Two simulations with the same number of turns and different distances. The magnetic field strength near the central turns does not change with the distance.*

When these distorted flux lines cross nearby conductors, eddy currents are induced, generating extra losses and temperature rise. The strength of this field does depend on the distance to the producing air gap, so in this case the name "Proximity" is more adequate.

The 2D field produced by these two effects can be added up as vectors, and the resulting field in a given turn is the external field that turn sees.

Because we are focusing on analytical methods in this chapter, we will not discuss on finite-elements analysis, which is the typical approach for calculating magnetic fields, as it can calculate them with high accuracy in three dimensions.

In this section, we will focus on 2D methods [22–25], concretely on Binns–Lawrenson's [22] and Roshen's [24] models, which provide the values of the x and y components of the magnetic field strength for any point in a 2D representation of the winding window produced by a turn (round or rectangular) or an air gap, respectively.

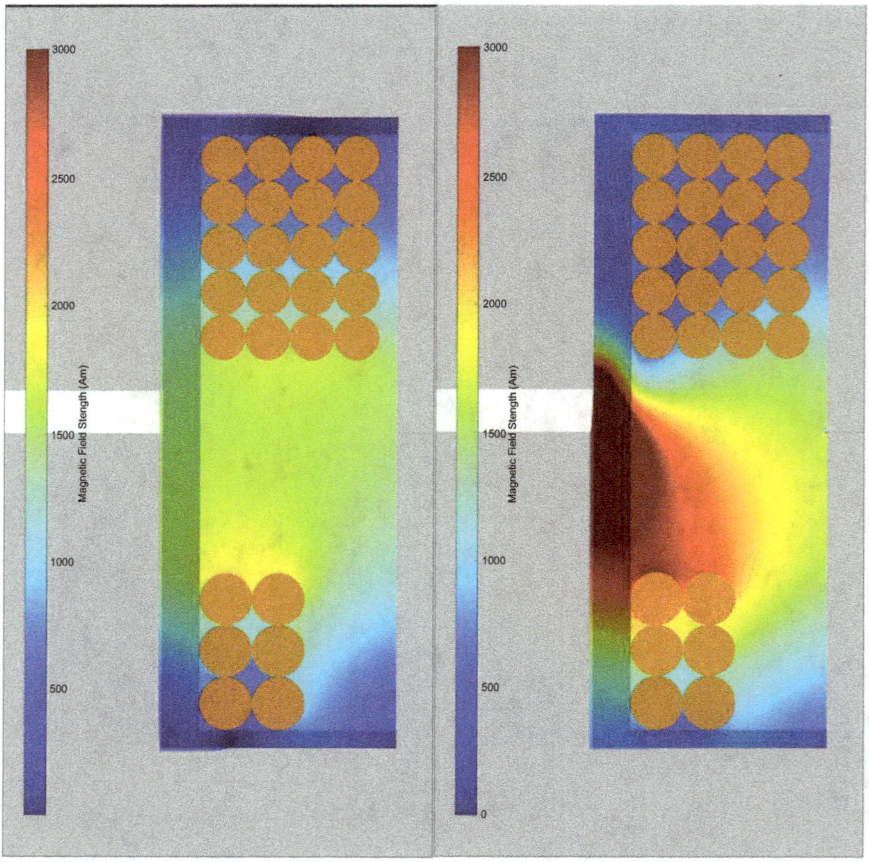

Figure 9.2 Same magnetics with the magnetic field produced by the turns but without fringing field (left) and with it (right)

The code for Binns–Lawrenson's model calculates the magnetic field strength between an inducting turn, either round or rectangular, and an induced point, which can be another turn or a point in the 2D space.

In the implementation, we differentiate between round wires and rectangular wires. In both cases, we calculate the x and y components of the magnetic field strength between the inducing wire and an induced point.

For round wires:

$$H_x = -I \frac{\left(y_{inducing} - y_{induced}\right)}{2\pi\left(\left(y_{induced} - y_{inducing}\right)^2 + \left(x_{induced} - x_{inducing}\right)^2\right)} \tag{9.11}$$

$$H_y = I \frac{\left(x_{inducing} - x_{induced}\right)}{2\pi\left(\left(y_{induced} - y_{inducing}\right)^2 + \left(x_{induced} - x_{inducing}\right)^2\right)} \tag{9.12}$$

where $x,y_{inducing}$ are the cartesian coordinates of the point that induces the magnetic field (the wire in this case) and $x,y_{induced}$ the coordinates of the point at which we wish to calculate the induced magnetic field. This induced point can be other wires in order to calculate the proximity loss or the point of a regular grid to plot the field.

For rectangular wires:

$$H_x = \frac{I}{8\pi ab}\left[(y+b)(\theta_1-\theta_2)-(y-b)(\theta_4-\theta_3)+(x+a)\log\left(\frac{r_2}{r_3}\right)-(x-a)\log\left(\frac{r_1}{r_4}\right)\right] \quad (9.13)$$

$$H_y = \frac{-I}{8\pi ab}\left[(x+a)(\theta_2-\theta_3)-(x-a)(\theta_1-\theta_4)+(y+b)\log\left(\frac{r_2}{r_1}\right)-(y-b)\log\left(\frac{r_3}{r_4}\right)\right] \quad (9.14)$$

with θ_i being the angle between corner i and the induced point, r_i being the distance between corner i and induced point, I being the current circulating through the rectangular conductor, and a and b the dimensions in horizontal and vertical axes, respectively. The center of the rectangular conductor is used as the origin of coordinates (Figure 9.3). The corners are numbered in the following way:

1. Bottom right corner, positive x and negative y coordinates.
2. Bottom left corner, negative x and y coordinates.
3. Top left corner, negative x and positive y coordinates.
4. Top right corner, positive x and y coordinates.

For the case of rectangular wires, we need to correct the angles, especially in the case of points vertically or horizontally aligned with the wire, making sure they are always between 0° and 360°. This would not be needed in normal trigonometric functions, but in this formula, we are subtracting and multiplying angles by distances, so a negative or larger than 360° can really throw away our calculations (Figure 9.4).

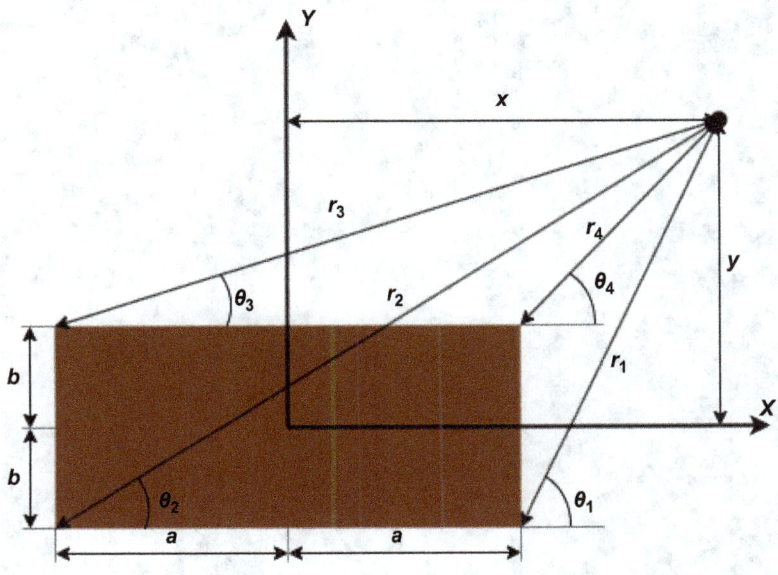

Figure 9.3 Geometrical diagram showing the coordinates system and the distances used in the calculation. Colored rectangle is a 2D section of the rectangular conductor

```
std::vector<double>
get_magnetic_field_strength_from_round_conductor_into_point(std::vector<d
ouble> inducingPoint, double inducingCurrent, std::vector<double>
inducedPoint) {
    double divisor = 2 * std::numbers::pi * (pow(inducedPoint[1] -
inducingPoint[1], 2) + pow(inducedPoint[0] - inducingPoint[0], 2));
    double Hx = -inducingCurrent * (inducingPoint[1] - inducedPoint[1]) /
divisor;
    double Hy = inducingCurrent * (inducingPoint[0] - inducedPoint[0]) /
divisor;

    return {Hx, Hy};
}

std::vector<double>
get_magnetic_field_strength_from_rectangular_conductor_into_point(std::ve
ctor<double> inducingPoint, double inducingCurrent, double inducingWidth,
double inducingHeight, std::vector<double> inducedPoint) {
    double a = inducingWidth / 2;
    double b = inducingHeight / 2;
    double x = inducedPoint[0] - inducingPoint[0];
    double y = inducedPoint[1] - inducingPoint[1];
    double r1 = hypot(y + b, x - a);
    double r2 = hypot(y + b, x + a);
    double r3 = hypot(y - b, x + a);
    double r4 = hypot(y - b, x - a);

    double tetha1 = atan((y + b) / (x - a));
    double tetha2 = atan((y + b) / (x + a));
    double tetha3 = atan((y - b) / (x + a));
    double tetha4 = atan((y - b) / (x - a));

    if (!(x > a && -b < y && y < b)) {
        if (x > a && y < -b) {
            tetha1 += 2 * std::numbers::pi;
        }
        else if (x < a || y < -b) {
            tetha1 += std::numbers::pi;
        }

        if (x > -a && y < -b) {
            tetha2 += 2 * std::numbers::pi;
        }
        else if (x < -a || y < -b) {
            tetha2 += std::numbers::pi;
        }

        if (x > -a && y < b) {
            tetha3 += 2 * std::numbers::pi;
        }
        else if (x < -a || y < b) {
            tetha3 += std::numbers::pi;
        }
```

```
        if (x > a && y < b) {
            tetha4 += 2 * std::numbers::pi;
        }
        else if (x < a || y < b) {
            tetha4 += std::numbers::pi;
        }
    }

    double common_part = inducingCurrent / (8.0 * std::numbers::pi * a *
b);
    double Hx = common_part * ((y + b) * (tetha1 - tetha2) - (y - b) *
(tetha4 - tetha3) + (x + a) * log(r2 / r3) - (x - a) * log(r1 / r4));
    double Hy = -common_part * ((x + a) * (tetha2 - tetha3) - (x - a) *
(tetha1 - tetha4) + (y + b) * log(r2 / r1) - (y - b) * log(r3 / r4));
    return {Hx, Hy};
}
```

Figure 9.4 Magnetic fields example of round and rectangular conductors

For the magnetic field produced by an air gap, we use Roshen's model, which provides some formulas to calculate the x and y components of the magnetic field in any point of the 2D space of the winding window using the geometry of the gap and the value of the magnetic field strength inside the gap, which can be obtained from the magnetic flux density through the magnetic core

$$H_x = -H_g \log\left(\frac{x^2 + (y - l_g)^2}{x^2 + (y + l_g)^2}\right), \quad x^2 + y^2 \geq l_g^2 \Rightarrow m = 0 \tag{9.15}$$

$$H_y = -\frac{H_g}{\pi}\left(\operatorname{atan}\left(\frac{2xl_g}{x^2 + y^2 - l_g^2}\right) + mx\right), \quad x^2 + y^2 < l_g^2 \Rightarrow m = 1 \tag{9.16}$$

where x and y are the distance of the induced point to the center of the gap where it is exposed to the winding window, l_g is half the total length of the gap, and H_g is the magnetic field strength inside the gap (Figure 9.5).

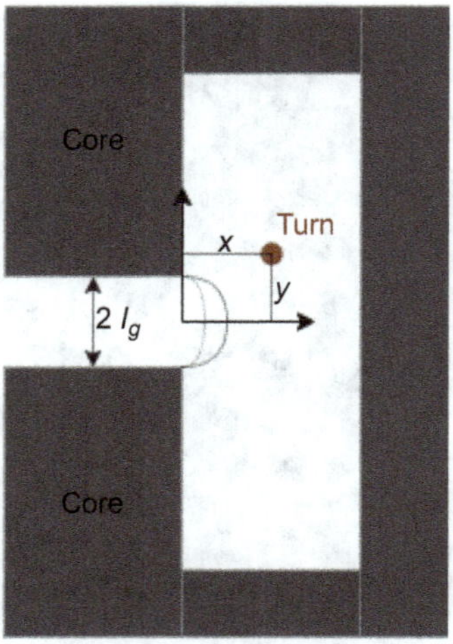

Figure 9.5 Geometrical diagram showing the coordinates system and the distances used in the calculation. Colored rectangle is a 2D section of the rectangular conductor

```
std::vector<double>
get_magnetic_field_strength_from_gap_into_point(std::vector<double>
gapCoordinates, std::vector<double> gapSectionDimensions, double
gapLength, double coreMagneticFluxDensityPeak, std::vector<double>
inducedPoint) {
    double vacuumPermeability = 1.25663706212e-6;
    double magneticFieldStrengthGap = coreMagneticFluxDensityPeak /
vacuumPermeability;
    double distanceFromCenterEdgeGapX;
    if (gapCoordinates[0] == 0) {
        distanceFromCenterEdgeGapX = inducedPoint[0] - (gapCoordinates[0]
+ gapSectionDimensions[0] / 2);
    }
    else {
        distanceFromCenterEdgeGapX = inducedPoint[0] - (gapCoordinates[0]
- gapSectionDimensions[0] / 2);
    }
    double distanceFromCenterEdgeGapY = inducedPoint[1] -
gapCoordinates[1];
    double halfGapLength = gapLength / 2;

    double magneticIntensityXDividend = pow(distanceFromCenterEdgeGapX,
2) + pow(distanceFromCenterEdgeGapY - halfGapLength, 2);
    double magneticIntensityXDivisor = pow(distanceFromCenterEdgeGapX, 2)
+ pow(distanceFromCenterEdgeGapY + halfGapLength, 2);
    double Hx = -magneticFieldStrengthGap / 2 / std::numbers::pi *
log(magneticIntensityXDividend / magneticIntensityXDivisor);
```

```
    double m;
    if (pow(distanceFromCenterEdgeGapX, 2) +
pow(distanceFromCenterEdgeGapY, 2) >= pow(halfGapLength, 2)) {
        m = 0;
    }
    else {
        m = 1;
    }

    double x = distanceFromCenterEdgeGapX * halfGapLength /
(pow(distanceFromCenterEdgeGapX, 2) + pow(distanceFromCenterEdgeGapY, 2)
- pow(halfGapLength, 2));
    double Hy = -magneticFieldStrengthGap / std::numbers::pi * (atan(x) +
m * std::numbers::pi);

    return {Hx, Hy};
}
```

9.4.1.2 Proximity effect losses

Why is this important for the proximity effect? Because this external field (H_e) is what will produce the proximity effect losses in our conductors. Several authors have proposed formulas for calculating the loss produced by a given H_e [16,21,23,26,27], though in this section, we will focus on Albach's [16] and Ferreira's [21].

As we did for the skin effect losses, we use functions to calculate the loss in one turn, which must be repeated for every turn in our coil.

For rectangular wires, including foil, we use Albach's model, which uses a solution to a simplification of the Maxwell equations that is solved using hyperbolic trigonometric functions with the dimensions of the wire

$$P_{proximity} = c\rho H_e^2 \cdot \mathrm{Re}\left\{\alpha d \cdot \tanh\left(\frac{\alpha d}{2}\right)\right\}, \quad \alpha = \frac{1+j}{\delta} \tag{9.17}$$

where d is the horizontal dimension of the wire, c is the vertical one, ρ is the resistivity of the copper, δ is the skin depth, and H_e is the external magnetic field in the center of the turn.

The equivalent code is

```
double vacuumPermeability = 1.25663706212e-6;
double copperResistivity = 0.00000001678;
double copperTemperatureCoefficient = 0.004041;

double calculate_proximity_losses_rectangular_wire(double
conductingWidth, double conductingHeight, double frequency,
std::vector<std::vector<double>> magneticFieldPoints, double temperature)
```

```
{
    double resistivity = copperResistivity * (1 +
copperTemperatureCoefficient * (temperature - 20));
    double skinDepth = calculate_skin_depth(frequency, temperature);

    double d = conductingWidth;
    double c = conductingHeight;

    std::complex<double> alpha(1, 1);
    alpha /= skinDepth;

    double He = 0;

    for (auto& datum : data) {
        He += hypot(datum[0], datum[1]);
    }
    He /= data.size();
    double turnLosses = c * resistivity * pow(He, 2) * (alpha * d *
tanh(alpha * d / 2.0)).real();
    return turnLosses;
}
```

For round and Litz wires, we use Ferreira's model, whose approach calculates a factor based on the geometry of the wire that converts the external magnetic field in turn to ohmic losses, doing an equivalence with Ohm's law

$$P_{proximity} = G \cdot H_{emax}^2 \cdot N$$
$$G = \frac{-2\pi\gamma}{\sigma} \frac{\left(ber_2(\gamma) * ber'(\gamma) + bei_2(\gamma) * bei'(\gamma)\right)}{\left(ber^2(\gamma) + bei^2(\gamma)\right)}, \quad \gamma = \frac{d}{\delta\sqrt{2}} \tag{9.18}$$

where σ is the conductivity of copper, $H_{e\ max}$ is the maximum magnetic field strength the turn is withstanding, N is the number of conductors or strands in a wire, and δ is the skin depth.

Ferreira's solution to Maxwell's equation for round wires makes use of the real and imaginary Kelvin functions and their derivatives, which are also implemented for completeness:

```
double vacuumPermeability = 1.25663706212e-6;
double copperResistivity = 0.00000001678;
double copperTemperatureCoefficient = 0.004041;

std::complex<double> bessel_first_kind(double order, std::complex<double>
z) {
    std::complex<double> sum = 0;
    std::complex<double> inc = 0;
    std::complex<double> aux = 0.25 * pow(z, 2);
    size_t limitK = 1000;
    for (size_t k = 0; k < limitK; ++k)
```

```
    {
        double divider = tgammaf(k + 1) * tgammaf(order + k + 1);
        if (std::isinf(divider)) {
            break;
        }
        inc = pow(-1, k) * pow(aux, k) / divider;
        sum += inc;
        if (std::abs(inc) < std::abs(sum) * 0.0001){
            break;
        }
    }
    auto bessel = sum * pow(0.5 * z, order);
    return bessel;
}

double kelvin_function_real(double order, double x) {
    return bessel_first_kind(order, x * exp(3.0 / 4 * std::numbers::pi *
std::complex<double>{0.0, 1.0})).real();
}

double kelvin_function_imaginary(double order, double x) {
    return bessel_first_kind(order, x * exp(3.0 / 4 * std::numbers::pi *
std::complex<double>{0.0, 1.0})).imag();
}

double derivative_kelvin_function_real(double order, double x) {
    return (kelvin_function_real(order + 1.0, x) +

kelvin_function_imaginary(order + 1.0, x)) / sqrt(2) + order / x *
kelvin_function_real(order, x);
}

double derivative_kelvin_function_imaginary(double order, double x) {
    return (kelvin_function_imaginary(order + 1.0, x) -
kelvin_function_real(order + 1.0, x)) / sqrt(2) + order / x *
kelvin_function_imaginary(order, x);
}

double calculate_proximity_factor(double conductingDiameter, double
frequency, double temperature) {
    double factor;
    double resistivity = copperResistivity * (1 +
copperTemperatureCoefficient * (temperature - 20));
    double skinDepth = calculate_skin_depth(frequency, temperature);

    double gamma = conductingDiameter / (skinDepth * sqrt(2));
    factor = - 2 * gamma * resistivity * (kelvin_function_real(2, gamma)
* derivative_kelvin_function_real(0, gamma) +
kelvin_function_imaginary(2, gamma) *
derivative_kelvin_function_imaginary(0, gamma)) /
(pow(kelvin_function_real(0, gamma), 2) +
pow(kelvin_function_imaginary(0, gamma), 2));
```

```
    return factor;
}

double calculate_proximity_losses_round_wire(double conductingDiameter,
double frequency, std::vector<ComplexFieldPoint> data, double
temperature, double numberStrands=1) {
    double proximityFactor =
calculate_proximity_factor(conductingDiameter, frequency, temperature);

    double He = 0;

    for (auto& datum : data) {
        He = std::max(He, hypot(datum[0], datum[1]));
    }

    double turnLosses = proximityFactor * pow(He, 2);
    turnLosses *= numberStrands;

    return turnLosses;
}
```

9.5 Winding losses for non-sinusoidal waveforms

We discussed calculating skin and proximity effect losses (or losses due to internal and external fields), but for sinusoidal current waveforms.

For non-sinusoidal waveforms, we can still use these models, but we need to decompose the waveform in its discrete Fourier transform, calculate the losses of each harmonic with their corresponding sinusoidal amplitude, and add them up.

References

[1] https://github.com/OpenMagnetics/MAS

[2] X. Zhang, F. Xiao, R. Wang, X. Fan and H. Wang, "Improved Calculation Method for Inductance Value of the Air-Gap Inductor," *2020 IEEE 1st China International Youth Conference on Electrical Engineering (CIYCEE)*, Wuhan, China, 2020, pp. 1–6, doi: 10.1109/CIYCEE49808.2020.9332553.

[3] J. Muhlethaler, J. W. Kolar and A. Ecklebe, "A Novel Approach for 3D Air Gap Reluctance Calculations," *8th International Conference on Power Electronics – ECCE Asia*, Jeju, Korea (South), 2011, pp. 446–452, doi: 10.1109/ICPE.2011.5944575.

[4] A. Balakrishnan, W. T. Joines and T. G. Wilson, "Air-Gap Reluctance and Inductance Calculations for Magnetic Circuits Using a Schwarz-Christoffel Transformation," in *IEEE Transactions on Power Electronics*, vol. 12, no. 4, pp. 654–663, July 1997, doi: 10.1109/63.602560.

[5] E. Stenglein and M. Albach, "The Reluctance of Large Air Gaps in Ferrite Cores," *2016 18th European Conference on Power Electronics and Applications (EPE'16 ECCE Europe)*, Karlsruhe, Germany, 2016, pp. 1–8, doi: 10.1109/EPE.2016.7695271.

[6] Colonel William T. McLyman, Transformer and Inductor Design Handbook, 3rd edn. CRC Press, 2004. ISBN: 978-0824753931

[7] C. P. Steinmetz, "On the Law of Hysteresis," in *Proceedings of the IEEE*, vol. 72, no. 2, pp. 197–221, 1984, doi: 10.1109/PROC.1984.12842.

[8] K. Venkatachalam, C. R. Sullivan, T. Abdallah and H. Tacca, "Accurate Prediction of Ferrite Core Loss with Nonsinusoidal Waveforms Using Only Steinmetz Parameters," *2002 IEEE Workshop on Computers in Power Electronics, 2002. Proceedings.*, Mayaguez, PR, USA, 2002, pp. 36–41, doi: 10.1109/CIPE.2002.1196712.

[9] S. Barg and K. Bertilsson, "Core Loss Calculation of Symmetric Trapezoidal Magnetic Flux Density Waveform," in *IEEE Open Journal of Power Electronics*, vol. 2, pp. 627–635, 2021, doi: 10.1109/OJPEL.2021.3133929.

[10] M. Albach, T. Durbaum and A. Brockmeyer, "Calculating Core Losses in Transformers for Arbitrary Magnetizing Currents a Comparison of Different Approaches," PESC Record. 27th Annual IEEE Power Electronics Specialists Conference, n.d., https://doi.org/10.1109/pesc.1996.548774

[11] A. Van den Bossche, V. C. Valchev and G. B. Georgiev, "Measurement and Loss Model of Ferrites with Non-Sinusoidal Waveforms," *2004 IEEE 35th Annual Power Electronics Specialists Conference (IEEE Cat. No. 04CH37551)*, Aachen, Germany, 2004, pp. 4814–4818, Vol. 6, doi: 10.1109/PESC.2004.1354851.

[12] J. Reinert, A. Brockmeyer and R. W. A. A. De Doncker, "Calculation of Losses in Ferro- and Ferrimagnetic Materials Based on the Modified Steinmetz Equation," in *IEEE Transactions on Industry Applications*, vol. 37, no. 4, pp. 1055–1061, 2001, doi: 10.1109/28.936396.

[13] T. Jieli Li Abdallah and C. R. Sullivan, "Improved Calculation of Core Loss with Nonsinusoidal Waveforms," *Conference Record of the 2001 IEEE Industry Applications Conference. 36th IAS Annual Meeting (Cat. No. 01CH37248)*, Chicago, IL, USA, 2001, pp. 2203–2210, vol. 4, doi: 10.1109/IAS.2001.955931.

[14] https://www.mag-inc.com/Design/Design-Guides/Curve-Fit-Equation-Tool/Thank-You/MagneticsCurveFitEquation.aspx

[15] https://s3.amazonaws.com/micrometals-production/filer_public/62/c0/62c0d9d3-96f3-41c6-830e-35d5a8aab07e/mmcurvefitcoefficientsall-sept21.xlsx

[16] Manfred Albach, Induktivitäten in der Leistungselektronik: Spulen, Trafos und ihre parasitären Eigenschaften. Springer Vieweg Wiesbaden, 2017. ISBN: 978-3-658-15080-8.

[17] R. P. Wojda and M. K. Kazimierczuk, "Winding Resistance and Power Loss of Inductors with Litz and Solid-Round Wires," in *IEEE Transactions on Industry Applications*, vol. 54, no. 4, pp. 3548–3557, 2018, doi: 10.1109/TIA.2018.2821647.

[18] Alan Payne, The AC Resistance of Rectangular Conductors, 2021.
[19] N. H. Kutkut, "A Simple Technique to Evaluate Winding Losses Including Two-Dimensional Edge Effects," in *IEEE Transactions on Power Electronics*, vol. 13, no. 5, pp. 950–958, Sept. 1998, doi: 10.1109/63.712319.
[20] A. W. Lofti and F. C. Lee, "Two-Dimensional Skin Effect in Power Foils for High-Frequency Applications," in *IEEE Transactions on Magnetics*, vol. 31, no. 2, pp. 1003–1006, March 1995, doi: 10.1109/20.364775.
[21] J. A. Ferreira and J. D. van Wyk, "A New Approach to Analyze Conduction Losses in High Frequency Magnetic Components," in *Transactions of the South African Institute of Electrical Engineers*, vol. 80, no. 1, pp. 61–67, Sept. 1989.
[22] Kenneth John Binns and P. J. Lawrenson, Analysis and Computation of Electric and Magnetic Field Problems, 2nd edn. Elsevier Ltd, 1973. ISBN: 978-0-08-016638-4.
[23] Jiří Lammeraner, Eddy Currents. CRC Press, 1966. https://archive.org/details/eddycurrents0000lamm
[24] W. A. Roshen, "Fringing Field Formulas and Winding Loss due to an Air Gap," in *IEEE Transactions on Magnetics*, vol. 43, no. 8, pp. 3387–3394, 2007, doi: 10.1109/TMAG.2007.898908.
[25] W. A. Roshen, "High-Frequency Fringing Fields Loss in Thick Rectangular and Round Wire Windings," in *IEEE Transactions on Magnetics*, vol. 44, no. 10, pp. 2396–2401, 2008, doi: 10.1109/TMAG.2008.2002302.
[26] H. Rossmanith, M. Doebroenti, M. Albach and D. Exner, "Measurement and Characterization of High Frequency Losses in Nonideal Litz Wires," in *IEEE Transactions on Power Electronics*, vol. 26, no. 11, pp. 3386–3394, 2011, doi: 10.1109/TPEL.2011.2143729.
[27] X. Wang, L. Wang, L. Mao and Y. Zhang, "Improved Analytical Calculation of High Frequency Winding Losses in Planar Inductors," *2018 IEEE Energy Conversion Congress and Exposition (ECCE)*, Portland, OR, USA, 2018, pp. 4336–4340, doi: 10.1109/ECCE.2018.8558397.

Part III

Concepts on magnetic flux coupling

Chapter 10

Use of permanent magnets in power electronics applications

Jens Friebe[1] *and Lennart Hoffmann*[1]

This chapter focuses on how the extension of the usable flux density range of soft magnetic materials for magnetic devices due to a premagnetization with a permanent magnet (PM) can be realized. It contains a short introduction to the principle of passive premagnetization as a distinction to current-based premagnetization. Therefore, the different aims of these two kinds of premagnetizations are derived by an inductor design for a converter stage. In the following sections, analytical calculation methods are presented for a combined magnetic circuit with a PM inserted into the air gap of a standard core geometry. Furthermore, Section 10.3 describes two different concepts for the realization of passive premagnetization, addressing the limiting factors of each concept such as stress on the hard magnetic material and losses in the PM. Additionally, the protection and application-specific choice of hard magnetic materials are discussed based on standardized parameters. Finally, a summary of the different concepts of the insertion of PMs in combined magnetic circuits will mark the benefits of passive premagnetization for the magnetic utilization of inductors and transformers.

10.1 Introduction to the benefits of passive premagnetization

Premagnetization of magnetic circuits with the aid of an additional winding is used, for example, in transducers for controlling AC loads [1]. The areas of application here are, for example, amplifier circuits or the stabilization of currents for certain applications [2]. Premagnetization can be achieved by a separate winding and the injection of a current into it. The example of a power electronic converter stage can be used to illustrate the disadvantage of current-based premagnetization. Considering V_a and V_b as voltages of a transformer with the windings a and b having the number of turns N_a and N_b with ideal coupling ($\phi_a = \phi_b$), as it is shown

[1]Department of Power Electronics, University of Kassel, Germany

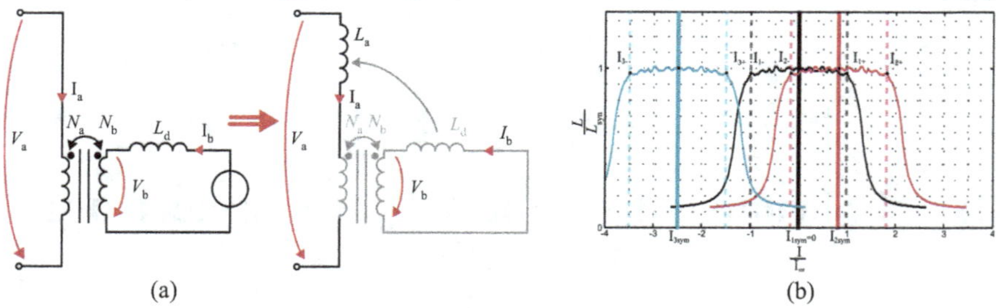

Figure 10.1 Electrical equivalent circuit for a current-based premagnetization considering an ideal transformer with a damping inductance in series with the output stage (a). Definitions of the symmetry current I_{sym} as well as the positive and negative saturation current using the example of a non-biased inductor (black), a biased inductor with a symmetry current close to the saturation current of the original inductance (red) and a very highly premagnetized inductance with a symmetry current greater than the saturation current of the original inductance blue. For illustrative purposes, the original measurement was shifted to show both cases with applied bias current.

in Figure 10.1(a), the following applies for the ac voltages:

$$V_b = \frac{N_a}{N_b} \cdot V_a \tag{10.1}$$

as well as for the impedances

$$\frac{Z_a}{Z_b} = \frac{N_a^2}{N_b^2}. \tag{10.2}$$

To obtain a desired constant premagnetization by the current through the second winding N_b, this current has to be constant; thus, the current should contain no ac components. By considering the output stage of the power electronic converter to be low impedance for switching frequency components f_s, this results in a short circuit in a small signal equivalent circuit. With the addition of a damping inductance in series to the output stage, it applies that

$$I_{b,fs} = \frac{N_b}{N_a} \frac{V_{a,fs}}{\omega \cdot L_d} \ll \frac{N_a}{N_b} \cdot I_{a,fs}. \tag{10.3}$$

This, however, contradicts the requirement that

$$\frac{V_{a,fs}}{V_{b,fs}} = \frac{I_{b,fs}}{I_{a,fs}}. \tag{10.4}$$

Taking the considerations for a low impedance output stage into account, the damping inductance on the secondary side L_d can be transformed to the primary side of the ideal transformer, resulting in a virtual inductance L_a with a measurable inductance of

$$L_a = \frac{N_a^2}{N_b^2} \cdot L_d. \tag{10.5}$$

Therefore, the design constraint for the damping inductance yields in

$$L_d = \frac{N_b^2}{N_a^2} \cdot L_a \tag{10.6}$$

whereas the current on the secondary side can be written as

$$I_b = \frac{N_a}{N_b} \cdot I_{a,dc} + \frac{N_a}{N_b} \cdot I_{a,fs}. \tag{10.7}$$

To find an estimation for the costs and box volume of the inductor, the indicator $PF_{=} = LI_{rms}I_{max}$ is introduced, with I_{rms} representing the thermal limits of the device and I_{max} the influence on the core in the form of the magnetic saturation flux density. With the simplification of $I_{rms}I_{max} = (I_{b,dc} + I_{b,fs})^2$, the indicator P_F of the inductance L_d is calculated as follows [3].

$$\begin{aligned} PF &= L_a \frac{N_b^2}{N_a^2} \cdot \left(\frac{N_a}{N_b} (I_{a,dc} + I_{a,fs}) \right)^2 \\ &= L_a (I_{a,dc} + I_{a,fs})^2 \end{aligned} \tag{10.8}$$

Under the neglections of an ideal transformer, the term in (10.8) equals the value for a non-premagnetized inductance. A real transformer does not transmit dc current, so at least the dc current must be taken into account according to the maximum saturation flux density of the transformer's core material. Since the magnetizing inductance L_m occurs in the electrical equivalent circuit in parallel to the inductance L_d, L_m must either be chosen to be very large compared to the inductance L_d, or L_d must be designed so that the measurable inductance L_a reaches the value of the desired inductance. This makes it clear that the current-based premagnetization is not applicable for volume reduction, but for targeted alteration of the magnetic circuit, for example, to reduce the inductance value. It has to be added that if the premagnetization is carried out with current, the current source must have a high impedance over a wide bandwidth.

In contrast to the control target for transducers, the purpose of permanent magnetic premagnetization is to shift the characteristic curve of an inductive component over the entire operating range, as shown in Figure 10.1(b). At this point, the symmetrical current I_{sym} should be introduced as a relative quantity as an indicator of premagnetization. In Figure 10.1(b), the required designations are shown using the example of three shifted LI-curves. Without premagnetization, the LI curve is symmetrical with respect to the ordinate. Thus, the symmetric current is $I_{1,sym} = 0$ with an identical value for the saturation currents $|I_{1+}| = |I_{1-}|$.

Introducing a permanent magnet (PM) into the main flux path of the magnetic device, the premagnetization causes a symmetrical shift by the value of the symmetrical current I_{sym} of the saturation currents I_+ and I_- and it applies that, for example, $I_{2+} = I_{2,sym} + I_{1+}$ and likewise $I_{2-} = I_{2,sym} + I_{1-}$. By using the symmetry current, the displacement caused by a bias magnetization can be designated by a single characteristic parameter. In principle, the effectiveness of a shift through various measures or through the use of different hard magnetic materials can be expressed. As the previous section examined the different aims of the current-based premagnetization and the permanent magnetic premagnetization, the focus of the following section lies on different methods to integrate the hard magnetic material into the main flux of the magnetic device. Therefore, Section 10.2 provides an analytical approach for a combined magnetic circuit with a PM integrated in the air gap of a standard core.

10.2 Analytical description of a combined magnetic circuit

For the simplified magnetic circuit shown in Figure 10.2, a relationship between the magnetic quantities can be determined. It is assumed here that the geometric dimensions, as well as the basic design parameters such as the number of turns, selection of the winding material and the soft magnetic material, are determined according to a standard inductance design but with double saturation flux density. The magnetic circuit also takes the reluctance of the stray field into account without addressing the saturation of the soft magnetic material. The depicted constellation in Figure 10.2 is referred to as a serial concept and is further analyzed in Section 10.3. Standard mathematical calculation methods for network analysis can be applied to derive the parameters that are relevant to the choice of the PM. Using Kirchhoff's second law applied to the magnetic circuit, it applies

$$N \cdot I = R_{m,fe} \cdot \phi_{m,fe} + R_{m,pm} \cdot \phi_{m,pm} \tag{10.9}$$

$$R_{m,pm} \cdot \phi_{m,pm} = R_\sigma \cdot \phi_\sigma. \tag{10.10}$$

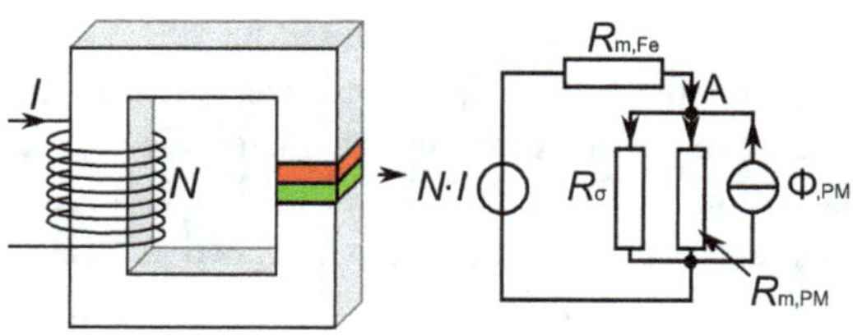

Figure 10.2 Combined magnetic circuit considering the stray field reluctance and a PM of constant polarization without addressing saturation behaviour of the soft magnetic material.

By applying Kirchhoff's first law to node A, the flux ϕ_{pm} yields in

$$\phi_{pm} = \phi_{\sigma} + \phi_{m,pm} - \phi_{m,fe}. \tag{10.11}$$

Taking the definition of the symmetric current I_{sym} into account, considering $I = I_{sym}$ and $\phi_{fe} = 0$, the flux density of the PM can be calculated as follows:

$$N \cdot \frac{I_{sat-} + I_{sat+}}{2} = R_{m,pm} \cdot \phi_{m,pm} = R_{\sigma} \cdot \phi_{\sigma} \tag{10.12}$$

$$\phi_{pm} = B_{pm} \cdot A_{pm} = N \cdot \frac{I_{sat-} + I_{sat+}}{2} \cdot \left(\frac{1}{R_{m,pm}} + \frac{1}{R_{\sigma}} \right) \tag{10.13}$$

$$B_{pm} = N \cdot \frac{I_{sym}}{A_{pm}} \cdot \underbrace{\left(\frac{1}{R_{m,pm}} + \frac{1}{R_{\sigma}} \right)}_{\Lambda_L} \tag{10.14}$$

Under the simplified assumption that $R_{m,fe} \ll R_{m,pm} || R_{\sigma}$, the same term results as for a model without consideration of the stray reluctance:

$$B_{pm} = \frac{L \cdot I_{sym}}{N \cdot A_{pm}}. \tag{10.15}$$

However, without neglecting the stray reluctance, the remanence flux density of the hard magnetic material has to be increased to obtain the same inductance for an identical geometry, as it can be derived by the inverse of the quotient of the necessary flux density without stray field to the flux density with stray field:

$$p_{s,Br} = \frac{R_{\sigma} + R_{m,pm}}{R_{\sigma}}. \tag{10.16}$$

To define requirements for the intrinsic coercive field strength of the PM, the maximal occurring magneto-motive force (MMF), $\theta_{max,pm}$, is calculated at maximal current I_{max} first without consideration of the stray field as follows:

$$\begin{aligned} \theta_{max,pm} &= R_{m,pm} \cdot \phi_{m,pm} = N \cdot I_{max} - R_{m,fe} \cdot \left(\frac{\theta_{max,pm}}{R_{m,pm}} - \phi_{pm} \right) \\ &= \frac{N \cdot I_{max} + R_{m,fe} \cdot \phi_{pm}}{1 + \dfrac{R_{m,fe}}{R_{m,pm}}} \end{aligned} \tag{10.17}$$

By inserting (10.13) without the stray reluctance in (10.17), the maximal MMF $\theta_{max,pm}$ results in

$$\theta_{max,pm} = \frac{N \cdot I_{max} + \frac{R_{m,fe}}{R_{m,pm}} \cdot N \cdot \frac{I_{sat-} + I_{sat+}}{2}}{1 + \frac{R_{m,fe}}{R_{m,pm}}}. \tag{10.18}$$

Considering a design with maximal unipolar excitation ($I_{sat-} = 0$ and $I_{sat+} = I_{max}$), the maximal MMF $\theta_{max,pm}$ can be written as

$$\theta_{max,\,pm} = N \cdot I_{max} \cdot \left(1 - \frac{1}{2} \cdot \frac{R_{m,fe}}{R_{m,fe} + R_{m,pm}}\right) \tag{10.19}$$

Utilizing the double flux density range, taking the condition $\theta_{max,pm} = H_{max} \cdot l_{pm}$ into account, the maximal field strength $H_{max,pm}$ results in

$$H_{max,pm} = \frac{2 \cdot B_{sat}}{\mu_0 \mu_{r,pm}}. \tag{10.20}$$

This stress places the highest demands on the intrinsic coercive field strength of the permanent magnetic material that can occur when an inductance is excited up to the saturation current of the soft magnetic material. Taking into account irreversible magnetization changes from approx. 80% of the intrinsic coercive field strength [4]:

$$H_{max,pm} = \frac{1}{0.8} \frac{2 \cdot B_{sat}}{\mu_0 \mu_{re,pm}} = 2.5 \cdot H_{cB}. \tag{10.21}$$

The field strength that a permanent magnetic material must possess is an intrinsic field strength in order to neglect all other protective effects and insert the PM directly in the main flux path as it applies to a serial concept. To extend the analysis for the stray field, the maximal MMF $\theta_{max,pm}$ is calculated as follows:

$$\theta_{max,pm} = \frac{N \cdot I_{max} + R_{m,fe} \cdot N \cdot I_{sym}\left(\frac{1}{R_{m,pm}} + \frac{1}{R_\sigma}\right)}{1 + R_{m,fe} \cdot \left(\frac{1}{R_{m,fe}} + \frac{1}{R_\sigma}\right)}. \tag{10.22}$$

Considering the same operational point ($I_{sat-} = 0$ and $I_{sat+} = I_{max}$), as for the calculations without stray field, the maximal MMF $\theta_{max,pm}$ results in

$$\theta_{max,pm} = N \cdot I_{max} \cdot \left(1 - \frac{1}{2} \frac{R_{m,fe}}{(R_\sigma || R_{m,pm}) + R_{m,fe}}\right). \tag{10.23}$$

It can be seen that the stray field leads to a reduction in the maximum MMF or field strength load. As a quotient of the MMF with consideration of the stray field to the MMF without consideration of the stray field results in

$$p_{s,\theta} = \frac{R_{m,fe} + R_{m,pm}}{R_{m,fe} + 2 \cdot R_{m,pm}} \cdot \frac{R_{m,fe} + 2 \cdot (R_\sigma || R_{m,pm})}{R_{m,fe} + (R_\sigma || R_{m,pm})}. \tag{10.24}$$

However, this only applies to a saturated magnetic material, as otherwise, the assumption that $R_{m,fe} \gg (R_\sigma || R_{m,pm})$ leads to the factor $p_{s,\theta}$ being equal to 1. With identical inductance and taking the stray field into account, the result is

$$p_{s,Hmax,L} = \frac{\Lambda_L - \Lambda_\sigma}{\Lambda_L} \tag{10.25}$$

as a factor for the maximal field strength stress. As this section provides the analytical basics to derive design determinating parameters for a PM being integrated into a combined magnetic circuit, the analytical approach can be adapted for any combined magnetic circuit, and therefore, different concepts of premagnetization are shown in the next section.

10.3 Concepts of passive premagnetization

Soft magnetic components are mainly characterized by their saturation flux density as well as their specific losses, described by the Steinmetz parameters k, α and β. Regarding the similar behaviour of these parameters of soft magnetic materials, ferrite materials with low saturation flux density tend to have lower specific losses. Therefore, materials with high saturation flux density are limited by their thermal behaviour due to the core-specific losses of the component. Likewise, materials with low specific losses come with the drawback of a decreased saturation flux density. To address this issue, it is possible to reduce the specific losses of materials with high saturation flux density by, e.g. sheeting iron cores. However, passive premagnetization aims to increase the saturation flux density of materials with low specific losses with regard to their application. The next section introduces two concepts of passive premagnetization as they are widely known in the literature. The presented analytical approach can be adapted for each of the concepts, as the derived terms of the previous section are universally applicable.

10.3.1 Concept of serial premagnetization

The insertion of the hard magnetic material directly into the main path of the magnetic circuit is referred to as a serial concept. Therefore, the hard magnetic material completely or partially fills an existing or specially designed air gap. Figure 10.3(a) shows a PQ core with two PMs inserted into the air gap of the outer legs. As depicted, the PM is used to insert a negative flux with respect to the main flux directly into the main path of the core. For unipolar excitation, the flux ϕ_{PM} of the PM will always be opposite to the main flux ϕ_{H}, shifting the effective flux density of the component to higher values [5]. This concept can effectively provide a reasonable amount of bias flux to the main path. However, the arrangement imposes several design constraints [6]:

1. The maximum bias flux is limited, as the magnetic cross section A_{PM} cannot be larger than the cross section of the core A_{core}.
2. There is a constant risk of demagnetization of the PM by the flux ϕ_{H} of the coil due to high currents, e.g. during inrush events [7].
3. Eddy currents will occur in the PM, which will lead to significant losses at high frequencies [6]. Therefore, the selection of polymer bonded hard magnetic material becomes crucial [8].

Figure 10.4(a) depicts simulation results of the fluxes ϕ_{PM}, ϕ_{fe} and $\phi_{\sigma,PM}$ for different MMFs. It can be seen that when the hard magnetic material and the

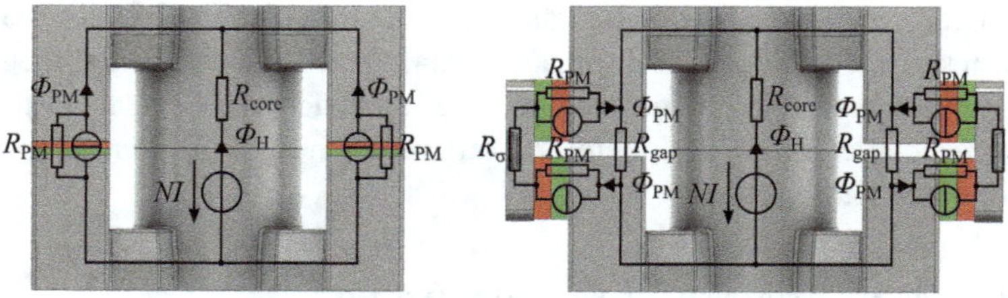

(a) (b)

Figure 10.3 Depiction of a concept of serial premagnetization with the PM being inserted directly into the main flux path (a) and the concept of parallel premagnetization with the PM bypassing the main flux path (b). The concepts are applied to a PQ core, showing the magnetic equivalent circuit [4].

winding are excited in the same direction, the soft magnetic material is saturated. However, as soon as the MMF is sufficiently positive and in opposite direction to the premagnetization, the soft magnetic material desaturates and leads to an increase in the flux through it and the PM. Simultaneously, due to the increase of the MMF, the flux through the leakage reluctance starts to rise. The resulting flux in the PM ϕ_{PM} can be calculated as follows:

$$\phi_{PM} = \phi_{fe} - \phi_{\sigma,PM}. \tag{10.26}$$

Since the resulting inductance is proportional to the derivative of the flux with regards to the exciting MMF, a constant inductance results for an MMF, shifting the LI curve of the inductor linearly towards higher saturation currents. As the serial concept appears to be the most simple approach to evoke a constant premagnetization by inserting the PM directly into the main flux path, several drawbacks come with this concept, as described. To overcome these limitations, the next section introduces the concept of parallel premagnetization.

10.3.2 Concept of parallel premagnetization

In contrast to the serial concept, the hard magnetic material is arranged parallel to the main path of the magnetic circuit, as depicted in Figure 10.3(b). The leakage reluctance now describes the main magnetic path parallel to the hard magnet. This arrangement has the disadvantage of a larger box volume but the advantage of a free air gap. For example, thermally conductive material or sensors for measuring the magnetic flux density can be inserted. The hard magnetic material takes up the space where the stray field lines primarily occur. This implies an optimization regarding the protection of the hard magnetic material, as it can be derived by (10.27), as the leakage reluctance is reduced and the reluctance of the hard magnetic material is increased with this parallel arrangement

$$p_\sigma = \frac{\Lambda_\sigma}{\Lambda_\sigma + \Lambda_{m,pm}}. \tag{10.27}$$

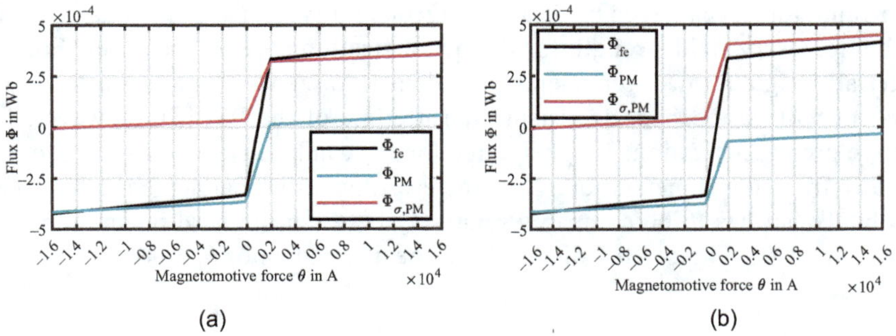

Figure 10.4 Magnetic fluxes in the soft magnetic material ϕ_{fe}, in the hard magnetic material ϕ_{PM} and in the stray field on the hard magnetic material $\phi_{\sigma,PM}$ using the example of the serial concept (a) and parallel concept (b).

In other words, this arrangement reduces the stress on the PM for the same geometry of the soft magnetic material. Furthermore, the size of the magnets and, according to that, the maximal bias flux is no longer limited by the cross-sectional area of the core [5,6]. Therefore, eddy currents can be reduced significantly in comparison to a serial concept. However, due to the derivation of the PM reluctance by changing the PM operating point, eddy currents are still existing in the PM but much smaller than directly generated eddy currents [5,6]. Using the same simulation methods as for the serial concept, Figure 10.4(b) shows the fluxes ϕ_{PM}, ϕ_{fe} and $\phi_{\sigma,PM}$ for different MMFs. As depicted, in contrast to the serial concept, the flux ϕ_{PM} does not become negative with respect to the PM, decreasing the maximal occurring MMF of the PM and reducing the stress of the hard magnetic material. As the previous sections describe in detail how to derive requirements on the hard magnetic material for different concepts of premagnetization, the next section provides an indicator-based selection guide for the hard magnetic material.

10.4 Selection of hard magnetic material

As described in Section 10.2, the premagnetization defines special requirements for the material of the PM. Typically, specifications like (BH_{max}) and H_{cJ} are indicated in the material description. If the material is polymer bonded, usually a p is placed after the label. For example, the material description x/yp indicates a polymer bonded hard magnetic material with $(BH_{max}) = x$ kJm^{-3} and $H_{cJ} = y \cdot 10$ kAm^{-1}. Since the definition of the maximal energy product (BH_{max})

$$(BH_{max}) = \frac{B^2}{4\mu_r\mu_0} \tag{10.28}$$

shows no dependence on the coercitive field strength H_{cJ}, selecting $H_{cJ} > 1.25 \cdot H_{cB}$ has no benefit for most other applications. Actually, choosing a high value for H_{cJ} can even be disadvantageous for typical applications of a PM, as the necessary field strength for the magnetization process increases. For the material design, a lower remanence flux density B_r as well as a lower viscosity constant will be reached, decreasing the lifetime of the PM [3]. However, for passive premagnetization, a higher value for H_{cJ} can be advantageous to prevent the PM from demagnetization. Nonetheless, this value cannot be determined absolutely but has to be compared relatively between the hard magnetic materials. Therefore, Figure 10.5 displays the specific indicator h_{rel}

$$h_{rel} = \frac{H_{cJ}}{H_{cB}} \tag{10.29}$$

for different hard magnetic materials. A clear differentiation between the different hard magnetic materials can be seen in Figure 10.5, whereas polymer-bonded materials show a reduction in B_r but not in h_{rel}. This indicates a theoretical

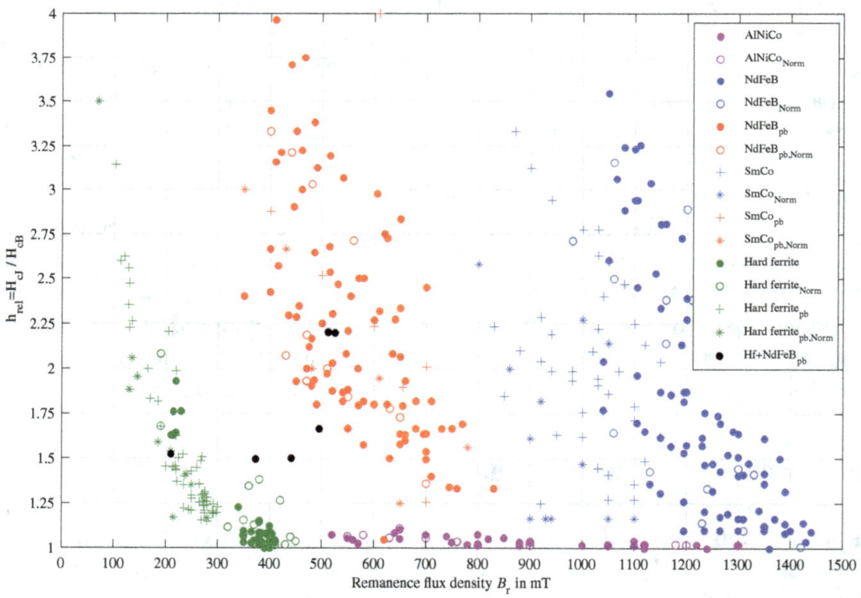

Figure 10.5 *h_{rel} of 367 commercially available hard magnetic materials and the respective specified material parameters are plotted over the remanence flux density. Rare-earth magnets (blue), plastic-bonded rare-earth magnets (red), AlNiCo magnets (violet) and hard ferrite, with and without plastic bonding (green), each forming their own areas. In addition, a hybrid material consisting of hard ferrite and NdFeB is shown (black) [9].*

possibility of different combinations of H_{cJ} and B_r. Due to the high electrical conductivity of non-polymer-bonded rare-earth materials as well as AlNiCo, these materials will be neglected for further investigations. With regard to (10.21), it can be recognized that there are materials that can be used even without protection by stray fields or saturation of the soft magnetic material if the maximum saturation current in the application is not exceeded, or only slightly exceeded, depending on h_{rel}. Those materials are indicated by a factor of $h_{rel} > 2.5$. According to Figure 10.5, even materials with a remanence flux density of above 650 mT fulfill these requirements. Decreasing the requirements on h_{rel} as described in previous sections, the selection of materials can be extended to hard magnetic materials with higher remanence flux density. This applies to combinations of soft and hard magnetic ferrites as well as soft magnetic materials with higher saturation flux density and polymer-bonded rare-earth materials.

10.5 Summary

There are various applications and topologies with magnetic devices suffering from poor magnetic utilization due to a unipolar excitation of the device [5]. If, for example, the current through the inductor only flows in one direction, for a non-premagnetized inductor, only 50% of the actual flux density range will be utilized. Therefore, shifting the LI curve of the inductor seems reasonable in order to adapt the usable flux density range to the operating point of the magnetic component. A practical approach to extend the usable flux density of soft magnetic materials is the injection of an additional constant flux in parallel to the main flux path. This can either be realized by an additional current-carrying winding, having the drawback of a damping inductance necessary to ensure a constant premagnetization. Thus, due to the feedback on the additional winding, no reduction in box volume can be expected for a current-based premagnetization. However, inserting a PM into the magnetic circuit offers the opportunity to constantly shift the LI curve of the inductor. Thus, a reduction in overall core volume can be achieved considering different concepts of passive premagnetization. Anyhow, the insertion of hard magnetic material in the flux path of the magnetic device places certain requirements on the PM. The stress on the PM relies on factors such as placement within the flux path, losses in the hard magnetic material due to different loss mechanisms as well as on the behaviour of the soft magnetic material. Therefore, it becomes inevitable to define design criteria for the PM based on factors such as coercitive field strength and necessary flux density. A comprehensive selection guide for the appropriate hard magnetic material based on specific indicators is presented in this chapter. This provides a generally applicable approach for a combined magnetic circuit, which can be adapted for various concepts of passive premagnetization as the two most widely known concepts are shown in this chapter.

References

[1] Mali P. Magnetic Amplifiers Principles and Application. J. F. Rider; 1963.

[2] Lenk A, Ballas RG, Werthschützky R, *et al. Magnetic Transducers*. Berlin: Springer; 2011. p. 247–311.

[3] Friebe J, and Zacharias P. Review of Magnetic Material Degradation Characteristics for the Design of Premagnetized Inductors. *IEEE Transactions on Magnetics*. 2014;50(3):40–48.

[4] Haavisto M, Kankaanpää H, and Paju M. Estimation of Time-Dependent Polarization Losses in Sintered Nd–Fe–B Permanent Magnets. *IEEE Transactions on Magnetics*. 2011;47(1):170–174.

[5] Brinker T, Gräber H, and Friebe J. GaN HEMT and SiC Diode Commutation Cell based Dual-Buck Single-Phase Inverter with Premagnetized Inductors and Negative Gate Driver Turn-off Voltage. In: *2022 24th European Conference on Power Electronics and Applications (EPE'22 ECCE Europe)*; 2022. p. 1–8.

[6] Aguilar AR, and Munk-Nielsen S. Design, Analysis and Simulation of Magnetic Biased Inductors with Saturation-Gap. In: *2014 16th European Conference on Power Electronics and Applications*; 2014. p. 1–11.

[7] Lin S, Friebe J, Langfermann S, *et al.* Premagnetization of High-Power Low-Frequency DC-Inductors in Power Electronic Applications. In: *PCIM Europe 2019; International Exhibition and Conference for Power Electronics, Intelligent Motion, Renewable Energy and Energy Management*; 2019. p. 1–7.

[8] Friebe J, Lin S, Fauth L, *et al.* Premagnetized Inductors in Single Phase DC–AC and AC–DC Converters. In: *2019 IEEE 15th Brazilian Power Electronics Conference and 5th IEEE Southern Power Electronics Conference (COBEP/SPEC)*; 2019. p. 1–6.

[9] Friebe J. Permanentmagnetische Vormagnetisierung von Speicherdrosseln in Stromrichtern [PhD thesis]. University of Kassel. Kassel; 2014.

Chapter 11

Orthogonal magnetization in power electronics applications

Camilo Suarez Buitrago[1] *and Wilmar Martinez*[1]

Orthogonal premagnetization in power electronics marks a significant advancement in magnetic components, particularly in the application of inductors and transformers. This approach deviates from traditional parallel premagnetization methods, offering a novel means of manipulating magnetic fields at orthogonal angles to enhance control over magnetic properties, which is crucial in developing controllable magnetics like inductors and transformers for power converters. Modelling and implementation studies of orthogonal premagnetized magnetic components indicate that dc bias fields can alter the hysteresis loop, influencing permeability and hysteresis losses. Additionally, the impact of an orthogonal premagnetization field on various core shapes is an open question. This chapter explores various configurations of controllable magnetics using orthogonal premagnetization, their trade-offs with conventional components, and their modelling and implementation implications. The validation of these concepts through finite-element simulations and experimental tests concludes the chapter, highlighting the practical applicability and transformative potential of orthogonal premagnetization in power electronics.

11.1 Introduction

Magnetic elements such as inductors and transformers are essential in electric systems, serving a wide range of applications from audio systems to power distribution networks, and are essential components of nearly all modern electronic devices. Traditionally, these components operate as passive elements with fixed characteristics such as voltage gain, turn ratio, and intrinsic inductance, exemplified by a standard 50/60 Hz distribution transformer. Historically, adjustability in such components has been mechanical, such as through the use of taps, and has seen significant use in domains including communication systems for tuning resonant circuits, aircraft control systems, motor control, and lighting dimming.

[1]KU Leuven, Department of Electrical Engineering (ESAT), ELECTA-Diepenbeek – Energy Ville, Belgium

However, conventionally, magnetic cores are utilized within their unsaturated or linear region, where the magnetization curve maintains a constant slope, reflecting a steady permeability and core reluctance. This stability is crucial for the majority of applications. Yet, the avoidance of operating beyond this linear region due to saturation effects limits the adaptability of these components. The knee or transition zone of the ***B–H*** curve, which lies between the linear and saturation regions, presents an opportunity as it exhibits a permeability that varies with the magnetic flux density B, resulting in a fluctuating reluctance. This variability is key to achieving controllability in magnetic components through electromagnetic methods, expanding their functionality beyond traditional fixed settings.

A simplified representation of a magnetic component is presented in Figure 11.1. The component has a coil made of conductive wire with N turns wound around a magnetic core characterized by a magnetic permeability (μ), an effective path length (l_e), and a cross-sectional area (A_e). Assuming a material with a constant permeability, the inductance (L) of the component is given by (11.1), and its core reluctance ($\Re$) by (11.2). In this context, regarding controllability, the modification of any of these parameters can be used as a tuning element of the magnetic component

$$L = \frac{N^2}{\Re} \tag{11.1}$$

$$\Re = \frac{l_e}{A_e \mu} \tag{11.2}$$

Moreover, when a dc magnetic flux (B_{dc}) is superposed to the main flux of such magnetic component, the B–H operating point in the curve of the core material can be shifted, as shown in Figure 11.2. Such a phenomenon is possible by adding

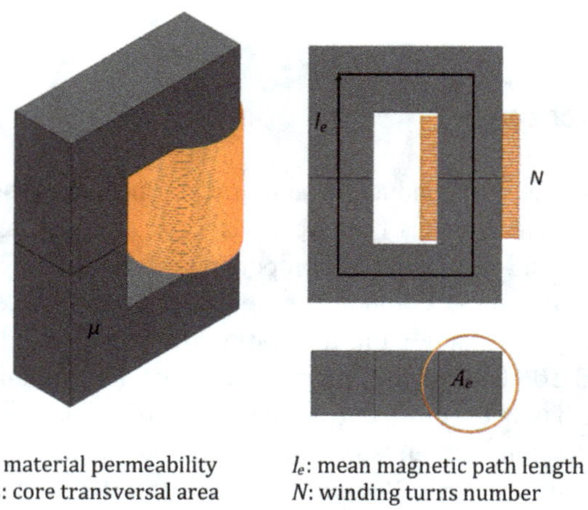

Figure 11.1 Simplified representation of a magnetic component with a ferrite magnetic core

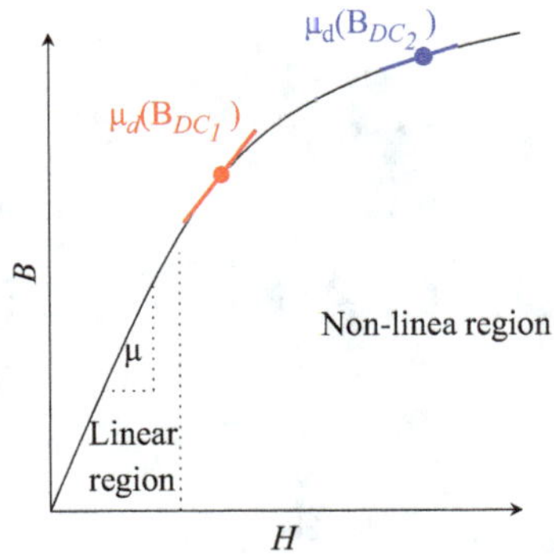

Figure 11.2 B–H curve of a ferromagnetic material

an auxiliary winding to the core connected to a constant current source (I_c). This addition allows for reluctance variations in the magnetic component, and, therefore, inductance values can be controlled without changing the number of turns of the main winding. In other words, working on the knee/transition region of the *B–H* curve provides a varying permeability dependent on the magnetic flux density *B*, and thus, a varying reluctance [1]. Consequently, controllability is possible in the magnetic component.

As is known, the small-signal ac reluctance of the magnetic core (γ) can be modelled using the concept of differential permeability, μ_d, as seen in (11.3) [2]. This approach enables the component to be analysed using the superposition principle and the equivalent reluctance circuit by linearizing the permeability around the control dc flux value

$$\gamma = \frac{l_e}{A_e \mu_d(B_{DC})} \tag{11.3}$$

The biasing method can be categorized as parallel, orthogonal, or radial based on the orientation of the generated dc magnetic flux in relation to the main magnetic flux, as depicted in [3] and Figure 11.3. The trade-offs among these active premagnetization methods are detailed in Table 11.1, and a thorough analysis can be found in [4].

Parallel premagnetization directly aligns the control and main fluxes, boosting magnetic coupling but also increasing non-linear current responses during large signal operations. Radial premagnetization concentrates flux interactions, providing accurate control, although more complex manufacturing due to core alterations is required. Finally, orthogonal premagnetization reduces direct coupling between fluxes, which decreases signal distortion, although it requires meticulous core geometry for successful implementation. The appropriateness of each method depends on particular needs regarding efficiency, signal integrity, and operational dynamics of the magnetic device.

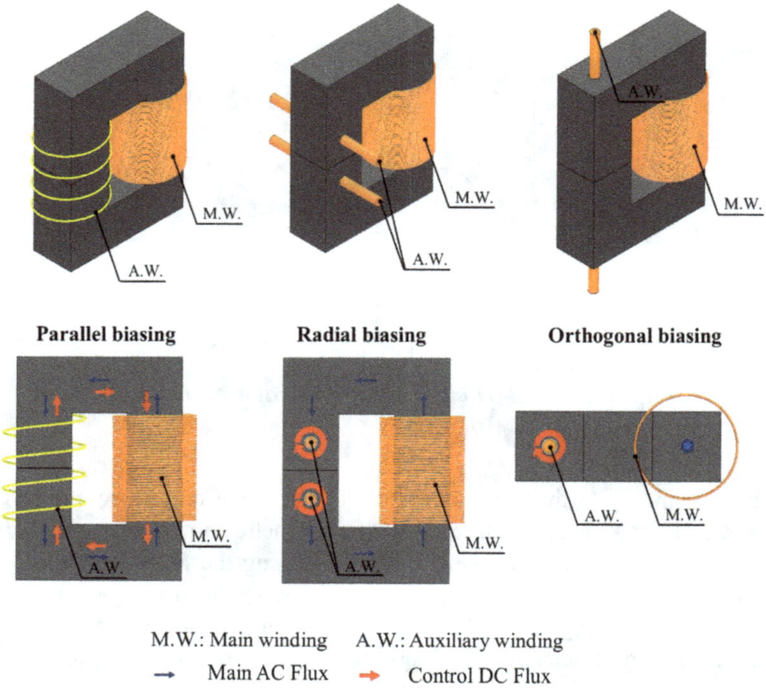

Figure 11.3 Implementation considering different active premagnetization methods Adapted from [3]

Although parallel premagnetization, employed in the traditional Magamps, is the most commonly used technique because of its maturity and easy implementability, radial and orthogonal variations offer unique characteristics. Moreover, radial/mixed premagnetization has a localized impact on the core. It concentrates the effect of the dc bias magnetic flux on a small region of the core, emulating the presence of an air gap. This characteristic has led to the technique being called a virtual air gap (VAG) in several publications.

Additionally, orthogonal premagnetization (also referred to as orthogonal biasing) shows a broad linear range in the material's magnetization curve across various control current levels, as presented in [4] and shown in the hysteresis curves of Figure 11.4. This characteristic leads to reduced distortion of the magnetic device's main current, even when experiencing substantial fluctuations in the magnetic flux, in comparison to other premagnetization techniques. However, it is important to note that under large-signal operation, the interactions between the control and main windings tend to be more pronounced, increasing the need for filtering on the control side or the use of high impedance control current sources.

Despite the comparatively higher coupling between the control and the main magnetic flux in parallel biasing, as well as the highly non-linear effects on the main current when operated under large-signal conditions, parallel biasing is still

Table 11.1 Trade-offs of the premagnetization methods

Method	Advantage	Disadvantage
Parallel	• Comparatively, the highest changes in the core's reluctance can be achieved. • Cost-effective, no complex manufacturing techniques are required.	• High coupling is present between the auxiliary/control winding and the main winding fluxes. • Highly non-linear current in the main winding under large-signal operation.
Radial	• The interaction between the control and main fluxes is concentrated in well-defined core sections.	• The hole-drilling process in ferrites can be difficult. • The hole distribution plays an important role in the device behaviour, being complex to analyse its impact on the main flux. • The holes reduce the effective area of the core, reducing the maximum main flux amplitude allowed.
Orthogonal	• Simple theoretical analysis • Small variation of the core permeability under large signal operation. Reduced distortion of the main current.	• Comparatively lower changes in the core's reluctance are achieved • Large impact on the material property variations due to the presence of air gaps in the core • Complex implementation, the core geometry plays a major role in obtaining an effective orthogonal flux interaction. (Non-uniform flux density distribution leads to a poor utilization of the core.)

the most commonly used technique in recent developments. Orthogonal biasing, with documented publications dating back to the 1970s [5] and earlier discussions provided in [6], along with radial biasing, introduced more recently in the late 1990s [7], could potentially address these issues, leading to a more compact and linear behaviour.

This chapter provides a detailed exploration of the orthogonal premagnetization method. It starts with a brief introduction to the operating principle. Then, in section 11.3, both simulation-based and analytical approaches to magnetic behaviour are discussed, followed by the Implementation section, which addresses practical core constraints and modelling challenges. The chapter also briefly reviews Applications of orthogonal biasing and concludes with a summary of key findings and future research directions. This structured approach offers a comprehensive understanding of theoretical concepts to practical implications and future possibilities.

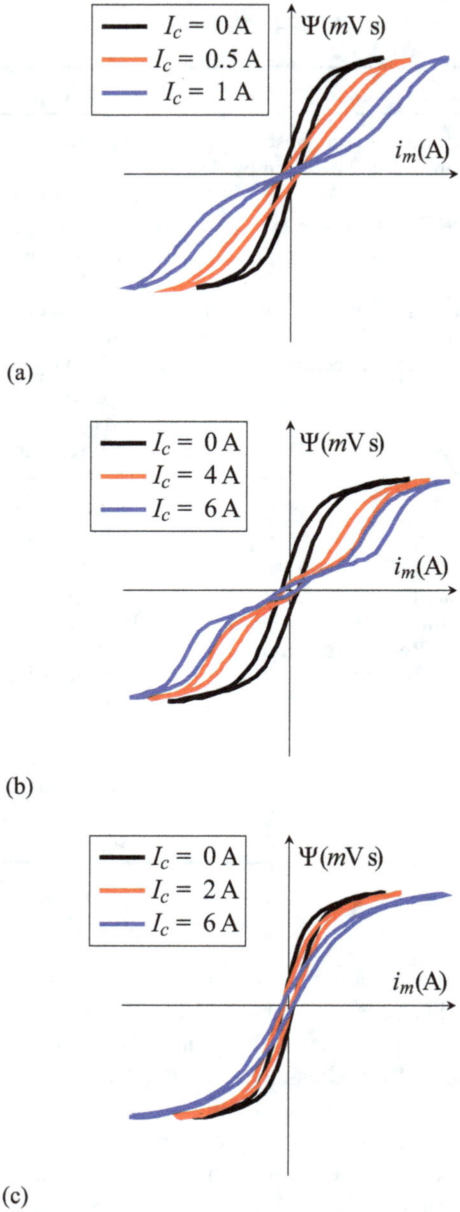

Figure 11.4 Characteristic hysteresis curves for different premagnetization methods (adapted from [4]). (a) Parallel: reduced linear region in the curves under the effect of high DC control currents. (b) Radial: highly nonlinear curve regions under the effect of DC control currents. (c) Orthogonal: wide linear regions on the curves under the effect of DC control currents.

11.2 Operating principle

Consider the hollow toroidal core shown in Figure 11.5, which consists of two windings. This magnetic device functions as a variable inductor through orthogonal flux interaction. It uses an auxiliary or control winding connected to a DC current source to achieve core saturation. This configuration allows for the adjustment of the inductance at the main winding.

For the purposes of analysis, the core is assumed to be made from non-oriented magnetic material with a uniform distribution of magnetic fluxes φ_c and φ_m, which are generated by currents I_c and i_m, respectively. Traditional linear circuit analysis would typically predict no coupling between the windings.

However, due to the non-linear relationship between flux and magnetomotive force in the magnetic core – as illustrated in Figure 11.6 –the flux generated by the control winding does interact with the flux linking the main winding. This phenomenon, known as parametric coupling, is explained further.

Assume the current through the main winding (i_m) is zero. Under this condition, the flux density in the core, denoted as B'_c, can be determined from the material's ***B–H*** curve by considering only the field intensity H_c associated solely with I_c, as shown in Figure 11.6 and represented by the square marker.

Now, consider I_m constant and equal to I_c. In this case, the net magnetic field intensity ***H*** must be calculated using (11.4), which assumes a purely orthogonal

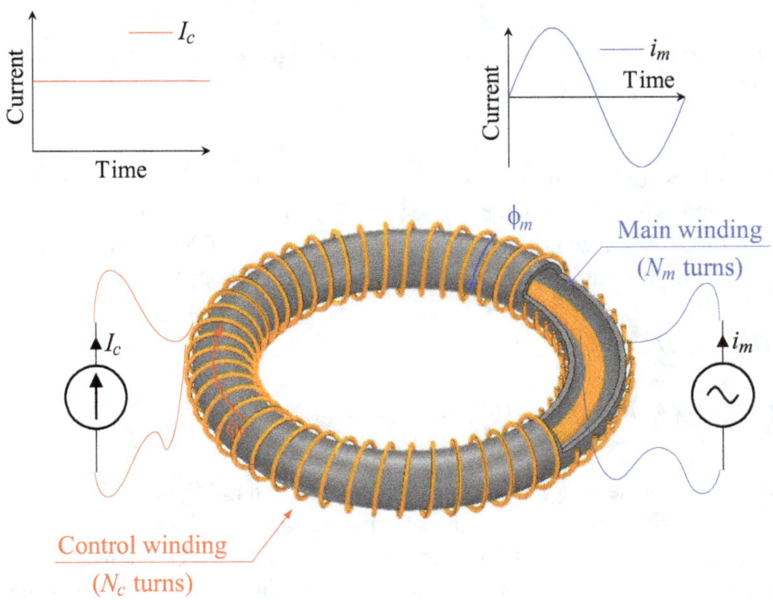

Figure 11.5 Hollow toroid with two windings and assumed current waveforms and magnetic flux directions

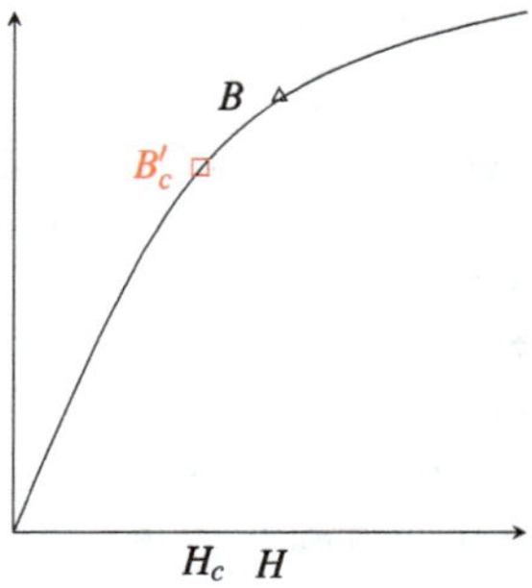

*Figure 11.6 Material **B–H** curve and derived flux density for i_m = 0, B′c (square marker) and net flux density for $i_m = i_c$, B (triangle marker)*

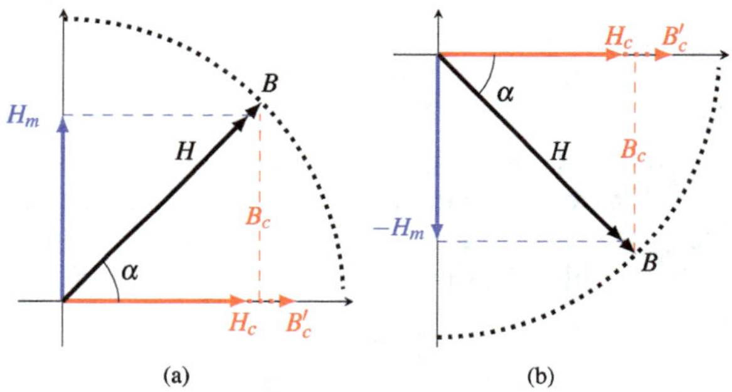

Figure 11.7 (a) and (b) Vector representation of the orthogonal flux interaction at the hollow

interaction of the fields along the core. This assumption is based on the specific core shape and winding arrangement. Similarly, to obtain the net magnetic flux density, ***B***, the ***B–H*** curve is again consulted, this time considering ***H***, as marked in Figure 11.6 with a triangle, and expressed mathematically in (11.5)

$$\|H\| = \sqrt{H_c^2 + H_m^2} \tag{11.4}$$

$$B(H_c, H_m) = B\left(\sqrt{H_c^2 + H_m^2}\right) \cdot \frac{H}{\sqrt{H_c^2 + H_m^2}} \tag{11.5}$$

The vector representation of this interaction is shown in Figure 11.7(a). It reveals that the flux density linked to the control winding, B_c, is lower than B'_c, as a consequence of the parametric coupling. Although this may seem counter-intuitive, it is important to note that when both magnetic flux intensities are combined, the net value falls into the non-linear region of the ***B–H*** curve. Here, the ratio between these quantities is lower compared to that in the linear region. As shown in Figure 11.7(b), this effect on the control flux density is independent of the polarity of H_m.

Following the same reasoning, if H_m exhibits a sinusoidal variation, it is possible to infer that, in this case, B_c would present a reduction of its mean value and fluctuations at double the frequency of the main magnetic flux intensity, as depicted in Figure 11.8.

The effects of the parametric coupling on the main flux density, B_m, can be derived following a similar approach. As a consequence of the orthogonal interaction between the fields, B_m is smaller than B'_m, which is the expected flux density value when considering the control current as zero. This is illustrated in Figure 11.9(a), where an additional point has been plotted in the ***B–H*** curve, considering the main flux intensity at half of its amplitude. As seen, the resultant magnetization curve, shown in blue, can be associated with a reduction of the material permeability in the direction of H_m. This can be analysed as a drop in the permeability in the direction of the main flux, μ_m, consequently leading to a drop in the inductance measured at the main winding, referred to as L_m. Further reductions can be achieved by increasing the control current I_c, as shown in Figure 11.9(b). In this way, the inductance can be controlled as a function of the DC current flowing through the auxiliary winding.

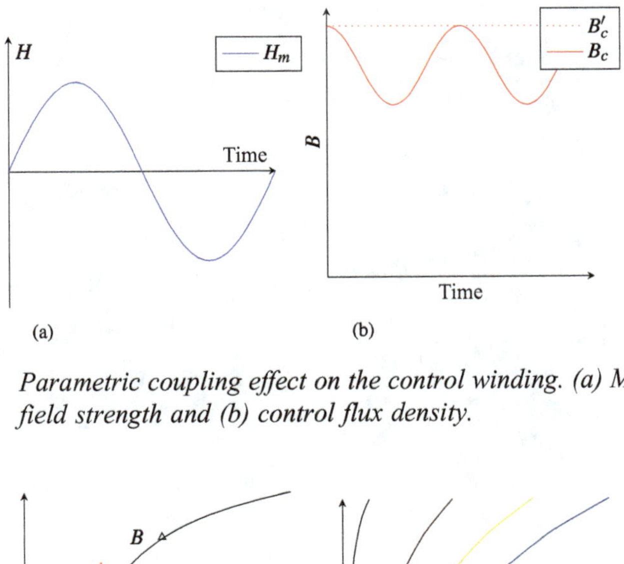

Figure 11.8 Parametric coupling effect on the control winding. (a) Main winding field strength and (b) control flux density.

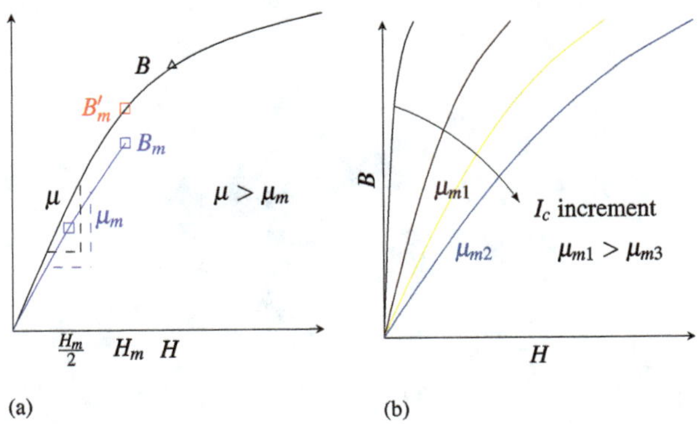

Figure 11.9 (a) and (b) Parametric coupling effect on the main winding

11.3 Modelling

Optimal conditions for purely orthogonal interactions between the main and control fields are restricted by the core shape, along with additional assumptions listed by the authors in [8], referring to the absence of losses in the magnetic material or stray fields, among others. When implementing alternative core designs, such as the one depicted in Figure 11.10(a), finite-element analysis often becomes

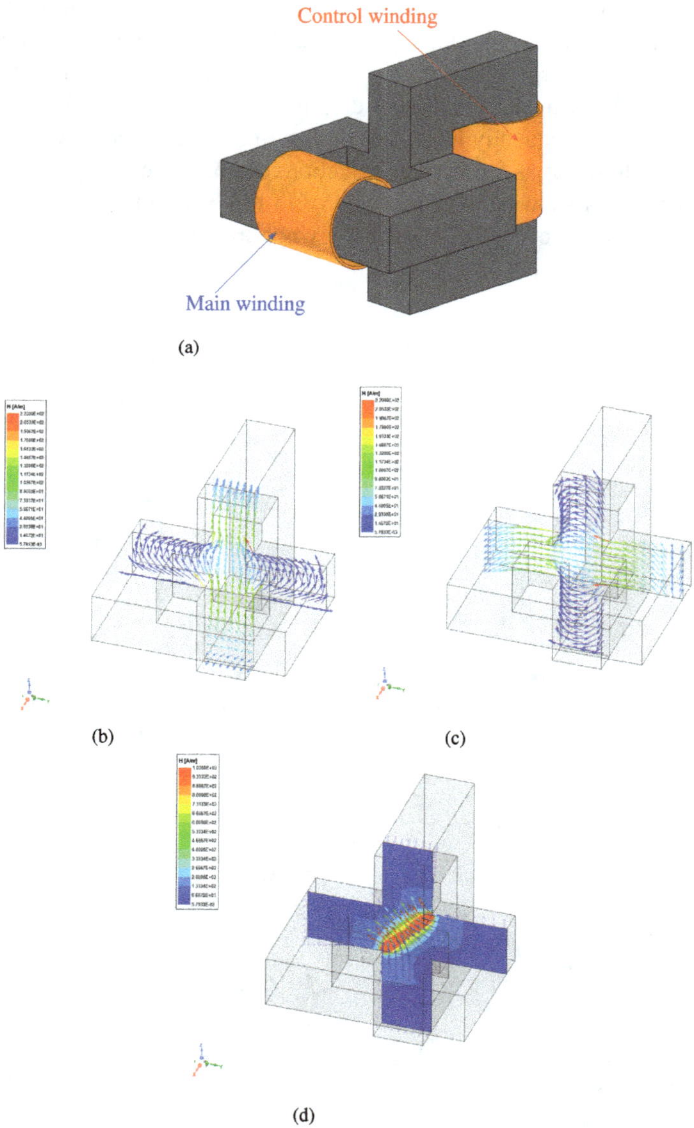

Figure 11.10 Orthogonal premagnetization, C, core implementation: finite-element simulations. (a) Orthogonal premagnetization, C core implementation, (b) ***H****, ($Ic \neq 0, i_m = 0$), (c)* ***H****, ($i_m \neq 0, I_c = 0$), and (d)* ***H****, ($i_m = I_c$).*

necessary. In practice, the region in the space of interaction extends beyond the intersection of the cores, and the absence of a uniform flux distribution leads to non-orthogonal interactions. Figure 11.10(b) and (c) illustrates the magnetic flux strength at the core intersection plane caused by the independent effects of I_c and i_m, respectively. As observed, the direction of both fields changes along the magnetic flux paths, creating regions where the interaction between the control and mains fields can be even parallel. These non-orthogonal interactions make it difficult to obtain analytical expressions that accurately model the behaviour of the coupling. Furthermore, since only a portion of the core is premagnetized, the core is not fully utilized, thereby limiting the variability of the inductance at the main winding [8]. Analysing the net field strength in this core arrangement, shown in Figure 11.10(d), it resembles more the behaviour of a VAG.

The modelling of the magnetic material is crucial, regardless of whether the approach is simulation-based or analytical. Accounting for factors such as hysteresis, eddy current losses, and temperature variations can significantly improve the precision of these models. However, this enhancement in accuracy demands extensive material characterization and detailed simulations, which are inherently time-consuming.

In other words, employing analytical models to describe the magnetization curve of the material simplifies the system modelling process and speeds up the analysis. In particular, (11.6), referred to as Brauer's model, provides an efficient method for determining the ***B–H*** curve of the material. This approach requires finding the material-dependant constants k_{1-3}, which are derived as detailed in [9]

$$H = (k_1 e^{k_2 B^2} + k_3)B \tag{11.6}$$

In order to illustrate how Brauer's model can be integrated into the modelling of magnetic devices with parametric coupling, the hollow core introduced in the previous section is employed to describe the process of obtaining the net flux density and field strength of the system. To simplify the procedure, it is assumed that the main winding is connected to a voltage source while the control winding is connected to a constant current source (Figure 11.11). Therefore, the magnitudes

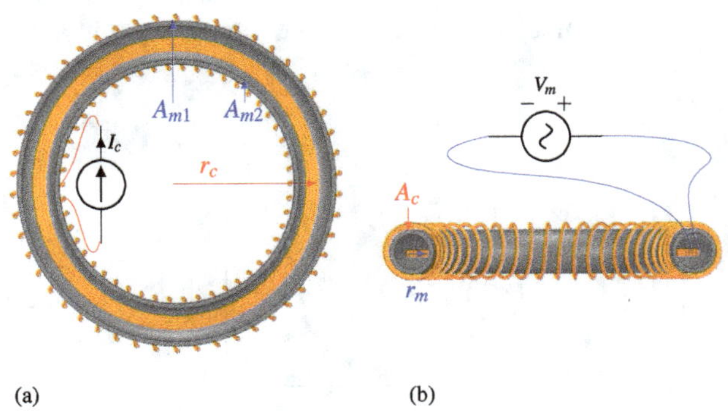

Figure 11.11 Hollow toroid dimensions and alternative current/voltage source connections. (a) Top view and (b) front view.

B_m and H_c are known and given by (11.9) and (11.10)

$$\|B_m\| = \frac{V_m}{N_m A_m} \int \sin(\omega t) dt \tag{11.7}$$

where

$$A_m = \frac{A_{m1} + A_{m2}}{2} \tag{11.8}$$

$$\|H_c\| = \frac{N_c I_c}{2\pi r_c} \tag{11.9}$$

By definition, the magnitude net flux density ***B*** can be obtained from (11.10). Where the angle formed by the components of the filed strength α can be obtained from (11.11)

$$\|B_m\| = \sin\,\alpha \|B\| \tag{11.10}$$

Therefore, considering (11.4) and (11.10), Brauer's model can be rewritten as follows:

$$\cos\,\alpha = \frac{H_c}{\sqrt{H_c^2 + H_m^2}} \tag{11.11}$$

$$\sqrt{H_c^2 + H_m^2} = \left(k_1 e^{k_2 \left(\frac{B_m}{\sin\,\alpha}\right)^2} + k_3\right) \frac{B_m}{\sin\,\alpha} \tag{11.12}$$

Then, the equation system formed by (11.10) and (11.11) with two unknown variables, H_m and B_c, can be easily solved numerically.

Following the principle of the small signal reluctance, *gama*, the analysis can be further simplified by specifically modelling the core reluctance as a parameter dependent only on the DC control current I_c, resulting in the circuit representation shown in Figure 11.12. This approximation is valid as long as the amplitude of B_m remains within the linear region of the ***B–H*** curve of the material.

Then, the reluctance of the main flux path, $\Re_m$, can be obtained from (11.13), where μ_m is directly obtained from the set B_m–H_m curves having I_c as parameter (Figure 11.9(b)). This calculation needs to be computed only once, therefore

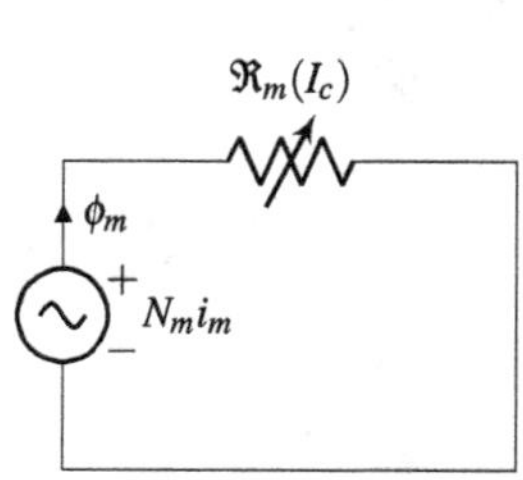

Figure 11.12 Inductor's small signal reluctance model

evaluating the impact of multiple sizes of an air gap present in the structure is simplified

$$\Re_m = \frac{2\pi r_m}{A_m \mu_m(I_c)} \tag{11.13}$$

When commercially available cores are used to implement magnetic devices featuring orthogonal premagnetization, which deviate from the ideal hollow toroid shape discussed previously, significant challenges arise. These challenges originate in part from irregularities in the transverse areas along the control flux path. Such irregularities lead to an uneven magnetic flux density, with some areas becoming highly saturated and others only partially so, thereby reducing the effectiveness of parametric coupling.

To comprehensively illustrate the effects of non-uniform transverse areas, FEM simulations are conducted to compare the ideal core shape with its nearest commercially available equivalent. It has been demonstrated that adding magnetic material in regions of higher saturation can modify the winding coupling behaviour, bringing it closer to the theoretically expected results.

Figure 11.13(a) illustrates a pot core designed to accommodate internal windings. This configuration closely resembles the ideal design for orthogonal magnetization, with the addition of two opposing holes to facilitate the insertion of

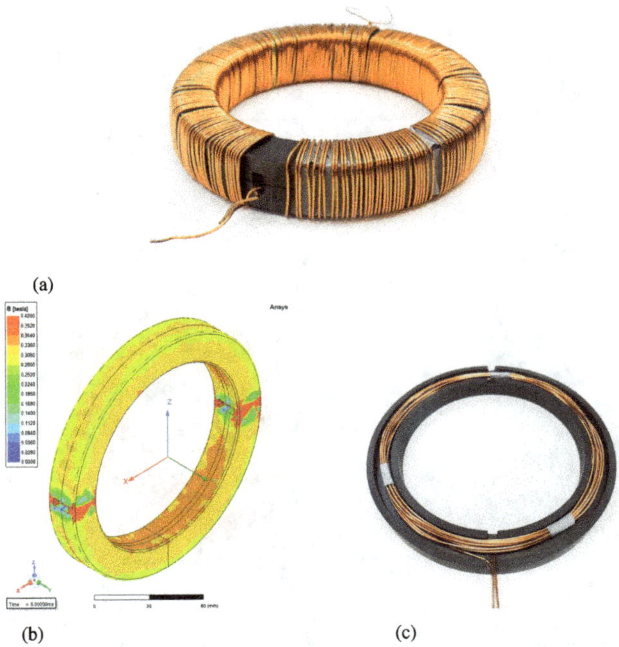

Figure 11.13 Variable inductor using a P148/112/10 core. (a) External control winding distribution and internal main winding terminals. (b) Control flux saturation profile. (c) Internal main winding. Adapted from [22]

internal windings. However, as depicted in Figure 11.13(b), this minor modification leads to localized saturation of the control flux due to the reduced cross-sectional area at these access points.

Note that a winding swap would lead to a similar situation, where the flux density in the inner section of the core would be larger than in the external part because of differences in the internal and external core radii and ultimately the transversal areas. However, regardless of the winding arrangement, it is shown that even small deviations from the ideal core shape significantly impact the achievement of an even control flux density.

Simulations of the core geometries under small-signal conditions ($H_m << H_c$) were conducted using FEA to analyse its characteristics. Figure 11.14(a) displays the normalized main inductance as a function of the control current, demonstrating its behaviour as a variable inductance. When compared to the same core structure without holes (ideal core shape), as shown in the same figure, it is clear that a lower

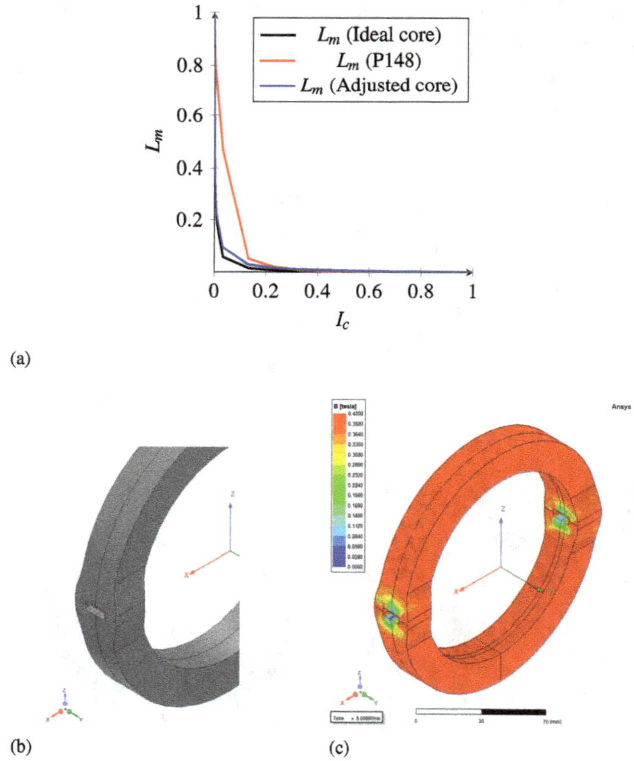

Figure 11.14 Proposed transversal core are modification on the P148 core – simulations and comparison with the standard core. (a) Simulations of the main inductance as function of the control current for the standard core, the core without holes and the proposed adjusted core. (b) Proposed core area adjustment. (c) Saturation profile of the adjusted core. Adapted from [22]

control current is needed to achieve the same inductance levels when a uniform cross-sectional area is ensured.

To further investigate this issue and confirm that it is caused by the non-uniform cross-sectional areas, an adjusted FEM model was created. This model includes additional ferrite material surrounding the holes, as shown in Figure 11.14(b) and (c). This addition of material compensates for the area reduction caused by the holes. As demonstrated in Figure 11.14(a), this adjustment successfully prevents local saturation, resulting in a significant improvement in the homogeneity of the magnetic field.

The experimental results, as shown in Figure 11.15, are consistent with the behaviour expected from the simulations. A uniform flux density along the control flux path results in larger variations in core reluctance compared to the standard core, as can be seen in Figure 11.15(b) and (c). Additionally, a wide linear region of the magnetization curve is preserved, as presented in Figure 11.15(d).

An important aspect of the implementation is the control current source, which requires a high impedance to prevent fluctuations in I_c due to the parametric coupling responsible for variations in the voltage at the control winding. The simplest approach is to use a larger air-core inductor in series with the control winding.

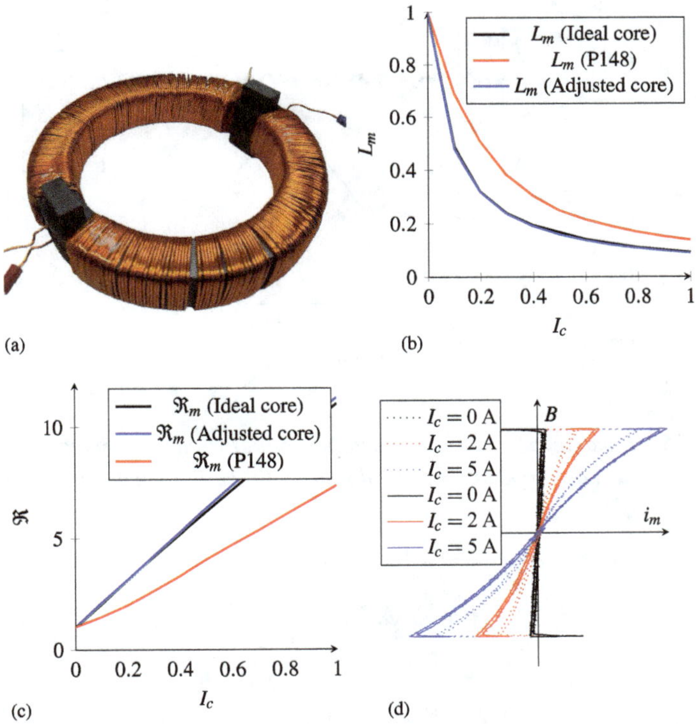

Figure 11.15 Experimental results P148 standard and adjusted cores core vs model. (a) Modified pot core by adding ferrite blocks at high saturation zones. (b) Small-signal characterization. (c) Normalized core reluctance variation. (d) Flux density curve as a function of the main current (dotted lines (P148 core), solid lines (adjusted core)). Adapted from [22]

Figure 11.16 illustrates the effects of parametric coupling on the control and main currents when a low-impedance source is connected to the control winding, compared to the behaviour when a current source is achieved by a series connection of an inductor. Although effective, this solution introduces an additional, expensive, and bulky element. Alternatively, the DC current source can be achieved using a bipolar BJT, as shown in Figure 11.17, or through DC–DC converters, referenced in [10,11].

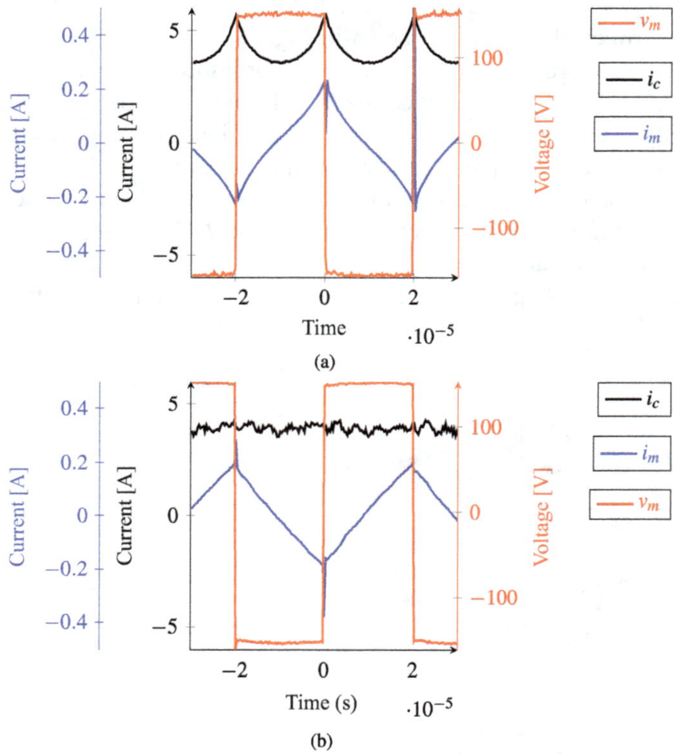

Figure 11.16 Parametric coupling: variable inductor current waveforms. (a) Low-impedance control source. (b) High-impedance control source.

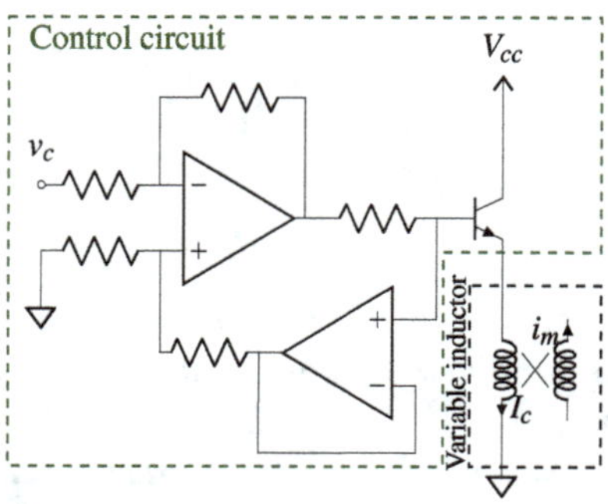

Figure 11.17 Control circuit, adapted from [12]

Nonetheless, the control circuit results in a reduction of the power density and increased complexity that have to be considered when integrating this type of magnetic device as part of a power converter.

11.4 Applications

The use of active premagnetization to achieve controllable magnetic devices in power electronics is mainly linked to variable inductors rather than transformers. This preference is driven by the growing importance of resonant and isolated converters in modern power electronic applications, such as electric vehicle charging. The ability to operate these converters at fixed or quasi-fixed switching frequencies and ensure efficient performance across a broad range of voltage and load conditions has renewed interest in the research of DC-biased magnetic components.

Over the past decade, variable inductors have been used in low-power applications (under 400 W) for switched-mode power supplies, particularly in the development of LED drivers. Key applications include their use in single-stage power factor correction (PFC) flyback converters and in unidirectional resonant switched capacitor step-up converters to enhance current regulation and response in OLED drivers.

VIs have added flexibility in circuits, enabling independent supply to multiple LED lamps. A complete list of references covering this topic can be found in [13].

Variable inductors have also found applications in higher power systems, such as in 2-kW and 3.3-kW converters used in inductive wireless power transfer (WPT). These systems utilize VIs to enhance efficiency at light loads and improve coupling in wireless EV charging structures. The use of VIs in these applications not only extends the zero-voltage-switching (ZVS) region but also adapts to a wider load range, benefiting both inductive and capacitive systems [14].

Recent advancements have explored novel control schemes for boosting PFC converters using VIs to regulate output voltage without traditional pulse-width modulation or frequency control, leading to more efficient systems with reduced voltage ripple [15,16]. Additionally, VIs are being used to maintain constant switching frequencies across varying input voltages and improve efficiency in isolated AC–DC converters by operating them at resonant frequencies [17,18].

Although parallel premagnetization is preferred in academic research mainly due to its simplicity in implementation, an orthogonal biasing method using variable inductors can also be utilized. However, it is important to note that this method typically results in comparatively lower variations in inductance.

Specific studies making use of orthogonal premagnetization include the implementation of a paralleled variable inductor with the high-frequency transformer of a dual active bridge (DAB) converter and a controllable transformer for the same topology, both aiming to reduce the switching losses of the converter under a wide range of load/voltage operating conditions [12,19]. A detailed description of the benefits of using orthogonal biasing in the weight reduction of non-isolated DC/DC converters is provided in [20]. Moreover, a comprehensive review of the different premagnetization methods and applications can be found in [4,21].

The ability to vary or control the inductance in inductors and transformers undoubtedly benefits the operation of modern power converters. However, to date, there is no clear quantification of the drop in efficiency and power density due to the use of active premagnetization and its required additional hardware.

11.5 Conclusions

While mathematical models for orthogonal flux magnetization are well-defined, they come with significant constraints, particularly concerning geometry, making practical implementation challenging. Therefore, it is imperative to develop modelling techniques that can be extended to non-ideal core geometries.

A thorough investigation into losses associated with orthogonal premagnetization in power converters is crucial. This includes analysing the control circuits required for operation and assessing the total power density of the solution. This information is essential for determining the application boundaries of this technology, which generally exhibits superior behaviour under large variations in the main flux compared to other premagnetization methods.

Although parallel premagnetization is more suitable for handling large variations in core reluctance and is less sensitive to the presence of air gaps, it suffers from limited use of the core as the main flux variation must remain small relative to the DC flux used for biasing. On the other hand, orthogonal premagnetization demonstrates almost linear behaviour under large flux variations, making it suitable for operations at high voltages, such as altering the parallel inductance of high-frequency transformers. Its application, for example, as a series inductance in DAB converters, may be affected by the presence of air gaps, which significantly impact core reluctance variation. However, a more in-depth analysis in this regard is warranted.

In general, numerous advantages of controlled magnetic devices in modern power electronics cannot be denied. In this context, the use of orthogonal premagnetization offers distinctive advantages. Nevertheless, further studies, especially concerning implementation and loss modelling, are needed.

References

[1] Perdigão MS, Menke M, Seidel AR, *et al.* A Review on Variable Inductors and Variable Transformers: Applications to Lighting Drivers. In: *2014 IEEE Industry Application Society Annual Meeting*; 2014. pp. 1–8.

[2] Alonso JM, Perdigão M, Dalla Costa MA, *et al.* Variable Inductor Modeling Revisited: The Analytical Approach. In: *2017 IEEE Energy Conversion Congress and Exposition (ECCE)*; 2017. pp. 895–902.

[3] Cobaleda DB, Tian F, Suarez C, *et al.* Linear Control Compensator for a Variable-Transformer in Wide-Voltage Power Converters. In: *2022 IEEE*

31st International Symposium on Industrial Electronics (ISIE); 2022. pp. 301–306.

[4] Pfeiffer J, Küster P, Schulz IEM, *et al.* Review of Flux Interaction of Differently Aligned Magnetic Fields in Inductors and Transformers. *IEEE Access*. 2021;9:2357–2381.

[5] Verma RP, Fam WZ. Theory and Performance of Parametric Transformers. *IEEE Transactions on Power Apparatus and Systems*. 1972;*PAS*-91 (6):2494–2504.

[6] Mumford WW. Some Notes on the History of Parametric Transducers. *Proceedings of the IRE*. 1960;48(5):848–853.

[7] Molcrette V, Kotny JL, Swan JP, *et al.* Reduction of Inrush Current in Single-Phase Transformer Using Virtual Air Gap Technique. *IEEE Transactions on Magnetics*. 1998;34(4):1192–1194.

[8] Meiksin ZH. Comparison of Orthogonal- and Parallel-Flux Variable Inductors. *IEEE Transactions on Industry Applications*. 1974;IA-10(3):417–423.

[9] Brauer J. Simple Equations for the Magnetization and Reluctivity Curves of Steel. *IEEE Transactions on Magnetics*. 1975;11(1):81–81.

[10] Choi S, Shin JW, Imaoka J, *et al.* Voltage-Controlled Variable Inductor for Fixed-Frequency Critical Conduction Mode Operation. *IEEE Transactions on Industrial Electronics*. 2023;70(6):5707–5716.

[11] Aganza-Torres A, Pfeiffer J, Zacharias P. Interleaved Buck Converter with Coupled Inductors as Auxiliary Bias-Current Source for Controllable Magnetic Devices. In: *PCIM Europe digital days 2020; International Exhibition and Conference for Power Electronics, Intelligent Motion, Renewable Energy and Energy Management*; 2020. pp. 1–8.

[12] Gao Y, Tang Y, Yao F, *et al.* Paralleled Variable Inductor Phase-Shifted Full-Bridge Converter with Full-Load Range ZVS and Low Duty Cycle Loss. *IEEE Transactions on Industrial Electronics*. 2024;71(5):4673–4684.

[13] Perdigão MS, Menke MF, Seidel AR, *et al.* A Review on Variable Inductors and Variable Transformers: Applications to Lighting Drivers. *IEEE Transactions on Industry Applications*. 2016;52(1):531–547.

[14] Buitrago CS, Cobaleda DB, Martinez W. Magnetically Controlled Transformer with Variable Turns Ratio and Low Series Inductance: Analysis and Implementation Toward Its Application in SMPS. *IEEE Transactions on Power Electronics*. 2023;38(11):14360–14374.

[15] Wei Y, Luo Q, Wang J, *et al.* Analysis and Design of the DCM Operation Boost PFC Converter with Magnetic Control. *IET Power Electronics*. 2019; 12(14):3697–3706.

[16] Yao K, Zhang Z, Yang J, *et al.* Quasi-Fixed Switching Frequency Control of CRM Boost PFC Converter Based on Variable Inductor in Wide Input Voltage Range. *IEEE Transactions on Power Electronics*. 2021;36(2):1814–1827.

[17] Zhang Z, Yao K, Ma C, *et al.* All-Fixed Switching Frequency Control of CRM Boost PFC Converter Based on Variable Inductor in a Wide Input Voltage Range. In: 2019 IEEE Energy Conversion Congress and Exposition (ECCE); 2019. pp. 1434–1441.

[18] Wei Y, Luo Q, Alonso JM, *et al.* A Magnetically Controlled Single-Stage AC–DC Converter. *IEEE Transactions on Power Electronics*. 2020;35 (9):8872–8877.

[19] Suarez C, Martinez W. Assessment of a DAB Converter with a Variable Transformer by Means of Orthogonal Premagnetization for Wide Voltage Operation and Enhanced Efficiency at Light-Load Conditions. In: *2023 IEEE 8th Southern Power Electronics Conference and 17th Brazilian Power Electronics Conference (SPEC/COBEP)*; 2023. pp. 1–7.

[20] Eichhorst D, Pfeiffer J, Zacharias P. Weight Reduction of DC/DC Converters Using Controllable Inductors. In: *PCIM Europe 2019; International Exhibition and Conference for Power Electronics, Intelligent Motion, Renewable Energy and Energy Management*; 2019. pp. 1–8.

[21] Zacharias P, Kleeb T, Fenske F, *et al.* Controlled Magnetic Devices in Power Electronic Applications. In: *2017 19th European Conference on Power Electronics and Applications (EPE'17 ECCE Europe)*; 2017. pp. P.1–P.10.

[22] J. Pfeiffer, *et al.* "Investigation of Variable Transformers Using Orthogonal Biasing," in IEEE Access, vol. 12, pp. 175375–175390, 2024, doi: 10.1109/ACCESS.2024.3501577.

Chapter 12

Multiple coupled phases using iron-based tape-wound cores

Christian P. Dick[1], *Valentin Wagner*[1] *and Patrick Deck*[1]

Electric and magnetic modeling of multi-phase coupled inductors is presented with the expected weight and volume reduction potential. The focus is on the technically relevant two-, three-, and four-phase coupled systems. Iron-based nanocrystalline ribbon is motivated and proposed as a favorable core material. A holistic, multi-objective design approach is proposed along with an exemplary design.

12.1 Motivation for coupled inductors based on tape-wound cores

Magnetic coupling of inverter phases is a well-known and widely used technology for reducing the overall size and weight of inductors. In particular, in high-current applications, where multiple inverter legs are necessary for current sharing, inversely coupled inductors are found, as depicted in Figure 12.1. Exemplary applications are voltage regulator modules in interleaved buck converters [1] or battery-connected bidirectional buck/boost converters [2,3].

In these approaches, the main mechanism of size and weight reduction comes from controlling the inverter-leg currents i_1 and i_2 to the same average value $\bar{i_1} = \bar{i_2}$, so that the DC flux component in parts of the magnetic core cancels out, as shown in Figure 12.2, resulting in a reduced core volume. Following this approach, for ferrite cores of identical volume, it is shown that the coupled system can carry more than twice the output current compared to a system with a single inverter leg and a standard inductor before saturation occurs [4].

A more far-reaching approach is the use of an iron-based ribbon, which features high saturation flux densities >1 T at very low losses, thus requiring a reduced core cross-section. A suitable material must have a reduced permeability in order to store the magnetic energy in a quasi-distributed air gap, which then leads to minimal copper losses due to a reduced proximity effect (no flux fringing at the air gap).

[1]TH Köln – University of Applied Sciences, Cologne Institute for Renewable Energy (CIRE) and Institute of Automation Engineering, Germany

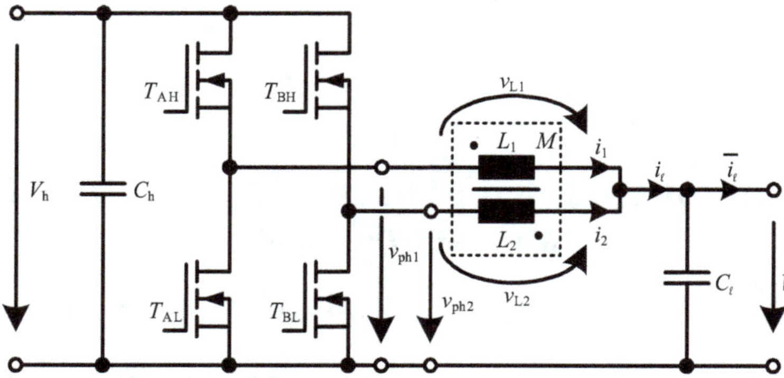

Figure 12.1 Power electronic building block comprising an inverse coupled inductor with two inverter-legs

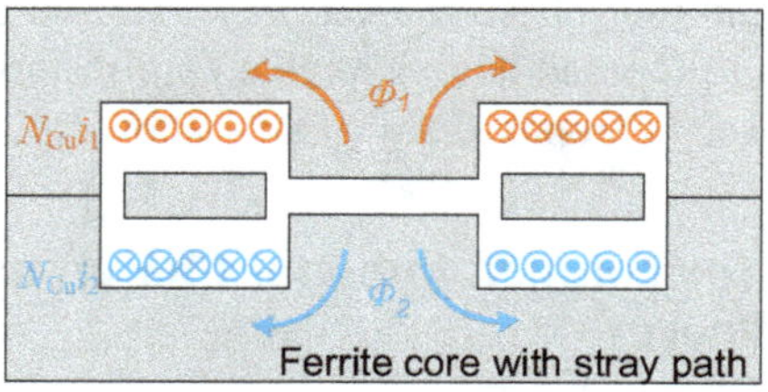

Figure 12.2 Basic principle of DC-flux cancelation due to superposition of main flux contributions Φ_1 and Φ_2; stray path features the DC-component (typ. low value due to increased reluctance)

However, the nanocrystalline ribbon known from common mode chokes has very high relative permeabilities ($\mu_r \gg 10,000$), so either core saturation occurs at low load currents or energy could only be stored in the air gap of a cut core, with the drawbacks of eddy currents at the cut and copper losses due to the proximity effect.

Recent innovations in soft magnetic materials have produced such highly efficient low-permeability ribbons. A stress- or creep-induced anisotropy is the basis for this property [5]. Relative permeabilities can be set as a design parameter down to a few hundred [6–8], which allows significant magnetic energy to be stored in a distributed air gap, and thus, the device can be used as a regular inductor, not only as a common mode choke. Before returning to the material itself, the focus is on how to arrange the ribbon to form a functional tape-wound coupled inductor, including electrical aspects.

The coupling of two inverter phases using a ribbon can be realized by an arrangement as depicted in Figure 12.3, or alternatively, where the inner stray cores and the outer coupling core could also be placed behind each other [9].

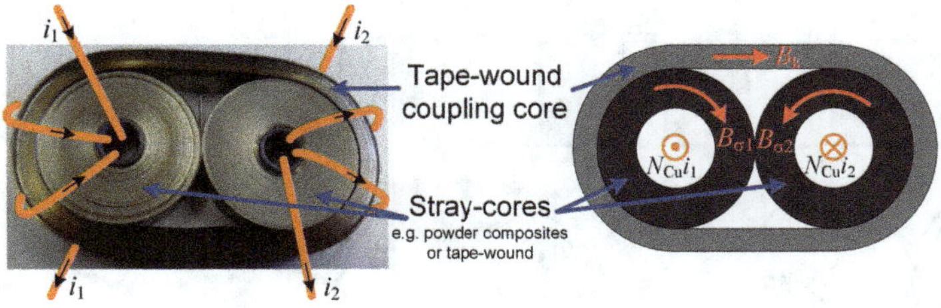

Figure 12.3 Possible arrangement of the two stray cores and a coupling core for a two-phase tape-wound coupled inductor

The magnetic device with two independent windings can be described with transformer equations as follows, where $L = L_1 = L_2$ is the self-inductance of both windings with the same number of turns, M is the mutual-inductance and $k = 0 \ldots 1$ is the positive coupling coefficient of the inversely coupled system (an alternative set of equations with negative k is also possible):

$$v_{L1} = L\frac{di_1}{dt} - M\frac{di_2}{dt} \qquad v_{L2} = L\frac{di_2}{dt} - M\frac{di_1}{dt}$$

$$k = \frac{M}{\sqrt{L_1 L_2}} = \frac{M}{L} L_k = L\left(1 - k^2\right) \tag{12.1}$$

By inserting L and M into the equations for v_{L1} and v_{L2} and substituting one of the current derivatives of the neighboring phase, one obtains the following relationship* directly:

$$v_{L1} = L_k\frac{di_1}{dt} - kv_{L2} \qquad v_{L2} = L_k\frac{di_2}{dt} - kv_{L1} \tag{12.2}$$

Obviously, the current-slope defining inductance is L_k and the switching state of the other coupled phase also affects the current. Therefore, in a 180° phase-shifted interleaved converter, both currents will feature a significant ripple at twice the switching frequency. Referring to Figure 12.3, the material in the inner stray path is stressed by a flux $B_{\sigma,i} \sim i_i$ also with twice the switching frequency. Furthermore, $B_k \sim (i_1 - i_2)$ applies in the coupling path. Together with the equations above, typical operating points are shown in Figure 12.4.

For the coupling path, it should be noted that, in addition to the zero DC flux $B_{k,dc} = 0$ T, the only significant frequency component seen by the material is the single switching frequency.

Regarding this qualitatively described alternating magnetic stress, the tape-wound material is again discussed for this aspect. Since such stress is not regularly given, stress analysis is performed on a special test bench that is designed and optimized for a fast, direct electrical measurement of the specific power losses [10].

*This relationship is furthermore derived for the general case in the electric model on multi-phase coupled magnetics in Section 12.2, including an intuitive equivalent circuit model of these equations.

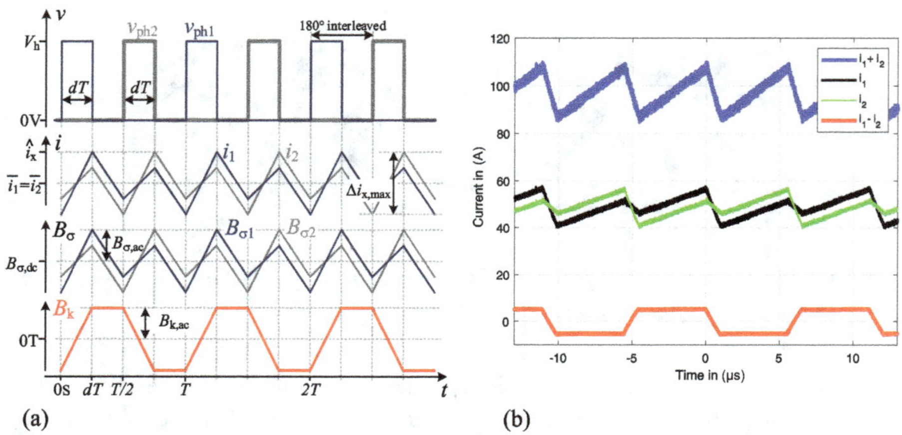

Figure 12.4 Exemplary operating points of the two-phase converter comprising a tape-wound coupled inductor (a) calculated currents and flux densities at d = 0.25 and (b) measured currents at d = 0.9

In this automated testbench, various techniques were applied to reduce measurement errors, such as (i) varying the measurement point such that measuring range switchover is favorable and (ii) decoupling the input voltage during the measurement to avoid common-mode interference on the measurement instruments. An example measurement based on the material VITROPERM® 850FF [11] featuring stress or creep-induced anisotropy [5] is depicted in Figure 12.5.

In contrast to ferrites, not only the saturation limit of the core material, but also the thermal limit is outstanding. The performance factor *PF* – which is the product of the peak flux density and the frequency at the thermal limit – is considered for both material classes at a frequency of 100 kHz. For ferrites at a typical thermal limit of 500 $\mathrm{mW/cm^3}$, specific losses result in $PF < 20,000$ HzT [12]. Sinusoidal excitation as required by standards is the basis for this figure. In contrast, the nanocrystalline iron-based tape-wound core is limited to ~ 250 W/kg. Using the numbers from Figure 12.5, this results in $PF \approx 100\ \mathrm{kHz} \cdot 350\ \mathrm{mT} = 35,000$ HzT. Although this comparison is in favor of the ferrites, since the results for the tape are obtained from non-sinusoidal curves as depicted in Figure 12.4, the difference in the performance factor is more than significant. Particularly for the stray core featuring a dc offset, a design can be carried out for both operating at thermal and saturation limitation.

Such high PF in combination with the high saturation flux density of the nanocrystalline ribbon results in a potentially huge reduction of core volume and, thus, in a reduction of the total weight and volume of the inductor itself.

With this motivation of a significant potential weight and volume reduction, this chapter is structured as follows. In Section 12.2, the modeling of the inductors is qualified to answer the question, which geometric quantities together with

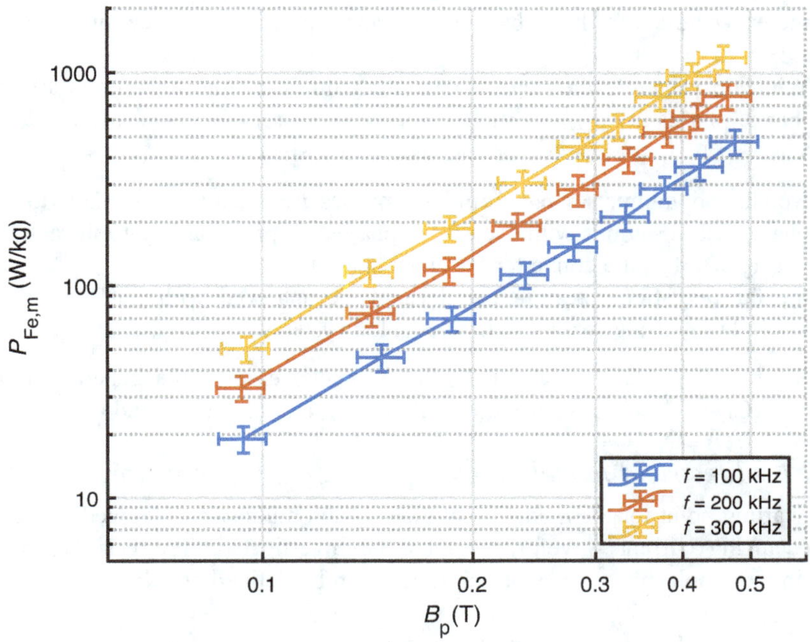

Figure 12.5 Core-loss densities of a tape-wound core with a permeability of μ_r =1500, stressed as coupling core at d = 0.35

material permeabilities lead to k and L_k being relevant for the electrical description on the one hand, and also lead to the dc and ac flux density components in both stray and coupling cores being relevant for the magnetic loss model on the other hand. This is not only conducted for two-phase systems but also for three- and four-phase systems.

Section 12.3 presents a detailed inductor design workflow. An example design for a 90 kW 48 kHz grid-connected converter is qualified to show the potential weight and volume reduction.

12.2 Modeling of multi-phase tape-wound-coupled inductors

In this chapter, the mathematical volume and weight potential of an interleaved coupled inductor system is derived. The modeling is divided into two parts. The first part is the description of the electrical model, including the electrical equivalent circuits and resulting current waveforms. The second part deals with the magnetic model of the tape-wound cores. The results of both parts lead to a combined model for the effective volume and weight. The ratios to the non-coupled case of two separate inductors are shown. With the knowledge of the sweet spots of potential maximum weight and volume reduction, the input parameter vectors for the subsequent Pareto optimization (Section 12.3) are limited.

While in principle the general case of multiple coupled phases is assumed (number of phases is N_{ph}), specific details are given for the technically more relevant cases of two, three, and four coupled phases, since

- weight and volume savings from 2→4 phases are much less compared to 1→2 phases, extra savings with even more phases are practically not relevant;
- wiring effort in the converter becomes high; and
- the fail-operational feature could easily be achieved with four phases, for example, by using two two-phase coupled systems.

Since two single inductors (SII) are a special case of the general coupled inductor (CI) with $k = 0$ the same model is used to compare SII and CI.

12.2.1 Electrical model of multi-phase coupled inductors

As a starting point for the modeling of tape-wound multi-phase coupled inductors, the occurring current waveforms should be investigated. The basic equations describing a multi-phase transformer system can be defined as follows:

$$v_{L_x} = L_x \frac{\mathrm{d}i_x}{\mathrm{d}t} + \sum_{s=1}^{N_s} \left(M_{x,s} \frac{\mathrm{d}i_s}{\mathrm{d}t} \right) - \sum_{r=1}^{N_r} \left(M_{x,r} \frac{\mathrm{d}i_r}{\mathrm{d}t} \right) \tag{12.3}$$

Here, the following definitions have to be considered:

- x: currently discussed phase;
- v_{L_x}: voltage applied to the CI's winding terminals of phase x, compare Figure 12.1;
- N_s: number of phases mutually coupled to current phase x;
- N_r: number of phases inversely coupled to current phase x;
- $N_{ph} = 1 + N_r + N_s$: total number of coupled phases.

Since only a positive coupling coefficient $k > 0$ and a positive mutual inductance are defined, the inverse coupling is represented by the index r and the negative sign in (12.3).

In the case of a symmetric magnetic system ($L_1 = L_2 = \ldots = L_x = L$, $M_{1,2} = M_{2,1} = \ldots = M_{x,s} = M_{x,r} = M$), it is now possible to replace the mutual inductances $M_{x,s}$ and $M_{x,r}$, since the coupling coefficient k describes the coupling between two phases and relates to the self- and mutual inductances with $M = kL$, see also (12.1) for the two-phase coupled system. Following [13], a usable equivalent electrical circuit is developed; in the first step, (12.3) is transformed to

$$v_{L_x} = L \frac{\mathrm{d}i_x}{\mathrm{d}t} + k\,L \left(\sum_{s=1}^{N_s} \frac{\mathrm{d}i_s}{\mathrm{d}t} - \sum_{r=1}^{N_r} \frac{\mathrm{d}i_r}{\mathrm{d}t} \right) \tag{12.4}$$

These equations for all phases $x = 1, \ldots, N_{ph}$ can be put into a matrix description, which is then inverted to obtain the current derivatives, resulting in

$$\frac{di_x}{dt} = \frac{1}{L} \cdot \frac{(N_{ph} - 2)k + 1}{1 + (N_{ph} - 2)k - (N_{ph} - 1)k^2} \cdot \left[v_{L_x} - \frac{k}{(N_{ph} - 2)k + 1} \cdot \left(\sum_{s=1}^{N_s} v_{L_s} - \sum_{r=1}^{N_r} v_{L_r} \right) \right] \tag{12.5}$$

and can be transformed into

$$v_{L_x} = \underbrace{L \cdot \frac{1 + (N_{ph} - 2)k - (N_{ph} - 1)k^2}{(N_{ph} - 2)k + 1}}_{=L_k} \cdot \frac{di_x}{dt} + \underbrace{\frac{k}{(N_{ph} - 2)k + 1} \cdot \left(\sum_{s=1}^{N_s} v_{L_s} - \sum_{r=1}^{N_r} v_{L_r} \right)}_{v_{k,x}} \tag{12.6}$$

The current derivative depends only on the voltages v_{L_x}, v_{L_s}, and v_{L_r} applied to the CI's winding, as shown in the equivalent circuit in Figure 12.6.

L_k is the relevant current limiting inductance and the voltage $v_{k,x}$ represents the influence of all coupled phases:

$$v_{k,x} = \frac{k}{(N_{ph} - 2)k + 1} \cdot \left(\sum_{s=1}^{N_s} v_{L_s} - \sum_{r=1}^{N_r} v_{L_r} \right) \tag{12.7}$$

$$L_k = L \cdot \frac{1 + (N_{ph} - 2)k - (N_{ph} - 1)k^2}{(N_{ph} - 2)k + 1} \tag{12.8}$$

Returning to the description using a transformer L-equivalent electrical circuit, L_k describes the serial inductance of one phase as depicted in Figure 12.7, which can be measured directly by short-circuiting the other phases.

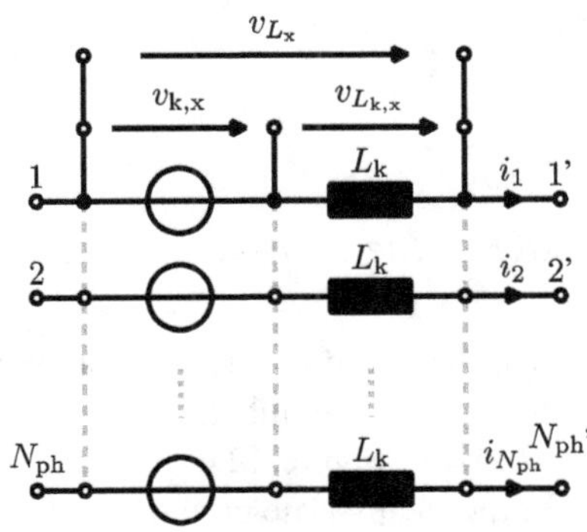

Figure 12.6 Equivalent electrical circuit of the multi-phase coupled inductor [13]

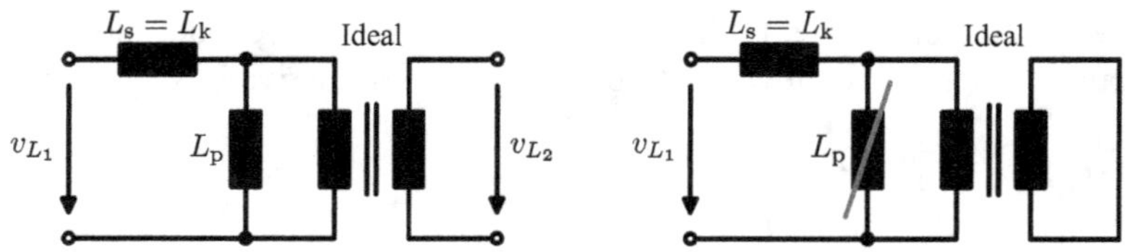

Figure 12.7 Transformer L-equivalent electrical circuit; figure from [14]

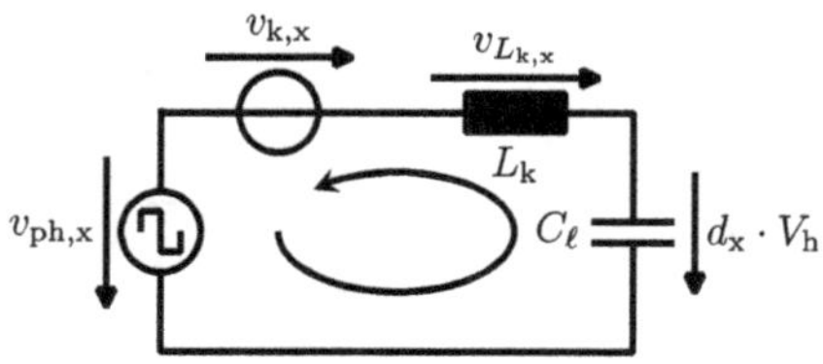

Figure 12.8 Voltage loop in the equivalent circuit of one phase; figure similar in [14]

For modeling the behavior, not only the current limiting inductance is of significance, but also the question of how the different phase voltages take influence on its voltage drop $v_{L_{k,x}}$. To evaluate this for arbitrary converter configurations, Figures 12.1 and 12.8 need to be considered.

This model describes the general case where the three phases do not feed into a common node where the currents sum up, see Figure 12.1, but are independent. Furthermore, the small resistive influence is not considered at this stage ($d_x \cdot V_h = \overline{v_{ph,x}}$). Of course, the case of a common node for all coupled phases is technically relevant and is discussed here as special case of the set of equations with $d_s = d_r = d$:

$$v_{L_{k,x}} = v_{ph,x} - v_{k,x} - d_x \cdot V_h \tag{12.9}$$

$$v_{k,x} = \frac{k}{(N_{ph} - 2)k + 1} \cdot \left[\sum_{s=1}^{N_s} v_{ph,s} - \sum_{r=1}^{N_r} v_{ph,r} - V_h \cdot \left(\sum_{s=1}^{N_s} d_s - \sum_{r=1}^{N_r} d_r \right) \right] \tag{12.10}$$

12.2.2 Electrical model of the two-phase inversely coupled inductor

This technically relevant topology from Figure 12.1, abbreviated with CI, is also qualified in Section 12.1. Inserting the numbers $N_{ph} = 2$, $N_r = 1$ and $N_s = 0$ into (12.10) directly leads to (12.2) and Figure 12.4. The curves for $v_{L_{k,1}}$, $v_{L_{k,2}}$ can be directly derived from this for the steady state period as well as the behavior for duty cycles greater than 0.5 [14]. With (12.2) and Figure 12.4 show the connection between the electrical quantities (voltages, frequency, and especially maximum phase ripple $\Delta i_{x,max}$ and the current peak value $\widehat{i}_x$) and the coupled inductors self and mutual inductances L an M.

12.2.3 Electrical model of the four-phase coupled inductor

In the four-phase coupled inductor $(N_{ph} = 4)$, two pairs of chopper phases are coupled. From the point of view of one phase, one other phase is mutually coupled ($N_s = 1$), and the remaining two are inversely coupled ($N_r = 2$), as depicted in Figure 12.9.

Two different winding configurations are possible with respect to the temporal order of the phases, as shown in Figure 12.10.

In the first configuration (CIV A), the mutually coupled phases follow each other in the temporal order of the interleaving structure, while in the second configuration order (CIV B), the mutually coupled phases alternate with the inversely coupled phases. Since CIV B shows no significant volume reduction [14], CIV A is the chosen configuration for the evaluation of the four-phase system and leads to the waveforms shown in Figure 12.11.

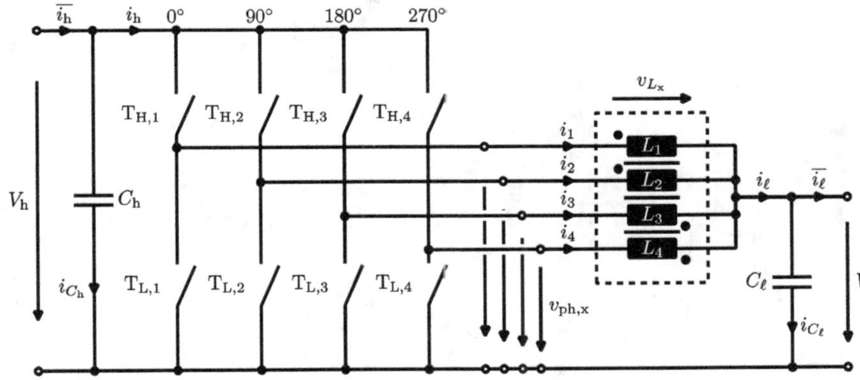

Figure 12.9 CIV in interleaved DC/DC converter: four-phase two-quadrant chopper with coupled inductor; figure from [14]

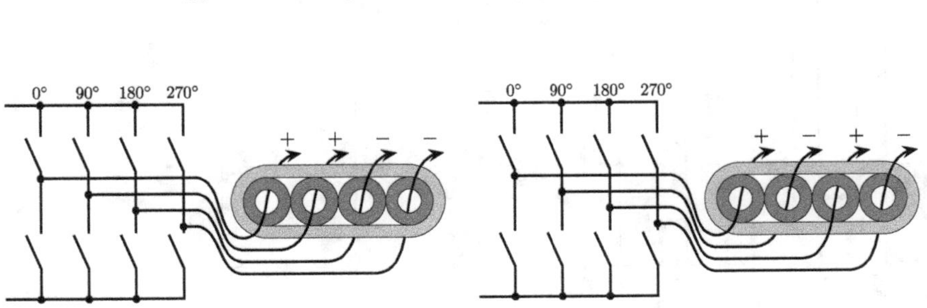

(a) (b)

Figure 12.10 Four-phase winding structure: (a) CIV A and (b) CIV B; figures from [14]

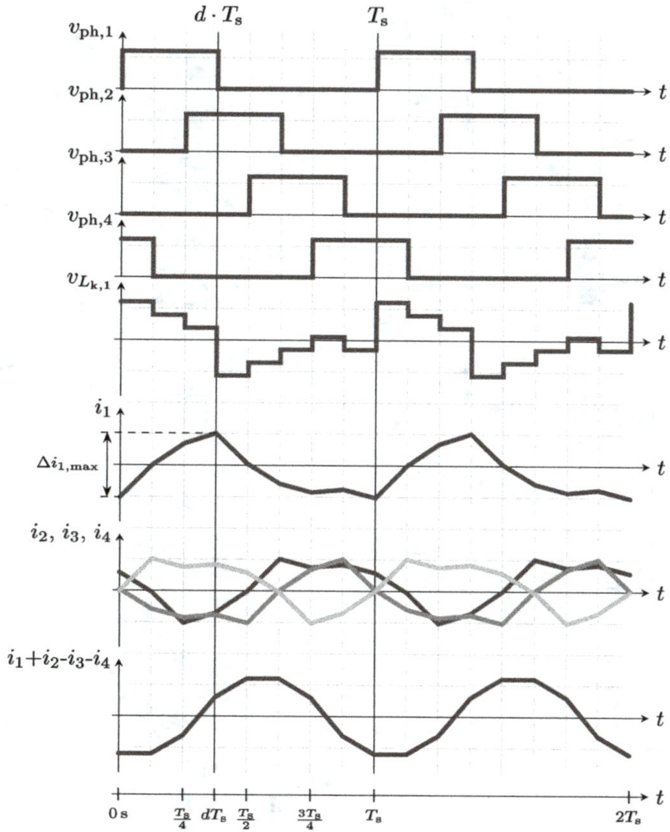

(a)

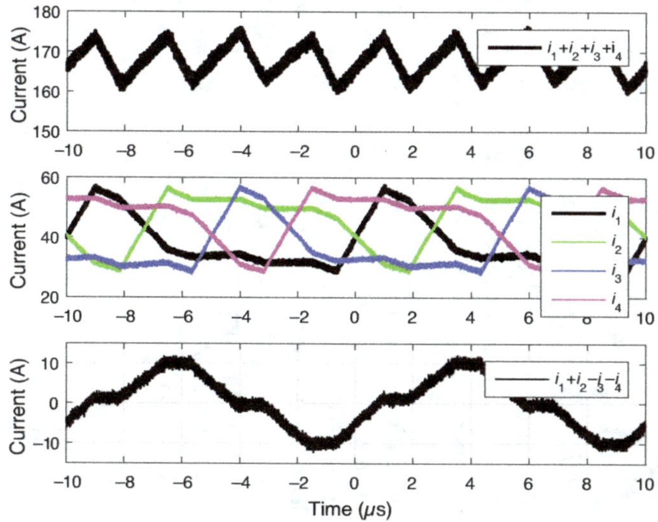

(b)

Figure 12.11 Current waveforms of the four-phase coupled inductor (a) d = 0.375 (valid for 0.25 < d ≤ 0.5) and (b) measured currents at d = 0.16, taken with coupled inductor depicted in Figure 2.15

Figure 12.11 shows two example sets of current waveforms. It should be noted that, when calculating the currents from the model described in (12.6)–(12.8), four duty cycle ranges must be distinguished, since when sweeping the duty cycle across $d = \{0.25; 0.5; 0.75\}$ the order of the applied voltage steps of $v_{L_{k,x}}$ changes on the time axis. Other curves like $v_{L_{k,x}}$ can be also directly derived from the model for the steady state period [14]. Equations (12.6)–(12.8) and Figure 12.11 show the direct relationship between the electrical quantities (voltages, frequency, and especially the maximum phase ripple $\Delta i_{x,\max}$ and the current peak value $\hat{i}_x$) and the coupled inductors self and mutual inductances L and M.

Similar to the two-phase system, the fundamental component of the current $(i_1 + i_2 - i_3 - i_4) \sim B_k$ proportional to the flux in the coupling path is again the single switching frequency.

In the four-phase system, negative and positive current peak values are not symmetrical around the average value of the current. This leads to a special current control as the four-phase system needs to be controlled in a way that the peak values and the peak-to-peak values of the four phases are the same for the best usage of the magnetic core. Average values in the four-phase system will differ depending on the phase number, coupling coefficient k, and duty cycle d.

12.2.4 Magnetic model of coupled inductors

In order to determine the values for the self-inductance L and the mutual inductance M, the core configurations wound around each other are examined in more detail and parameterized, as shown in Figure 12.12.

Low permeability and high saturation flux-density ribbon wound into ring cores as depicted leads to wound core configurations with separate elements for

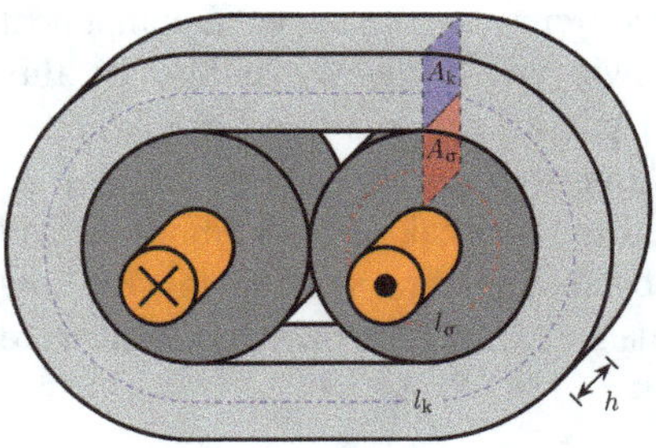

(a)

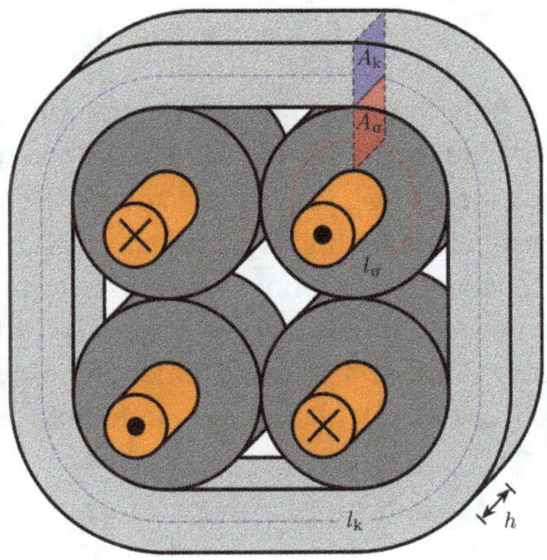

(b)

Figure 12.12 Core configurations of tape-wound cores with one exemplary electrical turn per phase: (a) two-phase and (b) four-phase; figures from [14]

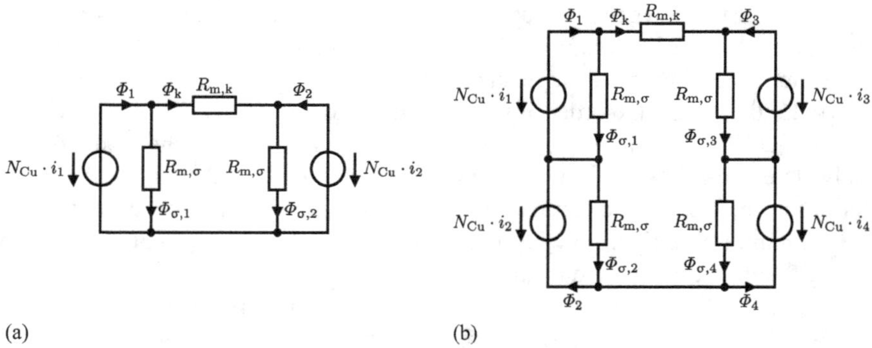

Figure 12.13 Reluctance models of tape-wound cores [14] (a) two-phase and (b) four-phase

leakage flux and mutual flux parts. Due to the preferred magnetic direction of the ribbon, the associated reluctances in the reluctance model in Figure 12.13 can be directly assigned to the different cores with $R_{\mathrm{m}} = \frac{l_{\mathrm{Fe}}}{\mu_0 \cdot \mu_{\mathrm{r}} \cdot A_{\mathrm{Fe}}}$, where the geometric numbers for l_{Fe} and A_{Fe} are the magnetically effective quantities, cmp. [5].

With the current waveforms modeled as illustrated in Figures 12.4 and 12.11, the various fluxes and flux densities are then calculated by

$$\phi_{\sigma,\mathrm{x}} = \frac{N_{\mathrm{Cu}}}{R_{\mathrm{m},\sigma}} i_{\mathrm{x}} \qquad B_{\sigma,\mathrm{x}} = \frac{N_{\mathrm{Cu}}}{A_{\mathrm{Fe}} \cdot R_{\mathrm{m},\sigma}} i_{\mathrm{x}} \tag{12.11}$$

$$\phi_k = \frac{N_{\mathrm{Cu}}}{R_{\mathrm{m,k}}} \left(i_{\mathrm{x}} + \sum_{\mathrm{s}=1}^{N_{\mathrm{s}}} i_{\mathrm{s}} - \sum_{\mathrm{r}=1}^{N_{\mathrm{r}}} i_{\mathrm{r}} \right) \qquad B_k = \frac{N_{\mathrm{Cu}}}{A_{\mathrm{Fe}} \cdot R_{\mathrm{m,k}}} \left(i_{\mathrm{x}} + \sum_{\mathrm{s}=1}^{N_{\mathrm{s}}} i_{\mathrm{s}} - \sum_{\mathrm{r}=1}^{N_{\mathrm{r}}} i_{\mathrm{r}} \right) \tag{12.12}$$

In both the two-phase and the four-phase systems, the fundamental component of the mutual flux ϕ_{k} shows the converter switching frequency, unlike the flux waveform of $\phi_{\sigma,\mathrm{x}}$ in the leakage flux section. Here, the flux has significantly higher harmonics resulting from all the phase-switching events.

Since the coupling core sees zero average flux, the peak-to-peak amplitude can be pushed to extreme values, only limited by temperature, leading to very low material usage and an extremely high-performance factor. The peak flux values are modeled to calculate the volume potential

$$N_{\mathrm{ph}} = 2 : \widehat{\Phi}_{\sigma,\mathrm{x,max}} = \frac{N_{\mathrm{Cu}}}{R_{\mathrm{m},\sigma}} \cdot \widehat{i}_{\mathrm{x,max}} = \frac{N_{\mathrm{Cu}}}{R_{\mathrm{m},\sigma}} \cdot \left[\frac{\bar{i}_\ell}{2} + \frac{V_{\mathrm{h}}}{8 \cdot L_k \cdot f_{\mathrm{s}} \cdot (1+k)} \right]$$

$$= \frac{N_{\mathrm{Cu}}}{R_{\mathrm{m},\sigma}} \cdot \left[\frac{\bar{i}_\ell}{2} + \frac{\Delta i_{\ell,\mathrm{max}}}{(1+k)^2} \right] \tag{12.13}$$

$$N_{\mathrm{ph}} = 2: \widehat{\Phi}_{\mathrm{k,max}} = \frac{N_{\mathrm{Cu}}}{R_{\mathrm{m,k}}} \cdot \widehat{i}_{\mathrm{k,max}} = \frac{N_{\mathrm{Cu}}}{R_{\mathrm{m,k}}} \cdot \left[\frac{1}{4} \cdot \frac{V_{\mathrm{h}}}{L_{\mathrm{k}} \cdot f_{\mathrm{s}}} \cdot (1-k)\right]$$

$$= \frac{N_{\mathrm{Cu}}}{R_{\mathrm{m,k}}} \cdot \left(2 \cdot \Delta i_{\ell,\mathrm{max}} \cdot \frac{1-k}{1+k}\right) \tag{12.14}$$

$$N_{\mathrm{ph}} = 4: \widehat{\Phi}_{\sigma,\mathrm{x,max}} = \frac{N_{\mathrm{Cu}}}{R_{\mathrm{m},\sigma}} \cdot \left(\frac{\bar{i}_{\ell}}{4} + \frac{\Delta i_{\mathrm{x}}}{2}\right)$$

$$= \frac{N_{\mathrm{Cu}}}{R_{\mathrm{m},\sigma}} \cdot \left(\frac{\bar{i}_{\ell}}{4} + \frac{V_{\mathrm{h}}}{8 \cdot L_{\mathrm{k}} \cdot f_{\mathrm{s}}} \cdot \frac{2k+1}{3k+1}\right) \tag{12.15}$$

$$= \frac{N_{\mathrm{Cu}}}{R_{\mathrm{m},\sigma}} \cdot \left[\frac{\bar{i}_{\ell}}{4} + 2 \cdot \Delta i_{\ell,\mathrm{max}} \left(\frac{2k+1}{3k+1}\right)^2\right]$$

$$N_{\mathrm{ph}} = 4: \widehat{\Phi}_{k,\mathrm{max}} = \frac{N_{\mathrm{Cu}}}{R_{\mathrm{m,k}}} \cdot \widehat{i}_{k,\mathrm{max}} = \frac{N_{\mathrm{Cu}}}{R_{\mathrm{m,k}}} \cdot \left[\frac{1}{4} \cdot \frac{V_{\mathrm{h}}}{L_{\mathrm{k}} \cdot f_{\mathrm{s}}} \cdot \frac{1-k}{2k+1}\right] \tag{12.16}$$

$$= \frac{N_{\mathrm{Cu}}}{R_{\mathrm{m,k}}} \cdot \left(4 \cdot \Delta i_{\ell,\mathrm{max}} \cdot \frac{1-k}{3k+1}\right)$$

The magnetically effective volume $V_{\mathrm{Fe}} = l_{\mathrm{Fe}} \cdot A_{\mathrm{Fe}}$ of the different cores is included in the square of the flux density when calculating $B = \Phi / A_{\mathrm{Fe}}$, using the flux Φ from the above equations and substituting $N_{\mathrm{Cu}} = \sqrt{L \cdot R_{\mathrm{m},\sigma} \cdot (1-k)} = \sqrt{L \cdot k \cdot R_{\mathrm{m,k}}}$ from $L = (\Phi_{\mathrm{k}} + \Phi_{\sigma}) N_{\mathrm{Cu}} / i = N_{\mathrm{Cu}}^2 \left(R_{\mathrm{m,k}}^{-1} + R_{\mathrm{m},\sigma}^{-1}\right)$. After the transformation, the following equations describe the volume for the two-phase coupled inductor:

$$N_{\mathrm{ph}} = 2: \quad V_{\mathrm{Fe},\sigma} = l_{\mathrm{Fe},\sigma} \cdot A_{\mathrm{Fe},\sigma}$$

$$= \frac{V_{\mathrm{h}} \cdot \mu_0 \cdot \mu_{\mathrm{r},\sigma}}{8 \cdot f_{\mathrm{s}} \cdot \left(\widehat{B}_{\sigma,\mathrm{max}}\right)^2 \cdot \Delta i_{\ell,\mathrm{max}}} \cdot \left[\frac{\bar{i}_{\ell}}{2} + \frac{\Delta i_{\ell,\mathrm{max}}}{(1+k)^2}\right]^2 \tag{12.17}$$

$$N_{\mathrm{ph}} = 2: V_{\mathrm{Fe,k}} = l_{\mathrm{Fe,k}} \cdot A_{\mathrm{Fe,k}} = \frac{V_{\mathrm{h}} \cdot \mu_0 \cdot \mu_{\mathrm{r,k}}}{2 \cdot f_{\mathrm{s}} \cdot \left(\widehat{B}_{\mathrm{k,max}}\right)^2} \cdot \frac{k \cdot (1-k)}{(1+k)^2} \cdot \Delta i_{\ell,\mathrm{max}} \tag{12.18}$$

To calculate the total effective volume $V_{\mathrm{Fe}} = 2 \cdot V_{\mathrm{Fe},\sigma} + V_{\mathrm{Fe,k}}$ is used. The possible volume saving can be calculated by comparing V_{Fe} with $V_{\mathrm{Fe}}(k=0)$, i.e., two uncoupled inductors, by the relation $\Gamma_V = V_{\mathrm{Fe}} / V_{\mathrm{Fe}}(k=0)$. To simplify the mathematical expression for this relation, additional relations are introduced in (12.19). This yields an expression for the volume relation Γ_V, which is the basis for the subsequent optimization:

$$\Gamma_B = \frac{\widehat{B}_{\mathrm{k,max}}}{\widehat{B}_{\sigma,\mathrm{max}}}; \Gamma_I = \frac{\Delta i_{\ell,\mathrm{max}}}{\bar{i}_{\ell}}; \Gamma_A = \frac{A_{\mathrm{Fe,k}}}{A_{\mathrm{Fe},\sigma}}; \Gamma_l = \frac{l_{\mathrm{Fe,k}}}{l_{\mathrm{Fe},\sigma}}; \Gamma_\mu = \frac{\mu_{r,\mathrm{k}}}{\mu_{\mathrm{r},\sigma}} \tag{12.19}$$

$$N_{\text{ph}} = 2 : \Gamma_V = \left(\left[\frac{1}{2} + \frac{\Gamma_I}{(1+k)^2} \right]^2 + \frac{2 \cdot \Gamma_l}{\Gamma_A \cdot (\Gamma_B)^2} \cdot \frac{k^2}{(1+k)^2} \cdot (\Gamma_I)^2 \right) / \left(\frac{1}{2} + \Gamma_I \right)^2 \tag{12.20}$$

The corresponding set of equations for the four-phase inductor using $V_{\text{Fe}} = 4 \cdot V_{\text{Fe},\sigma} + V_{\text{Fe,k}}$ is depicted below:

$$N_{\text{ph}} = 4 : \quad V_{\text{Fe},\sigma} = l_{\text{Fe},\sigma} \cdot A_{\text{Fe},\sigma}$$

$$= \frac{V_{\text{h}} \cdot \mu_0 \cdot \mu_{\text{r},\sigma}}{16 \cdot f_{\text{s}} \cdot \left(\widehat{B}_{\sigma,\max} \right)^2 \cdot \Delta i_{\ell,\max}} \cdot \left[\frac{\bar{i}_\ell}{4} + \left(\frac{4k+2}{3k+1} \right)^2 \frac{\Delta i_{\ell,\max}}{2} \right]^2 \tag{12.21}$$

$$V_{\text{Fe},k} = l_{\text{Fe,k}} \cdot A_{\text{Fe,k}} = \frac{V_{\text{h}} \cdot \mu_0 \cdot \mu_{\text{r,k}}}{f_s \cdot \left(\widehat{B}_{\text{k,max}} \right)^2} \cdot \frac{k \cdot (1-k)}{(3k+1)^2} \cdot \Delta i_{\ell,\max} \tag{12.22}$$

$$\Gamma_V = \frac{\left[\frac{1}{4} + \left(\frac{4k+2}{3k+1} \right)^2 \cdot \frac{\Gamma_I}{2} \right]^2 + \frac{4 \cdot \Gamma_l}{\Gamma_A \cdot (\Gamma_B)^2} \cdot \frac{k^2}{(3k+1)^2} \cdot (\Gamma_I)^2}{\left(\frac{1}{4} + 2 \cdot \Gamma_I \right)^2} \tag{12.23}$$

To illustrate the possible volume savings, these expressions can be plotted against the current relation $\Gamma_I = \Delta i_{\ell,\max} / \bar{i}_\ell$ the coupling coefficient as a parameter, see Figure 12.14. As a realistic example, the following figure illustrates the

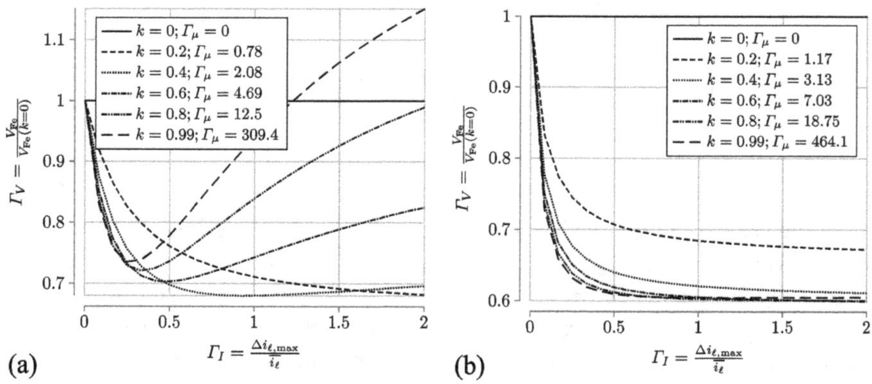

Figure 12.14 Volume relation Γ_V of effective magnetic core volumes depending on Γ_I, Γ_l, Γ_A, Γ_B, and k: (a) two-phase ($\Gamma_l = 2$; $\Gamma_B = 0.8$; $\Gamma_A = 1$) and (b) four-phase ($\Gamma_l = 3$; $\Gamma_B = 0.8$; $\Gamma_A = 1$). Figures from [14].

possible volume savings compared to non-coupled interleaved inductors ($k = 0$), shows dependencies, and compares two-phase and four-phase inductors.

It can be seen that the volume savings of the coupled inductor device depend on several parameters. Even higher volumes are possible compared to the non-coupled interleaved single inductors if these parameters are poorly chosen.

Furthermore, this set of equations is also used for optimization purposes. The knowledge of these sweet spots is used to reduce the input parameter range for the subsequent Pareto optimization to perform the design at the thermal and/or the flux density limit.

12.2.5 Special discussion on the three-phase coupled inductor

Another special case of the coupled inductor is the three-phase coupled inductor, e.g., for grid-tied inverters with $N_{ph} = 3$, $N_r = 0$ and $N_s = 2$. In contrast to the two- and four-phase applications where the phase currents sum up in a common node, the three phases feed into the different grid phases at different voltage levels, as shown in Figure 12.15.

The chosen topology features a DC-link-connected filter capacitor resulting in a non-floating DC-link with respect to the mains neutral point. The latter is important, for example, in photovoltaic converters or in interface converters between a DC microgrid and the regular AC grid. There are also applications with no connection between the DC-link and the filter capacitors, including coupled (but not tape-wound) cores, where the sum of the inductor currents is zero by design (apart from HF-CM noise, which is not discussed here, the focus is on filtering of the switching frequency). Without DC-link connected filter capacitors, the coupling core would be without/only with minor HF-CM filter functionality. This approach addresses the topology with a DC-link connected filter capacitor, with the coupling core significantly being stressed as qualified below. It also contributes to the HF-CM filter.

- The low main frequency component of, e.g., 50 Hz of the three currents i_1, i_2, and i_3 sum up to zero. Therefore, the coupling core is not stressed at the main frequency. However, the stray cores see the main frequency and, thus, exhibit full load excitation.

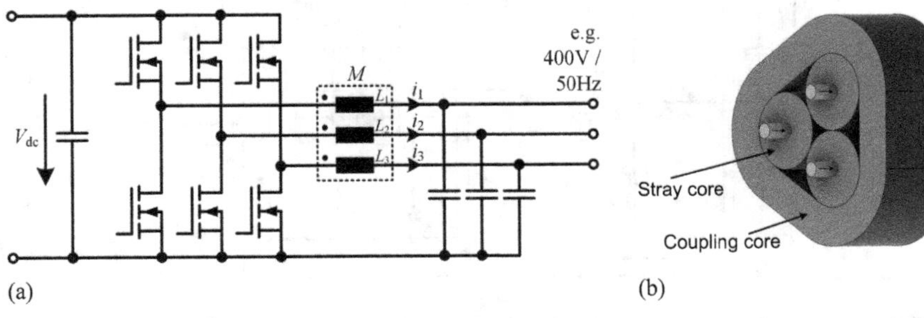

Figure 12.15 Three-phase tape-wound coupled inductor approach: (a) topology with DC-link connected filter capacitors and (b) magnetic core structure

- When, for example, space vector PWM is used, a third harmonic (150 Hz) plus other harmonics are driven through the coupled inductor as a common mode current. The coupling core will see these quite low-frequency charging currents of the filter capacitors. This effect will drive the coupling core to higher excitation; a design to the saturation limit is beneficial.
- At the same time, a design of the coupling core at the thermal limit can be realized. Current ripples caused by PWM switching contribute most significantly to core losses such that solutions can be found by varying switching frequency and current ripple being interlinked with geometry parameters, permeability, and thus, coupling coefficient and self-inductance.

The following design workflow can be applied analogously to the three-phase coupled inductor system.

12.3 Workflow of inductor optimization based on the example of a 90 kW, three-phase, three-level, grid-connected inverter operating at 48 kHz

One application for coupled inductors is grid-connected inverters. In this example, each line-phase is using two three-level interleaved inverter legs sharing one coupled inductor. Figure 12.16 shows a simplified schematic of the inverter discussed. This inductor is optimized to have the lowest possible weight while staying within the thermal limit of the materials.

A number of design parameters need to be identified. Before going into the details of the inductor itself, the first step is to define the boundary conditions, such as space constraints and maximum allowable losses, as well as the optimization targets (e.g., inductor losses, total inductor weight, core weight, total volume, etc.). For grid-connected inverters, this often includes a fixed switching frequency.

The next step is to select a current ripple in the output filter capacitor along with a corresponding switching frequency. In the first multi-objective Pareto

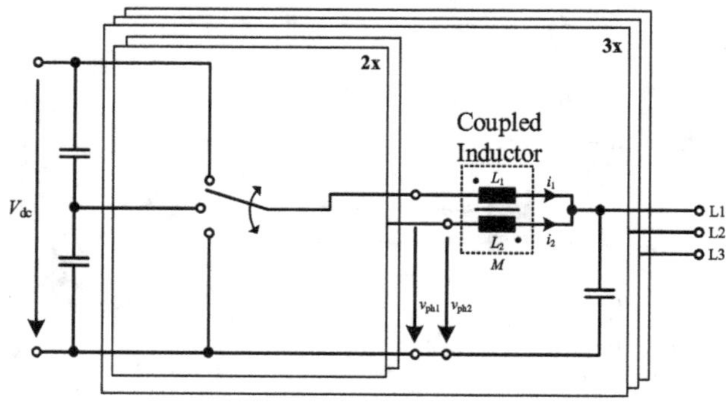

Figure 12.16 Topology of the three-level DC/AC converter comprising coupled inductors

optimization, a wide range of the following parameters at low resolution is chosen to calculate different inductors:

- Coupling factor k,
- Maximum flux density in the leakage cores $B_{p,\sigma}$,
- Maximum flux density in the coupling core $B_{p,\mathrm{k}}$,
- Number of conductor windings per phase N_{Cu},
- Relative permeability of the leakage cores $\mu_{r,\sigma}$,
- Relative permeability of the coupling core $\mu_{r,\mathrm{k}}$,
- Core height of the leakage cores h_{σ},
- Core height of the coupling cores h_{k}.

Together with the fixed input parameters, an inductor can be calculated for all permutations of these parameters. Of particular interest are the two optimization targets selected, e.g., total inductor losses and total inductor weight. All results that violate the previously selected boundary conditions or are impossible to manufacture (e.g., the conductors must fit inside of the cores) need to be discarded. All remaining results that are worse than any other result at both optimization targets are also discarded. This leaves a so-called "Pareto front" with only the optimal results of the calculation. Figure 12.19 shows an example of a Pareto front.

This optimization is then iteratively refined to get closer to the optimization targets by reducing the range of the input parameters and increasing the resolution based on the previous results.

One result is then selected, and a simplified CAD model is created from it. Together with the losses calculated in the previous step, a thermal simulation of the inductor in the environment of the converter (convection or forced air cooling, temperature of the environment) is conducted. Figure 12.17 shows a thermal simulation of such a simplified model.

The thermal simulation data can be used to estimate the maximum possible losses for this application. For this example, the power limit is estimated to be 35 W per coupled inductor.

The next step is to determine a "sweet spot" of the maximum output current ripple. This is done by performing a Pareto optimization for different current ripples and plotting the inductor weight at the previously determined loss value against the output current ripple. Figure 12.18 shows this plot for the example application. From this graph, a good compromise between weight and amount of filtering can be selected. A current ripple of 35 amperes peak to peak is chosen to further refine the design.

With the new value for the output current ripple, the Pareto optimization is iterated again, this time with special attention to the desired loss range. The resolution of the input parameters is further increased, while the range is decreased to reduce the computation time. A final design can now be selected from the Pareto front. For this design, another thermal simulation based on a detailed CAD model is conducted.

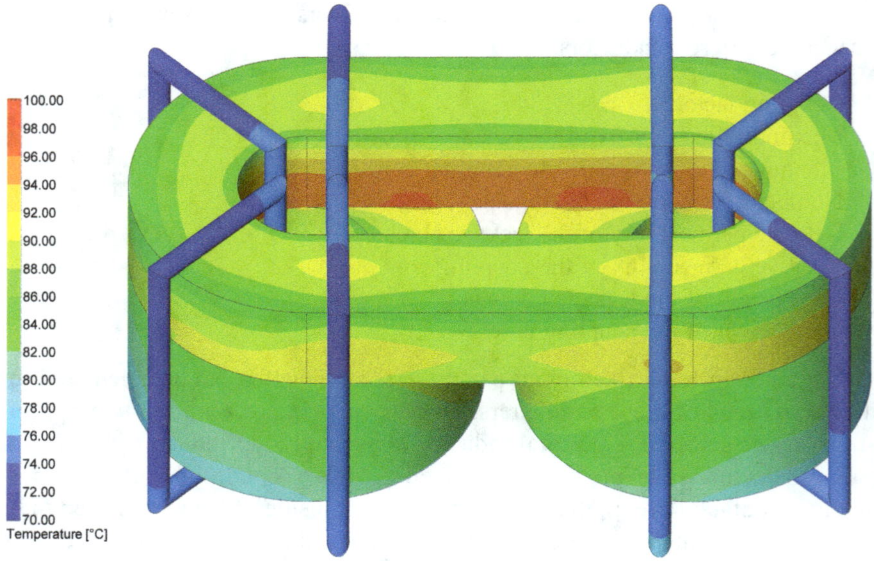

Figure 12.17 Thermal simulation of an exemplary coupled inductor (simplified model, case, and potting not shown)

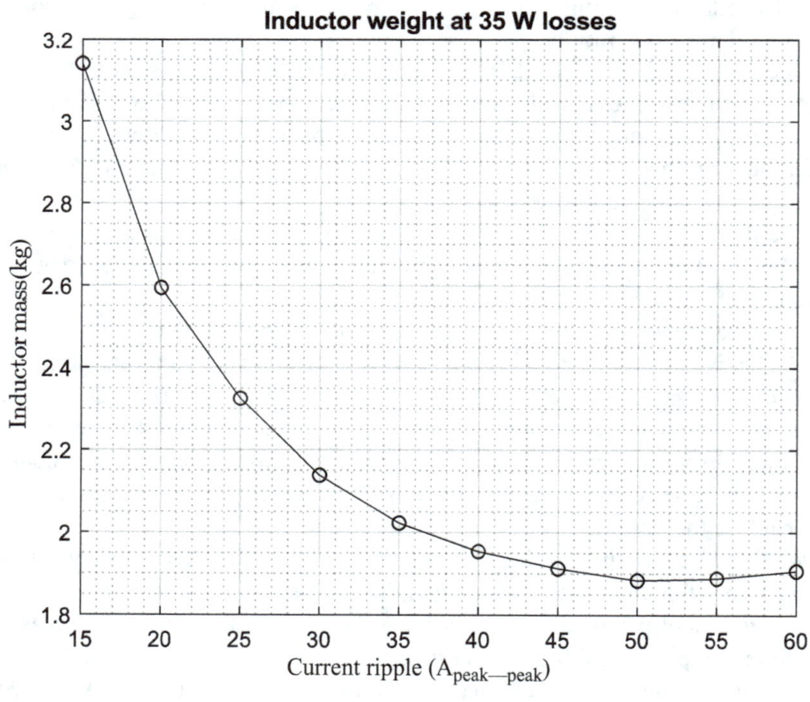

Figure 12.18 Mass of the coupled inductor over output current ripple

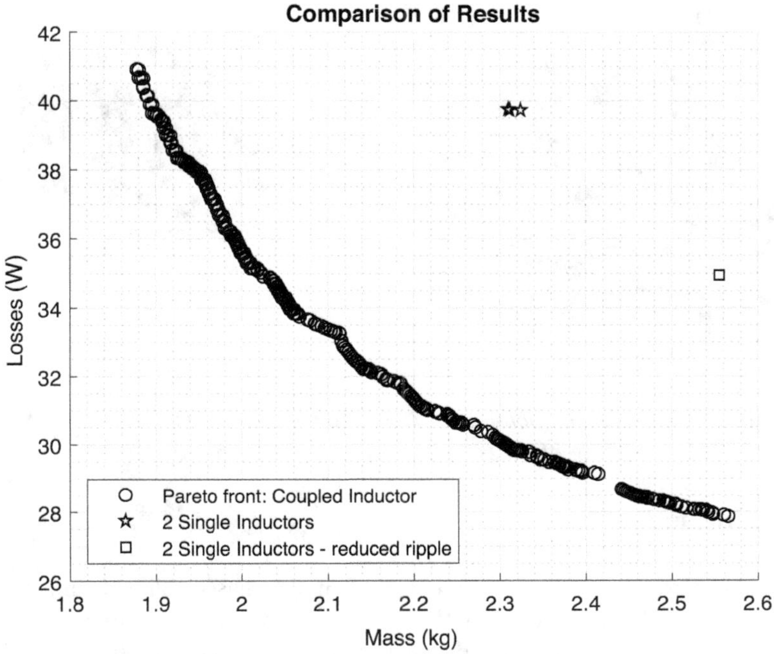

Figure 12.19 Pareto-calculation of the coupled inductor compared to two different single-inductor solutions

Finally, the electrical and thermal behavior of a prototype is to be validated in the final product or in a stress test bench as proposed in [15].

To compare the coupled inductor with two single inductors, such a solution is calculated with the same boundary conditions using the same optimization workflow. The two single inductors are set up in a similar way to the coupled inductor: with the cores in one common case and potting. Two completely separate inductors would profit from better cooling due to the increased surface area but would also be heavier. This again yields a Pareto front with only very few solutions very close to each other, as indicated with stars in Figure 12.19.

Compared to the solutions for the coupled inductor, two single inductors are either 22% heavier or produce 33% more losses. To reduce the losses in the single inductors to the previously defined maximum, the output current ripple has to be reduced from 35 amperes to 29 A peak to peak. The solution with two single inductors is then 26% heavier than the coupled solution. The two single inductors are also based on the nanocrystalline tape material, which features remarkably low loss behavior, see Figure 12.5, allowing extreme performance factors. Hence, there is another weight/volume/loss benefit compared to ferrite cores.

Figure 12.20 shows an image of a prototype coupled inductor next to a CAD rendering of the internal arrangement of the cores and windings. This inductor was designed for a higher power level at a different switching frequency.

(a) (b)

Figure 12.20 Coupled inductor for DC/AC converters, from [16] (a) CAD rendering of cores and copper windings and (b) prototype in the lab

12.4 Summary

The electrical and magnetic modeling of multi-phase coupled inductors is presented in detail, with a focus on the technically relevant two-, three-, and four-phase coupled systems. Iron-based nanocrystalline ribbon is motivated and proposed as a favorable core material due to its remarkably low loss behavior. A holistic, multi-objective design approach is proposed along with an exemplary design. Comparing two single inductors with a coupled system, both based on nanocrystalline core material only, the two single inductors are either 22% heavier or produce 33% more losses compared to the coupled version. In addition, another weight or efficiency benefit is given when compared to inductors comprising alternative core materials.

Acknowledgement

This work was supported in part by the German Federal Ministry for Economic Affairs and Climate Action through the publicly funded research project "GaN-HighPower" (Grant No. 03EE1111F) and by the German federal state of North Rhine-Westphalia using funds from the European Regional Development Fund (Coupled Tape-Wound Cores, grant number EFRE-0400028). The authors express their gratitude for this opportunity. Responsibility for this chapter's content lies with the authors.

References

[1] P.-L. Wong, P. Xu, P. Yang and F. C. Lee, Performance improvements of interleaving VRMs with coupling inductors, *IEEE Transactions on Power Electronics*, vol. 16, no. 4, pp. 499–507, July 2001, doi:10.1109/63.931059.

[2] M. Hirakawa, M. Nagano, Y. Watanabe, K. Andoh, S. Nakatomi and S. Hashino, High power density DC/DC converter using the close-coupled inductors, *IEEE Energy Conversion Congress and Exposition, San Jose, CA, USA*, Nr. pp. 1760–1767, 2009, doi:10.1109/ECCE.2009.5316389.

[3] X. Huang, F. C. Lee, Q. Li and W. Du, High-frequency high-efficiency GaN-based interleaved CRM bidirectional buck/boost converter with inverse coupled inductor, *IEEE Transactions on Power Electronics*, vol. 31, no. 6, pp. 4343–4352, 2016, doi:10.1109/TPEL.2015.2476482.

[4] P. Deck and C. P. Dick, Material savings using coupled inductors in hard switched power-electronic building blocks: modeling and experimental validation. *2016 18th European Conference on Power Electronics and Applications (EPE'16 ECCE Europe), Karlsruhe, Germany*, 2016, pp. 1–10, doi: 10.1109/EPE.2016.7695346.

[5] R. Hilzinger and W. Rodewald, *Magnetic Materials: Fundamentals, Products, Properties, Applications*, VACUUMSCHMELZE GmbH & Co. KG, Hrsg., Publicis Publishing Erlangen, 2013, ISBN: 978-3-89578-352-4.

[6] J. Petzold, Advantages of soft magnetic nanocrystalline materials for modern electronic applications. *Journal of Magnetism and Magnetic Materials*, vol. 242–245, Part 1, pp. 84–89, 2002, ISSN 0304-8853.

[7] Vacuumschmelze GmbH & Co.KG, Material Information "Nanocrystalline VITROPERM" – relative permeability down to 400 available. https://vacuumschmelze.com/products/soft-magnetic-materials-and-stamped-parts/nanocrystalline-material-vitroperm; retrieval date: 12 November 2023.

[8] Magnetec GmbH, Nanoperm® LM Cores for DC biased applications. *Datasheet information*.

[9] C. P. Dick, Converter device and coil arrangement for a voltage regulator. Patent EP3295464B1; US10068700B2; CN107787515B; WO2016180406A1, 11 May 2016.

[10] D. Bohne, V. Wagner, P. Deck and C. P. Dick, Loss characterization methodology for soft magnetic nano-crystalline tape materials in coupled inductors. *2022 24th European Conference on Power Electronics and Applications (EPE'22 ECCE Europe), Hanover, Germany*, pp. 01–11, 2022.

[11] G. Herzer, V. Budinsky and C. Polak, Magnetic properties of nanocrystalline FeCuNbSiB with huge creep induced anisotropy. *Journal of Physics: Conference Series*, vol. 266, p. 012010, 2011, doi:10.1088/1742-6596/266/1/012010.

[12] W. G. Hurley, T. Merkin and M. Duffy, The performance factor for magnetic materials revisited: the effect of core losses on the selection of core size in transformers. *IEEE Power Electronics Magazine*, vol. 5, no. 3, pp. 26–34, Sept. 2018, doi:10.1109/MPEL.2018.2848738.

[13] G. Zhu, B. A. McDonald and K. Wang, Modeling and analysis of coupled inductors in power converters. *IEEE Transactions on Power Electronics*, vol. 26, no. 5, pp. 1355–1363, May 2011, doi:10.1109/TPEL.2010.2079953.

[14] P. Deck, Gekoppelte Induktivitäten auf Basis niederpermeabler, nanokristalliner Bandmaterialien – Anwendung in getakteten Gleichspannungswandlern. *Wuppertaler Schriftenreihe zu Elektromobilität und Energiespeichersystemen*, 2003,

Band 3, Shaker Verlag, ISBN 978-3-8440-8990-5, doi: 10.25926/BUW/0-70, figure reproduced by permission of Shaker Verlag.

[15] V. Wagner, L. Geiger and C. P. Dick, *Stress Testbench for Two-Phase Interleaved Coupled Inductors in Three-Level Inverters using Cascaded Control, PCIM Europe 2023*; International Exhibition and Conference for Power Electronics, Intelligent Motion, Renewable Energy and Energy Management, Nuremberg, Germany, 2023, pp. 1–5, doi: 10.30420/566091186.

[16] Vacuumschmelze GmbH & Co. KG (VAC), *CAD Rendering and Prototype of Coupled Inductor, Copyright (c)*, Reproduced figures by permission of VAC, 2023.

Part IV

Integrated inductive components and design for high frequencies

Chapter 13

Design of integrated inductive components for high-power converters with and without insulation

Olympio Cipriano da Silva Filho[1] *and Demercil de Souza Oliveira Junior*[2]

Integrating magnetic devices is of great interest for different power converter topologies. The goals of reducing cost, dimensions, volume, and losses, as well as increasing power density, are always pursued by researchers to meet the requirements of a variety of applications. In the context of high-frequency and high-power converters, these aspects are very important.

This chapter introduces methods to reduce the overall magnetic component count through the integration of magnetic devices. There are different paths to achieve magnetic device integration; some are straightforward, while others require more analysis. For instance, simple integration can be achieved by adjusting the leakage inductance of transformers to eliminate the need for current-limiting inductors in converters like dual active bridge (DAB) or inductors that form resonant tanks in resonant converters.

As many converters employ interleaving techniques, it is common to find interleaved transformers associated in series with a filter inductance, resulting in two different magnetic devices. In such cases, with the integration procedures presented in this chapter one can generate only one non-insulated magnetic device with reduced cost, volume, and improved efficiency.

More complex integration solutions are also possible. One can, for instance, integrate an inductor, an interleaved transformer, and a transformer to create a single isolated magnetic device that accomplish the tasks performed by the elements individually. The integration of different magnetic devices is possible when they are connected in the circuit. For the aforementioned case, one DAB transformer isolating the primary and secondary sides of an AC–AC converter can be integrated with its current limit inductor, the filter input inductor, and the interleaved inductor if they are present in the converter topology. The integration method and the possible integration

[1]Department of Technology and Engineering, Federal University of Semi-arid Region, Brazil
[2]Department of Electrical Engineering, Federal University of Ceara, Brazil

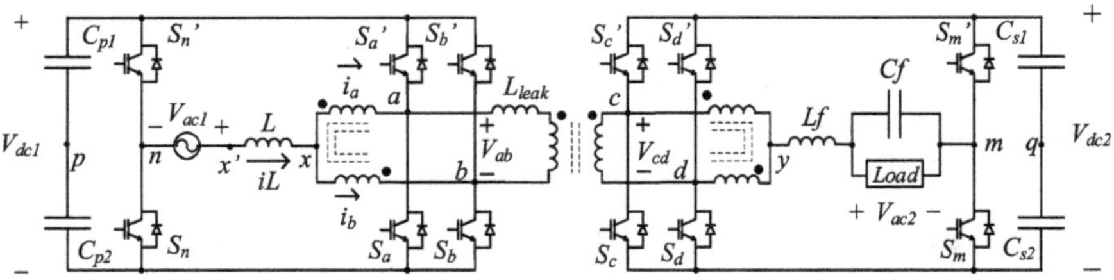

Figure 13.1 AC–AC converter topology with multiple magnetic devices

solution depend on the magnetics arrangement and are case-specific, which means that each scenario requires one solution.

The magnetic integration techniques presented in this chapter, and mentioned in the examples above are used to reduce the magnetic component count from six to two, taking an isolated symmetrical full-bridge converter topology presented in Figure 13.1, defined as a use case. The theoretical analysis of this converter is presented, from which a laboratory prototype was implemented using integrated magnetics. The result delivers higher efficiency and a lower component count when compared with the use of individual elements.

13.1 The problem of the number of magnetic devices

Power converters are essentially produced using semiconductor switches, passive elements, and electronics (usually involving sensors, controllers, etc.). Among the passive components, the most common are capacitors and magnetic components. There is a myriad of magnetic components, and they play many different roles in the converter, such as energy storage, split current, galvanic isolation, and filtering. Due to the numerous applications, it is common, in certain topologies, to find a great number of magnetic devices. The problem is that magnetic devices need to be designed, manufactured, are heavy, occupy a significant volume, are expensive, may increase the losses of the converter, and can generate noise. For these reasons, many designers try to limit their use.

One strategy to overcome these problems is to employ high-frequency converters, which allow the use of smaller and lighter magnetic devices. However, this may increase the losses in some cases and may cause electromagnetic interference. Another possibility is to develop planar magnetic devices on the printed circuit board (PCB). In [1], the authors integrate a high-frequency inductor into a PCB using a toroid shape and magnetic sheets. However, this strategy requires high frequency to reduce the required inductor size to fit on the PCB. Besides, the low cross-sectional areas of the magnetic device sheet make it difficult to achieve high inductance values. One can eliminate inductors completely by using switched-capacitor solutions [2]. The converters are usually more compact, lightweight, and cheaper, but they present some limitations, such as low efficiency and higher ripple.

For high-power converters, magnetic devices are usually present, and one path to deal with the problems cited before is magnetic integration.

13.2 AC–AC converter with multiple magnetic devices

Some topologies, such as those in [3–5], for instance, require a high number of magnetic components. In [3], a single-phase isolated AC–AC converter is presented, requiring four different magnetic components, including a transformer, one filter inductor, and two interleaved transformers. Similarly, in [4], a Z-source AC–AC converter is introduced, also necessitating four magnetic components. The use of a DAB in [5] to isolate the source from the load also requires four magnetic components.

The work presented in [6] employs the DAB in an AC–AC single-phase topology but includes interleaving operation as well. In this topology, as depicted in Figure 13.1, there are six different magnetic elements: two filter inductors, two interleaved transformers, one transformer, and one current-limiting inductor in series with it. This topology will be used in this chapter to illustrate how magnetic integration can be performed at different levels.

The first-level integration to perform is to integrate the inductance (L_{Leak}) with the transformer. The second-level possible integration in this topology is an option without isolation, in which the filter inductor and the interleaved transformer share the same magnetic core. The third-level integration consists of an option with isolation, and it integrates the second-level integration with a transformer. With this last integration, one magnetic device would performs the task of four different magnetic devices. In the following sections, each integration level is explained in detail.

13.3 Storing energy is leakage flux

The leakage magnetic flux is present in transformers since not all the flux of the primary winding links to the secondary winding. The leakage flux stores energy and is welcomed for the application mentioned previously since it produces a series leakage inductance. In the case of the study, there is a current limiting inductance in series with the transformer, so instead of designing an inductor to match the required series inductance, the energy of such an inductor can be stored in the leakage flux of the transformer.

The window area and the winding arrangement between the primary and secondary sides of the transformer have a great impact on the coupling coefficient. The window area is tied to the magnetic core, making it a less flexible parameter to change when adjusting the leakage. The winding arrangement, conversely, can be used to achieve greater or smaller leakage flux since different arrangements result in different coupling coefficients. The choice of the suitable winding arrangement is performed using empirical knowledge and experimentation.

The transformer of the DAB in Figure 13.1 was rated for 1 kW power with a 400 V DC link on both sides. The transformer employed an E55/28/21 core, with a 0.39 occupation factor and 0.19 of peak magnetic flux density inside the core, which was sufficient to avoid saturation in the N87 material employed. After

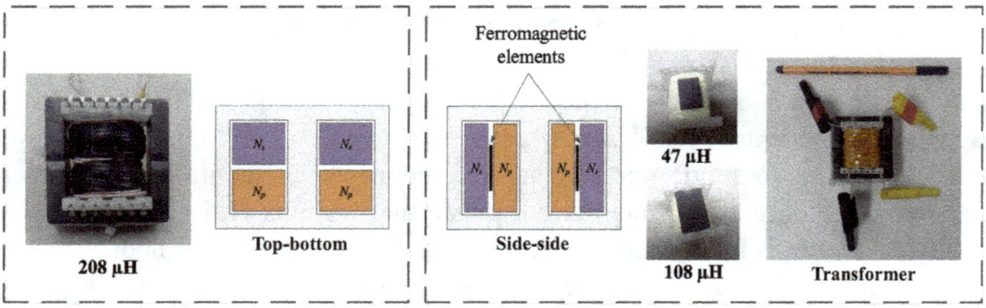

Figure 13.2 Experimental setup and results for two winding arrangement possibilities to adjust the leakage flux of a transformer

designing the transformer and determining the number of turns, it is possible to identify the winding arrangement that best suits the required inductance. For the DAB, a 108 μH inductance was required for the rated power and a phase shift of 15°. Given the requirements, Figure 13.2 shows the different winding arrangements and the obtained leakage inductances.

Figure 13.2 shows two winding arrangement possibilities, although there are many others. The top-bottom configuration results in a higher leakage flux and consequently higher inductance. This winding arrangement was tested, resulting in 208 μH of leakage inductance. A small leakage flux is obtained when the primary and secondary windings are interleaved. A medium leakage flux is obtained by the side-side configuration. This last possibility can be constructed with or without ferromagnetic elements between the windings. The ferromagnetic element is used to ease the leakage flux path and increase the resulting inductance. Notice that considering the side-side configuration without a ferromagnetic element results in very small inductance (not presented in the figure). To increase this value, two different ferromagnetic elements were tested. The smaller segment, with dimensions 20.16 mm × 14.02 mm × 2.26 mm, resulted in an inductance of 47 μH. The larger ferromagnetic element, measuring 31.95 mm × 20.50 mm × 2.99 mm, resulted in an inductance of 103 μH, which is a value very close to what was designed.

13.4 Integration of the interleaved transformer and the filter inductor

One interleaved transformer can be manufactured using toroid cores, taking advantage of the fact that the low-frequency fluxes are canceled due to the winding orientation. Toroid cores are known to minimize the leakage flux because of their geometry. To integrate the filter inductor with the interleaved transformer, one needs to store magnetic energy. Although there are some options to increase the energy stored in toroid cores, an E core was chosen. In an E core, one needs to take all leakage paths into account to obtain a more precise design. According to Kleeb [7], the most important leakage reluctances are presented in the equivalent circuit of an EE core, as shown in Figure 13.3(a).

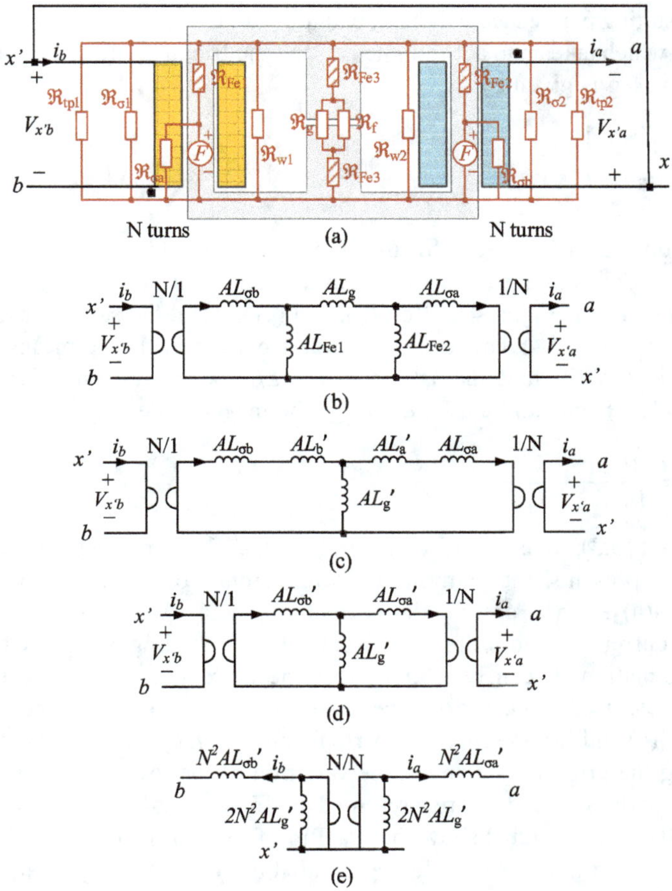

Figure 13.3 Integration of one inductor with an interleaved transformer (based on [8] and modified): (a) leakage reluctances of an EE core, (b) equivalent model, (c) Δ–Y conversion, (d) equivalent resulting circuit, and (e) element rearrangement to determine the longitudinal equivalent inductance of the integrated magnetic device

Essentially, there is leakage flux between the top and bottom surfaces of the EE core (translated to the reluctances R_{tp1} and R_{tp2}), leakage flux besides the outer legs (respective reluctances $R_{\sigma 1}$ and $R_{\sigma 2}$), inside the window (reluctances R_{w1} and R_{w2}), and in the windings (reluctances $R_{\sigma a}$ and $R_{\sigma b}$). Finally, one takes into account the reluctance of the air gap (R_g) and the fringing effect (R_f). As R_{tp1}, R_{tp2}, $R_{\sigma 1}$, $R_{\sigma 2}$, R_{w1}, and R_{w2} are in parallel, with the results association of the center leg, one can sum all the permeances to obtain the resulting leakage permeance 1

$$AL_{leak} = \frac{1}{R_{tp1}} + \frac{1}{R_{tp2}} + \frac{1}{R_{w1}} + \frac{1}{R_{w2}} + \frac{1}{R_{\sigma 1}} + \frac{1}{R_{\sigma 2}} \tag{13.1}$$

The resulting permeance will be in parallel with the resulting permeance of the central leg, which takes into account the gap and the fringing effect. The permeance of the central leg is given by

$$AL_{cl} = \frac{1}{\frac{1}{2 \cdot R_{Fe}} + \frac{1}{\frac{R_g \cdot R_f}{R_g + R_f}}} \tag{13.2}$$

Based on the reluctances defined in Figure 13.3(a) and the notation presented in [9], one derives the equivalent circuit presented in Figure 13.3(b), where $AL_g = AL_{cl} + AL_{leak}$. Figures 13.3(c), (d), and (e) present the steps to simplify the model performing a Δ–Y conversion and the isolation of the permeances, which form the inductor. The longitudinal inductance L_{LC}, obtained from Figure 13.3e, is detailed in (13.3) and can be used to design the magnetic device.

$$L_{LC} = \frac{AL_{\sigma a}{}' \cdot AL_{\sigma b}{}'}{AL_{\sigma a}{}' + AL_{\sigma b}{}'} N^2 \tag{13.3}$$

Besides (13.3), one needs to determine the flux and the current ripple. Figure 13.4 presents the current and flux ripple of the magnetic device. Figures 13.4(a) and (b) present the current ripple for two different duty cycle levels. The currents i_a and i_b are the currents in each winding, while the current iL is their summation. It can be noticed that the currents i_a and i_b have slightly different rates. This is caused by the reduced coupling between the windings. Figures 13.4(c) and (d) present the flux ripple on legs a, b, and in the central leg g. Considering the grid voltage angle, the variation of the low-frequency flux, and the ripple for legs a and b are presented in Figure 13.4(e), and for leg g in Figure 13.4(f). Both figures show the resulting flux in a traced line. The current ripple variation with the grid angle is presented in Figure 13.4(g) for the winding currents and in Figure 13.4(h) for the resulting current. The maximal ripple for the resulting current is given by

$$\Delta iL_{max} = \frac{V_{dc}}{8 \cdot L \cdot fs \cdot (1 - k)} \tag{13.4}$$

where L is the self-inductance, k is the coupling coefficient, fs is the switching frequency, and V_{dc} is the DC-bus voltage.

The maximum flux inside the core is given by

$$\Phi_{a(b),max} = \frac{L \cdot (1 - k) \cdot I_{a(b)pk}}{N} + \frac{(1 - M_f) \cdot M_f \cdot V_{dc}}{fs \cdot N} \tag{13.5}$$

$$\Phi_{g,max} = \frac{L \cdot (1 - k) \cdot I_{a(b)pk}}{N} + \frac{(M_f - 1) \cdot (1 - 2 \cdot M_f) \cdot V_{dc}}{fs \cdot N} \tag{13.6}$$

where M_f is the modulation index, N is the number of turns and $I_{a(b)pk}$ is the peak current of the winding.

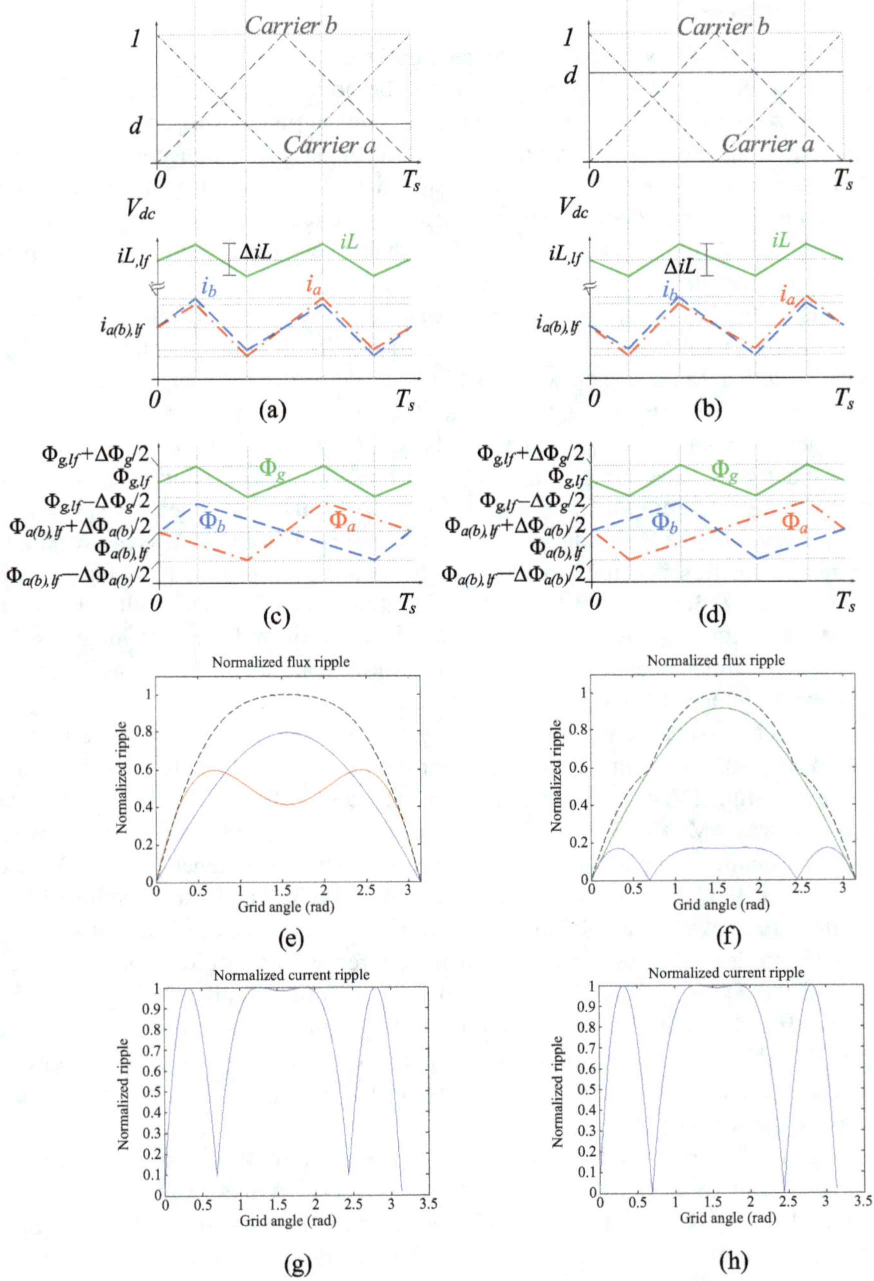

Figure 13.4 Current and flux swing: (a) and (b) present the detail of the current ripple in the windings and in the resulting current iL; (c) and (d) show the flux ripple for the core legs a, b, and g; (e) and (f) present the variation of the normalized ripple with the grid angle for legs a, b, and g; and (g) and (h) show the variation of the ripple of the windings and the resulting current with the grid angle.

13.4.1 Design and results

The integration of this section can replace one inductor and one interleaved transformer. Taking a look at Figure 13.1, it can be noticed that there are two occurrences of such type: the inductor L and the interleaved transformer connected to the point x' and the inductor Lf and the interleaved transformer connected to point y'.

Given this power rate of 1 kW assigned to this converter, the inductance required to achieve 20% of the current ripple is 640 μH. However, considering the design of the LC filter at the load side, the inductance was adjusted to 683 μH. It is on this side that this integration will be placed.

Based on the design requirements, one can determine the number of turns isolating the variable N in (13.3). The core chosen for the implementation was EE 55/28/21 with a 2 mm air gap with MF102 material from Tri Delta, which presents a saturation of 0.5 T. Based on the core, all the reluctances of Figure 13.3(a) can be determined mathematically or experimentally. The values found are 3.112 μH for AL_g', and 176.509 nH for AL_a' and AL_b'. The values for $AL_{\sigma a}$ and $AL_{\sigma b}$ are very small when compared with the other values (2.971 nH), but when considered in (13.3), it leads to one turn reduction in the design, resulting in 87 turns per winding. The maximum flux densities for legs a and b are 0.36 and 0.34 T for leg g were obtained from (13.5) and (13.6) and the core geometry. The manufactured device can be observed in Figure 13.5(a), and it reached a measured boxed volume of 120 065 mm^3. As this device integrates one inductor into an interleaved transformer, it is named an integrated interleaved transformer.

In order to compare the resulting integrated magnetic device with the individual counterparts, it is important to present the design of such parts. The inductor Lf was designed using amorphous alloy 2605SA1 from Metglas. The core employed was AMCC-6.3 with a permeance of 385 nH, which requires 41 turns of a 1.4 mm diameter copper wire to generate 645 μH of inductance. The peak flux obtained was 0.67 T, with a material saturation of 1.5 T. The boxed volume of the manufactured device was 58 976 mm^3, and it can be observed in Figure 13.5(b).

The interleaved transformer was manufactured with a toroidal core of model B64290L0082X087 from EPCOS® with N87 material. With a linkage flux of 3.33 mWb and 128 turns for each winding, a 0.13 T flux density was determined in the core. The winding used 0.8 mm copper non-magnetic wire. The resulting manufactured magnetic device can be observed in Figure 13.5(c), and the boxed volume obtained was 83 513 mm^3.

Figure 13.5(d) presents the volume comparison of the involved magnetic devices. It can be seen that the integrated interleaved transformer has a smaller volume than the summation of the volumes of the inductor and interleaved transformer, which are the devices it replaces. The overall volume of the interleaved transformer and the inductor is 142 489 mm^3, while the integrated interleaved transformer presents a volume of 120 065 mm^3 as stated before. This means that the integration performer resulted in a volume reduction of 15.74%.

Experimental results were obtained by applying the integrated magnetic device to the secondary side of the converter presented in Figure 13.1. Figure 13.5(e) shows the

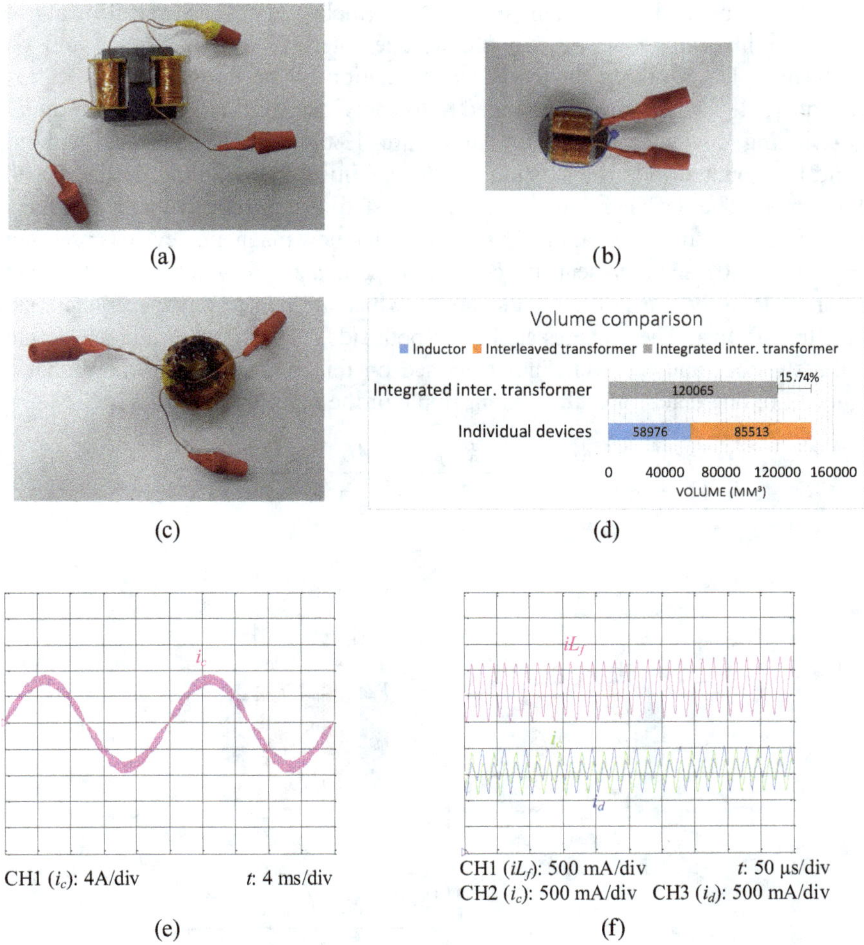

Figure 13.5 Integration results: (a) integrated interleaved transformer, (b) individual inductor, (c) individual interleaved transformer, (d) volume comparison, (e) winding current of the integrated device, and (f) ripple of both winding currents and the resulting current.

current in one winding (i_c) of the integrated interleaved transformer, while Figure 13.5(f) presents the currents i_c, i_d, and iL_f, which is the summation of the winding currents. The obtained inductance estimated by the current ripple was 724 µH.

13.5 Integration of the interleaved transformer, filter inductor, and transformer

In the previous section, the integration of the inductor and the interleaved transformer using an E core and two windings was demonstrated. The advantages are reduced volume and component count. Nevertheless, one can perform another integration involving one filter inductor, one interleaved transformer, and one transformer. The resulting device is a combination of the knowledge of Sections 13.3 and 13.4. Two windings are added to the previous configuration, resulting in a four-winding device, as presented in Figure 13.6(a). Each additional winding should be mounted over the existing one to maximize the coupling needed for the transformer. The leakage inductance is adjusted using ferromagnetic segments between this winding association. The model of the new magnetic device is presented in Figure 13.6(b), and it presents the connecting point *a* and *b*, which are related to the primary side of the transformer, and *c* and *d*, which are related to the secondary side of the transformer. The voltages applied to both sides of the transformer are defined by the switches and are well known. Based on this, and analyzing the circuit of Figure 13.6, one determines the current ripple of the input current *iL* in

$$\frac{\Delta iL}{\Delta t} = \frac{(v_{x'a} + v_{x'b}) \cdot [AL_g \cdot (1 - k_{bc}) - 2 \cdot AL_\sigma \cdot (1 + k_{ab})]}{L \cdot (1 - k_{bc}) \cdot (AL_g + 2 \cdot AL_\sigma) \cdot (k_{ab} - 2 \cdot k_{ab} + 1)} \tag{13.7}$$

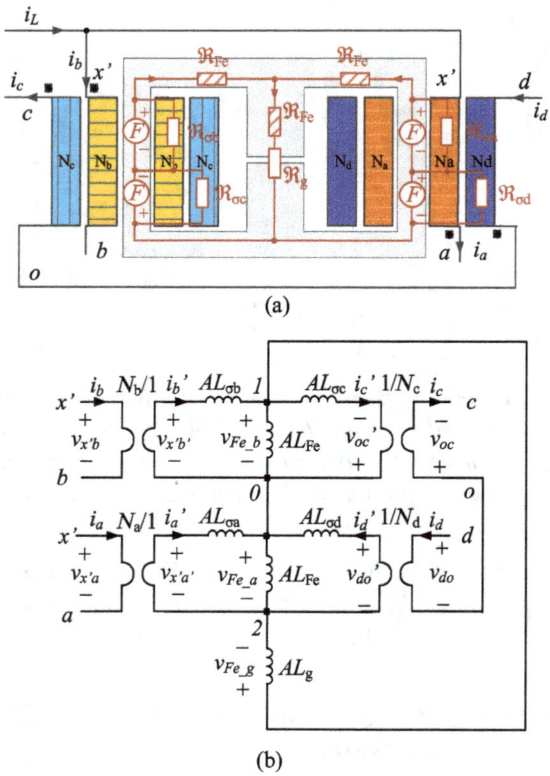

Figure 13.6 Integration of the magnetic elements [8]: (a) four winding magnetic integrated magnetic device and (b) respective model.

where AL_σ is the permeance of the leakage path of each winding, k_{ab} is the coupling coefficient between the windings of the primary side, which is related to the filter inductance, and k_{bc} is the coupling coefficient between the primary and secondary sides of the transformer.

Taking the variable nature of the duty cycle along the grid period, one determines the maximum current ripple achieved by the designed magnetic device, which occurs when $\omega_g \cdot t = sin^{-1}[3/(4 \cdot M_f)]$ and it is given in

$$\Delta iL_{max} = \frac{V_{dc} \cdot [AL_g \cdot (1 - k_{bc}) - 2 \cdot AL_\sigma \cdot (1 + k_{ab})]}{8 \cdot L \cdot fs \cdot (1 - k_{bc}) \cdot (AL_g + 2 \cdot AL_\sigma) \cdot (k_{ab} - 2 \cdot k_{ab} + 1)} \tag{13.8}$$

From Figure 13.6(b), the magnetic flux rate can be determined by calculating the voltage applied to the core legs, more specifically v_{Fe_a}, v_{Fe_b}, and v_{Fe_g}. These voltages can be written as a function of the winding applied voltages, which are known. Considering the leakage permeance the same for all windings, i.e., $AL_{\sigma a} = AL_{\sigma b} = AL_{\sigma c} = AL_{\sigma d} = AL_\sigma$, applying circuit law, considering $AL_g \gg AL_\sigma$, $AL_{Fe} \rightarrow \infty$, and after some other steps, the flux rate in each core leg is given by

$$\begin{aligned} v_{Fe\_a} &= \frac{\Delta\phi_a}{\Delta t} = \frac{1}{4} \cdot \frac{(3 \cdot v_{x'a} + v_{x'b} - v_{cd})}{N}, \\ v_{Fe\_b} &= \frac{\Delta\phi_b}{\Delta t} = \frac{1}{4} \cdot \frac{(v_{x'a} + 3 \cdot v_{x'b} + v_{cd})}{N}, \\ v_{Fe\_g} &= \frac{\Delta\phi_g}{\Delta t} = \frac{v_{x'a} + v_{x'b}}{N} \end{aligned} \tag{13.9}$$

The flux variation described above is combined with the flux defined by the low-frequency input current, which is present in windings a and b. The resulting flux can be used to determine the maximum flux achieved during one voltage grid cycle, which is obtained when $\omega_g \cdot t = \pi/2$, resulting in the following for $M_f > 0.75$:

$$\Phi_{a(b),max} = \frac{L \cdot (1 - k_{ab}) \cdot I_{a(b)pk}}{N} + \frac{(1 - M_f) \cdot M_f \cdot V_{dc}}{e \cdot fs \cdot N} - \frac{(1 - M_f) \cdot V_{dc} \cdot \phi}{4 \cdot \pi \cdot fs} \tag{13.10}$$

13.5.1 Design and results

In order to adjust the leakage inductance needed for the DAB, ferromagnetic segments are inserted between the windings of the primary and secondary sides of the transformer, as it is presented in Section 13.3. This device employs an E 55/28/25 core with a 2.6 mm gap to store energy for the inductance in a similar way as implemented in Section 13.4. A 2.36 A current ripple was defined (35%), resulting in a 352 μH inductance. For the leakage inductance of the DAB, the target value of 108 μH was maintained. To achieve this, one ferromagnetic segment I 20/14/2.3 – 20.1 mm/14.07 mm/2.26 mm was used in each leg of the device, placed between the primary and secondary windings. The obtained value of leakage inductance achieved is 112 μH, while 412 μH was obtained for the inductance.

Figure 13.7 presents all magnetic devices involved. The first row of devices presents the individual solutions, considering the integration mentioned in Section 13.3 to manufacture the transformer with its leakage inductance adjusted to the value required by the DAB. The second row of devices presents the integration results. The integrated transformer on the left side of this row represents the union of the inductor, interleaved transformer, and transformer, as presented. The second device, labeled integrated interleaved transformer, represents the union of the interleaved transformer and the inductor. Formerly, the converter presented in Figure 13.1 would need six different magnetic devices, but only five are shown due to the integration of the current limit inductor (L_{leak}) with the leakage inductance of the transformer.

A volume comparison between the individual magnetic devices and their integrated counterparts is presented in Figure 13.8(a). From the graphic, one notices that the integration has led to a volume reduction of 17.52% comparing the integrated with the individual devices. Figure 13.8(b) shows how the magnetic integration affected the efficiency of the converter. For the tested power levels, the efficiency was equal to or higher than the tests with the individual magnetic devices. Figure 13.8 (c) presents the winding current of the primary side of the integrated transformer, while Figure 13.8(d) presents the current on the secondary side of the integrated

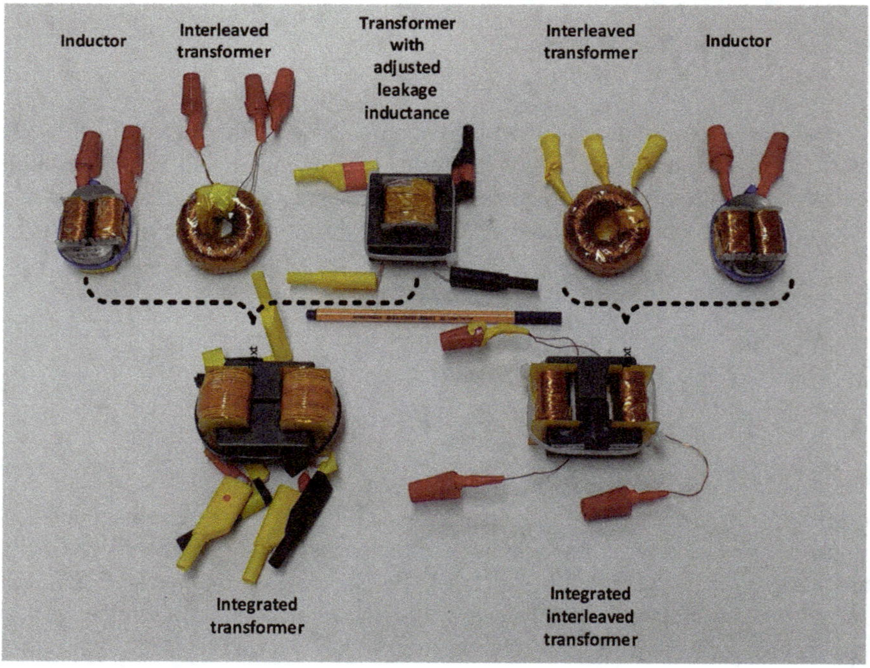

Figure 13.7 Designed magnetic devices. In the first row, the individual magnetic devices are presented, while in the second row of devices, the integrated solutions are shown, relating to the devices they replace.

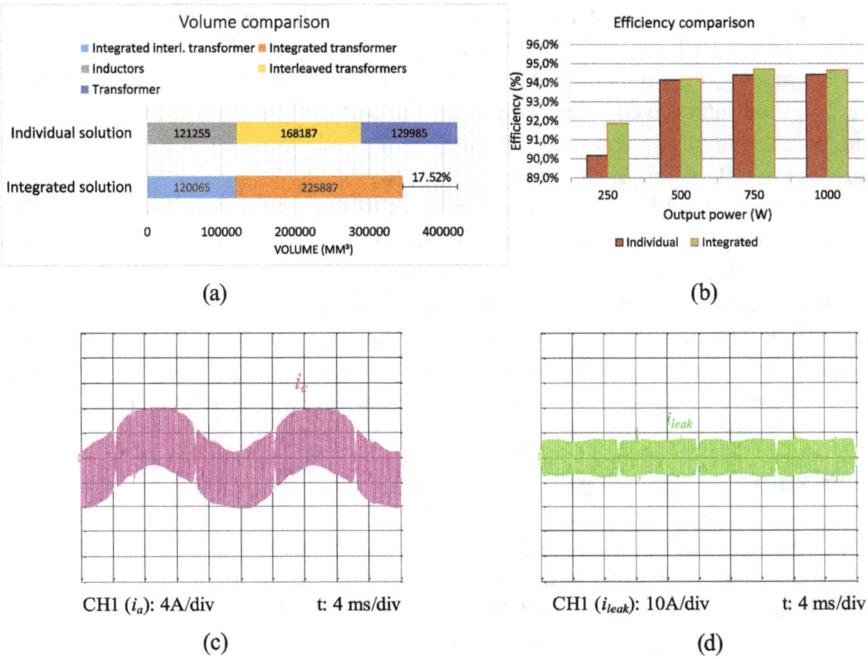

Figure 13.8 Experimental results: (a) graphic presenting the volume comparison between the individual magnetic devices and their integrated counterparts, (b) converter efficiency comparison with individual magnetic devices and their integrated versions, (c) current in the winding of the primary side of the integrated transformer, and (d) current in the secondary side of the integrated transformer.

transformer. As expected, the winding current of the primary side is the combination of the transformer current, represented in Figure 13.8(d), and the winding current of the integrated interleaved transformer, as presented in Figure 13.5(e).

13.6 Summary

The number of magnetic devices in power converters can be high in some topologies. The main problems related to this are heavy magnetic devices, which occupy great volume being usually expensive and increasing the losses of the converter. Considering that reducing costs, volume, weight, and increasing efficiency are common goals for any power converter designer, it is clear that establishing integration for magnetic devices is of great interest.

There are different strategies to reduce the number of magnetic devices in a converter. Understanding the leakage paths and how the leakage flux can be adjusted is very important for different integration types. The knowledge of the impact of different winding arrangements on the leakage inductance is essential, as well as the use of ferromagnetic segments.

Integrating one inductor with one interleaved transformer requires storing energy in the interleaved transformer, and this is mainly performed using air gaps. The core geometry plays an important role, and knowing the leakage paths increases the precision of the design.

Different strategies can be combined, as in the case of the integrated transformer, where the leakage adjustment of the transformer and the inductor energy storage manipulation were employed to design this device.

The design approach presented in this chapter, among other things, brings the magnetic model close to the real implementation and eases the simulation simply by using transformers and inductors as elements and allows the seamless integration between the magnetic model and the electric simulation model.

References

[1] Murphy R, McCloskey P, Cao Z, *et al.* High Frequency Magnetic Sheet Materials – Performance Factor Comparisons and Design of Toroidal Inductors Embedded in PCB. In: 2021 *IEEE Applied Power Electronics Conference and Exposition (APEC)*; 2021. p. 2897–2904.

[2] Chung H, and Ioinovici A. Development of a Generalized Switched-Capacitor DC/DC Converter with Bi-Directional Power Flow. In: *2000 IEEE International Symposium on Circuits and Systems (ISCAS)*. vol. 3; 2000. p. 499–502 vol.3.

[3] Ahmed HF, Cha H, Khan AA, *et al.* A Highly Reliable Single-Phase High-Frequency Isolated Double Step-Down AC–AC Converter With Both Non-inverting and Inverting Operations. *IEEE Transactions on Industry Applications*. 2016;52(6):4878–4887.

[4] Ahmed HF, Cha H, Khan AA, *et al.* A Family of High-Frequency Isolated Single-Phase Z-Source AC–AC Converters with Safe-Commutation Strategy. *IEEE Transactions on Power Electronics*. 2016;31(11):7522–7533.

[5] Huang AQ, Crow ML, Heydt GT, *et al.* The Future Renewable Electric Energy Delivery and Management (FREEDM) System: The Energy Internet. *Proceedings of the IEEE*. 2011;99(1):133–148.

[6] da Silva Filho OC, Tofoli FL, Barreto LHSC, *et al.* Single-Phase Isolated AC–AC Symmetrical Full-Bridge Converter. *IEEE Journal of Emerging and Selected Topics in Power Electronics*. 2022;10(1):846–855.

[7] Kleeb T. *Investigation on Performance Advantage of Functionally Integrated Magnetic Components in Decentralised Power Electronic Applications*. Kassel: Kassel University press; 2016.

[8] da Silva Filho OC. *A Family of High-Frequency Isolated Single-Phase AC–AC Converters [PhD thesis]*. University of Ceara. Fortaleza, CE; 2019. Available at http://www.repositorio.ufc.br/handle/riufc/48291.

[9] Zacharias P. *Magnetic Components – Basics and Applications*. Springer; 2022.

Chapter 14

Considerations for design optimization of high-power medium-frequency transformers

Drazen Dujic[1]

There are numerous ways to tackle the design optimization problem of inductive devices, being it a transformer or an inductor. The problem is generally nonlinear, nonconvex, and comprises a mix of continuous and discrete variables which makes it rather a difficult task. The design optimization can be carried out considering many different objectives, such as efficiency, weight, volume, and cost, to name a few, and trade-offs between them must be accepted. The power density, expressed as the volumetric power density (Watts per liter) or the gravimetric power density (Watts per kilogram), is a popular figure of merit adopted in many works.

As the power ratings and insulation requirements increase, as is the case with medium-voltage high-power galvanically isolated converters, scaling laws and trade-offs tend to be significantly different from the design problem encountered in low-voltage applications, such as switched-mode power supplies. This chapter provides an overview of the design considerations, taking the medium frequency transformer typically found inside all solid-state transformer topologies, as an example. Power density-driven design optimization must carefully consider all relevant domains: electrical, magnetic, dielectric, and thermal, to reach a feasible design.

14.1 Solid-state transformers

To keep the discussion limited and yet illustrative enough, let us consider the problem of the design of the medium-frequency transformer (MFT) for the solid-state transformer (SST). The SST is nowadays considered promising and emerging conversion technology for various MW-level medium-voltage applications and conversion needs [1]. The SST is, in essence, a galvanically isolated modular converter, which uses a plurality of MFTs, operated at high frequency, to achieve galvanic isolation between its input and output terminals, equipped with power semiconductor conversion stages. While the SST can be topologically arranged for

[1]Power Electronics Laboratory, Swiss Federal Institute of Technology in Lausanne - EPFL, Switzerland

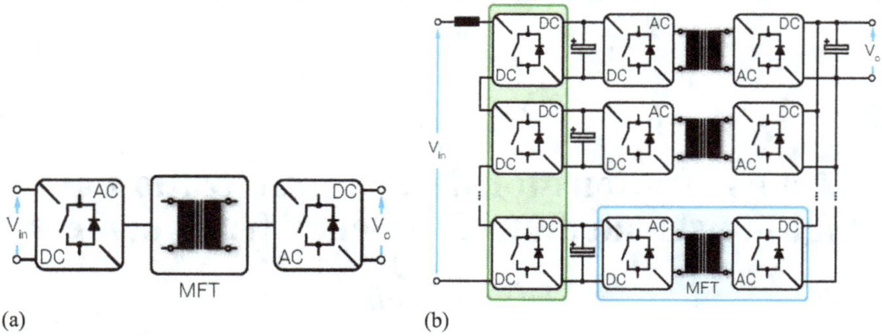

Figure 14.1 Exemplary DC–DC SST topologies: (a) bulk (monolithic) power conversion with single MFT rated for full power and (b) fractional (modular) power conversion with multiple MFTs of smaller power ratings

any conversion needs (AC–DC, AC–AC, DC–DC), Figure 14.1 illustrates two exemplary and generic variants of SST for DC–DC conversion.

Figure 14.1(a) illustrates the monolithic DC–DC SST, where a single MFT, rated for full power, is used to provide voltage adaptation and galvanic isolation. The actual topologies of power conversion stages are irrelevant, and a single MFT must be designed for the overall converter ratings in terms of power (e.g., 1 MW) and highest system voltage (e.g., 10 kV). On the contrary, modular SST shown in Figure 14.1(b) uses many converter stages of smaller power ratings compared to the overall SST ratings. The presented example is shown as a double-stage SST in input-series output-parallel configuration, but many other topological variations are possible. Assuming the use of five converter stages and the same exemplary ratings mentioned before, each MFT would be rated for 200 kW but still must withstand the same highest voltage between primary and secondary winding (e.g., 10 kV), even of the actual winding voltages are much smaller and defined by converter stages and not terminal voltages of the SST.

Since the modular SST topologies are much more popular nowadays, MFT design considerations elaborated in this chapter rely on this topological variant to highlight design trade-offs. Material presented hereafter is a brief overview of the challenges associated with design optimization of high-power medium-voltage MFTs, and many more details and considerations are available in the Ph.D. thesis [2,3].

14.2 MFT design considerations

In contrast to line frequency transformers operated at grid frequency (50 or 60 Hz), MFTs are directly driven by power conversion stages at some arbitrary switching frequency (few or tens of kHz), defined by the overall SST design optimization. Considering that MFTs inside the SST are essential for the converter operation and have a big influence on the overall performance, their design is subject to various requirements as well as constraints. Addressing these requirements, through design optimization, requires understanding design degrees of freedom, the range of influencing parameters, and the impact of the various technologies and materials on the final design. Both these are briefly presented next.

14.2.1 Scaling laws

While not extremely precise, the scaling law based on area product, A_p (m^4) is commonly used for line frequency transformer design [4,5]. It describes the dependency of the product of the winding (core window) area A_w and the core cross-sectional area A_{core} on the apparent power P_{MFT} (W), the current density in the windings J_m (A/m^2), the operating frequency f_{sw} (Hz), and the maximum flux density B_m (T). The derivation is rather simple and for an MFT with only two windings, the necessary core window area is given by the following:

$$A_w = \frac{N_1 A_{con1} + N_2 A_{con2}}{K_u} = \frac{N_1 I_1 + N_2 I_2}{K_u J_m} \tag{14.1}$$

where the core window utilization coefficient K_u (p.u.) quantifies the maximum space occupied by winding conductors in the core window, relative to its total available area. Furthermore, the N_1 primary and N_2 secondary number of turns are with the conductor's cross-sectional areas, A_{con1} (m^2) and A_{con2}, respectively, having a current density J_m (assumed identical for both windings).

The primary and secondary number of turns are derived from Faraday's induction law under the assumption of no flux bias and no core lamination (the effective core cross-sectional area corresponds to the actual core area). In case the core material is laminated, the utilization factor needs to be introduced to obtain the effective core cross-section area. The following expressions are acquired:

$$N_1 = \frac{V_1}{K_f A_{core} f_{sw} B_m}, \; N_2 = \frac{V_2}{K_f A_{core} f_{sw} B_m} \tag{14.2}$$

The parameter K_f is the waveform coefficient, defined by the excitation waveform (4 for square wave, 4.4 for sinewave). Combining (14.1) and (14.2) yields

$$A_w = \frac{V_1 I_1 + V_2 I_2}{K_u K_f A_{core} f_{sw} J_m B_m} \tag{14.3}$$

Acknowledging the input/output power expressions $P_1 = V_1 I_1$, $P_2 = V_2 I_2$, the area product Ap, representing scaling law, is derived as

$$V_{MFT} \approx A_p = A_w A_{core} = \frac{P_1 + P_2}{K_u K_f f_{sw} J_m B_m} = \frac{P_{MFT}}{K_u K_f f_{sw} J_m B_m} \tag{14.4}$$

Note that the core area product scaling law from (14.4) is based on the apparent power P_{MFT}, a power-handling capability of the MFT, which can be approximately up to two times larger than the output power P_2. The logic behind this considers that the primary side needs to handle the input power P_1, whereas the secondary processes the output power P_2 at the same time; therefore, the transformer per design needs to accommodate the sum of the two powers. Moreover, the core area product A_p is proportional to the transformer volume V_{MFT} (m^3) and it offers a link between the power-handling capability of the core and its size. For cores of standard shapes and sizes, in low-power applications, the A_p value is assigned to the standard core sizes by the manufacturer and it summarizes the dimensional and electrical properties of cores. For high-power designs and nonstandard or custom core sizes, this must be determined case by case.

Several observations can be made from (14.4). The volume of the MFT is directly proportional to its power ratings; hence, the modularity of converter design to some extent influences the overall volume of all magnetic devices inside the considered topology. The peak flux density B_m is mostly limited by the selected core material properties or thermally manageable core losses at operating frequencies of interest. However, materials with high saturation values of flux density are used for lower operating frequencies and vice versa, indicating a necessary trade-off [6]. This can be traced back to the fact that both flux density magnitude and frequency directly impact the core losses, i.e., high B_m values imply higher core losses, which can be managed only up to certain frequencies. Vice versa, low-power transformer designs at very high frequencies (MHz range) operate at very low flux density values to manage the core losses. Furthermore, it should be noted that the thermal conductivity of the core material also has a strong impact on the thermal aspect of the transformer design and that this has not been considered in the core area product expression given above. Therefore, this relation provides only a simplified estimation of the transformer size, and it does not show effects on the internal characteristics of the transformer (e.g., temperature rise) nor effects such as skin, proximity, fringing, and many more.

The current density J_m, similar to flux density, is limited by conductor properties and winding thermal design, e.g., the ability to remove winding losses effectively. In other words, by investing in a highly efficient cooling system (likely costly as well) for windings, J_m could be increased for the MFT volume to be decreased.

The switching frequency f_{sw} is predominantly affected only by the semiconductor devices used to provide excitation for the MFT. In that respect, considering today's power semiconductor devices and in contrast to flux density and current density range of values, the frequency is the only parameter that can be significantly changed (within several orders of magnitude, from hundreds of Hz to tens of kHz), thus having the most significant influence on the MFT size. Hence, the logical observation from (14.4) is that with the increasing operating frequency, the MFT volume decreases, which directly implies an increase in the volumetric power density. Even though it is not explicitly shown in the equations, volume reduction can entail weight reduction as well, which can improve gravimetric power density, as well. This has been a driving force for power density for many years in low-voltage applications, such as switched-mode power supplies.

A volume reduction of high-power MFTs is often simply not possible due to insulation coordination for medium voltage designs. With derived creepage and clearance distances, characteristic of the MFT design, to further decrease any of those (windings to winding, winding to the core), more advanced and complex insulation materials (with higher dielectric strengths) need to be used. The use of oil insulation reduces the volume dedicated to insulation but increases the overall weight compared to a design with air insulation. For a selected insulation system, after a certain frequency, the distance reduction is no longer feasible, and the optimal frequency exists where the lowest weight and volume can be achieved.

14.2.2 Requirements

Any MFT design would have to meet certain requirements and standards derived from the application and converter design itself.

Electrical requirements would typically define rated power level (P_n) and targeted MFT efficiency (η), allowing for a certain range of possible operating frequencies (f_{sw}), with converter stages defining primary (V_p) and secondary winding voltages (V_s). Often, MFT integrates required inductances for the converter operation, and the design must deliver needed leakage inductance ($L\sigma$) and magnetizing inductance (L_m).

Dielectric requirements need to consider and winding-to-winding voltages (typically the highest system voltage of the SST – V_{in} in Figure 14.1), winding and turn-to-turn voltages of windings (V_1), (V_2), windings-to-core, windings-to-enclosure (if any), etc. The slew rate of voltages applied to the winding must be taken into account (*dv/dt*), and various types of medium or insulating materials can be used for different parts of the MFT assembly. In general, insulation coordination is challenging and has a big impact on the final MFT design. In contrast to low-voltage MFT, where frequency increases driven power density is often limited by the losses and temperature, in the case of medium-voltage MFT design, dielectric requirements often pose the limit beyond which power density cannot be further increased with the increase of operating frequency.

Thermal requirements, namely allowed temperature rise on the winding (T_w) and the core (T_c), must be respected through a dedicated design of the cooling system. Due to the close integration of the MFT with power electronics stages, many cooling methods typically used for line frequency transformers are often not feasible. Challenges are further exaggerated due to the drive to improve the power density through high-frequency operation, effectively reducing the surfaces for heat removal.

Mechanical constraints are ultimately related to the desired form factor of the MFT and its total width, length, height ($W \times L \times H$), volume (V), and mass (M).

All of the mentioned requirements or constraints can be satisfied in infinite ways of combinations of windings and core arrangements, leading to many possible MFT designs. To understand the complexity of the design optimization, available technologies or the design space are presented next.

14.3 Design space and technologies

Having defined required features, parameters, and constraints, the design space defines boundaries within which all of the possible design alternatives exist. Often, it is next to impossible or unpractical to explore a complete design space and the designer would have to make some compromises and narrow it down to a subset of possible or feasible combinations of available technologies.

14.3.1 MFT construction types

Among numerous possible ways of organizing windings around the core, there are a few dominant construction arrangements (types), as shown in Figure 14.2. Each construction type has its pros and cons and is more or less suited for a specific set of requirements.

Shell-type: This type is usually assembled of two or more E-shaped magnetic cores, with the primary and secondary winding placed concentrically around the middle core leg. To achieve the required cross-sectional area, several core sets are typically stacked together. In the case of a shell-type MFT, a rather good coupling can be achieved between the windings, minimizing the leakage flux and hence yielding a low leakage inductance. Compared to the core type, it offers a larger core surface available for convective cooling. Conversely, the cooling surfaces of the windings are less accessible, increasing the challenge of thermal coordination. Some examples are reported in [6–8].

Core-type: It consists of two or more rectangular or C-cut magnetic cores, with the windings concentrically arranged around each of the core legs. This is a typical way to construct this type of MFT, but with adequate winding technique, noncut C-cores can be used as well. Improved winding insulation can be achieved, in case the windings are separately placed around the legs, simply due to the larger distances involved. However, such an arrangement requires a more complex analytical model of the leakage inductance due to unknown magnetic field distribution outside of the windings. According to She *et al.* [9], this construction type is the most convenient for manufacturing and offers a more compact transformer design than the one realized with the shell type. Primary and

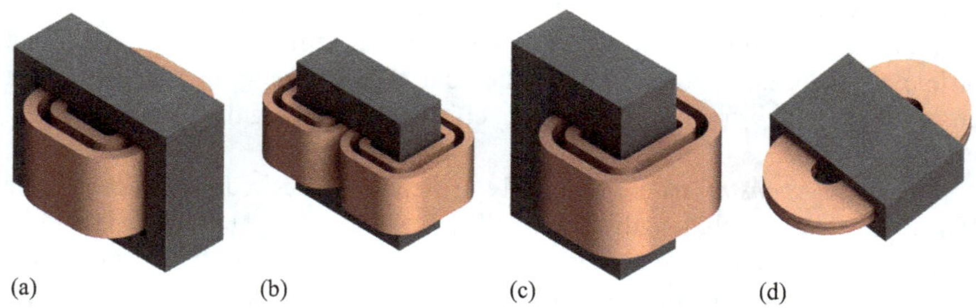

Figure 14.2 Common MFT construction types: (a) shell-type, (b) core-type, (c) C-type, and (d) planar-type

secondary windings can also be split, and their portions can be wound around each leg. These variations of winding placement offer a wide range of achievable leakage inductances. Representative examples are reported in [10–12].

C-type: This type is, by construction, a combination of the core-type and the shell-type. The windings are arranged in the same way as for the shell-type, except that C-cut cores are typically used. Additionally, the core can be constructed from two U-cores or a combination of U- and I-cores. The core window area can be reduced compared to the core-type but at the cost of limited winding arrangement possibilities. Thermally, this arrangement allows relatively good access to the windings (like core-type) and to the core (like shell-type). A good example of the MFT design with C-type construction can be found in [13].

Planar type: This particular construction type is employed when the converter design requires MFT with a low flat profile. Due to its characteristic construction, which features longer MLT of the windings, this type occupies a bigger volume and mass for the same core and winding cross sections. The resulting benefits are larger cooling surfaces for the same volume due to the flat shape, which offers more efficient convective cooling. In combination with the disc or printed circuit board (PCB) winding construction, the cooling medium can easily access the windings and the inner core surfaces. However, this type is mostly restricted to lower power ranges.

Presented variants are by no means an exhaustive list of options, and many winding-core arrangements are possible, as the high-power MFT design rarely can be served with commercial core sizes

14.3.2 Winding conductors

There are several popularly used conductor technologies for MFT windings, as illustrated in Figure 14.3. Frequency-dependent conduction losses, due to the skin and proximity effect, are one of the main parameters governing the choice of conductors used. Due to the relatively low cost and good mechanical and thermal properties, foil and bar conductors are usually preferred at moderate frequencies up to several kHz. In practice, the minimum realizable thickness of foil or bar wire is limited, and depending on the winding construction, these may experience

(a) (b) (c) (d)

Figure 14.3 Common conductor types: (a) foils, (b) bar wire, (c) Litz wire, and (d) hollow conductors (pipes).

significant additional losses due to the edge effect – especially in the case of foil windings at higher frequencies. While not illustrated, continuously transposed conductor (CTC) is another winding type derived from bar wires and commonly employed for high-power transformer windings at low-frequency applications. The conductor is usually made of an odd number of enameled bar wires, which together assume rectangular-shaped cables, eventually covered by a paper insulation layer. Thereby, the bars inside the CTC are radially transposed and can, thus, reduce the eddy current effects compared to a single solid conductor of the same cross-sectional area.

Nevertheless, at higher frequencies, typically above 5 kHz, fine-stranded litz wire represents a dominant technology of choice found in many MFT designs. Litz wires are available in various shapes and sizes, such as flat, round, and square pressed, with a cooling jacket in the middle, etc. Due to the very small diameter of the isolated strands, skin and proximity effects are greatly mitigated. Moreover, the edge effect is practically canceled, due to the isotropic skin and proximity mitigating properties, from any radial direction. Hollow conductors or pipes can offer good mechanical and very good thermal properties, due to the ability to use water cooling through an inner opening. As the wall thickness cannot be made extremely small, this type of technology is more interesting for lower frequency high-power (high current density) MFT designs. Coaxial cables are sometimes used as a part of the special coaxial transformer design, typically with unity turn ratios [7].

14.3.3 Winding arrangement

In addition to choosing the type of conductor, another important design choice is the arrangement of the windings and conductors within them, as illustrated in Figure 14.4. The cylindrical windings are concentrically stacked, whereas the disc windings are stacked axially, one on top of another, as shown in Figure 14.4(a) and Figure 14.4(b). This has implications on the electric, dielectric, and mechanical properties. Depending on the winding geometry ratios, the cylindrical windings typically result in a lower leakage flux density and the corresponding leakage

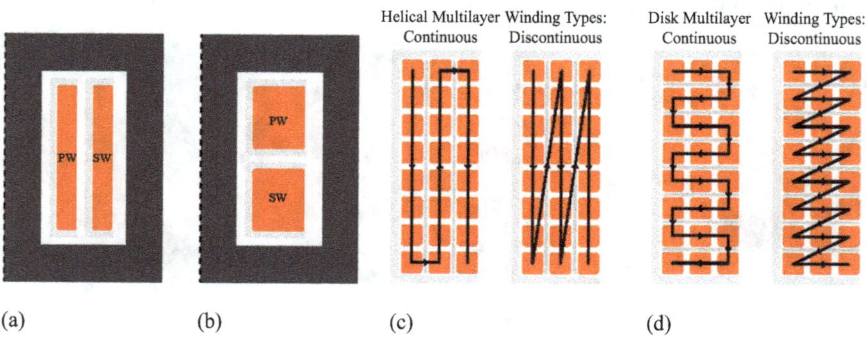

Figure 14.4 Winding arrangement: (a) cylindrical, (b) disc, (c) multilayer helical windings, and (d) multilayer disc windings

inductance compared to the disc arrangement with the same number of turns. Moreover, higher leakage flux density in disc windings results in a more pronounced proximity effect and an increase of the frequency-dependent AC resistance. Nevertheless, it is usually easier to implement good inter-winding insulation in the case of disc windings, as they are inherently better separated. The aforementioned conclusions do not necessarily hold for the planar transformer construction, as a special case of the disc winding transformer with extreme ratios—very low profiled windings. From the 2D electromagnetic point of view, the planar transformer construction can be seen as equivalent to the cylindrical, just rotated for 90°.

Interleaving of the windings is another interesting possibility but rather avoided in medium voltage-rated MFT designs. While this is a very powerful technique for reducing the leakage inductance and AC resistance of the low-voltage transformers operating on high frequencies, the added volume, necessary to facilitate medium voltage inter-winding insulation between each of the interleaved layers, is usually prohibitive. The insulation problem also extends within the winding, particularly in the case of multi-layer construction.

Two main conductor winding/stacking techniques are illustrated in Figure 14.4(c) and Figure 14.4(d). In the case of continuous winding, the potential difference between the two neighboring conductors in successive layers is two times higher compared to the discontinuous case. Nevertheless, the discontinuous winding requires a return conductor after each layer, thus complicating the design and increasing the length of the conductor and the resulting DC resistance. The choice of the appropriate alternative boils down to the trade-off between the complexity of adding the return conductor versus reinforcing the inter-layer insulation.

14.4 Materials

In addition to the spatial arrangement of the core and winding, the characteristics of the MFT are strongly governed by the characteristics of the utilized materials for the core, winding, and insulation coordination. Considering the loosely defined medium frequency range of 0.5–50 kHz, almost all core materials are suited to a certain range, as discussed next. The choice of material for the windings is rather simple, while the insulation coordination requires much more attention.

14.4.1 Core materials

High-power MFT designs can be found with almost all relevant core materials available: grain-oriented electrical steel (GOES), iron amorphous (Metglas), nanocrystalline, and ferrites. Even though each of these materials has its preferred frequency range, they can also be classified based on the nature of their construction into three groups, as displayed in Figure 14.5.

Low-frequency iron alloy-based cores used in the traditional line frequency transformers are usually made out of axially stacked lamination sheets, as shown in Figure 14.5(a). The same approach is used with GOES, with improved core loss

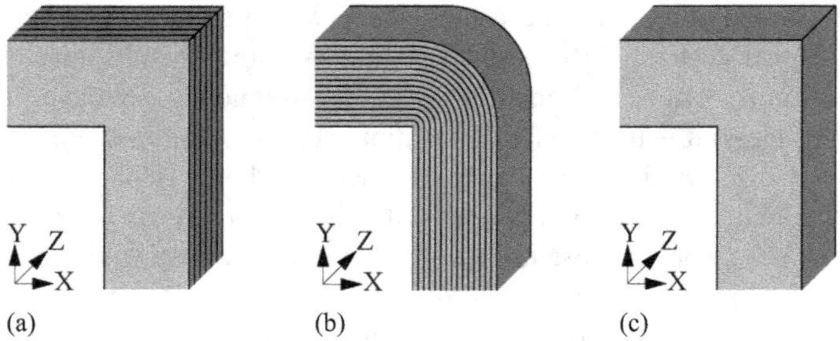

Figure 14.5 Typical construction of the cores: (a) axially laminated sheets, (b) radial laminated (wound) tapes, (c) isotropic, powder-based cores.

characteristics. The lamination thickness of the lamination goes down to 150–180 μm offering relatively acceptable eddy current loss suppression at frequencies up to 1–2 kHz [14]. Due to the high saturation induction (up to 2 T), excellent mechanical robustness, and resilience to very high temperatures (up to 400 °C), this material represents a very interesting alternative for MFTs designed for operation in hard conditions at modest frequencies.

At higher frequencies, eddy current and excess losses of the GOES materials are usually prohibitive. Another type of cores that are radially laminated are wound strip core materials (Figure 14.5(b)) such as iron amorphous (Metglas) and nanocrystalline, with a very thin strip width of 15–18 μm allowing to keep the eddy current losses at a reasonable level at elevated frequencies, roughly up to 50 kHz [15]. These materials however are characterized by lower saturation inductions of 1.56 and 1.23 T, respectively, and a maximum operating temperature of 155 °C. Due to the very low thickness and material properties, the strips are relatively brittle and require care during handling. While iron amorphous features a higher saturation induction compared to nanocrystalline, its losses are roughly three to five times higher within the frequency range from 1 to 50 kHz.

Finally, at high frequencies, above 50 kHz, isotropic ferrite cores are usually used, as shown in Figure 14.5(c). Ferrites are nonconductive ceramic materials with still good ferromagnetic properties, made by mixing large proportions of iron oxide with small proportions of one or more additional metallic elements, such as barium, manganese, nickel, and zinc [16]. Due to the very low conductivity of the material, the eddy current portion of the losses is practically negligible, thus allowing relatively efficient operation even at very high frequencies. However, regarding the ferromagnetic properties, their relative magnetic permeability is an order of magnitude lower compared to the aforementioned laminated materials, generating the so-called distributed air-gap effect. They also feature a relatively low saturation induction of around 0.3–0.4 T. Moreover, ferrite cores are mechanically brittle due to the ceramic structure. This poses restrictions on the core size and proportions and ultimately the application area.

It can be seen that similar to the trade-off in insulation materials (dielectric versus thermal properties), in the case of core materials, the main trade-off lies

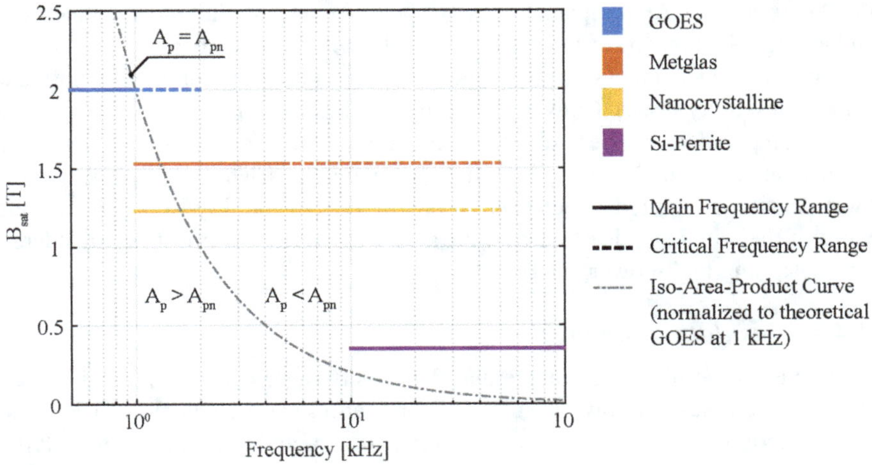

Figure 14.6 A qualitative map of the core materials in the saturation induction versus the operating frequency.

between the electric conductivity and magnetic properties, relative permeability, and saturation flux density. Moreover, it is important to mention the influence of the core construction on the thermal conductivity. Same as for the electromagnetic properties, laminated cores feature an anisotropic thermal conductivity: (i) relatively high along the lamination plane, defined by the strip material properties; and (ii) relatively low in a direction normal to the lamination, affected by the low conductivity of the necessary inter-strip insulation. This should be taken into account at the design stage to achieve optimal thermal coordination.

A chart of the theoretical magnetic saturation of the available core materials, with respect to the operating frequency, is given in Figure 14.6. The rough estimation of the critical frequency range where the core losses become thermally prohibitive is depicted with dashed lines. An iso-area-product curve, normalized to the theoretical area product of the MFT made with GOES and operating at 1 kHz is constructed to illustrate the influence of the frequency and the core material on the MFT size. As generally known, it can be seen that the materials with higher B_{sat} yield a more compact design for a given frequency. Moreover, it can be seen that even though the core materials for higher frequencies have a lower B_{sat} they will still yield higher power density at elevated frequencies. This is an important qualitative conclusion as it shows that the inherent technological curve of the core materials does not bind the potential for the MFT size reduction with the increase of the operating frequency.

14.4.2 Conductor materials

While many shapes of conductors are possible, as shown in Figure 14.3, there are two main conductor materials to be chosen from copper and aluminum. A selection of one or the other will impact the MFT design differently, mainly due to the inherent properties of these materials. Copper, having slightly lower electrical conductivity, features lower DC resistance of the windings compared to aluminum. At high frequencies, however, the copper conductors are more prone to additional skin and proximity losses. Conversely, aluminum has a significantly lower relative weight compared to copper, which makes aluminum conductors an interesting design alternative for MFTs, especially for applications where high gravimetric power density is required. Mechanical properties and the ability to realize various shapes of the windings, due to their size, are other details that must be considered when designing high-power MFTs.

14.4.3 Insulating materials

Depending on the blocking voltage and the target power density of the designed MFT, there are several ways insulation coordination can be achieved. The three main insulation types are introduced in section 14.3.5, and according to each type, the blocking voltage, and the selected insulation material, appropriate distances can be determined.

Note that the term oil is used to describe all the insulation liquids, including esters as well, which are not oil substances. The suitable insulating material is usually selected based on its dielectric and mechanical strength, which are interdependent properties. The parameter that affects both properties is the temperature. When exposed to high temperatures above the allowed values for a long time, the insulation experiences aging and a decline in both insulating and material, e.g., mechanical properties.

Note that high insulation voltage levels, common for medium-voltage and high-voltage applications, require larger distances for the insertion of the insulation material, which implies a reduction of the MFT power density. This can be compensated and improved by selecting a material with enhanced dielectric properties. Still, the main insulation concepts rely on air, solid, or oil as an insulating medium, and they are discussed in the following:

Air: In general, it is considered to be a good electrical insulator with a breakdown voltage of $\approx$3 kV. However, considering the humidity percentage and polluting particles in the air, this value usually reduces to 1 kV. Both values are defined for 1 mm distances. Hence, in case air is the only insulating material employed, large clearance distances are needed, which reduces the power density of the MFT. Nevertheless, an obvious advantage of this medium is its availability, no additional cost, and mass to the design. Additionally, it allows for combined insulation and cooling with the help of natural or forced convection.

Oil: Dielectric fluids used for transformer insulation can be divided into mineral-based and vegetable-based oils, which are used to produce natural ester liquids. Mineral oils have been in use for a very long time; however, they suffer from low fire resistance (flammable by nature) and have a damaging effect on the environment. Therefore, ester oils were introduced to improve these aspects, and they additionally exhibit higher breakdown voltages [17]. Regarding PD occurrence, oil, similar to air, possesses a self-healing ability, particularly when in motion, and it offers combined cooling and insulation capabilities. Nevertheless, oil degradation can occur after a long period of use. As mentioned in the previous section, the use of oil for insulation and cooling requires additional equipment, which adds weight and volume to the design [12]. Moreover, there are considerable integration issues, since it is not a typical power electronics technology. Typical values for this type of insulation medium are 5–47 kV/mm, regarding the dielectric strength, and 2–5 regarding the dielectric constant.

Solid: Designs that do not use any type of liquid insulation are known as dry-type designs. There are three main possibilities, namely, vacuum-pressure impregnation (VPI), resin-impregnated, and cast-resin transformers, which differentiate from each other based on the insulation procedure and the temperature class. VPI transformers use insulating materials, such as Nomex, Glastic, Kapton, Glass tape, and similar for winding insulation. After application of the material, the windings are dried to remove moisture, impregnated with varnish under vacuum pressure and additionally baked to cure the varnish. In this way, the resulting windings are both electrically and mechanically well made. The second method includes the insulating material, which is pre-impregnated with resin or pulled through a resin bath during winding, after which the windings are baked to cure the resin. Lastly, the cast resin dry type was developed as an alternative to the resin impregnation method, and it includes encapsulation of the windings with epoxy resin to prevent moisture from reaching them. The windings are placed into molds where the resin is introduced under vacuum, followed by baking for curing, cooling, and removing the molds. Both casting methods are susceptible to cracking, due to the use of epoxy-based materials, and require replacement, in case the insulation gets damaged. Note that all three solid insulation designs are usually air-cooled with natural or forced convection.

One of the key differences between solid type and air and liquid insulation is the potential presence of air pockets or voids between the conductor and the insulating material. This can lead to the accumulation of high electric fields that can discharge into the insulation and cause irreparable damage. Conversely, the complete casting of the transformer windings yields cooling problems, due to the low thermal conductivity of cast resin. Furthermore, thermal stresses can cause delamination, i.e., material fracturing into layers, as well as internal partial discharge (PD). Nevertheless, cast resin windings improve the power density compared to air due to an increase of breakdown voltage to around 45 kV [18] and reduced insulation distances. A list of frequently used insulating materials together with some dielectric properties is given in Table 14.1. Lastly, for insulation of large power and distribution transformers, the common practice includes solid insulation for the windings, in the form of paper and pressboards, and the whole component is immersed in the dielectric fluid, i.e., oil. As already emphasized before, the thermal and insulation coordination must be considered together since they impact each other significantly.

Table 14.1 Dielectric properties of common insulating materials

Dielectric material	Dielectric strength (kV/mm)	Dielectric constant ε_r (p.u.)
Air	1–3	1
Oil	5–47	2–5
Mica tape	60–230	5–9
NOMEX 410	18–27	1.6–3.7
PTFE	60–170	2.1
Mylar	80–600	3.1
Paper	16	3.85
PE, XLPE	35–50	2.3
KAPTON	118–236	3.9

14.5 Thermal and insulation coordination

Having selected core and winding geometries and materials, the choices need to be made among technologies for the cooling system and insulation coordination and its realization. Often, these two design aspects are closely linked as many cooling mediums are either good or bad insulators and vice versa.

14.5.1 Thermal design

Regardless of the MFT design, all of the generated winding and core losses and heat need to be exchanged with the surrounding ambient. All three modes of heat transfer, conduction, convection, and radiation are almost always present, with one of them often being dominant and maximized through design optimization. There are three main cooling methods or mediums for both core and windings, which are employed depending on the cooling effect that the designer wants to achieve. Those are namely, air, oil, and water.

Natural convection with air, also known as the self-cooled, is the simplest, cheapest solution, but not efficient for high powers and is often not feasible due to high voltages. The cooling effect can be improved by adding fans, resulting in forced air convection methods. Some MFT designs use heat sinks attached to the core as a means to remove core losses through conduction before convection (natural or forced) would remove it to the ambient. The location of heat sinks needs to carefully consider the thermal properties of the core in different directions due to thermal anisotropy of laminated cores, as shown in Figure 14.5.

The oil is a well-known and utilized solution in large power distribution transformers, for both thermal and insulation coordination, and it offers stronger convective cooling compared to air. However, this option comes with increased complexity and cost due to additional heat exchange apparatus (containers with oil, pumps to achieve force convection, and heat exchangers), and it is seldom used for MFTs. Nevertheless, examples can be found in [10,12].

Lastly, forced convection with water is very efficient, either as a direct or an indirect cooling method, but it cannot contribute to the insulation design. To use it for active cooling of the windings, the designs need to use either hollow conductors or have the conductors encompassed with cooling jackets. Considering that the cooling medium comes in contact with components at various voltage potentials, it must have low electrical conductivity, which is achieved with de-ionized water [3,12]. For core cooling special ducts or panels can be inserted for the water to circulate, which represents an indirect method. Additionally, it also relates to using water for cooling the oil in heat exchangers or aluminum coolers with water channels that can be placed on transformer surfaces.

14.5.2 Insulation coordination

Ensuring electric insulation of all the components within the MFT that are on different potentials is of paramount importance for reliable operation, as well as for the safety of users. The insulation material and the corresponding distances need to be selected carefully. Insulation coordination is the process of defining insulation levels by considering voltage levels and determining the dielectric strength of the equipment by considering the environment and characteristics of the available protection devices. In general, there are three main insulation types [19]:

Functional insulation: Its purpose is to ensure the correct functioning of the component. However, it does not protect or isolate against electrical shock. For an MFT it includes turn-to-turn insulation, layer-to-layer insulation, interwinding insulation, and winding-to-core insulation.

Basic insulation: Its task is to ensure protection against electrical shock. Furthermore, it needs to make sure that the failure of the functional insulation does not endanger life as a consequence of the grounding of the exposed parts.

Protective separation: It includes supplementary, double, and reinforced insulation. Supplementary insulation mitigates the risk of electrical shock in case of basic insulation failure. Double insulation is a redundant insulation layer, which doubles the basic insulation and increases the protective separation. Lastly, the reinforced insulation increases the distance prescribed by the basic insulation requirement (additional air clearance or insulation thickness of solid or liquid basic insulation).

The presented insulation types are illustrated in Figure 14.7. Thereby, the clearance provides the shortest air distance, whereas the creepage provides the shortest distance along an insulator surface between two conductive parts. The creepage distance is regarded as supplementary insulation. When it comes to the insulation design of traditional grid-frequency operated transformers, this is well studied and understood. However, operation at elevated frequencies and nonsinusoidal excitation bring various challenges. Thereby, several points need to be taken into consideration. Namely, the square voltage excitation that the MFTs experience has a much smaller rise time compared to the sinusoidal voltage, i.e., high-time derivatives (*dv/dt*). This in turn can lead to overvoltages between winding turns and turns-to-ground due to existing parasitics coming from the capacitive coupling. The result is uneven voltage distribution across the windings and

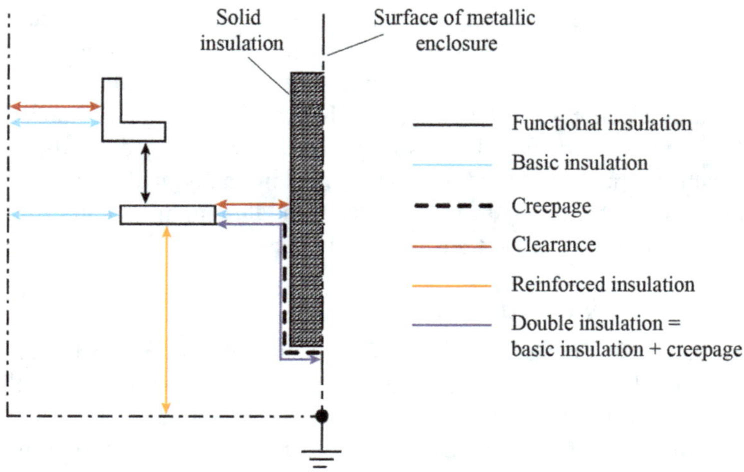

Figure 14.7 Illustration of different insulation types

increased electrical stress [20]. This in turn leads to a higher probability of PD occurrence, which, if it is continuously repeated, causes premature aging and insulation failure. Secondly, increased operating frequency implies increased core and winding losses, as well as dielectric losses, which altogether contribute to higher operating temperatures and higher thermal stress. As a result, many insulation materials experience a decrease in insulating capability, which eventually can lead to thermal breakdown.

Safety standards exist for different types of transformers, depending on their electrical specifications, construction type, and similar, and they give minimal creepage and clearance distances that need to be obeyed by the manufacturer. The values are generally determined depending on several criteria [19]. The Over-Voltage Protection Category describes the electrical conditions the component is installed. It makes the differentiation regarding indoor and outdoor installation, exposure to lightning strikes, and availability of protection against it. Pollution Degree describes the level of dust deposit and humidity, which influences the conductivity of the particles present in the environment. The material group, through a Comparative Tracking Index, describes the ability of the insulating material to resist degradation. Finally, system voltage, depending on the specific standard could be the maximal or RMS of the working voltage, impulse voltage, or temporary overvoltage, is used to define operating stresses and higher voltage levels that equipment must be tested against.

Additional influencing factors are altitude, the shape of the insulating surface, time under voltage, and others. However, considering the increase of the operating frequency into the MF range, there are still no concrete standards dealing with the insulation coordination of MFTs developed for the SSTs. Consequently, MFT designers are obliged to follow different standards. Some examples are IEC 60092-503 [21], which provides requirements for systems operating with voltages above

1 kV and up to 15 kV. EN 50124 [22] is the installation coordination standard for railway systems with voltages up to 25 kV, often followed in the work on SST and MFT for railways. IEC 61800 [23] is a popular standard for systems up to 35 kV and is most commonly applied to power semiconductor-based power conversion and medium voltage drive systems, while EN 60071 [24] provides details of the insulation coordination for systems with voltages above 1 kV.

There are several standard insulation tests used to verify the insulation design performance in normal operating conditions and for occasional overvoltage pulse incidences. Namely, those are the PD test, basic lightning impulse test, and the dielectric withstand test, also known as the high voltage, i.e., Hi-pot test. The PD test is used to ensure that the solid insulation is PD-free in a certain voltage range, i.e., to test the level of gradual insulation material degradation due to repeated localized PDs. The impulse test is a type of safety test, which aims to simulate overvoltages of atmospheric origins and the ones generated due to equipment switching. Lastly, the Hi-pot test verifies the ability of safe operation of the component during the rated electrical conditions, e.g., the effectiveness of the designed insulation. All the tests should be set according to recommendations stated in the corresponding safety standards.

14.6 Design optimization

Designing an MFT with desired characteristics implies solving a multi-physical, multi-objective optimization problem, in the sense that various domains and aspects need to be considered, such as electrical, magnetic, dielectric, and thermal. They all make an integral part of the problem and contribute to the challenges, which need to be properly addressed during the design phase. The mentioned domains are reflected in many transformer characteristics that indirectly have an impact on the converter performances. Those are namely magnetizing and leakage inductance, parasitic capacitances, voltage and current ratings, insulation voltages, operating frequency, power losses, and so on. Thus, achieving an optimal design requires considering all the relevant effects and influences, as well as selecting a suitable combination of technologies, some of which are discussed in this chapter.

Computer-aided design of power transformers was developed end of the 1960s and the beginning of 1970s of the last century and quickly became standard practice [25]. Considering the computation capabilities of the computers at the time, the governing equations and models needed to be as simple as possible, yet, capable of accurately describing the relevant effects and design trends and were usually analytical [26]. Based on the relevant phenomena discussed in the previous sections, one can recognize that the problem of the MFT design is highly complex, nonlinear, nonconvex and it is a multi-objective problem. To apply any optimization technique to this problem, one needs to define objectives, specifications, variables, and constraints. Therefore, as the first step, it is necessary to define an objective function, also known as a cost function, which can reflect a single design criterion or make a combination of criteria with weighting factors. Finally, the selected optimization algorithm needs to find the (global) optimum, maximum, or minimum of the defined cost function. Different optimization methods can be compared based on their execution time, accuracy, robustness, sensitivity to variations, and so on.

There are various optimization algorithms reported in the literature. First, genetic algorithms rely on operators such as mutation, crossover, and selection that are inspired by natural biological evolution to find optimal solutions. Second, design algorithms that use machine learning and neural networks have been reported. From given specifications and goals, a trained artificial neural network generates Pareto fronts and selects the optimal designs. A drawback here is that large data sets of good quality need to be available to train the network before use. Some numerical methods, such as Newton-Raphson, which is a gradient-based method can also be employed. Lastly, there are brute force or parametric sweep algorithms which look into all possible combinations of optimization variables. In case a large number of optimization variables is defined, this method can be very computationally intensive, but on the other hand, it is easy to implement. Deeper classification with prominent algorithms can be found in [27].

Figure 14.8 illustrates the flowchart of the brute force MFT design optimization method [2]. It combines the functionality from design optimization techniques described in [28–30] and offers additional degrees of freedom (e.g., different

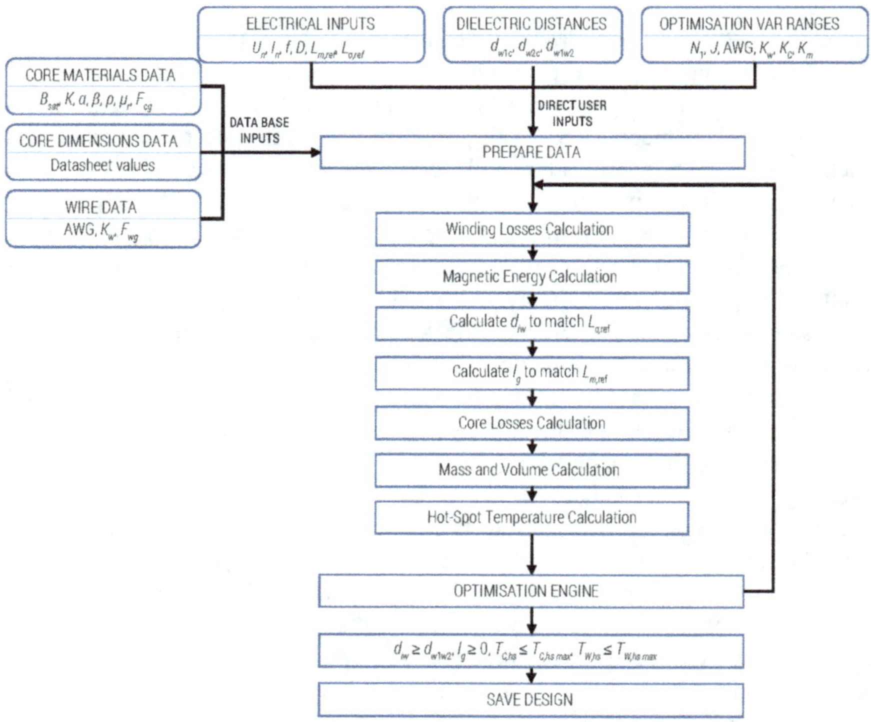

Figure 14.8 An example of the MFT design optimization algorithm using the brute force approach

characteristics and geometric proportions of the primary and secondary winding), thus generating a much more versatile design population and uncovering more design possibilities. This is enabled by proper transformations and the use of novel sophisticated data-driven models [31] or fast analytic models [32–35], providing improved accuracy and lower computational cost—both paramount for the quality of the overall optimization. It is used here just as an example to illustrate the process and the outcome of the optimization.

There are three distinctive parts of the optimization algorithm. (i) Direct user inputs directly specify requirements and constraints. Desired MFT parameters are given as fixed values or range values over which the search will be performed. (ii) Database inputs containing relevant datasheet parameters of all used materials. (iii) All the relevant models, corresponding to the set of selected geometries and technologies, that are executed over the optimization space, defined by the expert user.

The proposed MFT optimization methodology is applied for the design of shell-type MFT, rated for 100 kW, with a 1:1 turns ratio with ±750 V rated windings, operated at 10 kHz and having target leakage inductance of 3.3 μH and magnetizing inductance of 750 μH. For the core material, the Ferrite N87 is considered, and rectangular Litz wire is used for the winding. Air cooling and air insulation are applied for both core and winding. The maximum temperatures of the core and the windings are set according to the given material properties to 100 and 150 °C.

The proposed methodology consists of two steps. In the first step, all optimization variables are varied freely in a mathematically arbitrary way, thus exposing all mathematically feasible designs, as shown in Figure 14.9 with circle markers, where the color code corresponds to the highest hot-spot temperature within the windings (calculated during the evaluation of that particular design). The upper boundary of the feasibility set plots represents the Pareto front, related to the trade-off between

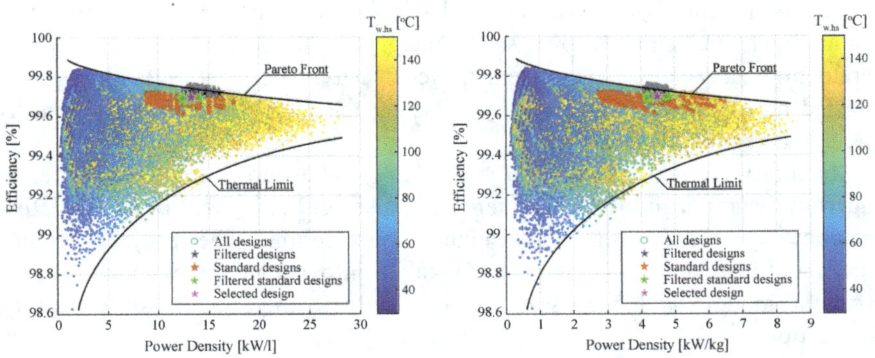

Figure 14.9 Plot of all possible shell-type MFT designs and filtered optimal: (a) efficiency vs. volumetric power density and (b) efficiency vs. gravimetric power density

maximum achievable efficiency and power density. Its descending trend shows how efficiency must be sacrificed for the increase in power density. The lower boundary represents the thermal limit. It shows the implications of the size decrease on the cooling conditions, and therefore, the maximum acceptable losses. In other words, it can be seen that an increase in the power density (decrease in size) of a design that is on a thermal limit is thermally feasible only if the losses are decreased, –increased efficiency.

These designs are interactively filtered in terms of key performance indicators (maximum weight, volume and hot-spot temperatures, efficiency, etc.) using performance filters, allowing to arbitrarily shift and narrow down the filtered design domain by relaxing and tightening the corresponding constraints. This is the strength of the brute force approach, but it also assumes expert users with knowledge of acceptable trade-offs for a given application.

In the second step, the filtered designs are analyzed against the available standard core and wire geometries, and the design is re-optimized with fixed discrete values corresponding to the selected (most similar) components, as displayed in Figure 14.9 with red star markers. These designs are filtered in the same way as before, using the interactively defined performance filters, as highlighted with green star markers. The final design is selected from this set, based on various additional designer preferences, such as construction simplicity, experience, etc., as highlighted with the purple star marker.

While the discussion about various Pareto optimal fronts is important, as it gives a good insight into the limitations of what is the maximum theoretically achievable performance and provides a good platform for comparison of different combinations of design choices and materials, when it comes to prototyping, it is important to take into account the safety margins, especially for the phenomena with high stochasticity such as natural convection cooling.

In that respect, it can be seen that the optimal design is not placed on the boundary of any of the parameter ranges. In order to achieve a thermally robust design, the maximum allowed core and winding hot-spot temperature rise constraint (filter) are tightened by a safety margin of 20 °C, thus generating a new Pareto front between efficiency and power density. The most optimal design is selected on this new Pareto front by adjusting the efficiency and power density filters in a desired manner.

In this way, it is possible to individually take into account the different uncertainties for each MFT characteristic, caused either by the limited modeling accuracy or manufacturing and assembly imperfections. The result is a robust Pareto optimal solution that properly takes into account the uncertainty of different models and processes relative to their importance for the given application.

Finally, by applying the described optimization process to an MFT with presented electric specifications, an optimal design is selected, marked with a star in Figure 14.9. The realized MFT prototype is an air-insulated and air-cooled shell-type MFT utilizing square Litz wire for windings and N87 SIFERRIT U-cores, as shown in Figure 14.10(a) [36].

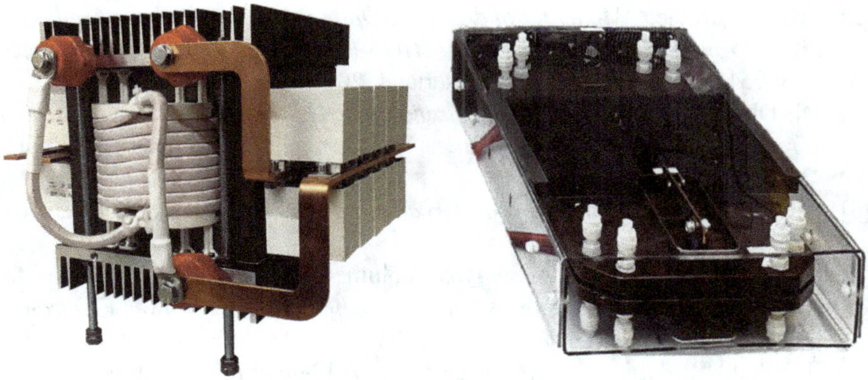

(a) (b)

Figure 14.10 Prototyped MFTs rated for 100 kW and operated at 10 kHz: (a) shell-type and (b) planar-type.

Additionally, the design optimization is repeated with the same electrical specifications (power, operating frequency, leakage, and magnetizing inductance) but with a different combination of technologies. Another prototype, of a planar-type, having also square Litz wire windings, but casted in epoxy and nanocrystalline core, as shown in Figure 14.10(b), is manufactured. It is important to note that these two MFTs are electrically absolutely identical while relying on a completely different set of technologies for their realization.

14.7 Summary

Increased interest in emerging SST conversion technology brought the design optimization of the high-power MFT into the spotlight. Even though the design of line frequency transformers has been mastered for more than a century, executing this at elevated operating frequencies while meeting high-power medium-voltage requirements is still a challenging task. A plethora of technologies are available to designers, and an infinite number of combinations can lead to a functional design. It is hence important to understand design space and utilize it as efficiently as possible. Whether the realized design is optimal or not will greatly depend on the sought-after optimality criterion, and trade-offs are inevitable and can be well controlled during the design optimization process. This chapter has provided some design considerations that designers of high MFT should consider and have in mind.

References

[1] J. E. Huber and J. W. Kolar, "Solid-State Transformers: On the Origins and Evolution of Key Concepts," *in IEEE Industrial Electronics Magazine*, vol. 10, no. 3, pp. 19–28, 2016, doi:10.1109/MIE.2016.2588878

[2] M. Mogorovic, *"Modeling and Design Optimization of Medium Frequency Transformers for Medium-Voltage High-Power Converters,"* Ph.D. thesis No: 7511, EPFL, Lausanne, Switzerland, 2019.

[3] N. Djekanovic, *"Modeling, Optimization and Design of High Power Medium Frequency Transformers,"* Ph.D. thesis No: 10238, EPFL, Lausanne, Switzerland, 2023.

[4] C. W. T. McLyman, *Transformer and Inductor Design Handbook*. CRC Press, 20014.

[5] W.-J. Gu and R. Liu, "A Study of Volume and Weight vs. Frequency for High-Frequency Transformers," *in Proceedings of IEEE Power Electronics Specialist Conference – PESC '93*, 1993, pp. 1123–1129.

[6] I. Villar, *"Multiphysical Characterization of Medium-Frequency Power Electronic Transformers,"* Ph.D. thesis No: 4622, EPFL, Lausanne, Switzerland, 2010.

[7] L. Heinemann, "An Actively Cooled High Power, High-Frequency Transformer with High Insulation Capability," *APEC. Seventeenth Annual IEEE Applied Power Electronics Conference and Exposition (Cat. No. 02CH37335)*, Dallas, TX, USA, 2002, pp. 352–357 vol.1, doi:10.1109/APEC.2002.989270.

[8] T. Kjellqvist, S. Ostlund, S. Norrga and K. Ilves, "Thermal Evaluation of a Medium Frequency Transformer in a Line Side Conversion System," *2009 13th European Conference on Power Electronics and Applications*, Barcelona, Spain, 2009, pp. 1–10.

[9] X. She, A. Q. Huang and R. Burgos, "Review of Solid-State Transformer Technologies and Their Application in Power Distribution Systems," *in IEEE Journal of Emerging and Selected Topics in Power Electronics*, vol. 1, no. 3, pp. 186–198, 2013, doi:10.1109/JESTPE.2013.2277917.

[10] H. Hoffmann and B. Piepenbreier, "Medium Frequency Transformer in Resonant Switching Dc/Dc-Converters for Railway Applications," *Proceedings of the 2011 14th European Conference on Power Electronics and Applications*, Birmingham, UK, 2011, pp. 1–8.

[11] I. Villar, L. Mir, I. Etxeberria-Otadui, J. Colmenero, X. Agirre and T. Nieva, "Optimal design and experimental validation of a Medium-Frequency 400kVA power transformer for railway traction applications," *2012 IEEE Energy Conversion Congress and Exposition (ECCE)*, Raleigh, NC, USA, 2012, pp. 684–690, doi:10.1109/ECCE.2012.63427514.

[12] N. Djekanovic and D. Dujic, "Design Optimization of a MW-level Medium Frequency Transformer," *PCIM Europe 2022; International Exhibition and Conference for Power Electronics, Intelligent Motion, Renewable Energy and Energy Management*, Nuremberg, Germany, 2022, pp. 1–10, doi:10.30420/565822101.

[13] C. Zhao, D. Dujic, A. Mester, *et al.*, "Power Electronic Traction Transformer—Medium Voltage Prototype," *in IEEE Transactions on Industrial Electronics*, vol. 61, no. 7, pp. 3257–3268, 2014, doi:10.1109/TIE.2013.2278960.

[14] T. Belgrand and R. Lemaître, "Grain Oriented Electrical Steel for Solid State Transformers," *in ECPE Workshop: New Technologies for Medium-Frequency Solid-State Transformers in Lausanne*, Switzerland, 2019.

[15] FINEMET F3CC Series Cut Core and material properties data sheet, Hitachi Metals, Ltd. https://www.hitachi-metals.co.jp.

[16] C. C. Barry and N. M. Grant, *Ceramic Materials*. Springer, 20114.

[17] V. A. Thiviyanathan, P. J. Ker, Y. S. Leong, F. Abdullah, A. Ismail, and M. Zaini Jamaludin, "Power Transformer Insulation System: A Review on the Reactions, Fault Detection, Challenges and Future Prospects," *Alexandria Engineering Journal*, vol. 61, no. 10, pp. 7697–7713, 2022.

[18] T. B. Gradinger, U. Drofenik and S. Alvarez, "Novel Insulation Concept for an MV Dry-Cast Medium-Frequency Transformer," *2017 19th European Conference on Power Electronics and Applications (EPE'17 ECCE Europe), Warsaw, Poland*, 2017, pp. P.1–P.10, doi:10.23919/EPE17ECCEEurope.2017.8099006.

[19] S. Ozdemir, N. Altin, A. Nasiri and R. Cuzner, "Review of Standards on Insulation Coordination for Medium Voltage Power Converters," *in IEEE Open Journal of Power Electronics*, vol. 2, pp. 236–249, 2021, doi:10.1109/OJPEL.2021.3065813.

[20] C. Zheng, Q. Wang, H. Wang, C. L. Bak and Z. Shen, "Uneven Inter-turn Voltage Distribution among Windings of Medium-Voltage Medium/High-Frequency Transformers," *2020 IEEE International Conference on High Voltage Engineering and Application (ICHVE)*, Beijing, China, 2020, pp. 1–4, doi:10.1109/ICHVE49031.2020.9280046.

[21] IEC Electrical Installations in Ships – Part 503: Special Features – AC Supply Systems With Voltages in the Range of Above 1 kV up to and Including 15 kV, IEC Standard 60092-503:2007, 2007.

[22] EN Railway Applications – Insulation Coordination – Part 1: Basic Requirements – Clearances and Creepage Distances for All Electrical and Electronic Equipment, EN Standard 50124-1, 2017.

[23] EN Adjustable Speed Electrical Power Drive Systems – Part 5-1: Safety Requirements – Electrical, Thermal and Energy, EN Standard, 61800, 5–1, 2007.

[24] EN Insulation Co-Ordination – Part 1: Definitions, Principles and Rules, EN Standard 60071–1, 2006.

[25] F. Judd and D. Kressler, "Design Optimization of Small Low-Frequency Power Transformers," *in IEEE Transactions on Magnetics*, vol. 13, no. 4, pp. 1058–1069, July 1977, doi:10.1109/TMAG.1977.1059509.

[26] W. G. Hurley, W. H. Wolfle and J. G. Breslin, "Optimized Transformer Design: Inclusive of High-Frequency Effects," *in IEEE Transactions on Power Electronics*, vol. 13, no. 4, pp. 651–659, July 1998, doi:10.1109/63.704133.

[27] A. Garcia Bediaga, *"Optimal Design of Medium Frequency High Power Converters," Ph.D. thesis No: 6205, EPFL*, Lausanne, Switzerland, 20114.

[28] M. Bahmani, *"Design and Optimization Considerations of Medium-Frequency Power Transformers in High-Power DC-DC Applications,"* PhD thesis, Chalmers University of Technology Gothenburg, Sweden, 2016.

[29] G. Ortiz, *"High-Power DC-DC Converter Technologies for Smart Grid and Traction Applications,"* PhD thesis, ETH Zurich, Switzerland, 20114.

[30] U. Drofenik, "A 150kW Medium Frequency Transformer Optimized for Maximum Power Density," *in 2012 7th International Conference on Integrated Power Electronics Systems (CIPS)*, 2012, pp. 1–6.

[31] M. Mogorovic and D. Dujic, "Computationally Efficient Leakage Inductance Estimation of Multi-Winding Medium Frequency Transformers," *in PCIM Europe 2019; International Exhibition and Conference for Power Electronics, Intelligent Motion, Renewable Energy and Energy Management*, 2019, pp. 1–6.

[32] M. Mogorovic and D. Dujic, "Thermal Modeling and Experimental Verification of an Air-Cooled Medium Frequency Transformer," *in 2017 19th European Conference on Power Electronics and Applications (EPE'17 ECCE Europe)*, 2017, P.1–P.9.

[33] M. Mogorovic and D. Dujic, "Medium Frequency Transformer Leakage Inductance Modeling and Experimental Verification," *in 2017 IEEE Energy Conversion Congress and Exposition (ECCE)*, 2017, pp. 419–4214.

[34] M. Mogorovic and D. Dujic, "Modeling and Experimental Verification of Geometry Effects on Core Losses," *in 10th International Conference on Power Electronics – ECCE Asia (ICPE 2019-ECCE Asia)*, 2019, pp. 1–7.

[35] M. Mogorovic and D. Dujic, "Computationally Efficient Estimation of the Electric-Field Maximums for the MFT Insulation Coordination," *2019 IEEE Energy Conversion Congress and Exposition (ECCE)*, Baltimore, MD, USA, 2019, pp. 7118–7123, doi:10.1109/ECCE.2019.8912195.

[36] M. Mogorovic and D. Dujic, "100 kW, 10 kHz Medium-Frequency Transformer Design Optimization and Experimental Verification," *in IEEE Transactions on Power Electronics*, vol. 34, no. 2, pp. 1696–1708, 2019, doi:10.1109/TPEL.2018.28355614.

Chapter 15

Multiwinding transformer-based solid-state transformers

Hamzeh Beiranvand[1]*, Fabian Groon*[1]*, Yoann Pascal*[2] *and Marco Liserre*[1,2]

Power distribution networks are characterized by their voltage level, i.e., medium voltage (MV) and low voltage (LV). Typically, the consumers need LV to feed their loads. In traditional distribution networks, the power flow was unidirectional, and power was delivered from the high-voltage substation to the distribution substation through radially connected overhead or underground MV lines. A distribution substation is basically a low-frequency transformer (LFT) that converts MV to LV with some additional protective devices and switchgear. Nevertheless, the steady growth of renewable energy sources has led to voltage violations originating from the bidirectional power flow. Moreover, intermittent high-power loads such as fast-charging stations cause current violations.

Solid-state transformers (SSTs) have been introduced as a remedy for the aforementioned challenges. SSTs typically employ different power converter topologies to convert MV to LV while providing additional functions such as bidirectional power flow control, galvanic isolation through medium-frequency transformers (MFTs), providing MV DC and LV DC ports, enabling hybrid AC and DC networks, enabling meshed networks of AC and DC lines at MV and LV levels. Voltage violation in the distribution network is solved by voltage control of the SST coupled with the use of DC lines. The current violation is solved by meshing the network, where the load demand can be supplied through different paths. Although SSTs are introduced aiming to increase the flexibility and connectivity of the network, their application is not limited to distribution networks. Many other applications operating at both MV and LV levels can benefit from the SST functionality. Examples of applications include the integration of renewable resources through isolated DC–DC conversion, fast charging stations, and electric transportation.

This chapter addresses the SST topologies that can be achieved by using medium to high-frequency multiwinding transformers (MWT). It is outlined that

[1]Power Electronics, Kiel University, Germany
[2]Fraunhofer Institute for Silicon Technology, Germany

the MWT topologies combined with traditional modular converter topologies lead to new SST topologies. The operation of the new SST topologies, which can operate as resonant or non-resonant converters, is discussed. Moreover, the applicability of voltage-controlled variable magnetics is extended to realize self-tuning SSTs to deal with the variable nature of loads. The transformer topologies resulting from different core and winding configurations are described, where planar transformers are highlighted. Emerging applications that benefit from MWT-based SST topologies are detailed.

15.1 Introduction to SSTs

SST terminology: Richard Rutz used the term solid-state transformer for the first time in a patent registered in 1967 [1]. However, the circuit topology, which is known as SST in today's literature, can be found in William McMurray's patents dating back to 1968 [2,3]. Nowadays, high-power isolated DC–DC converters, where the isolation is provided by medium- or high-frequency transformers, can be considered the core of an SST. SSTs can be used not only to connect DC networks of different voltage levels but also AC networks by means of additional DC-AC converter stages.

SST types: The primary motivator to select the SST topology is usually dictated by the application and the required level of functionality. However, the functions and topologies of the SST greatly depend on the location of the MFT in the conversion stage, as shown in Figure 15.1 [4]. SSTs can be classified into four types: A, B, C, and D. SST type A might be the key enabler for applications where isolated AC–AC conversion is required in a limited space. In applications where the connectivity and voltage control at MV-level is of interest, then SST type B is preferred. Similarly, SST type C ensures connectivity and controllability on the LV side. SST type A has the minimum functions, and type D has the most available control functions. SST types B and C can be considered as a compromise between types A and D. Therefore, this chapter focuses on SST type D owing to its higher control functions. Moreover, the isolated DC-DC converter within SST type D is the key to devise new SST topologies discussed in the chapter.

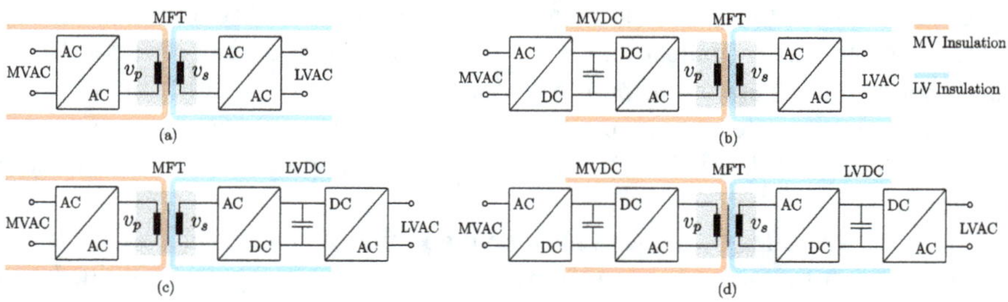

Figure 15.1 Different types of SSTs based on the location of the MFT in the power conversion chain: (a) Type A, (b) Type B, (c) Type C, and (d) Type D

SST functions: SST not only functions as the LFT and converts the MV to LV but also controls the power flow at the input and output ports. The front-end converter on the MV side can inject or absorb reactive power to control the voltage level to a desired value, exactly as Static Compensator (STATCOM). Differently from the STATCOM, the SST is able to regulate the active power provided by the possible renewable resources downstream of it. The front-end AC–DC converter converts the MVAC voltage to either MVDC or multiple LVDC ports. This means that new degrees of freedom are provided for the interconnection of renewable sources and new grid-forming possibilities [5]. SSTs functioning with additional support for the flow of power and communication have been regarded as smart transformers (ST) [6]. Other functions of SSTs include connectivity at DC links, providing reactive power for voltage support, creating a meshed grid, power quality improvement, islanding control, integration of EV charging stations, integration of energy storage for dispatching, reverse power flow control, and soft load reduction.

Isolation: Galvanic isolation is the main requirement to ensure safe interconnection for adjusting voltage levels. Moreover, efficient and effective voltage conversion can be achieved by medium- or high-frequency transformers. Modular topologies have been adopted such that the operating voltage of the transformers falls in the LV level while the input voltage of the front-end converter is still MV [7]. If the SST requires an MVDC port, then modular multilevel converter (MMC) based front-end converters can be used. This directly impacts the transformer design in both cascaded and single-stage conversion from DC to high-frequency AC. Regardless of the selected converter topology, the transformer must satisfy the MV insulation requirements.

Transformer frequency: In terms of frequency, various SST prototypes have been reported in the literature with transformer frequencies ranging from a few hundred Hz up to more than 100 kHz depending on the used power semiconductors and power level. Increasing the frequency and consequently shrinking the transformer volume and weight are the key concepts of the SST. Increasing the frequency and the voltage level might increase the power density and efficiency of the transformer, as shown in [8]. However, increasing the frequency comes with drawbacks as the switching losses of the semiconductor increase, and the overall efficiency might be decreased. The operating frequency of the transformer is considered from the limitations of the power converters with respect to the considered voltage and current ratings. For example, if MMC is used as a high-frequency inverter with a high-power IGBT, the switching frequency will be limited to a few kHz. On the other hand, using low-voltage SiC devices with lower current ratings, frequencies in the order of hundreds of kHz can be applied, achieving a promising system-level efficiency and power density [9].

Electric field stresses: It is worth noting that electric field stresses become considerable if the voltage of the SST is in the MV range. Electrical shields create an equipotential surface that can be used to confine the electric field to a specific volume of the device and to give it a determined topology, thereby relaxing the requirement in terms of dielectric design and electric field management in the rest of the device. This is, for instance, useful when only part of the entire device is encapsulated, the shield ensuring then that the electrical field remains confined within the dielectric material. However, eddy currents can be induced in the shields, leading to additional losses and heating. Furthermore, shields bring additional common mode capacitance. If shielding can help shape the electric field topology, precautions must be taken at the exit region, where the shield stops covering the live conductor, to avoid field concentration. This can be achieved by various techniques, such as prolonging the conductor shield with a semiconductor one, the resistivity of which will ensure a grading of the electric field. This technique, widely used in electrical machines, reduces the electrical stress and the risk of breakdown. Field plates and stress-relief cones can also be used [10].

15.2 MWT-based SST topologies

The topology of the SST is significantly influenced by the MV AC front-end converter. Three possible topologies with a high degree of modularity are the three-level neutral point converter (NPC), modular multilevel converter (MMC), and cascaded H-bridge (CHB) [5]. NPC and MMC topologies provide MVDC voltage levels in the SST while there is no MVDC link in CHB. Moreover, NPC requires large inductors and capacitors to satisfy the power quality requirements while it cannot reach very high voltage levels. Thereby, various SST topologies with and without MVDC links are discussed and compared.

Figure 15.2(a) shows a conventional system without SSTs where different loads and sources are integrated by isolated converters. By introducing SSTs, the LFT transformer can be eliminated, and the AC–DC front-end converters can be integrated into one, as illustrated in Figure 15.2(b). This evolution contributes to reducing the conversion stages without sacrificing the isolation. MWTs couple multiple ports, ensuring galvanic isolation between them. MWT-based SSTs could effectively improve energy efficiency by reducing the number of conversion stages in particular for local generation and loads. Figures 15.2(c) and (d) are two possible variants of SSTs that reduce the local power flow path from a source to a load or storage from six to three. In the MWT-based SST in Figure 15.2(d), the power flow path from the grid to the load center is reduced by one compared to a conventional SST. A complete comparison of the number of conversion stages is given in Table 15.1.

In the case of MWT-based SST topologies, power flow control between different ports becomes more complicated because the ports have a cross-coupling effect [11]. The power flow decoupling between the ports has been reported in two ways. The first method is to use control techniques in the software of the controller through a look-up table. This method is not scalable and adds complexity to the controller design. The second method is to decouple the power from the root, i.e., the inductance network of the MWT. This method can greatly reduce the complexity of the controller implementation. However, special considerations are required to proportionally design the inductance network and the relevant parasitic parameters.

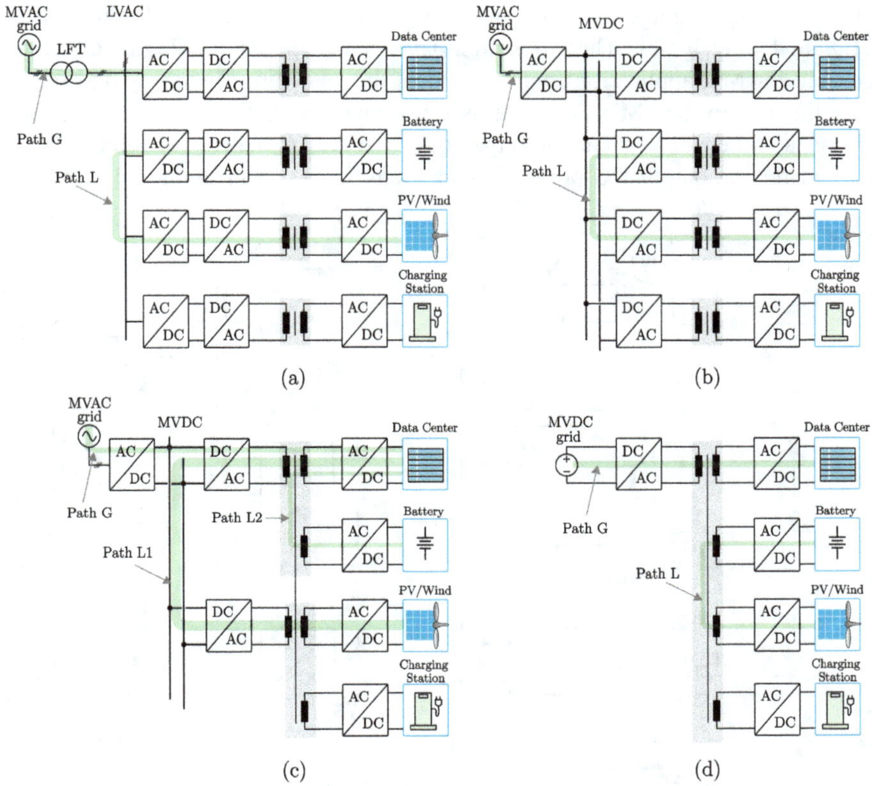

Figure 15.2 Evolution of the SST connectivity employing MWTs: (a) conventional energy conversion without SST, (b) energy conversion with SST, (c) multiple MWT-based SST with MVDC distribution, and (d) MWT-based DC SST

Table 15.1 Comparison of the number of conversion stages for different power flow paths

Circuit	**Path G**	**Path L**	**Path L2**
Conventional converters in Figure 15.2(a)	5	8	–
SST in Figure 15.2(b)	4	6	–
MWT-based SST in Figure 15.2(c)	4	6	3
MWT-based DCSST in Figure 15.2(d)	3	3	–

15.2.1 DC–DC SST topologies with MVDC link

High-power IGBT switches are usually built with standard blocking voltage, V_B, such as 600, 1200, 1700, 3300, 4500, and 6500 V. Considering an utilisation factor in the range of $u \in [0.5, 0.7]$, two-level converter topologies might be able to reach

4500 V. At this voltage level, the switching frequency is strongly limited by the IGBT switching losses. Therefore, multilevel and modular topologies are indispensable for MV applications and also boost the switching frequency and miniaturize the transformer. MMCs can be used to convert MVDC to medium-frequency MVAC, and then an MWT reduces the voltage to LVAC, where the voltage can be rectified for the output DC ports. This topology is shown in Figure 15.3 (a), and

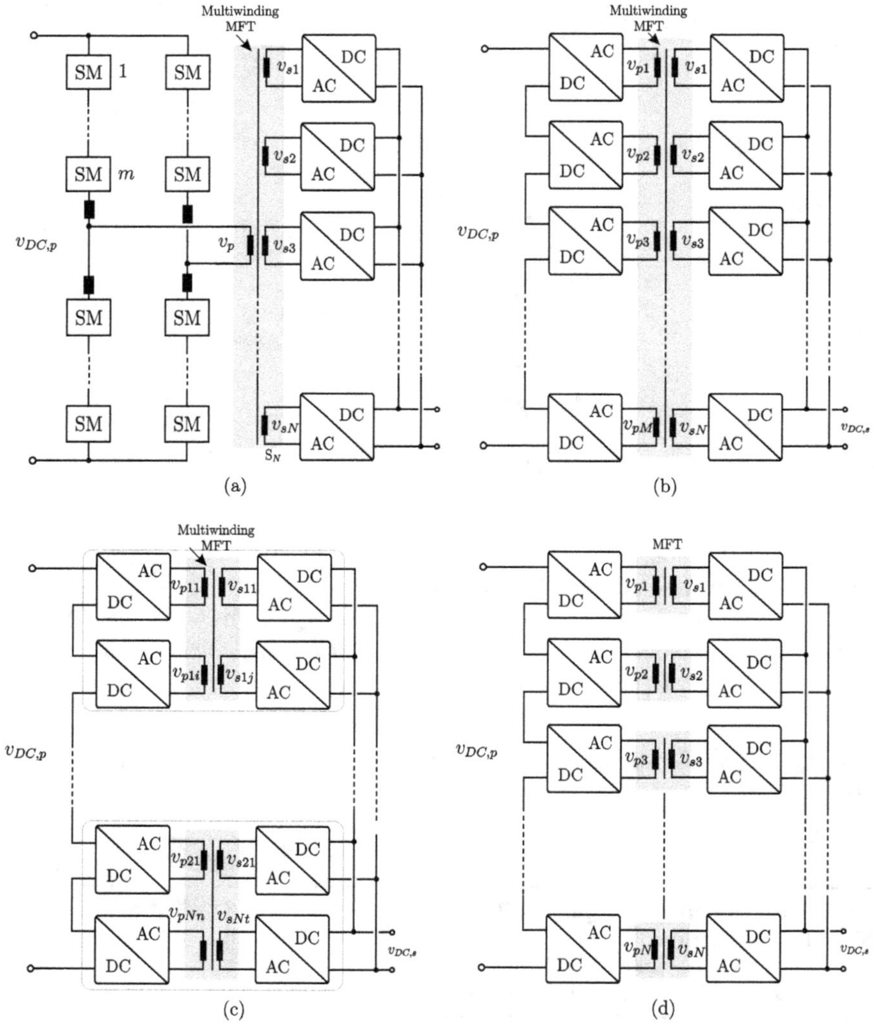

Figure 15.3 DC–DC SST topologies derived from the flexibility of the transformer design: (a) MMC-based DC–DC converters with one MV winding and many LV windings MFT, (b) DC–DC converters constructed from many input and output windings with serially connected inputs and parallel connected outputs, (c) DC–DC converters configured from multiple MWTs, and (d) DC–DC converters with individual transformers in ISOP configurations. The converter block can be either a DAB or a resonant converter.

hereafter, it is abbreviated to MW-MMC DC–DC, i.e., MWT-based MMC DC–DC converter. This topology reduces the current stress and voltage stress on both primary and secondary side semiconductors. The voltage of MMC can be controlled in staircase operation with trapezoidal or quasi 2-level (Q2L) waveform, minimizing the d*v*/d*t*, thereby reducing dielectric losses and electromagnetic (EM) emissions [12]. However, it might not necessarily guarantee soft switching for all the switches in the MF MMC. In the SST operation, the output ports should be paralleled; therefore, it is required to ensure equal current distribution in the output ports. If the MW-MMC DC–DC converter is operated as a fast charging station, multiple EVs can be charged at the same time while maintaining the galvanic isolation between all of them.

Figure 15.3(b) shows an MWT-based input-series output-parallel (ISOP) DC-DC converter (MW-ISOP). In this topology, the number of transformer windings on the primary side can be different from the secondary side. For the sake of simplicity, they are equated in this figure. MWT eases the power flow among the different ports; hence, the primary side converters are only required to maintain the balanced voltage, and the power will be naturally balanced. The secondary side windings might be treated similarly to the MW-MMC DC-DC converter, depending on the application as an SST or EV charging station.

To further improve the modularity, but also simplify the MWT design, the MWT can be split into multiple MWTs, as shown in Figure 15.3(c). An example of such a converter can be created from a quadruple-active bridge converter (symmetric or asymmetric) with ISOP configuration, i.e., multiple MWT ISOP (MMW-ISOP) DC–DC converter. The main drawback of this topology is the need for power and voltage balancing on the primary side if each converter processes different power, e.g., in fast charging applications. However, this might not be required in the SST application. The only requirement for the secondary side is to balance the output port currents.

The topology with the highest modularity in the transformer construction is the ISOP DC-DC converter with a separated transformer per module, as depicted in Figure 15.3(d). The independent operation of the output ports is very limited and almost impossible to use for multiple EV charging points. If configured for SST operation, the only requirement is to ensure equal voltage on the primary side, which will lead to equal power and, consequently, equal current distribution among the output ports.

15.2.2 AC-DC SST topologies without MVDC link

SSTs configured from CHB and followed by a dual active bridge (DAB) or resonant DC–DC converter are the most popular. Plenty of research has been carried out on this topology [13–16]. Similar to the topologies discussed in the previous subsection, MWTs lead to different topologies for the overall SST, as depicted in Figure 15.4. A single MWT can be used to interconnect the submodules of an AC-DC SST, as shown in Figure 15.4(a). Hereafter, acronym MW-SST will be used instead. The topology in Figure 15.4(b) employs a single-output MWT to configure the AC-DC SST, i.e., SMSO-SST. Multiple MWT-based AC-DC SSTs (MMW-SSTs) are shown in Figure 15.4(c), where MWTs with different numbers

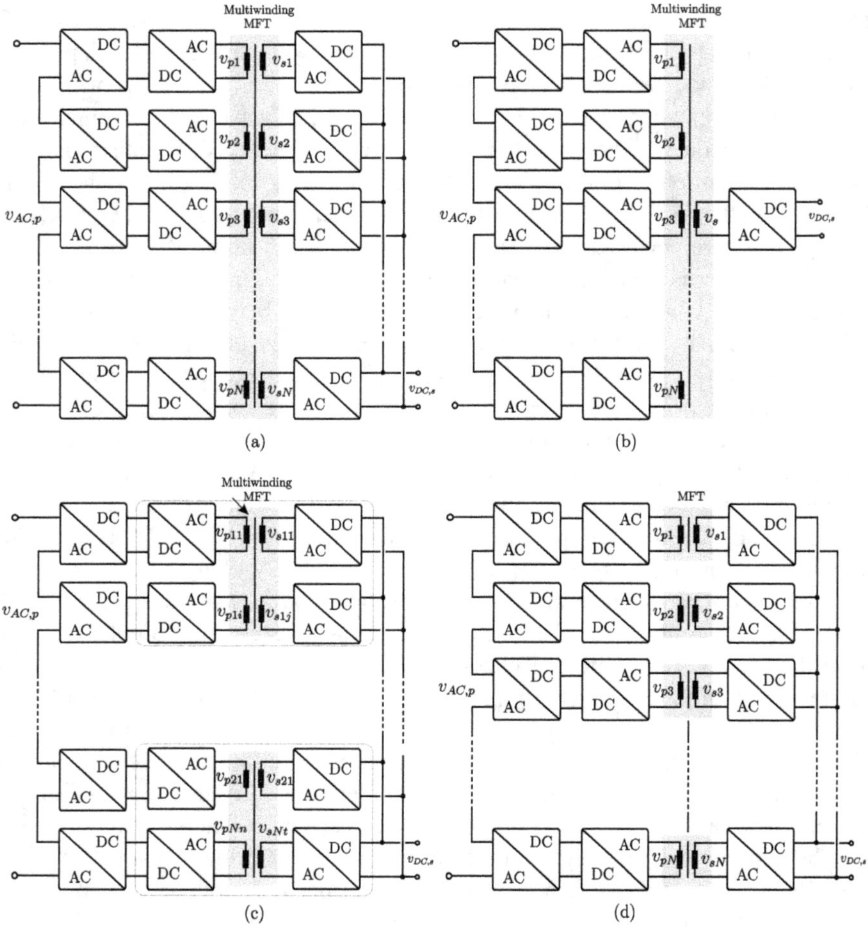

Figure 15.4 AC–DC SST topologies with distributed DC voltage: (a) an AC–DC SST configured from a single MWT, (b) an AC–DC SST configured from a single single-output MWT, (c) an AC–DC SST constructed from multiple MWT, and (d) an AC–DC SST constructed with two-winding transformers

of input and output winding can be used. A comparative study of MMW-SSTs configured with quad-active bridges can be found in [16]. Finally, Figure 15.4(d) illustrates a two-winding transformer-based AC-DC SST (TW-SST) where conventional two-winding transformers are considered.

The MWTs used in topologies shown in MW-SST and SMSO-SST reduce the modularity of the SST with respect to the transformer; instead, the power routing and fault tolerant capabilities improve significantly as MWT acts as a power buffer. For SMSO-SST topology, the current level in the secondary side is much higher than the primary side and it might require parallel switches or converters to fulfill the required rating. Although MMW-SST and TW-SST have the highest modularity for the transformer, the converters require balanced operation between the modules.

15.2.3 Power flow control

Depending on the configuration of the MF AC link passive components network, both resonant and non-resonant converters can be realized using the DC-DC converter topologies shown in Figure 15.3. In the case of non-resonant converters, such as dual active bridge (DAB) converters, sufficient inductance should be provided in series in the MF AC link to enable the power flow control for a given phase shift φ between primary and secondary side converters. The basic waveforms of the DAB converter operation with a two-winding transformer are shown in Figure 15.5 (a).

The power transfer in a DAB converter works analogously to a grid system with two voltage buses and a line in between. Power is transferred from the leading system to the trailing system. A square-wave voltage is generated by the switching components and applied to the system, resulting in the transfer of a trapezoidal current. The slope of the current ramp depends on the voltage difference between the primary and secondary sides and the value of the inductance. The power flow equation is as follows:

$$P = \frac{V_p V_s}{\omega L} \varphi (1 - \frac{\varphi}{\pi}) \tag{15.1}$$

where V_p and V_s are the DC voltages at primary and secondary sides, L is the total inductance in the AC link, and φ is the phase shift between two converters.

In this system, the power transfer is mainly dependent on the phase shift between the primary and secondary sides. During the design of the converter, the inductance and the frequency can further influence the maximum power of the system. By shifting the phase angle in both directions, the converter can be operated bidirectionally. The maximum power can be achieved with a phase angle of 90°.

A resonant converter consists of a resonant network with an inductor and a capacitor, as illustrated in Figure 15.5(b). A square wave voltage is generated by the switching components, whereby a sinusoidal current is transmitted. The fundamental harmonic is mainly responsible for power transmission. Through the resonant network and the applied frequency, the gain of the system can be determined. The gain of the resonant converter is the typical measure of power transfer.

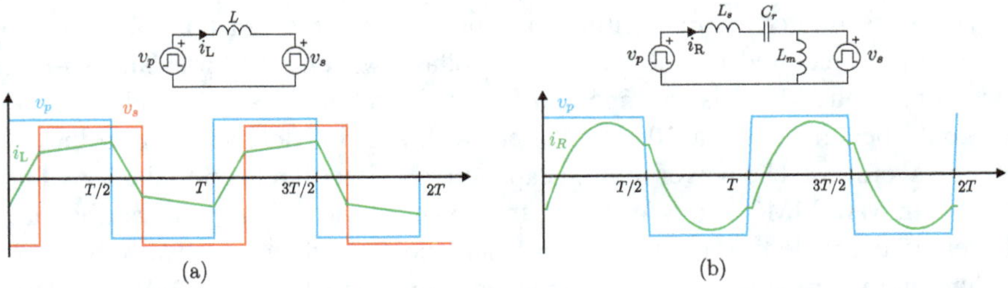

Figure 15.5 Basic current and voltage waveforms in the MF link transformer: (a) DAB converter and (b) series-resonant converter

By selecting the ratios of the resonant network, different gain bands can be generated, one of which is selected to meet the requirement of the system. The frequency of the converter determines the power transfer of the system, as it changes the impedance of the resonant network. There is no phase shift of the voltages between the primary and secondary sides. The current shape is dependent on whether the converter is operated above or below the resonant frequency and does not give a perfect sinusoidal shape in either region. The voltage gain is as follows:

$$G = \left| \frac{L_n f_n^2}{(L_n + 1)f_n^2 - 1 + j[(f_n^2 - 1)f_n Q_e L_n]} \right| \tag{15.2}$$

where L_n is the ratio of magnetizing inductance L_m to leakage inductance L_s, f_n is the normalized switching frequency, and Q_e is the quality factor of the resonant network.

15.2.4 Comparative study of diverse SSTs

Each MV DC-DC SST topology comes with some advantages and disadvantages compared to others. The presented topologies shown in Figure 15.3 are compared in the following. Starting from the MFT efficiency η_{MFT}, MWT-based topologies manifest fewer power losses as the size of the magnetic core can be reduced to a degree compared to other topologies. In the MW-MMC, the soft switching operation cannot be easily guaranteed for all the switches in MMC. Therefore, the power losses of the semiconductors are higher than the other topologies. According to the total efficiency criteria for the IGBT semiconductors, η_{IGBT}, the ISOP converter might suggest the highest efficiency. In MWT, reactive power can be more prominent. The MFT of the MW-MMC uses less core material compared to other topologies, and it has the highest efficiency. This is followed by MW-ISOP, MMW-ISOP, and finally, ISOP, which comes with the lowest η_{MFT}. In particular, the power flow is coupled among the different ports for MW-ISOP and MMW-ISOP converter topologies. In the MW-MMC, the cross-coupling effect between the ports is less severe than that in the other MWT topologies, and hence, the increased power losses due to circulating reactive power in the MVAC link are less. The fault-tolerant capability is very important in modular converters. It is associated with maintaining the operation under fault conditions in one or multiple submodules. Since MWTs act as a power buffer, MW-MMC and MW-ISOP outperform MMW-ISOP and ISOP. The operation of ISOP is much more impacted by a fault. The MW-MMC switching frequency, which determines the MFT operating frequency, is limited more significantly compared to the other topologies. The dwelling time required for the trapezoidal voltage waveform to prevent two-level operations of the MMC is an important factor. Switching losses depend on the V_B and could be similar for all the converters with slight differences due to the different waveforms. The power routing capability is directly related to the use of the MWTs. In MW-MMC and MW-ISOP, the power of the output ports can be independently controlled from the primary side ports leading to always balanced loading on the primary side IGBT semiconductors. Nonetheless, MMW-ISOP and ISOP are very limited as unequal power flow in the output ports will lead to

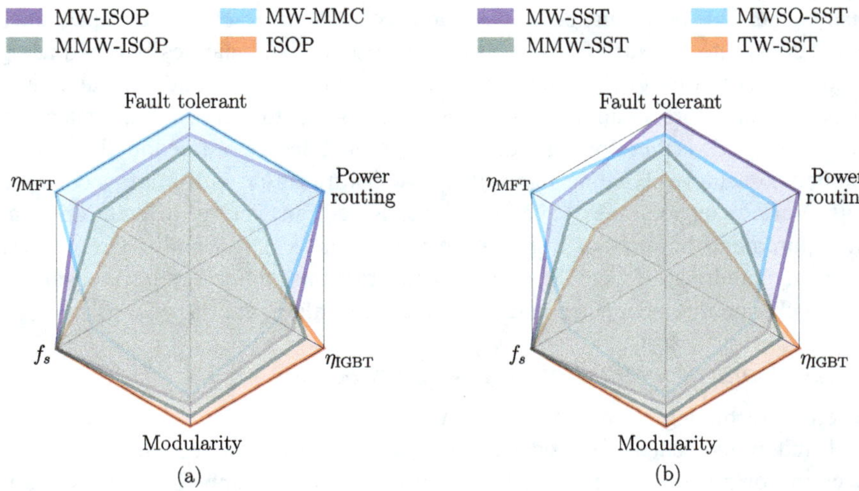

Figure 15.6 Qualitative comparison of diverse topologies (a) MW-MMC, MW-ISOP, MMV-ISOP, and ISOP, and (b) MW-SST, MWSO-SST, MMW-SST, and TW-SST. The AC-DC and DC-DC SST topologies are compared in terms of MFT efficiency (η_{MFT}), IGBT semiconductor efficiency (η_{IGBT}), fault-tolerant capability, capability to reach high switching frequency f_s, and modularity.

unequal voltage distribution and therefore power on the primary side. The maximum achievable switching frequency is limited by the trapezoidal output voltage of the MMC. ISOP offers the highest degree of modularity among others as each submodule has a dedicated MFT. Figure 15.6(a) shows a graphical representation of the previous discussion.

AC–DC topologies can also be compared for different indexes similar to DC–DC SSTs. Figure 15.6(b) shows this qualitative comparison. Considering that there is no MMC converter with inherent fault-tolerant capability in MW-SST topology, its fault-tolerant capability degrades. Moreover, a high-current level at low voltage might negatively impact its efficiency. Generally, the same comparisons and discussions outlined for DC-DC SSTs are also valid for AC-DC SSTs. For example, when comparing TW-SST and ISOP, the only difference is the integration of the front-end CHB converter to the input side of the DC-DC converters. Assuming similar submodules and operating conditions, then the two should lead to similar performance indexes. Therefore, further explanation is not required.

15.3 Self-tuning SST topologies

15.3.1 Voltage-controlled magnetics

Designing magnetic components for a specific application answers to a set of contradicting constraints, for instance, between response time, filtering capability, density, thermal performances, loss, and cost. Voltage-controlled magnetics (VCM) constitute a family of passive devices that can be dynamically tuned using an external control signal. Tuneable parameters include self and mutual inductances and therefore coupling coefficient. Compared to conventional magnetic components, VCMs therefore introduce an additional degree of freedom that can be exploited to achieve dynamic multi-objective optimization with improved Pareto fronts. Various technologies have been investigated, answering different application requirements in terms of response time (from μs to s), tuneability (from a few percent to hundreds of percent), power consumptions of the tuning mechanism (both to change the inductance and to maintain its value), etc.

15.3.2 Control method of magnetic properties

Numerous technologies, based on various physical phenomena, have been proposed and implemented to produce adjustable magnetic components for signal processing or power applications. For energy conversion, tapped windings, commuted using mechanical or solid-state relays, provide a simple solution offering discrete variation of characteristics. This technology is widely used in distribution LF transformers to maintain constant grid voltage.

Another approach, widely studied over the last few decades, consists of biasing the magnetic core in the non-linear part of its *B–H* magnetization curve using an auxiliary winding supplied with a current defined by the VCM control signal. This technology makes it possible to vary the effective permeability of a core or part of it, thereby acting on the inductance or coupling between windings [17].

Magnetostrictive materials feature a coupling between mechanical stress and magnetic characteristics. Conversely, piezoelectric materials exhibit an intrinsic coupling between electric field and stress. The mechanical combination of magnetostrictive and magnetoelectric layers results in a magnetoelectric composite with a coupling between the electrical field and magnetic characteristics. In such a stack, applying a voltage to the piezoelectric layers generates stress that is transmitted to the magnetostrictive layers, resulting in a variation in their permeability. Although this technology has mainly been used in small-signal applications, the approach appears to have high potential for power conversion applications [18].

Various more exotic methods of inductance control have been investigated, such as using an actuator to adjust the distance between the two halves of a core in order to adjust the length of its air gap or the position of a core in the air gap of a magnetic circuit, or the geometry of the winding. Ferrofluids can be used to control the shape of the magnetic core and the path of the field lines. However, these technologies remain laboratory curiosities or are applied in very specific applications.

15.3.3 Use case

VCMs have been proposed and implemented in various power electronics structures and applications, such as fault current limiting or voltage regulation by adjusting a series impedance or the turn-ratio of a transformer. Other possible

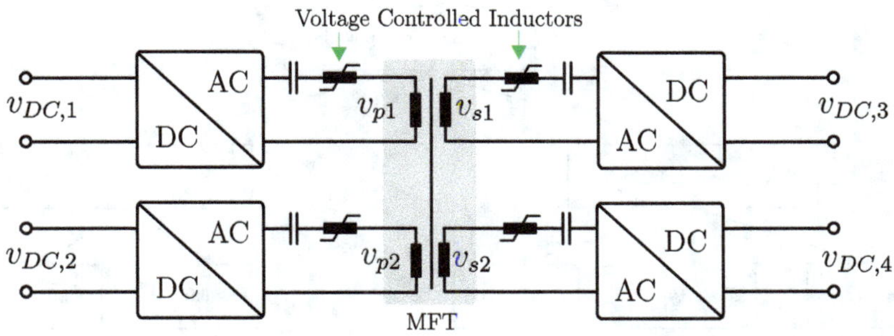

Figure 15.7 SST cell with voltage-controlled magnetics [19]

applications involve dynamically adjusting the value of the VCM to ensure operation at the converter's point of maximum efficiency. The use of VCMs also enables the control of a resonant converter by acting directly on its resonant frequency rather than its switching frequency—which is particularly beneficial in terms of conducted EM emission and, potentially, efficiency.

In the case of a hydrogen-powered electric bus, for example, the slow dynamics of the fuel cell means that other sources must also be integrated for peak-shaving and delivering the high pulse power required. These can be a combination of batteries and super-capacitors, for instance. These different sources and loads are then interconnected using an SST acting as a smart energy router, and based, for example, on a multiport resonant converter built around a MWT (Figure 15.7). In this case, the initial uncertainty over the value of the components, possibly exacerbated by ill-controlled parasitic elements and drifts due to temperature and aging, results in different resonant frequencies for the ports. This renders the operation and control of the system, to ensure good control of power flows, tricky to implement. By integrating voltage-controlled inductors in place of or in addition to traditional resonant elements, they can be used to control the flow of power between the ports [19].

15.4 Transformer topologies

15.4.1 Overview

Transformers can be mainly categorized into wire wound transformers and planar transformers. Both these transformer types have subgroups of different topologies. These topologies are shown in Figure 15.8. Each of these topologies has its own advantages and disadvantages that must be weighted during the design process. The different topologies are described below.

15.4.2 Wire-wound transformer

Wire-wound transformers typically use litz-wire for the windings. Copper foil is also a valid option, but insulation material must be placed between the windings,

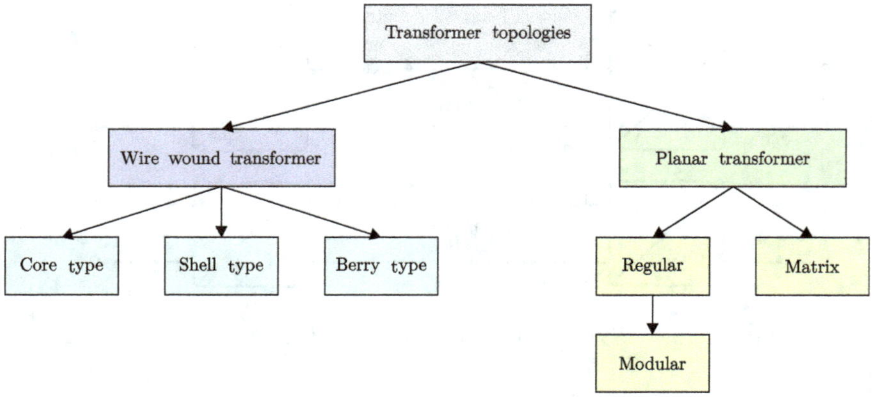

Figure 15.8 Overview of the different transformer topologies

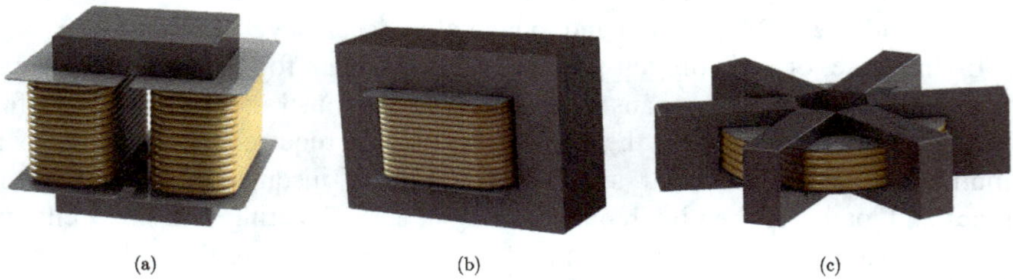

Figure 15.9 Different topologies of wire wound transformer: (a) core-type transformer, (b) shell-type transformer, and (c) Berry-type transformer

thereby complexifying manufacturing. Wire-wound transformer comes with a variety of different architectures, which typically use E, I, or U-cores.

Core-type transformer uses U-cores, where each of the two winding is wound around a separate leg of the core, as seen in Figure 15.9 (a). Generally, more windings could be added, where one at the top and one at the bottom of the core is placed to achieve a MWT. Due to the longer magnetic path, it has a higher magnetizing current and therefore lower efficiency. Mechanically, it has less strength, and magnetic forces from the windings are higher. The major advantages are easy maintainability for repair purposes and better cooling efficiency due to the exposed surface area of the windings. Core-type transformers are mainly used in high power high voltage applications.

Shell-type transformer uses E-core or two separate U-cores, where the windings are wound around the center leg of the core, as seen in Figure 15.9(b). This protects the winding mechanically from the outside. The middle leg typically has double the width and, with that, double the area of the legs on the outside. MWTs with this configuration are very easy to achieve since multiple windings can be wound around the middle leg until the window space of the core is filled. Additionally, windings can be wound around the outer legs for half the flux, but this leads to flux asymmetry, unbalance, and the mechanical protection of the windings would be lost. The advantages are lower magnetizing current and better mechanical support, but the manufacturing costs are higher. They are typically used for low-power low-voltage applications.

Berry-type transformer uses multiple U-cores placed in a circle, as shown in Figure 15.9(c). The windings are wound inside the inner and outer leg of the U-core, placing it in the middle of the core. The windings are both wound around the same leg as in the shell-type transformer, which makes it easy to add more windings to the transformer. The number of cores placed in a circle can be varied and this structure achieves low leakage. The big advantage of this transformer is that it allows high power transmission, and heat dissipation is greater than in other transformer types, since the windings can dissipate the heat to the transformer cores.

15.4.3 Planar transformer

Planar transformers (PTs) use PCBs for the windings instead of litz-wire. The windings are integrated into the PCB, and there can be multiple turns on one layer and multiple turns in one PCB if a multilayer PCB is used. In comparison to the litz-wire transformer, PT comes with a variety of advantages. PT is very reliable to manufacture, so there will be no issue during mass production with variety between them. They have a very low profile, which allows them to achieve very high power density at high frequencies. They have a greater heat removal capability, since the flat surface allows better heat dissipation, and the surface area when connected to a cold plate, heatsink, or even still air is greater. Two of the biggest advantages are the ability to interleave the windings and predict/control the parasitic in a PT. Interleaving the windings with a PT is very easy and allows to achieve low AC resistance. This interleaving also allows the PT to achieve very low leakage inductance compared to the wire-wound transformer.

A variety of architectures with the PT can be achieved. The typical PT uses a combination of E- or I-cores with two planar windings. The planar windings can be achieved by using PCBs with copper traces inside the PCB, but copper foil can also be used. When copper foil is used, an insulation material must be manually placed between the windings, since there is no FR4 like in the PCBs. Typical frequencies and power ranges of a PT is around several tens of kHz up to low MHz and around a few watts up to almost hundreds of kW [20–22].

Multiwinding planar transformer (MWPT) are built the same way as the typical transformer but with more than two planar windings (primary, secondary, and tertiary). This allows to connect multiple sources and loads together on one transformer. This makes them a great fit for SST applications. The windings can be connected in any kind of arrangement (series or parallel). The frequency range of MWPT is in the same range, but power can be up to several hundred kW [23–25]. An example of a MWPT with a turn ratio of 7:1:1 is given in Figure 15.10(a) with a primary winding and two secondary windings. The MWPT is used in a resonant

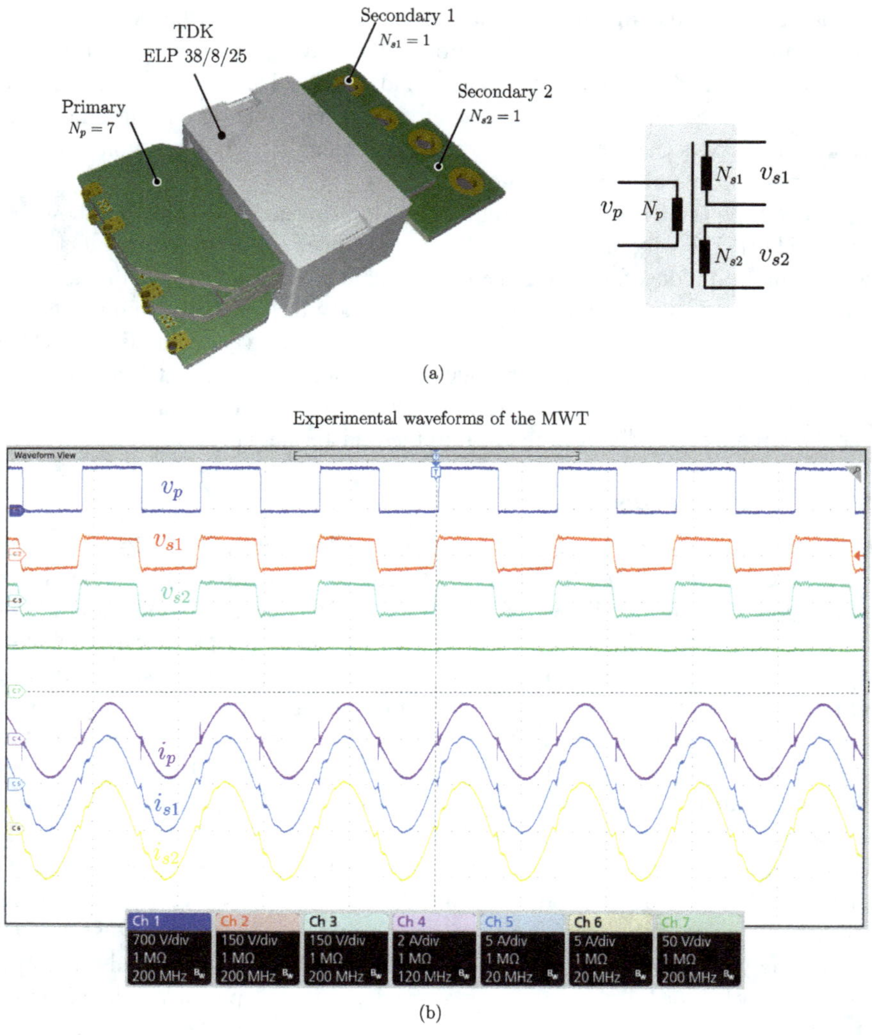

Figure 15.10 An example MWPT used in EVs [25]: (a) 3D-Rendering of the MWPT, and (b) waveforms of a resonant converter with the MWPT

converter, as shown in Figure 15.10(b), where both secondary windings are connected in parallel and share the same load. As it can be seen, one of the main advantages of MWPT is that it achieves good current sharing between both secondary windings, since PTs can achieve very reliable turn ratios [26].

Modular transformers are composed of multiple transformers that are connected in series or parallel, resulting in a total of four possible configurations. This allows us to scale the power of individual transformers to higher power. This was shown in [27] where a 200 kW IPOS transformer for an SST with a 1:1 turn ratio and low parasitic capacitance was built. There is a trade-off between efficiency and

Figure 15.11 Matrix planar transformer with the top core plane removed for visual

the number of transformers. The optimum number of transformers can be obtained at the maximum efficiency point.

Planar matrix transformer is composed of several individual transformers connected together like a matrix. The input is connected in series with N transformers and the output is connected in parallel with M transformers. Each of the transformers has the same turns ratio N_t. This means a matrix transformer typically has one input but several outputs that are connected in parallel. By rearranging the cores of the matrix transformer, a flux cancelation along its path can be achieved [28,29]. In combination with this approach, the individual transformers are also connected in the form of a pillar, where each pillar represents one winding, as seen in Figure 15.11. Matrix transformers are mostly used with frequencies above >500 kHz and can go up to 1 MHz [30,31]. The power level is typically in the low power range of 1 kW [30,32] but can also be used for higher powers [33].

15.5 Emerging applications of MWT-based SSTs

The conventional grid applications of SST and its potential are well known. In this context, some special applications are getting more importance with grid interface. For instance, public and mass transportation is also steadily moving toward electric power, where EVs, ships, aircrafts, and trucks have been the main targets. Thereby, fast charging stations (FCSs) should be proportionally installed to the EV growth rate. Moreover, the energy distribution paradigm is shifted toward net-zero nano- and micro-grids where multiple sources and loads can be interconnected with an MWT-based SST. More details are given in the following.

Fast charging stations (FCSs): A large number of FCSs are typically placed close to each other to support the charging of multiple vehicles at the same time. Based on the regulations, FCSs require galvanic isolation between the grid and EVs. By using an MWT-based SST, FCSs can be integrated with galvanic isolation between all the components as well as the reduced number of energy conversion stages as demonstrated in Section 15.2. MWT-based SST can handle the power distribution between the fast charging stations [34,35]. The SST also enables a very low-complexity integration of solar panels that can be placed on the roofs of the fast charging stations [34].

Electric transportation applications: Aircraft, electric ships, and electric vehicles are likely to have multiple energy sources in the future, such as batteries, hydrogen, and super-capacitors. Thereby, different sources and loads should be interconnected optimally with minimum weight, volume, and maximum energy efficiency to ensure the highest range possible. MWT-based SSTs make it possible to connect all sources together to power the electric transportation vehicle [36]. The peculiarities of MWT-based DC-DC SST guarantee the minimum number of energy conversion stages and therefore very high efficiency and power density.

Net-zero microgrids: MWT-based SSTs enable net-zero microgrids by connecting multiple sources, such as wind farms, photovoltaic systems, and battery storage, and multiple loads, such as fast charging stations, hydrogen electrolysis stations, public transportation and industrial facilities, and distributing the power as desired. Since everything is connected to the SST, no external power from a main grid is required to manage the energy demand within the microgrid [37]. Thereby, small-scale net-zero networks such as hospitals and data centers can have a secure supply independent from grid interruptions through central MWT-based SSTs.

15.6 Summary

Various SST topologies resulting from the different MWT transformer topologies have been discussed. Considering the influence of the DC link voltage level, two groups of topologies with and without MVDC link are compared, namely DC–DC and AC–DC SSTs. In AC–DC SSTs, the front-end CHB converter is followed by the isolated DC–DC converter, while in DC–DC SSTs, the MVDC link is used directly as the input of the isolated DC–DC converters. It is highlighted that the SSTs can be configured in a variety of forms with different levels of modularity. It is shown that MWT-based SSTs can reduce the number of conversion stages between the local loads and generations to half compared to a conventional SST topology which leads to great contribution to energy efficiency while maintaining the galvanic isolation between each port. The advantages and disadvantages of SST topologies were described and qualitatively compared. The diverse topologies can be operated as resonant or non-resonant converters, regardless of whether two windings or MWTs are used. Moreover, it is shown that voltage-controlled variable inductors enhance the tunability of the SST. Finally, transformer topologies and configurations for both wire-wound and planar transformers were described. The chapter concludes by surveying with an overview of the emerging applications of MWT-based SSTs.

References

[1] Rutz RF. Solid state transformer. *United States patent*. 1967; (US60913567A).

[2] McMurray W. Multipurpose power converter circuits. *United States patent*. 1968;(US72166468A).

[3] McMurray W. Power converter circuits having a high frequency link. *United States patent*. 1968;(US3517300A).

[4] She X, Huang AQ, and Burgos R. Review of solid-state transformer technologies and their application in power distribution systems. *IEEE journal of emerging and selected topics in power electronics*. 2013;1(3):186–198.

[5] Liserre M, Buticchi G, Andresen M, *et al.* The smart transformer: Impact on the electric grid and technology challenges. *IEEE Industrial Electronics Magazine*. 2016;10(2):46–58.

[6] Liserre M, Perez MA, Langwasser M, *et al.* Unlocking the hidden capacity of the electrical grid through smart transformer and smart transmission. *Proceedings of the IEEE*. 2022;111(4):421–437.

[7] Costa LF, Buticchi G, and Liserre M. Highly efficient and reliable SiC-based DC–DC converter for smart transformer. *IEEE Transactions on Industrial Electronics*. 2017;64(10):8383–8392.

[8] Beiranvand H, Rokrok E, and Liserre M. Vf-constrained $\eta\rho$-pareto optimisation of medium frequency transformers in ISOP-DAB converters. *IET Power Electronics*. 2020;13(10):1984–1994.

[9] Li Z, Hsieh E, Li Q, *et al.* High-frequency transformer design with medium-voltage insulation for resonant converter in solid-state transformer. *IEEE Transactions on Power Electronics*. 2023;38(8):9917–9932.

[10] Ye H, Fechner T, Lei X, *et al.* Review on HVDC cable terminations. *High voltage*. 2018;3(2):79–89.

[11] Beiranvand H, Hoffmann F, Pascal Y, *et al.* Multiwinding Transformer Leakage Inductance Optimization for Power Flow Decoupling in Multiport DC-DC Converters. In: *2021 IEEE 15th International Conference on Compatibility, Power Electronics and Power Engineering (CPE-POWERENG)*; 2021. p. 1–8.

[12] Czyz P, Papamanolis P, Guillod T, *et al.* New 40kV/300kVA Quasi-2-Level Operated 5-Level Flying Capacitor SiC "Super-Switch" IPM. In: *2019 10th International Conference on Power Electronics and ECCE Asia (ICPE 2019-ECCE Asia)*; 2019. p. 813–820.

[13] Huang AQ. Medium-voltage solid-state transformer: Technology for a smarter and resilient grid. *IEEE Industrial Electronics Magazine*. 2016;10 (3):29–42.

[14] Bifaretti S, Zanchetta P, Watson A, *et al.* Advanced power electronic conversion and control system for universal and flexible power management. *IEEE Transactions on Smart Grid*. 2011;2(2):231–243.

[15] Huber JE, Kolar JW. Optimum number of cascaded cells for high-power medium-voltage AC–DC converters. *IEEE Journal of Emerging and Selected Topics in Power Electronics*. 2016;5(1):213–232.

[16] Costa LF, Hoffmann F, Buticchi G, *et al.* Comparative analysis of multiple active bridge converters configurations in modular smart transformer. *IEEE Transactions on Industrial Electronics*. 2018;66(1):191–202.

[17] Vollin J, Tan FD, Cuk SM. Magnetic Regulator Modeling. In: *Proc. Eighth Annual Applied Power Electronics Conference and Exposition*; 1993. p. 604–611.

[18] Yan Y, Geng LD, Zhang L, *et al.* High-power magnetoelectric voltage tunable inductors. *IEEE Transactions on Industrial Electronics*. 2021;68 (6):5355–5365.

[19] Pereira T, Wei Y, Pascal Y, *et al.* Self-tuning multiport resonant DC/DC converter based on actively-controlled inductors for hybrid storage system integration. *IEEE Transactions on Power Electronics*. 2022;38(4):4787–4804.

[20] Ge Z, Wu H, Liu Y. Low-loss segmented shielding technique for PCB-winding planar transformers. *IEEE Transactions on Power Electronics*. 2022;38(1):12–16.

[21] Loef C, De Doncker RW, Ackermann B. On high frequency high voltage generators with planar transformers; 2014. p. 1936–1940.

[22] Groon F, Beiranvand H, Pereira T, *et al.* PCB Layer Optimization of Planar Medium Frequency Transformer for On-Board EV Chargers. In: *2022 24th European Conference on Power Electronics and Applications (EPE'22 ECCE Europe)*; 2022. p. P–1.

[23] Serban E, Saket MA, Ordonez M. High-performance isolated gate-driver power supply with integrated planar transformer. *IEEE Transactions on Power Electronics*. 2021;36(10):11409–11420.

[24] Zhang J, Hurley WG, Wölfle WH. Design of the Planar Transformer in LLC Resonant Converters for Micro-Grid Applications. In: *2014 IEEE 5th International Symposium on Power Electronics for Distributed Generation Systems (PEDG)*; 2014. p. 1–7.

[25] Groon F, Pereira T, Beiranvand H, *et al.* GaN-Based Multiport Resonant Converter for Automotive Applications. In: 2023 IEEE Applied Power Electronics Conference and Exposition (APEC); 2023. p. 892–899.

[26] Groon F, Pereira T, Schikowski S, *et al.* Resonant Multiport Converter with High Interconnection Capability and Lower Parameter Deviations. In: *PCIM Europe 2023; International Exhibition and Conference for Power Electronics, Intelligent Motion, Renewable Energy and Energy Management*; 2023. p. 1–8.

[27] La Ganga A, Re R, Guglielmi P. Input parallel output series structure of planar medium frequency transformers for 200 kW power converter: model and parameters evaluation. *Energies*. 2021;14(5):1450.

[28] Reusch D, Lee FC. High frequency bus converter with low loss integrated matrix transformer; 2012. p. 1392–1397.

[29] Huang D, Ji S, Lee FC. LLC resonant converter with matrix transformer. *IEEE Transactions on Power Electronics*. 2013;29(8):4339–4347.

[30] Fei C, Lee FC, Li Q. High-efficiency high-power-density LLC converter with an integrated planar matrix transformer for high-output current applications. *IEEE Transactions on Industrial Electronics*. 2017;64(11):9072–9082.

[31] Dai M, Zhang X, Li H, *et al.* LLC converter with an integrated planar matrix transformer based on variable width winding; 2019. p. 1–4.

[32] Tria LAR, Alam KS, Zhang D, *et al.* Comparative study of multicore planar transformers on printed circuit boards. *IET Power Electronics*. 2017;10 (12):1452–1460.

[33] Zhang Z, Liu C, Wang M, *et al.* High-efficiency high-power-density CLLC resonant converter with low-stray-capacitance and well-heat-dissipated planar transformer for EV on-board charger. *IEEE Transactions on Power Electronics*. 2020;35(10):10831–10851.

[34] Liserre M, Hoffman F, Pereira T. Multiwinding-transformer-based dc-dc converter solutions for charging stations [technology leaders]. *IEEE Electrification Magazine*. 2021;9(2):5–9.

[35] Hoffmann F, Person J, Andresen M, *et al.* A multiport partial power processing converter with energy storage integration for EV stationary charging. *IEEE Journal of Emerging and Selected Topics in Power Electronics*. 2021; 10(6):7950–7962.

[36] Buticchi G, Costa LF, Liserre M. Multi-port DC/DC converter for the electrical power distribution system of the more electric aircraft. *Mathematics and Computers in Simulation*. 2019;158:387–402.

[37] Bartolucci L, Cordiner S, Mulone V, *et al.* Towards net zero energy factory: a multi-objective approach to optimally size and operate industrial flexibility solutions. *International Journal of Electrical Power & Energy Systems*. 2022;137:107796.

Chapter 16

Design of high-frequency magnetic components for resonant converter

Feng Jin[1], Zheqing Li[1] and Qiang Li[1]

16.1 Introduction

The isolated converters, such as LLC converter, CLLC converter, etc., are widely used in many applications, including data centers [1,2], telecom, electric vehicle (EV) applications [3–5], solid-state transformer applications [6,7], and railway applications [8–12]. In all these converters, both a transformer and a minimum of one inductor are essential. For some converters that do not require large input/output voltage regulation capabilities, a small inductance is enough. The converter has a fixed ratio of input and output voltage, namely DC transformer (DCX), which has gained popularity [5,6,13,14]. Taking the DCX LLC converter as an example, the converter operates at a fixed switching frequency and maintains a unity gain for all load conditions. For applications with regulation requirements, incorporating an additional stage before/after the DCX converter is crucial, which inevitably results in higher costs and a more intricate system structure. Specifically, in the case of a DC/DC (D2D) LLC converter that possesses regulation capabilities, a specific amount of resonant inductance is necessary.

Magnetic components play a crucial role in a converter, and their associated magnetic loss accounts for a significant fraction of its overall loss. Thus, enhancing the performance of these magnetic components directly boosts the overall efficiency and power density of the DC–DC converter. One effective strategy to achieve enhanced efficiency and power density is through magnetic integration. Several methods of integrating leakage inductance with magnetizing inductance can be found in [4,11,15–28].

Some combine the inductor core and transformer core to share the same magnetic path to reduce flux density and core loss [15,16]. However, extra space and volume are required.

For simplicity of integrating the leakage inductance with the transformer, one way is to increase the leakage inductance by increasing the physical distance between primary and secondary windings [17,18]. In [18], two sets of primary

[1]Center for Power Electronics Systems (CPES), Virginia Tech, USA

windings and one set of secondary windings are arranged across the center leg of an EI-core. By maintaining a specific distance between the primary and secondary windings, the built-in leakage inductance for a CLLC converter is adjusted. However, using this technique, the leakage flux does not circulate within the magnetic core; instead, it leaks outside the core. This leakage flux can interact with nearby components and circuits, leading to undesirable voltages and currents in neighboring conductive materials, which might impact their operations. This design also lacks interleaving between the primary and secondary windings, causing increased winding losses. Implementing this method in mass production can be complicated due to the precision needed in controlling the leakage inductance tolerance.

To confine the leakage flux inside the magnetic core, a low-permeability magnetic material between the main transformer core legs is used to adjust the coupling between the primary and secondary sides to control the integrated leakage inductance [19,20]. The low-permeability magnetic material can be replaced by a low reluctance magnetic path with a small air gap to adjust the equivalent permeability [11,21–23]. A common example is an EI-core-based integrated transformer, where the primary and secondary windings are positioned on two legs of the transformer, and the third core leg is utilized to finetune the built-in leakage inductance [21]. This third core leg can be segmented into multiple pieces to facilitate flux distribution in the core plates [11,24]. The primary and secondary windings can be interleaved to reduce the AC winding loss due to smaller magnetomotive force (MMF).

Integrating leakage inductance with a transformer can simplify circuit complexity and enhance power density when designed appropriately. Using integrated cores prevents leakage flux from escaping into the air, mitigating the potential for significant fringing effects, and subsequently reducing winding losses. When the magnitude of the built-in leakage flux is comparable to that of the magnetizing flux, combining the paths of the leakage flux and magnetizing flux can be advantageous, allowing them to share the same magnetic path. While integration will lead to an increase in total flux, the flux density can be effectively reduced by enlarging the core's radius. Since the cross-sectional area is directly proportional to the square of the core radius, even a modest enlargement in radius can substantially increase the cross-sectional area. This leads to a slight increase in winding loss, corresponding to the expanded radius. However, for applications desiring larger inductance and higher power, the leakage flux becomes substantially larger than the magnetizing flux. This necessitates an additional cross-sectional area for integrated magnetics (IT), leading to a considerably higher winding loss than a discrete transformer due to the enlarged winding length.

The previous literature has not extensively explored the limitations and potential challenges associated with IT. This chapter will focus on the comparison of IT solution and discrete inductor and transformer (DIT) solution for a high-power LLC converter.

The high-power LLC converter adopted for this chapter is the first stage of the auxiliary power supply (APS) for railway applications. The APS consists of a D2D

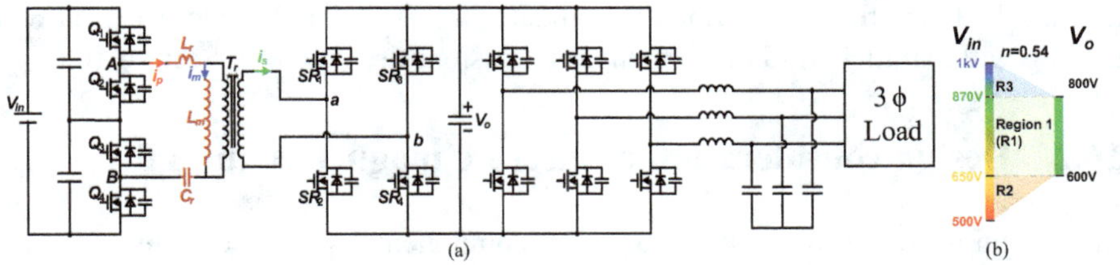

Figure 16.1 Topology of APS and the voltage range of LLC converter: (a). Topology. (b). Voltage range of the LLC

three-level LLC converter [29] and a three-phase inverter, as shown in Figure 16.1 (a) [10,11]. The input voltage range of the LLC converter is from 500 V to 1 kV, the output voltage is from 600 to 800 V, and the total output power is 30 kW. The input and output voltage relationship can be found in Figure 16.1(b). 1.2 kV silicon-carbide (SiC) devices can be adopted for both the primary side and the secondary side. This chapter focuses on the magnetic components design of the LLC converter. The details of the control method for this LLC converter can be found in [10].

The key contents of this chapter are summarized as follows:

1. Comprehensive design considerations for 4-leg-core-based IT and DIT solutions have been provided. This includes an in-depth derivation of inductance parameters, effective turns ratio, and flux calculations for each magnetic part.
2. A detailed exploration of flux distribution within a 4-leg-core-based IT is offered, illuminating both the merits and challenges associated with integrating built-in leakage inductance with magnetizing inductance. The analysis highlights specific limitations inherent to the IT solution when applied to high-power LLC converters. Consequently, the DIT solution with proper design emerges as a more advantageous alternative than the IT solution for these high-power scenarios.
3. A new DIT solution with parallel windings for the resonant inductor and transformer is proposed with good current-sharing capabilities. Different transformer winding configurations lead to varied current-sharing behaviors. The ideal case was identified and validated through experimental results, demonstrating successful current sharing between parallel windings for the inductor and the transformer.

The outline of the chapter is organized as follows: Section 16.2 delves into the general design considerations for both DIT and IT solutions, supplemented with a thorough assessment of a general 4-leg-core-based IT. Section 16.3 presents the flux waveforms and provides a comparison of IT and DIT solutions through finite-element analysis (FEA) simulation, facilitating a deeper understanding of flux distribution. Section 16.4 outlines the design of the inductor and transformer for the DIT solution, emphasizing the effective current sharing between parallel windings for a 30 kW LLC converter. In Section 16.5, experimental results of the IT solution and the DIT solution is provided to validate the design and explore the current sharing with parallel windings. Finally, conclusions are drawn in Section 16.6.

16.2 Design consideration of discrete magnetics and IT

In designing Litz-wire-based magnetic components within a converter, for example, discrete inductor and discrete transformer, the windings and the core need careful consideration. As for the selection of Litz-wire, the wire diameter for each strand can be determined based on the skin depth at the maximum switching frequency, and the total strand count can be calculated accordingly. The cross-sectional area of the core legs for the inductor and transformer can be determined based on the maximum flux going through the magnetics and the desired maximum flux density. For integrated transformers with built-in leakage inductance, the maximum flux is determined by the magnetizing flux and built-in leakage flux. In Section 16.2, the design consideration of the DIT and IT solutions is included.

16.2.1 Design consideration for a discrete inductor

To start with a discrete inductor, an EI-core-based inductor that has N_L turns across the center leg shown in Figure 16.2(a) is adopted for evaluation; its sectional-view and reluctance model are shown in Figures 16.2(b) and (c). In the reluctance model, the core's reluctance is assumed to be zero due to the large permeability of the ferrite core compared with that of free air. Therefore, only the air gap reluctance is considered, which is R_{gL}, as calculated in (16.1). This assumption remains valid for subsequent analysis of the transformer reluctance model unless otherwise specified.

$$R_{gL} = l_{gL}/(\mu_0 \cdot A_{eL}) \tag{16.1}$$

where A_{eL} is the cross-sectional area of the center leg, l_{gL} is the air gap length of the center leg, μ_0 is the permeability of free space, i_L is the inductor current, and N_L is the turns number.

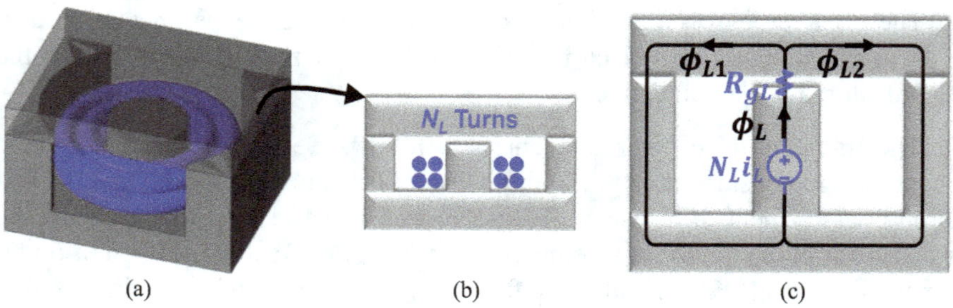

Figure 16.2 Discrete inductor structure and reluctance model: (a) inductor structure, (b) sectional view, and (c) reluctance model

The flux in the center leg, φ_L, can be calculated in (16.2), and the inductance can be calculated in (16.3):

$$\varphi_L = N_L i_L / R_{gL} \tag{16.2}$$

$$L = N_L \varphi_L(i_L)/i_L = N_L{}^2/R_{gL} \tag{16.3}$$

The flux of the center leg can be rewritten in (16.4) based on (16.2) and (16.3), and its flux density of the center leg can be calculated in (16.5)

$$\varphi_L = L \cdot i_L / N_L \tag{16.4}$$

$$B_L = \varphi_L / A_{eL} = L \cdot i_L / (N_L \cdot A_{eL}) \tag{16.5}$$

As expressed in (16.4), the flux generated is directly related to the inductor current and inductance value and inversely related to the number of turns. Given a specific inductance and inductor current, increasing the number of turns is a direct approach to decrease the flux. However, the larger turns number will result in a larger air gap as (16.6) points out that the required air gap is proportional to the square of the turns number. The larger air gap may bring a larger fringing effect and cause larger winding loss; therefore, a longer center leg is required to avoid the fringing effect and additional winding loss with the price of a larger volume. The necessity to mitigate fringing flux challenges has led to the use of a distributed air gap configuration in industrial applications [30,31]. Rather than utilizing a single long gap, this approach involves distributing the gap into several smaller segments, as shown in Figure 16.3(a). This strategy effectively reduces the fringing flux effects and minimizes the winding loss. However, it is worth noting that implementing a distributed air gap does reduce the final inductance of an inductor [30], which requires FEA evaluation to explore its impact. This reduction is constrained by manufacturing technology and the overall cost associated with producing such an inductor with the distributed air gap. For the ideal case, with the reluctance model shown in Figure 16.3(b), the total air gap reluctance can be summed to be one value, as shown in (16.7). l_{gLi} is the length of i_{th} air gap of the distributed air gaps. The calculation of inductance and generated flux can still be calculated based

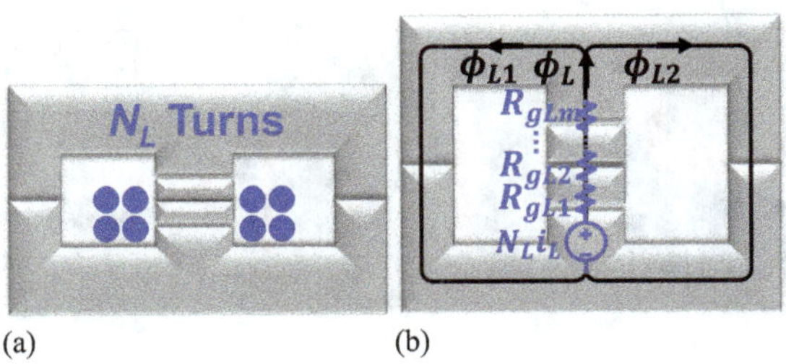

(a) (b)

Figure 16.3 Discrete inductor with distributed air gap: (a) sectional view and (b) reluctance model on (16.3) and (16.4):

$$l_{gL} = \mu_0 \cdot A_{eL} \cdot {N_L}^2/L \tag{16.6}$$

$$R_{gL} = \sum_{i=1}^{m} R_{gLi} = \sum_{i=1}^{m} \left(l_{gLi}/(\mu_0 \cdot A_{eL})\right) \tag{16.7}$$

For the inductor used for an LLC resonant converter, the inductor current, i_L (it equals to i_p in Figure 16.1(a)), is load-dependent and will increase with the output power. According to (16.4), for a given inductance, the flux inside the core becomes larger along with the increase of i_p. Therefore, the maximum flux occurs at the condition of lowest input voltage and maximum output power as the highest resonant current, I_{p_max}, can be expected. The maximum flux density, B_{L_max}, can be used for the selection of the required turns number as calculated in (16.8):

$$N_L = \frac{L \cdot I_{p\_\max}}{A_{eL} \cdot B_{L\_\max}} \tag{16.8}$$

As (16.8) points out, the cross-sectional area, A_{eL}, will also have an influence on determining the required turns number, N_L. If smaller A_{eL} is selected, more turns number will be necessary to reduce the $B_{\max}$, leading to a larger window area of the magnetic core and a larger air gap, and vice versa. The final selection of N_L and A_{eL} should consider the magnetic loss, magnetic core volumes, and its manufacturing constraints.

16.2.2 Design consideration for a discrete transformer

As for a discrete transformer, a UI-core-based transformer where has (N_{p1})P(N_{s1})S and (N_{p2})P(N_{s2})S winding arrangement across the transformer legs are used for evaluation as shown in Figure 16.4(a), its sectional view of the transformer and reluctance model are shown in Figure 16.4(b) and (c). To simplify the analysis, only the reluctance of the air gap is considered, and the reluctance of the core is assumed to be zero due to its relatively high permeability of ferrite material.

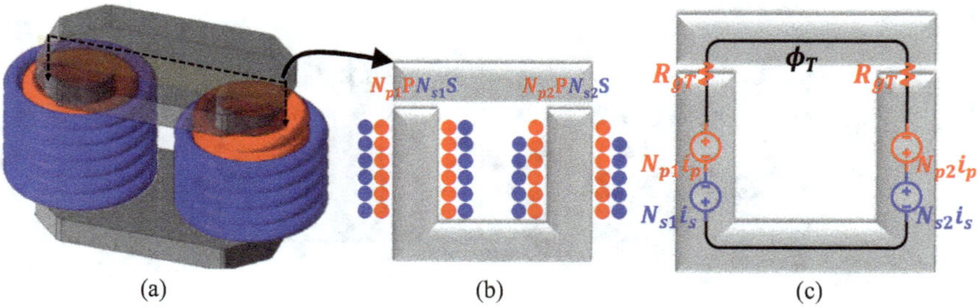

Figure 16.4 Transformer structure and reluctance model for DIT solution: (a) a UI-core-based transformer structure, (b) sectional view, and (c) reluctance model

Based on the reluctance model shown in Figure 16.4(c), the flux inside the core can be calculated as

$$\varphi_T(i_p, i_s) = \frac{(N_{p1} + N_{p2}) \cdot i_p - (N_{s1} + N_{s2}) \cdot i_s}{2R_{gT}} \tag{16.9}$$

$$R_{gT} = l_{gT}/(\mu_0 \cdot A_{eT}) \tag{16.10}$$

where R_{gT} is the reluctance of air gap in each transformer core leg, A_{eT} is the cross-sectional area of the core leg, l_{gT} is the air gap length, i_p is the primary current, and i_s the secondary current.

The magnetizing inductance, L_m, can be calculated as

$$L_m = \frac{(N_{p1} + N_{p2})\varphi_T(i_p, 0)}{i_p} = \frac{(N_{p1} + N_{p2})^2}{2R_{gT}} \tag{16.11}$$

In the ideal case, the flux generated by the primary windings will be fully linked to the secondary windings, resulting in zero leakage inductance in both the primary and secondary sides in a cantilever model, as shown in Figure 16.5(a), namely, $L_{kp} = 0$. In Figure 16.5(a), below KCL can be generated:

$$i_m = i_p - i_s/N_e \tag{16.12}$$

where N_e is the effective turns ratio between primary and secondary windings, here, it is $(N_{p1}+N_{p2})/(N_{s1}+N_{s2})$.

The generated flux inside the transformer core can be rewritten in (16.13). As indicated by (16.13), the magnetizing flux is directly proportional to the magnetizing current. This magnetizing current is determined by the voltage-second across the transformer and remains constant irrespective of the load conditions. Because it will not increase with the output power increase, the worst case can be designed as the maximum output voltage where the highest voltage-second is applied. The flux density of the transformer core can be calculated in (16.14)

$$\varphi_T = \frac{L_m}{N_{p1} + N_{p2}} i_m \tag{16.13}$$

$$B_T = \frac{\varphi_T}{A_{eT}} = \frac{L_m}{(N_{p1} + N_{p2})A_{eT}} i_m \tag{16.14}$$

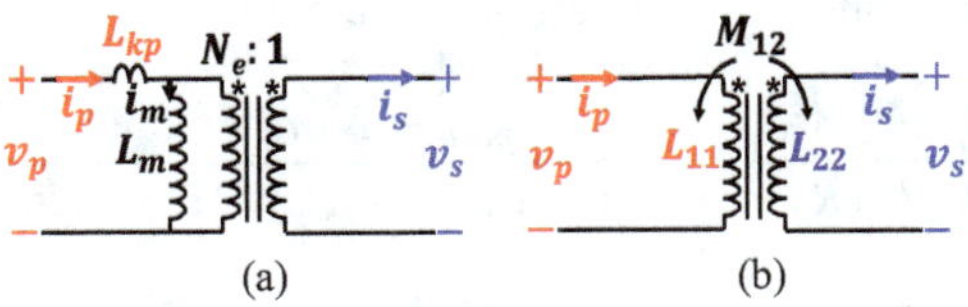

Figure 16.5 Transformer models: (a) cantilever model and (b) couple inductor model

In an LLC resonant converter, the maximum magnetizing current, I_{m_max}, occurs at the maximum output voltage, V_{o_max}, as follows:

$$I_{m\_\max} = \frac{N_e V_{o\_\max}}{L_m} \frac{T_s}{4} \tag{16.15}$$

where T_s is the switching period.

The maximum flux density, B_{T_max}, can be derived in (16.16) from (16.14) and (16.15)

$$B_{T\_\max} = \frac{V_{o\_\max} T_s}{4(N_{s1} + N_{s2}) A_{eT}} \tag{16.16}$$

As (16.16) reveals, both the cross-sectional area, A_{eT}, and secondary turns number, (N_{s1}+N_{s2}), will determine the maximum flux density, B_{T_max}. The turns number of the primary and secondary windings can be selected based on the designed turns ratio for the converter, and the cross-sectional area, A_{eT}, can be finalized based on the secondary turns number and maximum allowed flux density.

16.2.3 Design consideration for an integrated transformer

To reduce the number of magnetic components and achieve the benefits of integration, several integration techniques were proposed. By introducing (an) additional low reluctance path(s) to the UI core of the UI-core-based transformer shown in Figure 16.4(a), the controllable leakage inductance can be integrated with proper winding arrangement [11,21,23]. In this chapter, a UI-core with two additional side legs is adopted to achieve evenly distributed flux density in the core plates and smaller core loss, which is defined as a 4-leg-core. The 4-leg-core-based IT with generalized representations of winding arrangements, as shown in Figure 16.6, is adopted for evaluation.

In the reluctance model, R_{g1} is the air gap reluctance of core leg 1 with (N_{p1}PN_{s1}S) winding arrangement, R_{g2} is the air gap reluctance of core leg 2 with (N_{p2}PN_{s2}S) winding arrangement, and R_{g3} is the air gap reluctance of side leg. The two side legs are designed with the same cross-sectional area and air gap length to ensure mechanical stability after assembly.

$$R_{gi} = l_{gi}/(\mu_0 \cdot A_{ei}), i = 1,2,3 \tag{16.17}$$

where l_{gi} is the air gap length of the ith core leg and A_{ei} is the cross-sectional area of the ith core leg.

The flux of each core segment can be calculated as

$$\varphi_1 = \frac{N_{p1} i_p - N_{s1} i_s}{R_{g1} + R_{g2}//(R_{g3}/2)} + \frac{N_{p2} i_p - N_{s2} i_s}{R_{g2} + R_{g1}//(R_{g3}/2)} \frac{R_{g1}//(R_{g3}/2)}{R_{g1}} \tag{16.18}$$

$$\varphi_2 = \frac{N_{p2} i_p - N_{s2} i_s}{R_{g2} + R_{g1}//(R_{g3}/2)} + \frac{N_{p1} i_p - N_{s1} i_s}{R_{g1} + R_{g2}//(R_{g3}/2)} \frac{R_{g2}//(R_{g3}/2)}{R_{g2}} \tag{16.19}$$

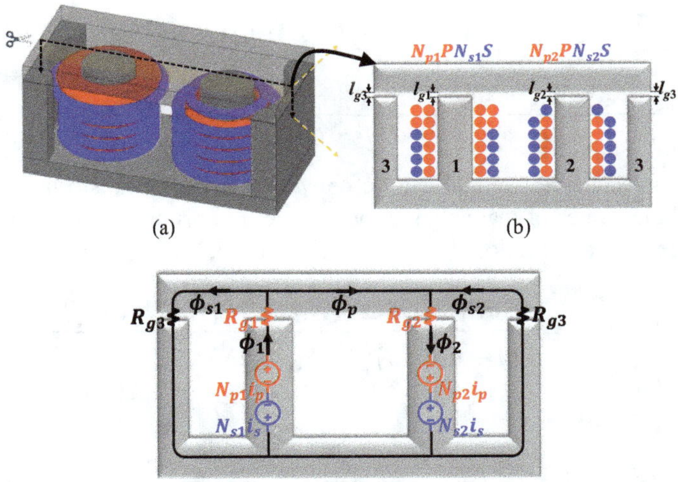

Figure 16.6 IT structure and its reluctance model: (a) transformer structure, (b) sectional view, and (c) reluctance model

$$\varphi_{s1} = -\varphi_{s2} = \frac{N_{p1}i_p - N_{s1}i_s}{R_{g1} + R_{g2}//(R_{g3}/2)} \frac{R_{g2}//(R_{g3}/2)}{R_{g3}} - \frac{N_{p2}i_p - N_{s2}i_s}{R_{g2} + R_{g1}//(R_{g3}/2)} \frac{R_{g1}//(R_{g3}/2)}{R_{g3}} \tag{16.20}$$

$$\varphi_p = \frac{N_{p1}i_p - N_{s1}i_s}{R_{g1} + R_{g2}//(R_{g3}/2)} \frac{R_{g2}//(R_{g3}/2)}{R_{g2}//R_{g3}} + \frac{N_{p2}i_p - N_{s2}i_s}{R_{g2} + R_{g1}//(R_{g3}/2)} \frac{R_{g1}//(R_{g3}/2)}{R_{g1}//R_{g3}} \tag{16.21}$$

To simplify the calculation, the reluctance R_{g2} is defined as α times of R_{g1}, and R_{g3} is defined as β times of R_{g1}. The inductor parameters of the couple inductor model shown in Figure 16.5(b) can be calculated as

$$L_{11} = (N_{p1}\varphi_1(i_p, i_s = 0) + N_{p2}\varphi_1(i_p, i_s = 0))/i_p = \frac{2\alpha N_{p1}{}^2 + 2N_{p2}{}^2 + (N_{p1} + N_{p2})^2\beta}{(2\alpha + \alpha \cdot \beta + \beta)R_{g1}} \tag{16.22}$$

$$L_{22} = (-N_{s1}\varphi_1(i_p = 0, i_s) - N_{s2}\varphi_1(i_p = 0, i_s))/i_s = \frac{2\alpha N_{s1}{}^2 + 2N_{s2}{}^2 + (N_{s1} + N_{s2})^2\beta}{(2\alpha + \alpha \cdot \beta + \beta)R_{g1}} \tag{16.23}$$

$$L_{12} = \left(-N_{s1}\varphi_1\left(i_p, i_s = 0\right) - N_{s2}\varphi_1\left(i_p, i_s = 0\right)\right)/i_p$$
$$= -\frac{2aN_{p1}N_{s1} + 2N_{p2}N_{s2} + \left(N_{p1} + N_{p2}\right)\left(N_{s1} + N_{s2}\right)\beta}{(2a + \alpha \cdot \beta + \beta)R_{g1}} \tag{16.24}$$

The couple inductor model can be converted into the transformer cantilever model, and the values of magnetizing inductance, L_m, primary leakage inductance, L_{kp}, and effective turns ratio, N_e, can be found in (16.25)–(16.27):

$$L_m = L_{12}{}^2/L_{22} = \frac{\left(2aN_{p1}N_{s1} + 2N_{p2}N_{s2} + \left(N_{p1} + N_{p2}\right)\left(N_{s1} + N_{s2}\right)\beta\right)^2}{(2a + a\beta + \beta)\left(2aN_{s1}{}^2 + 2N_{s2}{}^2 + \left(N_{s1} + N_{s2}\right)^2\beta\right)R_{g1}} \tag{16.25}$$

$$L_{kp} = L_{11} - L_m = \frac{2\left(N_{p1}N_{s2} - N_{p2}N_{s1}\right)^2}{\left(2aN_{s1}{}^2 + 2N_{s2}{}^2 + \left(N_{s1} + N_{s2}\right)^2\beta\right)R_{g1}} \tag{16.26}$$

$$N_e = \sqrt{L_m/L_{22}} = \frac{2aN_{p1}N_{s1} + 2N_{p2}N_{s2} + \left(N_{p1} + N_{p2}\right)\left(N_{s1} + N_{s2}\right)\beta}{2aN_{s1}{}^2 + 2N_{s2}{}^2 + \left(N_{s1} + N_{s2}\right)^2\beta} \tag{16.27}$$

where

$$\beta = \frac{N_e\left(2aN_{s1}{}^2 + 2N_{s2}{}^2\right) - \left(2aN_{p1}N_{s1} + 2N_{p2}N_{s2}\right)}{\left(N_{s1} + N_{s2}\right)\left(N_{p1} + N_{p2} - N_e\left(N_{s1} + N_{s2}\right)\right)} \tag{16.28}$$

As (16.26) points out, there are two requirements for integrating leakage inductance: (1) $N_{p1}N_{s2} \neq N_{p2}N_{s1}$ and (2) β is not infinitely large. As (16.27) points out, N_e does not equal the physical turns ratio and can be related to the selection of α and β (it equals the physical turns ratio when β is infinitely large, which means no leakage inductance is integrated, namely, it is a perfect transformer). As revealed in (16.25)–(16.27), the transformer parameters L_m, L_k, and N_e are not independent and should be designed through appropriate selection of N_{p1}, N_{p2}, N_{s1}, N_{s2}, α, and β, which can be a potential limitation of IT. By assuming L_m, L_{kp}, and N_e are selected based on circuits design, the flux inside each core segment, φ_1, φ_2, φ_{s1}, φ_{s2}, φ_p, can be rewritten as

$$\varphi_1 = \frac{N_{s2}}{N_{p1}N_{s2} - N_{p2}N_{s1}}L_{kp}i_p + \frac{N_{p2} - N_{s2}N_e}{N_e\left(N_{p2}N_{s1} - N_{p1}N_{s2}\right)}L_m i_m \tag{16.29}$$

$$\varphi_2 = \frac{N_{s1}}{N_{p2}N_{s1} - N_{p1}N_{s2}}L_{kp}i_p + \frac{N_{p1} - N_{s1}N_e}{N_e\left(N_{p1}N_{s2} - N_{p2}N_{s1}\right)}L_m i_m \tag{16.30}$$

$$\varphi_p = \frac{N_{s1} - N_{s2}}{2N_{p2}N_{s1} - 2N_{p1}N_{s2}}L_{kp}i_p + \frac{N_{p1} - N_{p2} - N_e\left(N_{s1} - N_{s2}\right)}{2N_e\left(N_{p1}N_{s2} - N_{p2}N_{s1}\right)}L_m i_m \tag{16.31}$$

$$\varphi_{s1} = -\varphi_{s2} = \frac{N_{s1} + N_{s2}}{2N_{p1}N_{s2} - 2N_{p2}N_{s1}} L_{kp} i_p + \frac{N_{p1} + N_{p2} - N_e(N_{s1} + N_{s2})}{2N_e(N_{p2}N_{s1} - N_{p1}N_{s2})} L_m i_m \tag{16.32}$$

As shown in (16.29)–(16.32), the flux inside each core segment is determined by both the leakage flux ($L_{kp}i_p$) and magnetizing flux ($L_m i_m$), which means that the total flux will become larger when larger leakage inductance is built-in.

For the LLC resonant converter, i_p is load dependent and will increase when the output power increases and i_m is load independent, which is determined by the voltage-second across the transformer. i_p is much higher than i_m for high high-power LLC converter, which will make the leakage flux much higher than the magnetizing flux. As for designing DIT, both the inductor and the transformer can be optimized individually due to the decoupled structure. However, the flux inside the inductor core and the flux inside the transformer core cannot share the same magnetic path, and no flux cancellation can be expected. As for designing IT, several design considerations of IT should be pointed out here:

1. L_m, L_{kp}, and N_e can be designed with proper selection of N_{p1}, N_{p2}, N_{s1}, N_{s2}, R_{g1}, α, and β. However, N_{p1}, N_{p2}, N_{s1}, and N_{s2} are integer numbers; the change in numbers will impact inductance parameters and effective turns ratio in a discrete step. The turns ratio across the two transformer legs, N_{p1}/N_{s1} and N_{p2}/N_{s2}, should be designed as closely as possible to ensure similar flux amplitude of the two transformer core legs with integrating certain leakage inductance. For more built-in leakage inductance, a larger difference in turns ratio of N_{p1}/N_{s1} and N_{p2}/N_{s2} are necessary. By considering easier manufacturing, the cross-sectional area and air gap length of the two transformer core legs can be designed exactly the same. Namely, α is set to 1.
2. The effective turns ratio of the IT will be smaller than the physical turns ratio, $N_e < (N_{p1}+N_{p2})/(N_{s1}+N_{s2})$, as built-in leakage inductance is integrated. Therefore, the physical turns ratio should be larger than the required turns ratio.
3. The leakage inductance is created through different turns ratios ($N_{p1}/N_{s1} \neq N_{p2}/N_{s2}$) across transformer core legs. This variation causes a lack of interleaving between primary and secondary windings, potentially leading to increased winding losses due to the accumulation of MMF.
4. The leakage inductance due to the physical distance between primary and secondary windings may not be negligible and can be evaluated through FEA tools.
5. The total flux will be predominantly influenced by the leakage flux, especially with larger integrated leakage inductance and higher output power. This leads to larger winding loss, caused by a longer winding length necessitated by the expanded core sectional area aimed at reducing the flux density. The magnetic loss of an extra inductor can be saved. However, the total magnetic core loss and winding loss should be evaluated thoroughly.

16.3 Flux distribution comparison between IT and DIT solutions

To compare the DIT and IT solutions, a 30 kW LLC converter with specification and circuit parameters defined in Table 16.1 was adopted for evaluation.

As specified in Figure 16.1(b), the LLC converter operates at resonant frequency when the input voltage range is from 650 to 870 V (Region 1, R1) to achieve the best efficiency for nominal conditions; when the input voltage is lower than 650 V (Region 2, R2), it will operate in boost operation mode to boost the gain of the converter for high power conditions [10]. When the input voltage exceeds 870 V (Region 3, R3), it will enter variable frequency modulation to reduce the output voltage. The key operation waveforms of different operation regions are shown in Figure 16.7.

The root-mean-square (RMS) current of the primary side and secondary side at the lowest input voltage, $V_{in} = 500$ V, and full load, $P_o = 30$ kW, are 134 and 70 A, respectively. Two transformers connected parallel on the primary side and series on the secondary sides are adopted for the magnetic components of this 30 kW LLC converter, as shown in Figure 16.8(a). For comparison, the DIT solution adopts two discrete transformers with input parallel output series connection and one series

Table 16.1 Specification of 30 kW LLC converter

Design parameters	Values
Input voltage, V_{in}	500–1000 V
Nominal input voltage, V_{in_nom}	750 V
Output voltage, V_o	600–800 V
Maximum output power, P_{omax}	30 kW
Resonant frequency, f_r	100 kHz
Transformer turns ratio	0.54
Magnetizing inductance	35.0 μH
Resonant inductance	3.8 μH
Resonant capacitors	672.0 nF

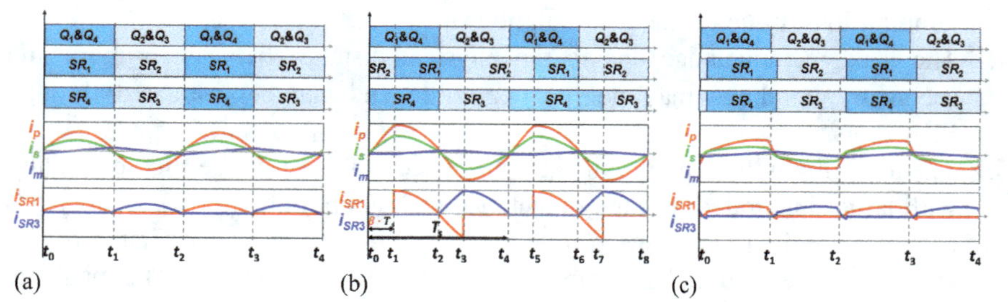

Figure 16.7 Operation waveforms of different regions: (a) Region 1, (b) Region 2, and (c) Region 3

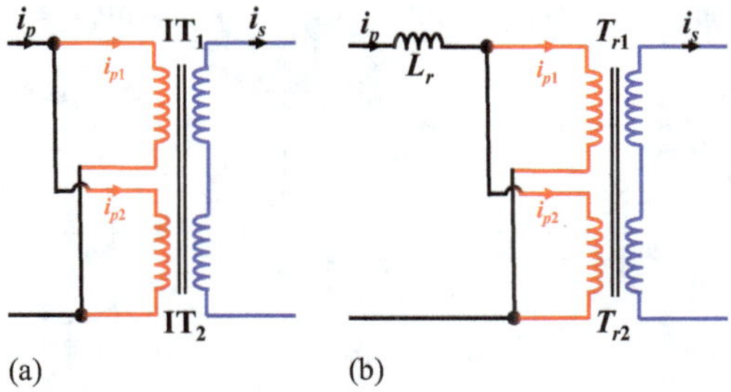

Figure 16.8 Transformer connections of IT and DIT solutions: (a) IT solution and (b) DIT solution

Table 16.2 Equivalent parameter of each transformer

Design parameters	**IT**	**DIT**
Magnetizing inductance, L_{m_each}	70.0 μH	70.0 μH
Built-in leakage inductance, L_{kp_each}	7.6 μH	~0 μH
Discrete resonant inductance, L_r	0 μH	3.8 μH
Transformer turns ratio, N_{e_each}	1.08	1.08

resonant inductor, as shown in Figure 16.8(b). Equivalent AWG 6 (2700 strands of AWG40) Litz wire is used for the primary and secondary windings to build the magnetic components according to NFPA 070 National Electrical Code. For each transformer, the parameters are listed in Table 16.2.

When it comes to the design of the magnetics, the worst case should be considered. For designing the discrete inductor, the maximum flux occurs at condition 1 of $V_{in} = 500$ V, $V_o = 600$ V, $P_o = 30$ kW as the highest primary current is expected. As for the design of a discrete transformer, the maximum magnetizing flux with the largest voltage-second across the transformer occurs at condition 2 of $V_{in} = 870$ V, $V_o = 800$ V, $P_o = 30$ kW. However, for the IT solution, both the magnetizing flux and the leakage flux will flow through the core, and it requires careful evaluation of flux distribution inside the core. Therefore, the abovementioned two conditions are used for evaluating the flux distribution inside the magnetic cores.

16.3.1 Comparison of flux waveforms for DIT and IT solutions

For discrete transformer, the primary to secondary turns number is selected to 10:9 (5P5S for one transformer leg and 5P4S for another transformer leg) for each transformer to achieve the effective turns ratio in Table 16.2, and these flux waveforms at condition 1 and condition 2 are shown in Figures 16.9(a) and (b). As depicted in Figures 16.9(a) and (b), the magnetizing flux, φ_T, of condition 1 is smaller than that of condition 2 as smaller voltage-second is expected, and the inductance flux, φ_L, of condition 1 is much larger than that of condition 2 as larger

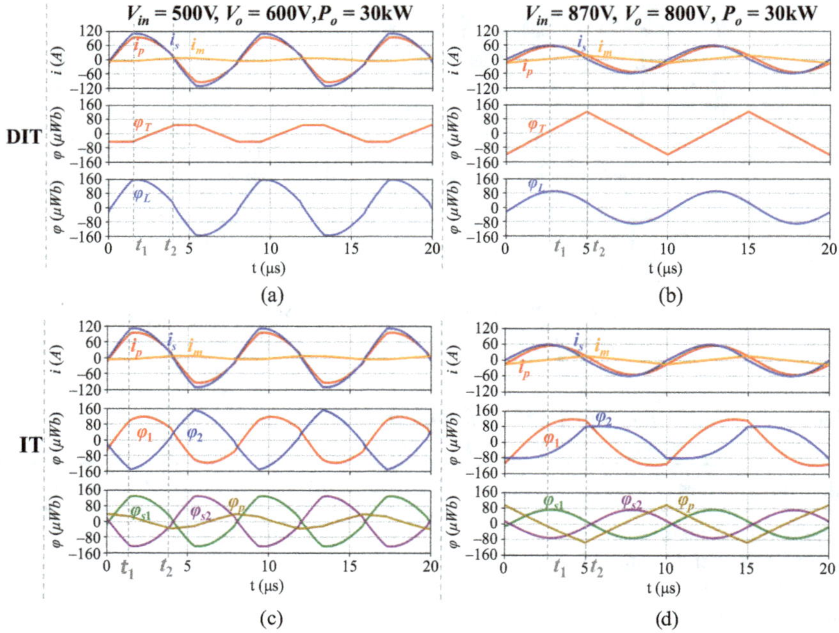

Figure 16.9 Flux waveforms of DIT and IT solutions: (a) DIT of condition 1, (b) DIT of condition 2, (c) IT of condition 1, and (d) IT of condition 2

resonant current is expected for lower input voltage. Based on Figure 16.9(b), the cross-sectional area of transformer core legs and plates is set to 615.8 mm^2, where its core leg's radius is set to 14.0 mm to keep the maximum flux density to 200 mT that the maximum flux amplitude (120.8 μWb) occurs at condition 2. For a discrete inductor, four turns of two parallel windings are adopted, and its structure to achieve current sharing among parallel windings will be discussed in Section 16.4. Based on Figure 16.9(a), the cross-sectional area of the inductor core leg is set to 855 mm^2 where its core leg's radius is set to 16.5 mm to keep the maximum flux density to 180 mT as the maximum flux amplitude (156.0 μWb) occurs at condition 1.

For the integrated transformer, the primary to secondary turns number is selected to 13:11 (8P5S for one transformer leg and 5P6S for another transformer leg) for each transformer to achieve the effective turns ratio and inductance parameters in Table 16.2, and the flux waveforms at condition 1 and condition 2 are shown in Figures 16.9(c) and (d). As Figures 16.9(c) and (d) point out, the flux inside each core segment is different and has a sinusoidal component ($L_{kp}i_p$ related) and a triangle component ($L_m i_m$ related). The distribution of both components differs along with the conditions; however, the maximum flux occurs at condition 1 as the leakage flux dominates. Based on Figure 16.9(c), the cross-sectional area of IT's cores and plates is set to 855 mm^2, where its core leg's radius is set to 16.5 mm to keep the maximum flux density to 180 mT as the maximum flux amplitude (151.9 μWb) occurs at condition 1.

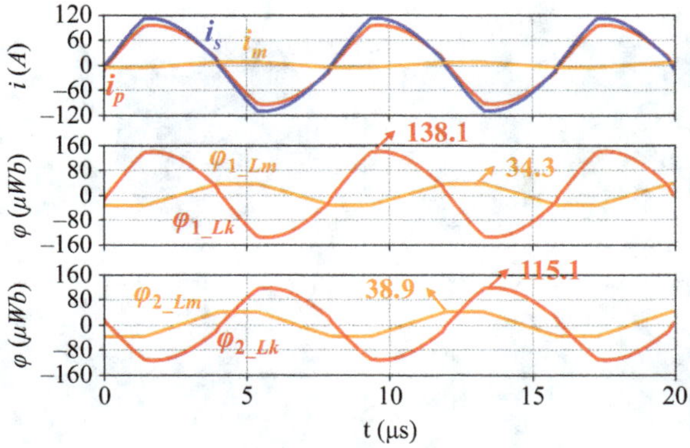

Figure 16.10 Flux distribution waveforms of IT under condition 1

Figure 16.10 shows the waveforms of magnetizing flux and leakage flux for condition 1 of IT. φ_{i_Lm} and φ_{i_Lk} are the magnetizing flux and leakage flux of core leg *i*, respectively; *i* can be 1 or 2. The summation of φ_{i_Lm} and φ_{i_Lk} will determine the total flux of core leg *i*, φ_i. In this high-power LLC converter, the primary resonant current, i_p, is much higher than that of the magnetizing current, i_m, as shown in Figure 16.10. As a result, the leakage flux, φ_{i_Lk}, primarily determines the overall flux amplitude within the core. This results in a larger cross-sectional area for the transformer core legs, leading to an increased core volume and winding loss. The φ_{i_Lk} will become even larger with more built-in leakage inductance and higher resonant current. As (16.25) and (16.26) indicate, the values of L_m and L_{kp} are interdependent, and the induced magnetizing flux and leakage flux cannot be adjusted independently. In contrast with IT, the magnetizing flux and leakage flux of DIT can be adjusted independently by changing the turns number of the inductor and transformer, as (16.4) and (16.13) reveal.

16.3.2 FEA comparison of flux distribution for DIT and IT solutions

Maxwell three-dimensional (3D) FEA is implemented to compare the flux distribution of IT and DIT solutions at the given conditions. Figure 16.11 depicts the flux distribution results of DIT at time instance t_1 (maximum resonant current amplitude) and t_2 (maximum magnetizing current amplitude) under the condition of condition 1 and condition 2, respectively.

In Figure 16.11(a), at the time instance t_1 (time instance with maximum primary resonant current) under condition 1, the flux within the discrete inductor distributes evenly from the center leg to both side legs. Simultaneously, the magnetizing flux within the discrete transformer circulates from one core leg to the other through the core plate. However, for the IT depicted in Figure 16.11(c), the flux, created by the built-in leakage inductance, flows from the transformer core leg directly to the nearest side leg, rather than equally distributed between the core

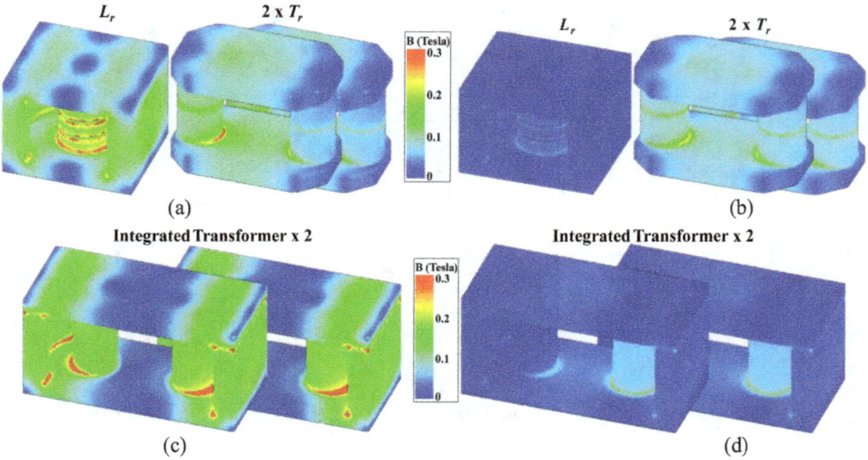

Figure 16.11 FEA results comparison of DIT and IT under condition 1: (a) t_1 of DIT, (b) t_2 of DIT, (c) t_1 of IT, and (d) t_2 of IT

plate and side legs. This flux behavior aligns well with the waveform illustrated in Figure 16.9(c). Such an uneven flux distribution can result in increased core losses. At the time instance of t_2 (time instance with maximum magnetizing current) of condition 1, the primary current amplitude equals the maximum magnetizing current, much smaller than the peak resonant current. As shown in Figure 16.11(b), the flux density inside the inductor core is very small, and the magnetizing flux is still relatively large as its magnetizing flux reaches its peak value, as Figure 16.9(a) points out. In Figure 16.11(d), the flux density inside the core of IT is relatively small due to a larger core-sectional area being used.

Under condition 2, the primary resonant current decreases, while the magnetizing current increases in comparison to that of condition 1. The flux density of the discrete inductor shown in Figure 16.12(a) will be smaller than that of condition 1 in Figure 16.11(a). Simultaneously, the magnetizing flux within the discrete transformer is close to zero due to the zero-magnetizing current at t_1. For the flux density distribution of IT depicted in Figure 16.12(c), the leakage flux will flow from the center core leg to the nearest side leg with a smaller value than that in Figure 16.11(c). At the time instance of t_2 of condition 2, the magnetizing current is significantly larger than it is under condition 1, owing to the increased voltage-second across the transformer. As illustrated in Figure 16.12(b), the flux density within the inductor core remains minimal. However, the flux density within the discrete transformer reaches maximum and surpasses the value observed at condition 1, as shown in Figure 16.11(b). From the flux distribution in Figure 16.12(d), the magnetizing flux dominates and flows through one core leg to another through the core plate, where the two sides legs do not help distribute the flux as the reluctance of the air gap is greater than the reluctance of the core plate. The flux density within the core of the IT being lesser than that in the DIT can be attributed to its larger cross-sectional area.

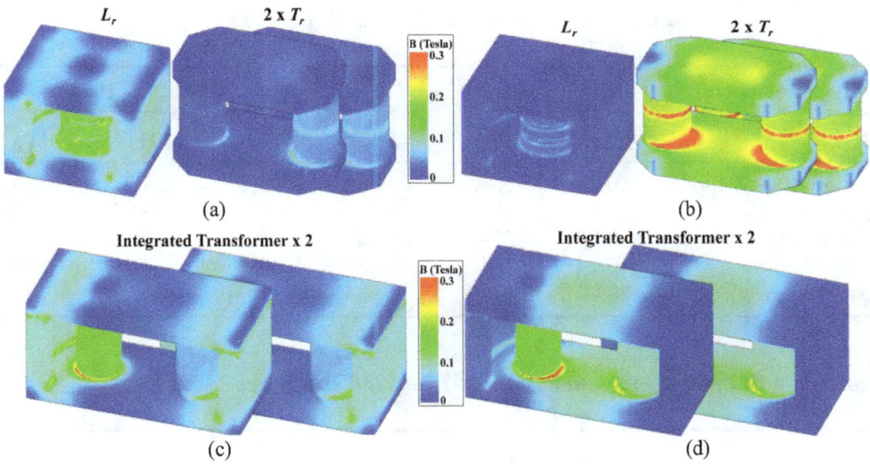

Figure 16.12 FEA results comparison of IT and DIT under condition 2: (a) t_1 of DIT, (b) t_2 of DIT, (c) t_1 of IT, and (d) t_2 of /IT

Based on the above analysis, the following can be summarized:

1. For a discrete inductor, the maximum flux density inside the core appears at the maximum resonant current at the lowest input voltage, and the inductor flux can be distributed evenly inside the core plates and side legs with an EI-core. The selection of the cross-sectional area of the inductor core should be based on the lowest input voltage and maximum output power.
2. For a discrete transformer, the maximum flux density appears at the maximum magnetizing current at the highest output voltage (expecting maximum voltage-second). The time instance with peak flux density inside the inductor core and the transformer core is not the same. The selection of the cross-sectional area of the transformer core of the IT solution should be based on the highest output voltage and lowest switching frequency.
3. For IT, the flux inside the magnetic core legs will be determined by both the built-in leakage flux and magnetizing flux. The flux inside the core side legs will be mainly determined by the built-in leakage flux. The flux inside the core plate will be mainly determined by the magnetizing flux. At the lowest input voltage, the higher resonant current is generated, which causes larger leakage flux and a larger cross-sectional area of the core center leg and side leg being used to reduce the flux density. At higher input and output voltage, the magnetizing flux becomes larger, and the leakage flux reduces due to a smaller resonant current. As the total flux is the summation of magnetizing flux and leakage flux, the selection of cross-sectional area should be based on the worst case with maximum flux within the whole voltage and load ranges.
4. The flux distribution in the core of IT is not evenly distributed when the built-in leakage flux is much larger than the magnetizing flux; thus, a larger cross-sectional area of the core will be necessary. The impact of the winding loss is discussed in Section 16.3.3.

Table 16.3 Magnetic loss comparison of IT and DIT solutions

Voltage parameter	500 V		750 V		1000 V	
	DIT[a]	IT	DIT[a]	IT	DIT[a]	IT
L_r wind. loss (W)	40.0	N/A	12.6	N/A	18.7	N/A
L_r core loss (W)	33.8	N/A	8.4	N/A	11.9	N/A
T_r wind. loss (W)	120.5	274.2	38.6	85.4	42.8	126.4
T_r core loss (W)	10.1	141.1	43.8	39.6	44.1	62.8
Total loss (W)	204.4	415.3	103.4	125.0	117.5	189.2

Table 16.4 Dimension comparison of IT and DIT solutions

	DIT[a]	IT
Dimension of L_r (mm)	70.9 × 82.9 × 60.9	N/A
Dimension of T_r (mm)	61.5 × 133.4 × 65 × 2 pcs	75 × 163.6 × 78.8 × 2 pcs
Total volume (L)	1.425	1.933

[a]The DIT solution includes one inductor with parallel windings and two transformers with input parallel and output series connections.

16.3.3 *Comparison of magnetic loss for IT and DIT solutions*

For a magnetic component, the magnetic loss includes the core loss and winding loss. In this chapter, the core loss will be evaluated based on Maxwell 3D FEA simulation by considering the core dimension, non-uniform flux distribution, dimension resonance, eddy current loss, and skin effect [32]. A two-dimensional (2D) Maxwell FEA simulation was implemented to evaluate the winding loss by considering the skin effect, proximity effect, and fringing effect.

The simulated winding loss and core loss for different input voltage ranges of both solutions are summarized in Table 16.3, and their dimension comparison is shown in Table 16.4. Compared with the IT solution, the DIT solution with one inductor and two transformers results in a volume reduction of 26%, as Table 16.4 points out. The increased cross-sectional area of the IT solution is necessary to decrease the flux density at the lowest input voltage. However, this results in a larger magnetic volume and increased winding loss. At the lowest input voltage, the total magnetic loss of the DIT solution significantly outperforms that of the IT solution, as Table 16.3 reveals.

16.4 Design of discrete magnetics for a 30 kw LLC resonant converter

To achieve the required gain range (input voltage from 500 V to 1 kV and output voltage from 600–800 V) for the LLC converter, the parameters of the circuits were provided in Table 16.1.

As discussed in Section 16.3, equivalent AWG 6 (2700 strands of AWG40) Litz wire is used for the primary and secondary windings to build the magnetic components according to NFPA 070 National Electrical Code. For the IT solution, there are two pieces of integrated transformers connected primary in parallel secondary in series to handle the total 30 kW output power. To make a fair comparison with the same quantity of magnetic components, a discrete inductor, and a discrete transformer were required for the 30 kW LLC converter. With one inductor and one transformer, the RMS current of primary windings becomes two times that of the two transformers solution, which requires an equivalent AWG 3 Litz wire to be used. The cost of AWG 3 Litz wire is much higher than that of the two pieces of AWG 6 Litz wire, and the manufacturing of transformer windings with AWG 3 Litz wire as primary windings and AWG 6 Litz wire as secondary side becomes more difficult than using same AWG 6 Litz wires for both primary and secondary windings. Considering these factors, parallel primary windings with AWG 6 Litz wire were adopted for the discrete inductor and discrete transformer. To guarantee current sharing among parallel windings, the winding arrangements of discrete inductors and discrete transformers were carefully designed.

16.4.1 Design of discrete transformer with parallel primary windings

Instead of the two transformers with input parallel and output series, parallel primary windings for one discrete transformer were adopted. As mentioned, the transformer winding was built with two parallel AWG 6 Litz wires. To achieve the required turns ratio of 0.54, the primary turns number is set to 6, which means two six turns of primary windings will be connected in parallel, and the turns number of secondary windings is set to 11. The physical turns ratio is set slightly higher than the required turns ratio due to inherent leakage inductance, which will reduce the effective turns ratio. The cross-sectional area of the transformer core leg can be designed in (16.33) based on (16.16):

$$A_{eT\_\min} = \frac{V_{o\_\max} T_{s\_\min}}{4(N_{s1} + N_{s2}) B_{T\_\max}} = 910 \text{ mm}^2 \tag{16.33}$$

where $V_{o\_\max}$ is 800 V, $T_{s\_\min}$ is 10 μs, $B_{T\_\max}$ is 200 mT.

The final core leg radius, R_T, is set to 17.5 mm with a certain margin by considering the leakage flux that goes through the core and the tolerance during manufacturing.

To avoid introducing the extra leakage inductance due to unbalanced windings for a 6:6:11 turns ratio transformer, the UI-core type transformer was selected. The equivalent circuits of the 6:6:11 transformer are shown in Figure 16.13(a). The total 11 turns of secondary windings (marked with blue solid) can be divided into two parts: six turns of secondary windings are placed across one transformer leg (left core leg) and another five turns of secondary windings are placed across another transformer core leg (right core leg) as shown in Figure 16.13(b). As

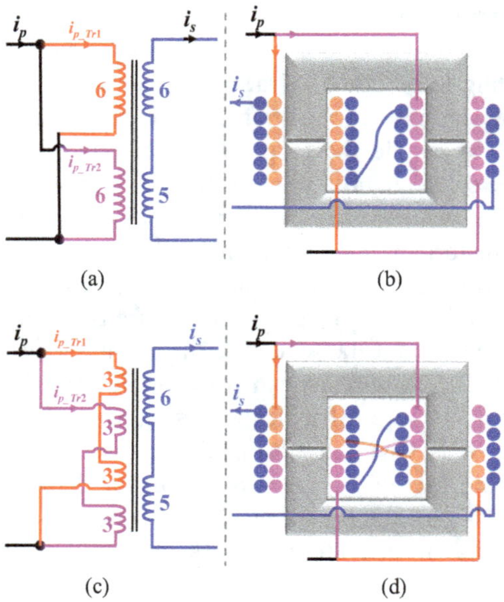

Figure 16.13 Proposed transformer with parallel primary windings: (a) circuits of structure 1, (b) winding arrangement of structure 1, (c) circuits of structure 2, and (d) winding arrangement of structure 2

Figure 16.13(b) depicts, the six turns of primary windings (marked with red slash) are placed across one transformer core leg, and another six turns of primary windings (marked with pink dot) are placed across another transformer core leg. The current sharing between parallel windings and their current density were evaluated in Maxwell 3D FEA simulations using the eddy current solver. The current excitation for 3D FEA simulation is based on the nominal input voltage, V_{in}=750 V, and 30 kW full load condition, where the primary current excitation, i_p, is 128.7 A and its phase is −11.1°, secondary current excitation, i_s, is 70.5 A and its phase is 0°. The current waveforms for the parallel primary windings of structure 1, depicted in Figure 16.13(b), are presented in Figure 16.13(a), exhibiting clear variations in amplitude due to the unsymmetrical structure of windings. To achieve good current sharing between parallel windings, structure 2 of parallel windings for the discrete transformer was proposed in Figures 16.13(c) and (d). In Figure 16.13(c), the transformer circuits are presented with an exchange of three turns for each parallel winding to achieve a symmetrical structure, while Figure 16.13(d) illustrates the winding arrangement of structure 2. The simulated current waveforms of i_{p_Tr1} and i_{p_Tr2} are shown in Figure 16.14(b), which shows good current-sharing results between parallel windings. The current density shown in Figure 16.15 aligns closely with the current sharing waveforms from Figure 16.14.

The final design parameters of discrete transformers with parallel windings are shown in Table 16.5.

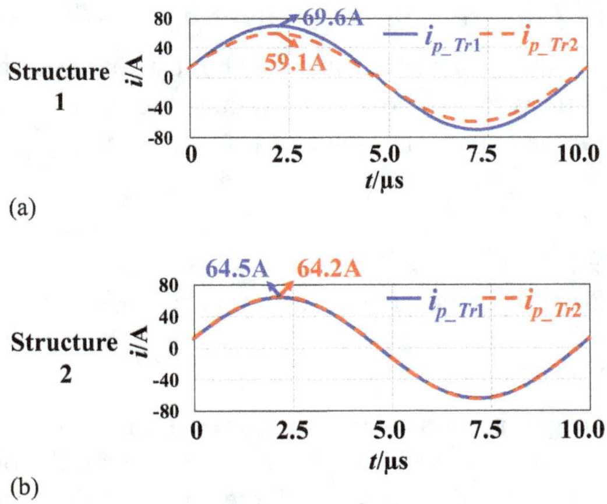

Figure 16.14 Simulated current sharing results of different structures: (a) structure 1 and (b) structure 2

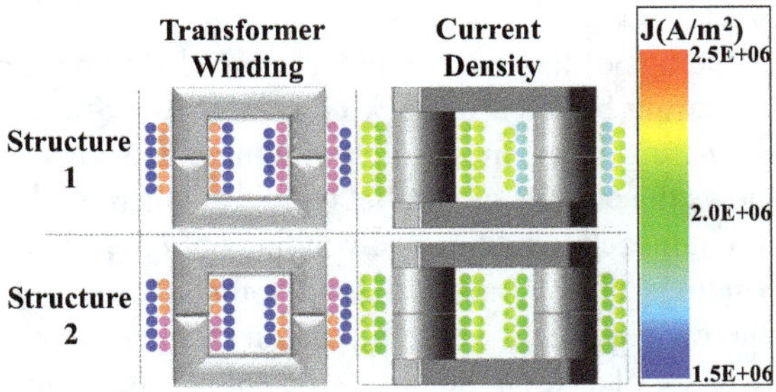

Figure 16.15 Simulated current density at peak primary resonant current of different structures

Table 16.5 Design parameters of discrete transformer

Design parameters	Values
Turns of primary windings, N_{p1}, N_{p2}	6 in Parallel
Turns of secondary windings, N_s	11
Core leg radius, R_T	17.5 mm
Litz wire	AWG 6 (2700 strands of AWG 40)
Core material	DMR96A
Magnetizing inductance, L_m	35.0 μH
Simulated leakage inductance, L_{kp}	0.5 μH
Maximum flux density, B_{T_max}	0.2 T
Window area	22 mm × 52 mm
Dimension	68.5 mm × 147.4 mm × 80.3 mm

16.4.2 Design of discrete inductor with parallel windings

As mentioned before, the inductor winding was built with two parallel AWG 6 Litz wires. The required turns number of each winding and cross-sectional area of the inductor core leg can be determined in (16.34):

$$N_L A_{eL} = L \cdot i_{p\_\max} / B_{L\_\max} \tag{16.34}$$

where i_{p_max} is 189.1 A and B_{L_max} is 180 mT.

The required air gap can be calculated in (16.35) based on (16.6) and (16.34):

$$l_{gL} = \mu_0 N_L i_{p\_\max} / B_{L\_\max} \tag{16.35}$$

The required air gap is proportional to the required turns number. More turns can be used to reduce the core loss; however, it comes with the price of increased core window area and winding loss and larger air gap that bring difficulty in building the distributed air gaps and increased cost. The turns number, N_L, was selected to be 4 with a total 6 mm air gap, and the total air gap will be divided into three same air gaps for easier manufacturing.

Based on (16.34), the cross-sectional area is set to 855 mm^2 with a core leg radius of 16.5 mm.

The parallel windings of the discrete inductor needed to be carefully designed to achieve current sharing between these windings. Three cases are used for evaluating the current sharing between parallel windings, as shown in Figure 16.16. Figure 16.16(a) shows the circuits of parallel windings where i_{p_L1} and i_{p_L2} are the currents that go through each winding. The windings of i_{p_L1} are blue dotted round shape, and the windings of i_{p_L2} are red solid round shape. In the first case of Figure 16.16(b), the windings of i_{p_L1} are placed on the inner side of all layers, and the windings of i_{p_L2} are placed on the outer sides of all layers, which makes these two parallel windings have different winding length and magnetic field distribution across the winding, thus, unbalanced current among parallel windings was generated. There are two symmetrical winding structures of the inductor with parallel windings, as shown in Figures 16.16(c) and (d), respectively. In Figure 16.16(c), the windings of case 2 are placed as intertwined, with their windings alternately positioned on every layer. In Figure 16.16(d), the windings occupy opposite sides of the window area, with the upper two layers designated for i_{p_L1} and the lower

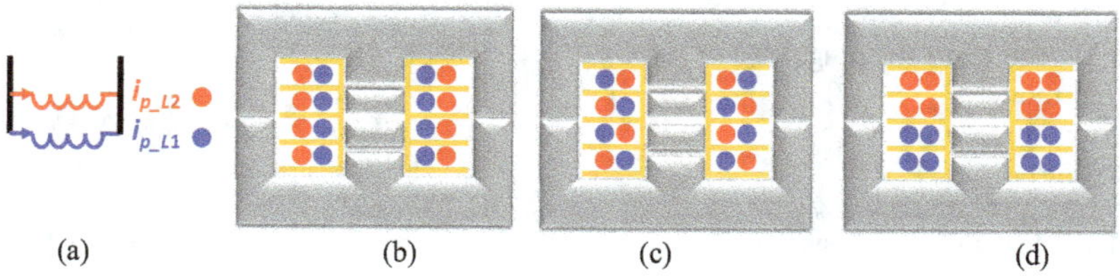

Figure 16.16 Inductor circuits and different structures: (a) circuits of parallel windings, (b) case 1, (c) case 2, and (d) case 3

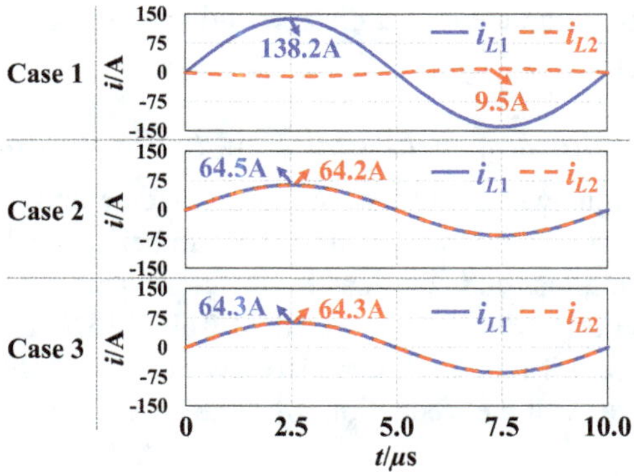

Figure 16.17 Simulated current sharing results of different cases

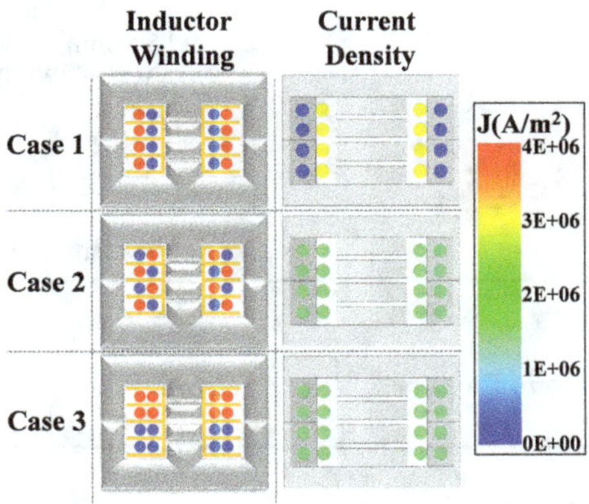

Figure 16.18 Simulated current density at peak resonant current of different cases

layers allocated to i_{p_L2}. These three cases were built and evaluated in Maxwell 3D FEA simulations to check the current sharing and current density of the windings. The current excitation for 3D FEA simulation is based on the nominal input voltage, V_{in} = 750 V, and 30 kW full load condition, which is 128.7 A peak resonant current, 100 kHz switching frequency, and zero phase. The simulated current-sharing waveforms of these three cases are shown in Figure 16.17. The current density at the time instance with maximum inductor current (one-fourth of the switching frequency) was shown in Figure 16.18. The current density shown in Figure 16.18 aligns closely with the current sharing waveforms from Figure 16.17. In this representation, there is a notable variance in current distribution among the parallel windings for case 1, while cases 2 and 3 demonstrate ideal current sharing. Compared with case 2, the inductor design of case 3 was selected because of its simpler manufacturing process and comparable current sharing capability.

The final design parameters of a discrete inductor with parallel windings are shown in Table 16.6.

16.4.3 Loss comparison of IT and New DIT solutions

The new DIT solution, which consists of one inductor and one transformer, is competitive with the IT solution regarding the number of magnetic components. The simulated winding loss and core loss for different input voltage ranges of both IT and new DIT solutions are summarized in Table 16.7, and their dimension comparison is shown in Table 16.8. Compared with the IT solution, the new DIT solution results in a volume reduction of 39.5%, as Table 16.8 points out.

Table 16.6 Design parameters of discrete inductor

Design parameters	**Values**
Turns number of each winding, N_L	4 in Parallel
Core leg radius, R_L	16.5 mm
Litz wire	AWG 6 (2700 strands of AWG 40)
Core material	DMR96A
Inductance, L	3.3 μH
Maximum flux density, B_{L_max}	0.18 T
Window area	22 mm × 41 mm
Dimension	70.9 mm × 82.9 mm × 60.9 mm

Table 16.7 Comparison of magnetic loss of IT and new DIT solution

Voltage parameter	**500 V**		**750 V**		**1000 V**	
	DIT[a]	**IT**	**DIT[a]**	**IT**	**DIT[a]**	**IT**
L_r wind loss (W)	40.0	N/A	12.6	N/A	18.7	N/A
L_r core loss (W)	33.8	N/A	8.4	N/A	11.9	N/A
T_r wind loss (W)	77.8	274.2	25.8	85.4	45.7	126.4
T_r core loss (W)	15.7	141.1	48.4	39.6	47.1	62.8
Total loss (W)	167.3	415.3	95.2	125.0	123.4	189.2

Table 16.8 Dimension comparison of IT and new DIT solution

	DIT[a]	**IT**
Dimension of L_r (mm)	70.9 × 82.9 × 60.9	N/A
Dimension of T_r (mm)	68.5 × 147.4 × 80.3 × 1 pcs	75 × 163.6 × 78.8 × 2 pcs
Total volume (L)	1.169	1.933

[a]The new DIT solution includes one inductor with parallel windings and one transformer with primary parallel windings.

According to Table 16.7, the total magnetic loss of the new DIT solution is significantly lower than that of the IT solution and surpasses the performance of the original DIT solution with one inductor and two transformers, as listed in Table 16.3.

16.5 Experimental results

To validate the comparison results of IT solution and new DIT solution, a 30 kW LLC resonant converter with 44 W/in^3 and 99% peak efficiency was built, as shown in Figure 16.19(a). The detailed circuit parameters are listed in Table 16.9. The magnetic components of IT and DIT solutions were built and can be

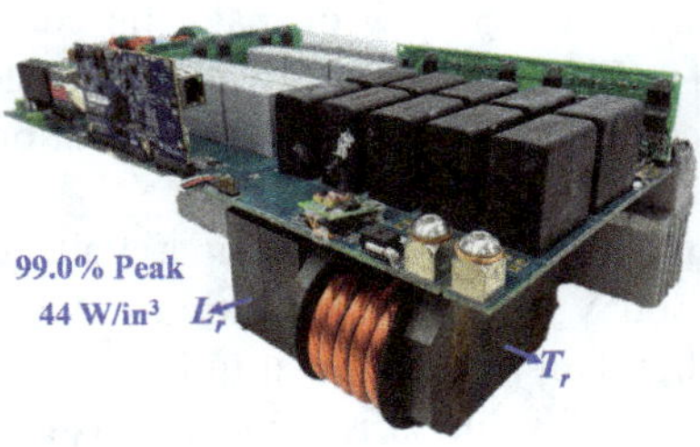

(a)

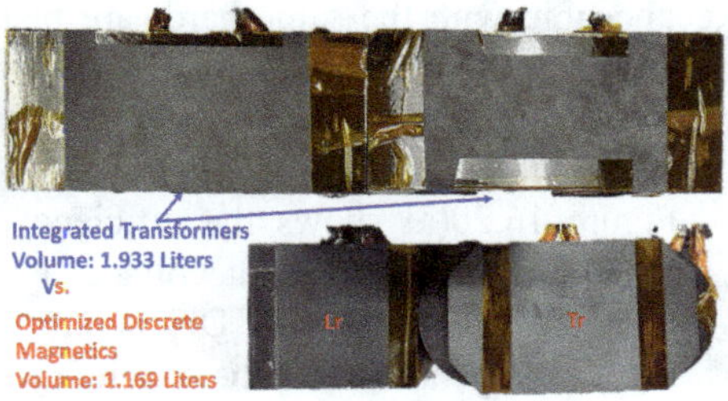

(b)

Figure 16.19 Hardware prototype: (a) prototype with DIT and (b) magnetics comparison of IT and DIT solutions

Table 16.9 Circuits parameters of 30 kW LLC prototype

Design parameters	**Values**
Input voltage, V_{in}	500–1000 V
Output voltage, V_o	600–800 V
Maximum output power, P_{omax}	30 kW
Primary SiC device	FF6MR12KM1P
Secondary SiC device, 2 pcs in parallel	C3M0016120K
Gate driver of pri. devices	IXDN630YI
Gate driver of sec. devices	NCP81074ADR2G
Core material	DMR96A
Resonant frequency, f_r	100 kHz
Transformer turns ratio, N_T	0.54
Magnetizing inductance, L_m	35.0 μH
Resonant inductance, L_r	3.8 μH
Resonant capacitors, C_r	672.0 nF
Input capacitor, C_{in1}, C_{in2}	120 μF, 6 pcs MKP1848620704P
Output capacitor, C_f	200 μF, 10 pcs MKP1848C62090JP4
Printed circuit board	Four layers 4 oz. Cu., 2 mm thick
Prototype dimension with heat sink	420 × 210 × 125 mm

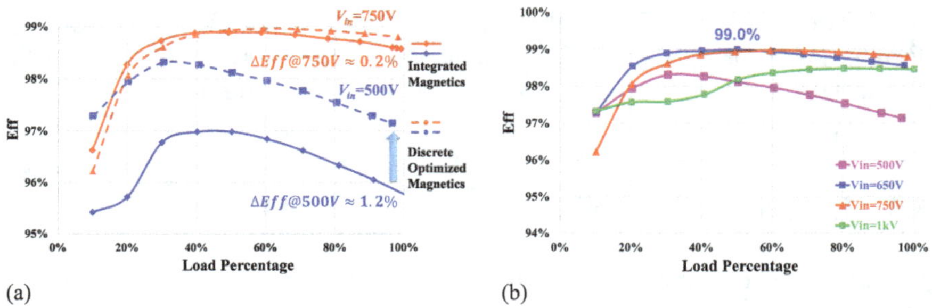

Figure 16.20 Efficiency curves: (1) efficiency comparison between IT and DIT solutions and (2) efficiency of DIT at different input voltages

exchangeable with the same hardware platform. The top view of magnetic components of the IT and DIT solutions is shown in Figure 16.19(b); over 40% footprint reduction and 39.5% volume reduction can be saved with the DIT solution compared with IT solution.

Figure 16.20(a) shows the efficiency comparison at nominal input voltage V_{in} = 750 V and lowest input voltage, V_{in} = 500 V, of the two solutions. When V_{in} = 750 V, the efficiency of DIT is 0.2% higher than IT solution at full load. For low input voltage at 500 V, the efficiency of the DIT solution is 1.2% higher than that of IT solution. Figure 16.20(b) shows the efficiency at different load conditions of the DIT solution; the peak efficiency can reach up to 99% at 750 V nominal input voltage and 50% load.

Figure 16.21 shows the key operation waveforms of 50% and 100% load conditions when the input voltage is 500 V (minimum input voltage), 750 V (nominal input voltage), and 1000 V (maximum input voltage). In each waveform, there are the drain-source waveforms of primary switches Q_2 and Q_4, v_{ds_Q2}, and v_{ds_Q4}, the gate-source waveform of primary switch Q_2, v_{gs_Q2}, the drain-source waveforms of secondary switches SR_2 and SR_4, v_{ds_SR2} and v_{ds_SR4}, the gate-source waveform of secondary switch SR_2, v_{gs_SR2}, primary resonant current, i_p, secondary resonant current, i_s, and magnetizing current, i_m, which is calculated from the primary current minus the reflected secondary current to the primary side with a factor of turns ratio. It can be seen from Figure 16.21 that the primary and secondary devices can achieve ZVS. The experimental waveforms in Figure 16.21 prove the effectiveness of DIT magnetics applied in this 30 kW LLC converter.

Figure 16.22 shows the current sharing waveforms of parallel windings in the resonant inductor and transformer of the DIT solution. Figure 16.22 includes the current sharing waveforms of 50% and 100% load conditions when the input voltage is 500, 750, and 1000 V, respectively. In each waveform, the drain-source waveforms of primary switches Q_2 and Q_4, v_{ds_Q2} and v_{ds_Q4}, the drain-source waveforms of secondary switches SR_2 and SR_4, v_{ds_SR2} and v_{ds_SR4}, each winding's current in the parallel windings of resonant inductor, i_{p_Lr1} and i_{p_Lr2}, each winding's current in the parallel windings of transformer, i_{p_Tr1} and i_{p_Tr2}, primary resonant current, i_p, secondary resonant current, i_s, and magnetizing current, i_m.

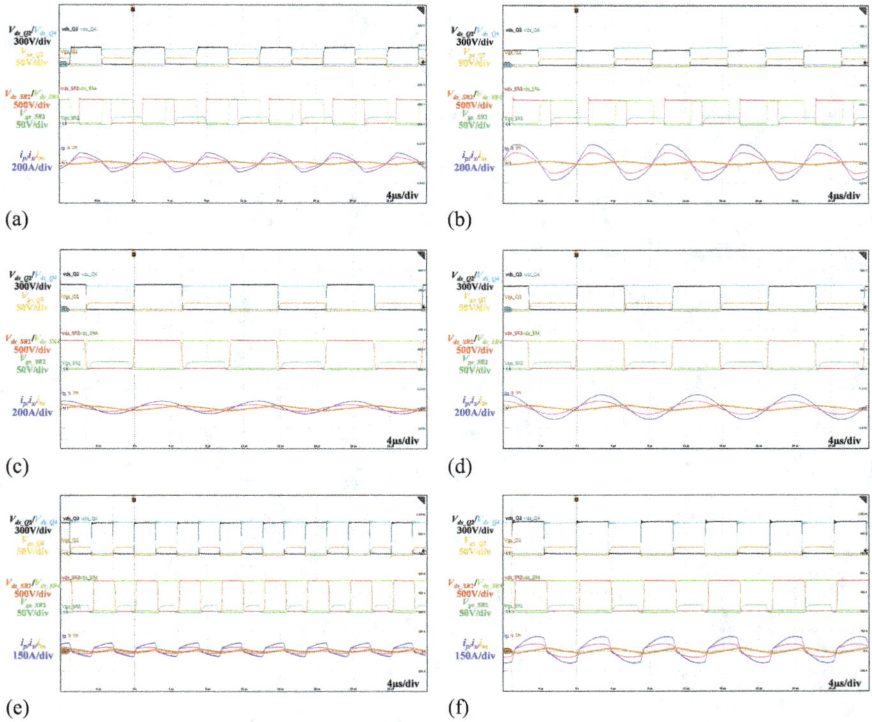

Figure 16.21 Key operation waveforms of DIT solution: (a) V_{in} = 500 V, V_o = 600 V, P_o = 15 kW; (b) V_{in} = 500 V, V_o = 600 V, P_o = 30 kW; (c) V_{in} = 750 V, V_o = 700 V, P_o = 15 kW; (d) V_{in} = 750 V, V_o = 700 V, P_o = 30 kW; (e) V_{in} = 1000 V, V_o = 800 V, P_o = 15 kW; and (f) V_{in} = 1000 V, V_o = 800 V, P_o = 30 kW

i_{p_Lr2} is calculated from (i_p - i_{p_Lr1}), and i_{p_Tr2} is calculated from (i_p - i_{p_Tr1}). As shown in Figure 16.22, the current waveforms of i_{p_Lr1} and i_{p_Lr2} are overlapped with each other, and the current waveforms of i_{p_Tr1} and i_{p_Tr2} are overlapped as well, which proves the good current sharing capability of the proposed parallel winding structures for both discrete inductor and discrete transformer. The data of current waveforms are extracted from the oscilloscope, and their RMS values are calculated and summarized in Table 16.10. The maximum difference in current sharing of the resonant inductor is 3.7%, and the maximum difference in current sharing of the transformer is 5.7%, which proves the good current sharing capability.

Figure 16.23 presents the thermal performance results of the DIT magnetics under specific conditions: 500 V input voltage, 30 kW output power, and 3 m/s air cooling at room temperature, 25 °C. As Figure 16.23 shows, the core temperature of the resonant inductor is 42.3 °C, the core temperature of the transformer is 37.5 °C, and the winding temperature of the transformer is 63.9 °C. The maximum

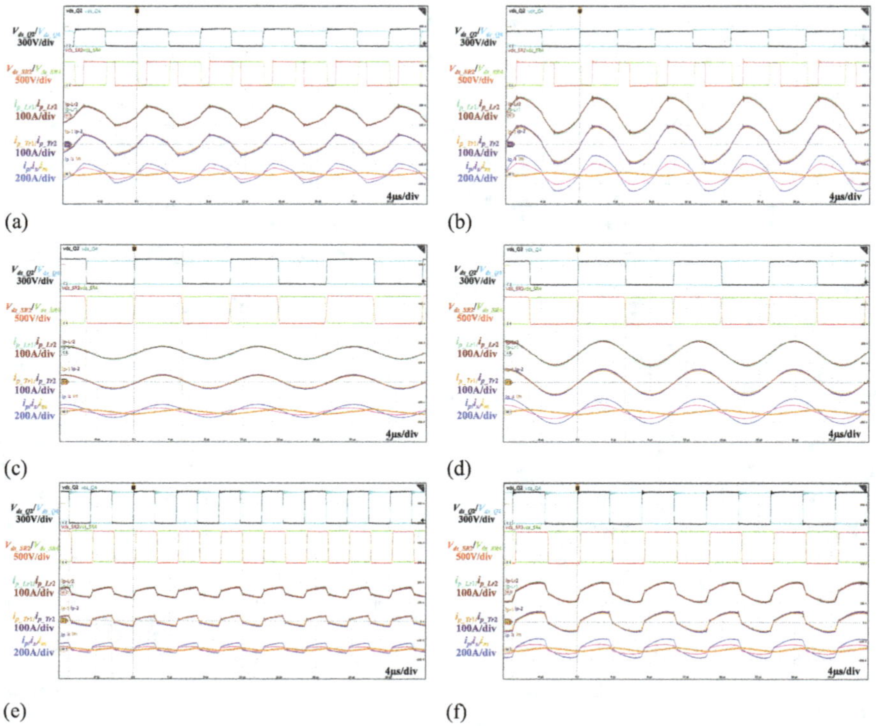

Figure 16.22 Current sharing waveforms of DIT solution: (a) V_{in} = 500 V, V_o = 600 V, P_o= 15 kW; (b) V_{in} = 500 V, V_o = 600 V, P_o = 30 kW; (c) V_{in} = 750 V, V_o = 700 V, P_o = 15 kW; (d) V_{in} = 750 V, V_o = 700 V, P_o = 30 kW; (e) V_{in} = 1000 V, V_o = 800 V, P_o = 15 kW; and (f) V_{in} = 1000 V, V_o = 800 V, P_o = 30 kW

Table 16.10 Current sharing results of DIT solution

Load	Input vol. (V)	I_{p_Lr1} (A)	I_{p_Lr2} (A)	Δi_{p_Lr} (%)	I_{p_Tr1} (A)	I_{p_Tr2} (A)	Δi_{p_Tr} (%)
50%	500	34.8	33.1	2.6	32.5	35.4	4.3
	600	27.9	26.5	2.6	25.9	28.5	4.7
	650	27.2	25.5	3.2	25.1	27.6	4.8
	750	23.7	22.5	2.5	22.0	24.2	4.8
	870	21.6	20.8	1.9	20.1	22.3	5.2
	1000	19.1	18.6	1.3	17.8	20.0	5.7
100%	500	67.5	62.8	3.6	62.7	67.6	3.8
	600	56.8	53.3	3.2	52.2	57.9	5.2
	650	53.1	49.3	3.7	48.6	53.7	5.0
	750	45.8	42.8	3.4	42.2	46.4	4.8
	870	38.9	36.7	2.8	35.9	39.7	5.1
	1000	39.2	36.9	3.0	36.2	39.9	4.8

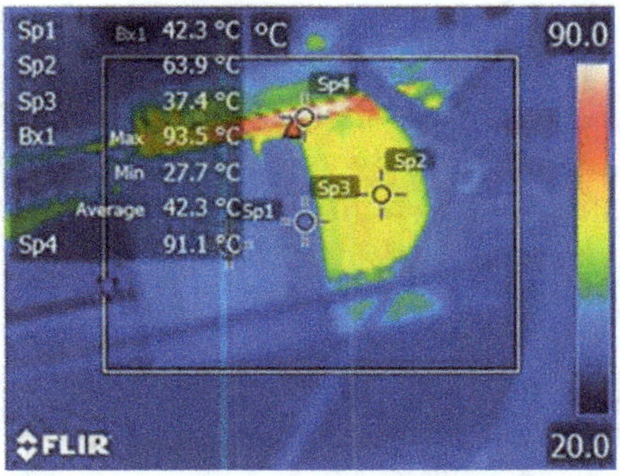

Figure 16.23 Thermal results of the DIT solution

temperature of the thermal image is 93.5 °C at the location of the resonant capacitors where there is no air flow. Other methods, such as adding thermal interface material or increasing the air speed across the resonant capacitors, can be beneficial in improving their thermal performance.

16.6 Conclusions

In this chapter, a comparison study of DIT solutions and IT solutions for high-power LLC converters was implemented. Both the 4-leg-core-based IT solution and the DIT solution have been thoroughly discussed in terms of design considerations. The limitations of IT solutions for high-power LLC converters are explained through a comparative analysis using FEA simulations and analytical models of flux distribution. Through Maxwell 3D and 2D FEA simulations, the magnetic losses of both solutions under various voltage conditions were evaluated, and the results of the DIT solution highlight a notable reduction in overall magnetic losses and offer a smaller volume compared with the IT solution.

The proposed DIT solution, featuring a parallel winding configuration for both the inductor and transformer, was evaluated and validated in a 30 kW LLC converter used in APS for railway applications, demonstrating efficient current-sharing capabilities. The proposed DIT solution prototype achieved a peak efficiency of 99% and a power density of 44 W/in^3.

Such findings reveal the limitation of IT solutions for high-power applications and emphasize the potential of DIT solutions in optimizing performance and efficiency in high-power LLC converter applications.

References

[1] A. Nabih, M. H. Ahmed, Q. Li and F. C. Lee, "Transient Control and Soft Start-Up for 1-MHz LLC Converter With Wide Input Voltage Range Using Simplified Optimal Trajectory Control," in *IEEE Journal of Emerging and Selected Topics in Power Electronics*, vol. 9, no. 1, pp. 24–37, 2021.

[2] R. Gadelrab, A. Nabih, F. C. Lee and Q. Li, "LLC Resonant Converter with 99% Efficiency for Data Center Server," 2021 *IEEE Applied Power Electronics Conference and Exposition (APEC)*, 2021, pp. 310–319.

[3] P. H. Pham, A. Nabih, S. Wang and Q. Li, "11-kW High-Frequency High-Density Bidirectional OBC with PCB Winding Magnetic Design," 2022 *IEEE Applied Power Electronics Conference and Exposition (APEC)*, Houston, TX, USA, 2022, pp. 1176–1181, doi:10.1109/APEC43599.2022.9773735.

[4] F. Jin, A. Nabih, C. Chen, X. Chen, Q. Li and F. C. Lee, "A High Efficiency High Density DC/DC Converter for Battery Charger Applications," 2021 *IEEE Applied Power Electronics Conference and Exposition (APEC)*, Phoenix, AZ, USA, 2021, pp. 1767–1774, doi:10.1109/APEC42165.2021.9487108.

[5] Y. Cao, M. Ngo, N. Yan, D. Dong, R. Burgos and A. Ismail, "Design and Implementation of an 18 kW 500 kHz 98.8% Efficiency High-density Battery Charger with Partial Power Processing," *in IEEE Journal of Emerging and Selected Topics in Power Electronics*.

[6] Z. Li, E. Hsieh, Q. Li and F. C. Lee, "High-Frequency Transformer Design With Medium-Voltage Insulation for Resonant Converter in Solid-State Transformer," in *IEEE Transactions on Power Electronics*, vol. 38, no. 8, pp. 9917–9932, 2023, doi:10.1109/TPEL.2023.3279030.

[7] C. Zhao, Y. -H. Hsieh, F. C. Lee and Q. Li, "Design and Analysis of a High-frequency CLLC Resonant Converter with Medium Voltage Insulation for Solid-State-Transformer," 2021 *IEEE Applied Power Electronics Conference and Exposition (APEC)*, 2021, pp. 1638–1642.

[8] M. Ocklenburg, M. DÃűhmen, X. Wu, and M. Helsper, "Next Generation Dc-Dc Converters for Auxiliary Power Supplies with SiC MOSFETs," in *2018 IEEE International Conference on Electrical Systems for Aircraft, Railway, Ship Propulsion and Road Vehicles International Transportation Electrification Conference* (ESARS-ITEC), Nov 2018, pp. 1–6.

[9] S. -W. Baek, H. -W. Kim and J. -W. Kang, "Auxiliary Power Unit for Railway Vehicles Using Three-Level LLC Converter with Pulse Width Modulation Control," 2019 *International Symposium on Electrical and Electronics Engineering (ISEE)*, 2019, pp. 232–236, doi:10.1109/ISEE2.2019.8921102

[10] C. Zhao, F. Jin, Z. Li, Y. -H. Hsieh, F. C. Lee and Q. Li, "Design and Control of a High-Power Wide-Gain-Range LLC Resonant Converter," 2022 *IEEE Applied Power Electronics Conference and Exposition (APEC)*, Houston, TX, USA, 2022, pp. 40–44.

[11] Z. Li, F. Jin, C. Zhao, E. Hsieh and Q. Li, "A High-Frequency Transformer Design with Leakage Integration for Auxiliary Power Supply in Railway Application," *2023 11th International Conference on Power Electronics and ECCE Asia (ICPE 2023-ECCE Asia), Jeju Island, Korea, Republic of, 2023*, pp. 3402–3409, doi: 10.23919/ICPE2023-ECCEAsia54778.2023.10213891.

[12] T. Yuan, F. Jin, Z. Li, C. Zhao and Q. Li, "Design of an Integrated Transformer With Parallel Windings for a 30-kW LLC Resonant Converter," *in IEEE Transactions on Power Electronics*, vol. 38, no. 11, pp. 14317–14333, 2023, doi:10.1109/TPEL.2023.3291954.

[13] Y. Cao, K. Ngo and D. Dong, "A Scalable Electronic-Embedded Transformer, a New Concept Toward Ultra-High-Frequency High-Power Transformer in DC–DC Converters," in *IEEE Transactions on Power Electronics*, vol. 38, no. 8, pp. 9278–9293, 2023, doi:10.1109/TPEL.2023.3279259.

[14] Y. Cao and D. Dong, "Electronic-Embedded Transformer Based DC Transformer With Trapezoidal Current and Natural Current Sharing," in *IEEE Transactions on Power Electronics*, vol. 38, no. 10, pp. 12348–12354, 2023, doi:10.1109/TPEL.2023.3295130.

[15] A. Nabih and Q. Li, "Design of 98.8% Efficient 400-to-48-V LLCLLC Converter With Optimized Matrix Transformer and Matrix Inductor," in *IEEE Transactions on Power Electronics*, vol. 38, no. 6, pp. 7207–7225, 2023, doi:10.1109/TPEL.2023.3244869.

[16] Y. Liu, H. Wu, Z. Ge and G. Ji, "Magnetic Integration for Multiple Resonant Converters," in *IEEE Transactions on Industrial Electronics*, vol. 70, no. 8, pp. 7604–7614, 2023, doi:10.1109/TIE.2022.3229381.

[17] R. Schlesinger and J. Biela, "Comparison of Analytical Models of Transformer Leakage Inductance: Accuracy Versus Computational Effort," in *IEEE Transactions on Power Electronics*, vol. 36, no. 1, pp. 146–156, 2021.

[18] G. Liu, D. Li, J. Q. Zhang, B. Hu and M. L. Jia, "Bidirectional CLLC Resonant DC-DC Converter with Integrated Magnetic for OBCM Application," 2015 *IEEE International Conference on Industrial Technology (ICIT)*, Seville, Spain, 2015, pp. 946–951, doi:10.1109/ICIT.2015.7125219.

[19] M. A. Bakar and K. Bertilsson, "An improved Modelling and Construction of Power Transformer for Controlled Leakage Inductance," 2016 *IEEE 16th International Conference on Environment and Electrical Engineering (EEEIC)*, Florence, Italy, 2016, pp. 1–5, doi:10.1109/EEEIC.2016.7555605.

[20] J. Biela and J. W. Kolar, "Electromagnetic Integration of High Power Resonant Circuits Comprising High Leakage Inductance Transformers," *2004 IEEE 35th Annual Power Electronics Specialists Conference (IEEE Cat. No.04CH37551)*, Aachen, Germany, 2004, pp. 4537–4545 Vol.6, doi:10.1109/PESC.2004.1354802.

[21] B. Li, Q. Li and F. Lee, "High-Frequency PCB Winding Transformer With Integrated Inductors for a Bi-Directional Resonant Converter," *IEEE Trans. Power Electron.*, vol. 34, no. 7, pp. 6123–6135, 2019.

[22] A. Nabih, F. Jin and Q. Li, "Efficient Integrated Transformer–Inductor With High PCB Utilization and Optimized Core," in *IEEE Transactions on Industrial Electronics*, doi:10.1109/TIE.2023.3294637.

[23] S. Mukherjee and P. Barbosa, "Design and Optimization of an Integrated Resonant Inductor With High-Frequency Transformer for Wide Gain Range DC–DC Resonant Converters in Electric Vehicle Charging Applications," in

IEEE Transactions on Power Electronics, vol. 38, no. 5, pp. 6380–6394, 2023.

[24] F. Jin, A. Nabih, Q. Li and F. C. Lee, "A Three Phase CLLC Converter with Improved Planar Integrated Transformer for Fast Charger Applications," *2021 IEEE Fourth International Conference on DC Microgrids (ICDCM)*, Arlington, VA, USA, 2021, pp. 1–5, doi:10.1109/ICDCM50975.2021.9504644.

[25] F. Jin, A. Nabih, Z. Li and Q. Li, "A Scalable Matrix Integrated Transformer with Controllable Leakage Inductance for a Bi-Directional Resonant Converter," in *IEEE Transactions on Power Electronics*, doi:10.1109/TPEL.2023.3270400.

[26] Y. Liu, Y. Song, D. Hu, Y. Li, Z. Zhang and H. Wu, "Overview of Planar Magnetics for High-Frequency Resonant Converters," in *Chinese Journal of Electrical Engineering*, vol. 8, no. 4, pp. 61–78, 2022.

[27] S. Wang, H. Wu, F. C. Lee and Q. Li, "Integrated Matrix Transformer with Optimized PCB Winding for High-Efficiency High-Power-Density LLC Resonant Converter," *2019 IEEE Energy Conversion Congress and Exposition (ECCE)*, Baltimore, MD, USA, 2019, pp. 6621–6627, doi:10.1109/ECCE.2019.8911885.

[28] F. Jin, Z. Li, A. Nabih and Q. Li, "Controllable Built-In Leakage Inductance for A Novel Matrix Integrated Transformer," *2023 IEEE Applied Power Electronics Conference and Exposition (APEC)*, Orlando, FL, USA, 2023, pp. 3248–3254, doi:10.1109/APEC43580.2023.10131277.

[29] C. Zhang and P. Barbosa, "Modulation Transition Methods Based on Trajectory Control for LLC Resonant Converters Operating in Wide Input-and/or Output-Voltage Range," in *IEEE Transactions on Power Electronics*, vol. 37, no. 12, pp. 14103–14114, 2022, doi:10.1109/TPEL.2022.3184631.

[30] R. Jez, "Influence of the Distributed Air Gap on the Parameters of an Industrial Inductor," in *IEEE Transactions on Magnetics*, vol. 53, no. 11, pp. 1–5, 2017, Art no. 8401605, doi:10.1109/TMAG.2017.2699120.

[31] https://www.tdk-electronics.tdk.com/en/2796388/products/product-catalog/ferrites-and-accessories/cores-with-distributed-air-gaps

[32] A. Nabih, F. Jin, R. Gadelrab, F. C. Lee and Q. Li, "Characterization and Mitigation of Dimensional Effects on Core Loss in High-Power High-Frequency Converters," in *IEEE Transactions on Power Electronics*, vol. 38, no. 11, pp. 14017–14036, 2023, doi:10.1109/TPEL.2023.3285633.

Part V

Selected Application Problems

Chapter 17

Magnetic materials and devices for EMI filtering

Marcin Kącki[1], Marek Ryłko[1], John G. Hayes[2] and Charles R. Sullivan[3]

Electromagnetic interference (EMI) protection is essential for electronic devices to reduce interaction with the environment and thus provide safe operation and conform with regulations. Differential- and common-mode chokes are the fundamental EMI-protection components. The common-mode choke operation requires a specific frequency characteristic that is matched to the electronic device. This requires careful selection of the magnetic material to fulfill the requirement and achieve the minimum size and manufacturing effort.

This chapter provides insight into magnetic components for EMI filtering within the conducted emission range. EMI disturbance is usually an effect of the commutation of the semiconductor-based switching cell, and it affects an electrical circuit by conduction, electromagnetic induction, or electrostatic coupling. The noise is further magnified and attenuated naturally by system parasitics and, finally, is mitigated in the EMI filter. The challenge of filter optimization is in the right selection of the magnetic core from the variety of available magnetic materials. The chapter guides the reader through the selected aspects of the complex permeability and its dependence on such frequency effects as eddy currents and dimensional resonance. The importance of a comprehensive and accurate dataset for the magnetic material is underlined as a key for design optimization. Finally, practical aspects based on examples are presented and discussed.

17.1 Challenges for EMI filtering

The continuous drive for the improvement of electrical energy conversion focuses on efficiency improvements and combines with higher switching frequencies to reduce the size and anticipated cost. This leads to new solutions and increases interest in new wide-bandgap (WBG) silicon-carbide (SiC) and gallium-nitride (GaN) semiconductors with superior switching performance over traditional silicon (Si)-based

[1]SMA Magnetics sp. z o.o., Department of Research and Development, Poland
[2]Power Electronics Research Laboratory School of Engineering University College Cork, Ireland
[3]Thayer School of Engineering at Dartmouth, USA

power semiconductors [1–3]. The benefits of using WBGs include higher switching frequencies, faster voltage, *dv/dt*, and current, *di/dt*, transients, and higher power densities that at the same time set a new frontier for EMI filtering and magnetic component design [4]. Regulations and industry standards for electromagnetic compatibility (EMC) stipulate limitations on radiated and conducted EMI noise generated by devices connected to the grid. In order to meet the regulations, there are a number of usual primary mitigation techniques, such as soft switching [5] and snubber circuits [6], backed with PCB layout guidelines [7–10], and topped with tailored EMI filters. The EMI filter is one of the last, but not the least, links on the EMI reduction chain to enable a power electronic system to conform with the regulations. EMI filters with attenuation characteristics matched to the EMI source suppress both common-mode (CM) and differential-mode (DM) noise. Therefore, the implementation of the right filter design based on high-accuracy modeling can determine the project's success and minimize iterations. Magnetic material specification is of key importance to achieve the goal of successful common-mode choke (CMC) design and the matching of the filter attenuation characteristic with the EMI source.

17.1.1 Impact of high-frequency pulse transitions on EMI generation

WBG semiconductors expand the frontier of power electronics beyond the frequencies achievable to date. Thus, higher switching frequencies set new requirements for EMI filter design as well as for magnetic material selection. Trapezoidal switching waveforms and related Fourier analysis are taken as an example to demonstrate the effect of switching frequency and fast voltage transients, *dv/dt,* on the EMI spectrum. The trapezoidal waveform with symmetric rise, τ_r, and fall times, τ_f, is used for the analysis is shown in Figure 17.1.

The signal duty ratio D is given by τ/T, where $T = 1/f_0$ is the period of the switching waveform with a frequency f_0, while τ is the pulse width. The calculated EMI spectrum has an envelope that is described by a series of linear asymptotes featuring gradients: 0, −20, and −40 dB/dec with increasing frequency [11]. Corner frequencies f_{c1} and f_{c2} where asymptotes intersect are determined by the

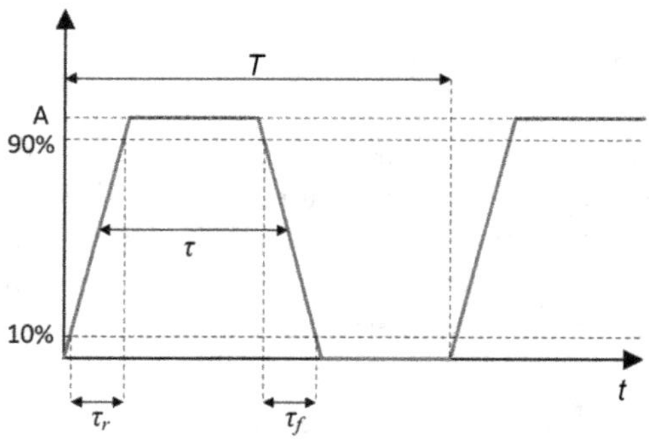

Figure 17.1 Trapezoidal waveform

waveform pulse width τ and rise/fall time τ_r/τ_f. The emission spectrum of the three switching signals, representing the successive semiconductor generations, was calculated to determine their impact on EMI filter requirements. The signal parameters are presented in Table 17.1 [2,4]. In the case of this analysis, only the waveform with $D = 0.5$ is considered. The calculated spectrum characteristics of the three switching signals are shown in Figure 17.2.

The most important observation from this analysis is that the high-frequency content of the examined signals is significantly affected by changing rise/fall time as well as the switching frequency. SiC semiconductors with a rise/fall time 6.5 times lower and a switching frequency 10 times higher than classic Si-based devices result in an increase of EMI noise by 30 dB at 1 MHz. Using GaN's semiconductors with even lower rise/fall times and a switching frequency of around 250 kHz further increases the high-frequency spectrum content. The Fourier

Table 17.1 Trapezoidal signal parameters

	Si	SiC	GaN
Amplitude (V)	500	500	500
Switching frequency f_0 (kHz)	10	100	250
Rise/fall time $\tau_r = \tau_f$ (ns)	100	15	8
First corner frequency f_{c1} (kHz)	6.36	63.66	159.15
Second corner frequency f_{c2} (MHz)	3.18	21.22	417.47

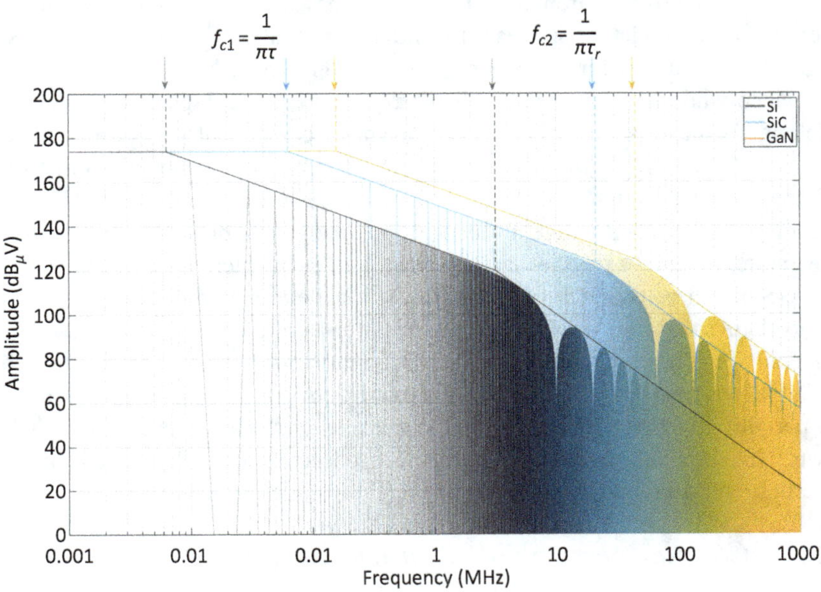

Figure 17.2 EMI emission spectrum comparison Si vs. SiC vs. GaN

transformation of the trapezoidal switching waveform shows that a switching frequency increase from 10 to 250 kHz has an effect only on the low-frequency spectrum envelope. High-frequency signal spectra are predominantly determined by the rise/fall times rather than the fundamental frequency. WBG semiconductors cause a shift of the EMI envelope to higher frequencies. This shift to higher frequencies creates a challenge for the magnetic component design.

17.2 Magnetic materials for EMI filters

Magnetic material selection is a key to successful magnetic component design. The ideal material for the EMI filter should feature high saturation flux density and high resistivity (as eddy-current losses are inversely proportional to the electrical resistivity). Moreover, the material should have a constant permeability characteristic over frequency and temperature, low permittivity, high mechanical robustness, and insensitivity to mechanical stress. The selection of materials for EMI filters is presented in Table 17.2. The main parameters, by default, are used for comparison of soft magnetic materials and include relative permeability, μ_r, saturation flux density, B_{SAT}, and electrical conductivity, σ. The table is based on the published manufacturers information [12–16].

Soft magnetic materials are divided into four distinctive material groups: amorphous, nanocrystalline, ferrites, and iron–nickel alloys [17]. Figure 17.3 presents a comparison of relative permeability vs. saturation of magnetic materials targeted for EMI applications.

Nanocrystalline and Fe-based amorphous materials have the highest saturation flux density, followed by iron–nickel alloys, Co-based amorphous, and ferrites. The material with high relative permeability and high saturation flux density reduces the CMC size as it allows for fewer turns and thus directly reduces copper losses and capacitance while improving the EMI filter performance at high frequencies. High switching frequencies raise the bar for magnetic component design and material selection. Due to the high-frequency capabilities, ferrites and nanocrystalline materials are of primary interest for high-performance magnetic components. The magnetic materials' properties are influenced by many variables such as frequency, temperature, physical size, mechanical stress, and even batch-to-batch material variations. Since there is a lack of standards for magnetic material characteristics and how to define them on datasheets, manufacturers strike a balance in presenting a sufficiently high level of information for most customers. For EMC applications, data is provided on complex permeability over the frequency range of interest, but complex permittivity over frequency is rarely provided and is also important, as will be shown. Not all filter parameters are determined solely by the magnetic material properties, and detailed tuning of designs requires in depth expertise.

17.2.1 Complex permeability characteristics

The analysis of complex permeability frequency characteristics is the primary method to assess material performance at a low magnetic flux amplitude. The

Table 17.2 Magnetic material properties

Manufacturer	Material	Type	Relative Permeability @10 kHz, 25°C	Saturation flux Density (T) @10 kHz, 25°C	Curie temp (°C)	Density (g cm^{-3})	Resistivity (Ω m)	Continuous operating temperature (°C)	Thermal conductivity (Wm^{-1} K^{-1})
Vacuumschmelze	Vitrovac 6030F	Amorphous	3500	0.82	365	7.80	1.3×10^{-6}	120	10.5**
Vacuumschmelze	Vitrovac 6025	Amorphous	70 000–100 000	0.58	240	7.70	1.4×10^{-6}	120	10.5**
Metglas Inc.	2605S3A	Amorphous	35 000	1.41	358	7.29	1.38×10^{-6}	150	10.0**
Hitachi Metals	Magnaperm 2714A	Amorphous	>72 000	0.57	225	7.59	1.42×10^{-6}	90	10.5**
Vacuumschmelze	Vitroperm 250F	Nanocrystalline	4000–6000	1.20	600	7.35	1.15×10^{-6}	180*	10.5**
Vacuumschmelze	Vitroperm 500F	Nanocrystalline	15 000–150 000	1.20	600	7.35	1.15×10^{-6}	180*	10.5**
Magnetec	Nanoperm	Nanocrystalline	5000 200 000	1.20	600	7.35	1.15×10^{-6}	200*	10.5**
Hitachi Metals	Finemet-1M	Nanocrystalline	70 000	1.23	570	7.40	1.1×10^{-6}	180*	10.5**
Hitachi Metals	Finemet-3L	Nanocrystalline	50 000	1.23	570	7.30	1.2×10^{-6}	180*	10.5**
Ferroxcube	3S5	MnZn Ferrite	3800	0.54	255	4.80	10.00	180*	3.5–5
Ferroxcube	3E27	MnZn Ferrite	6000	0.43	150	4.85	0.50	140	3.5–5
Ferroxcube	3E10	MnZn Ferrite	10 000	0.46	130	17.00	0.50	125	3.5–5
Ferroxcube	3E12	MnZn Ferrite	12 000	0.47	130	17.00	0.50	125	3.5–5
DMEGC	R15K	MnZn Ferrite	15 000	0.36	110	4.90	0.05	110	17.0
DMEGC	R10KZ	MnZn Ferrite	10 000	0.38	120	4.90	0.15	115	17.0
DMEGC	R10K	MnZn Ferrite	10 000	0.40	120	4.90	0.15	120	17.0
Ferroxcube	4B1	NiZn Ferrite	250	0.36	250	4.60	10^5	180	3.5–5
Ferroxcube	4S60	NiZn Ferrite	2000	0.26	100	17.00	10^5	100	3.5–5
MAGNETICS	Superalloy	Iron–nickel	70 000–10 0000	0.82	460	8.70	0.57×10^{-6}	180*	TBD
MAGNETICS	Permalloy 80	Iron–nickel	20 000–50 000	0.82	460	8.70	0.57×10^{-6}	180*	TBD
Vacuumschmelze	Vitrovac 6025	Amorphous	80 000	1.25	450	TBD	0.57×10^{-6}	180*	10.0**

*Limited by the box material.

**Along the lamination.

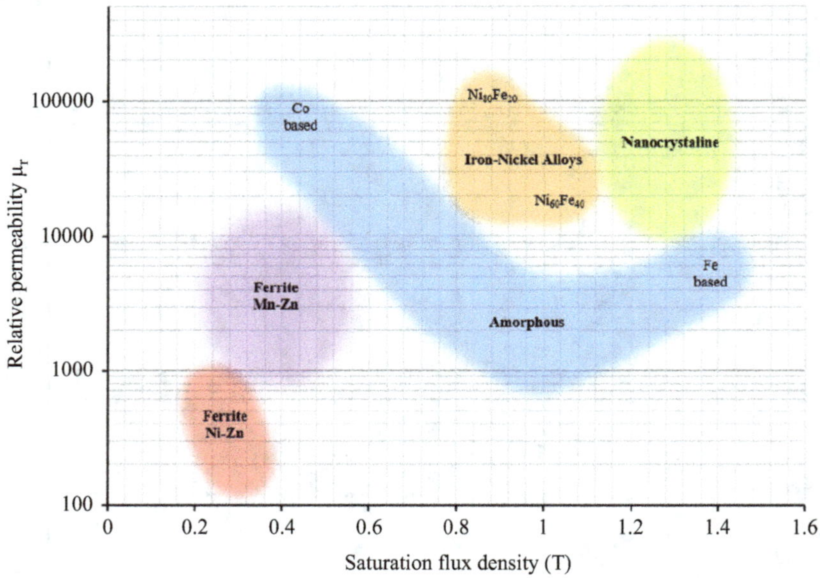

Figure 17.3 EMI material's relative permeability vs. saturation flux density

complex permeability is expressed by its real (reactive) component μ' and imaginary (dissipative) component μ''. Material losses are conventionally modeled as the imaginary component of complex permeability [18]. The relation of the complex impedance is represented by the series model, based on complex permeability, in the following manner:

$$\overline{Z} = j\omega L_S + R_S = j\omega L_0 \left(\mu_s' - j\mu_s''\right) \tag{17.1}$$

$$L_S = L_0 \mu_s' \tag{17.2}$$

$$R_S = \omega L_0 \mu_s'' \tag{17.3}$$

$$\tan\varphi = \frac{\mu_s''}{\mu_s'} \tag{17.4}$$

$$L_0 = \frac{\mu_0 N^2 A}{l} \tag{17.5}$$

where $\overline{Z}$ is the complex impedance, R_S and L_S are the series resistance and the series inductance, respectively, L_0 is the inductance of the core with unity permeability, ω is the angular frequency, j is the complex operator, μ' and μ'' are the real and imaginary permeabilities in the series model, φ is the loss angle (the phase angle between B and H), A is the core cross section, l is the magnetic path length in the core, and N is the turns number.

In order to determine the complex permeability frequency characteristics, caution must be taken as the ferrite core dimensions may affect the permeability

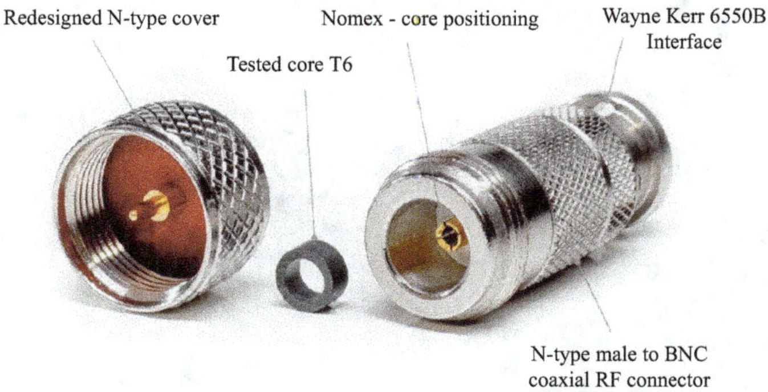

Figure 17.4 Short-ended coaxial line to measure the permeability characteristic

characteristics due to dimensional resonance and eddy currents; therefore, the selected core should have as small set of dimensions as feasible. Conversely, the small core dimensions directly translate to small values of measured resistance and reactance that may pose a challenge for the measurement system. As an example, a toroidal core T6/4/2.8 mm has been selected. It comes from the regular manufacturing process without machining, which may affect material parameters. In order to determine the complex permeability of the T6 core, a custom test fixture was used, as presented in Figure 17.4.

The test fixture provides one turn equally distributed around the tested core without any mechanical stress and a reliable connection to the measurement device via a concentric cable. Therefore, the material complex permeability characterization of the T6 core, made of various materials, can be easily repeated [19].

The complex permeability vs. frequency characteristics for 3F36 and 3E10 ferrites show the performance dependency on the core size, as shown in Figures 17.5 and 17.6, respectively [20].

In the presented Mn–Zn materials, the dimensions of the tested samples impact the measured complex permeability. A quality design of magnetic components for high-frequency applications requires detailed information on the magnetic material properties. Core characteristics show that the permeability depends not only on the magnetic material but also on the physical size. Therefore, the smallest cores provide the material data with the highest possible accuracy as long the size effects are negligible.

17.2.2 Complex permittivity characteristics

Complex permittivity, as in the case of complex permeability, is the second parameter that should be taken into account during the magnetic material selection process. The combination of the high permeability and high permittivity of MnZn ferrite affects the core high-frequency performance due to the skin effect and dimensional resonance. Therefore, accurate identification of the magnetic material

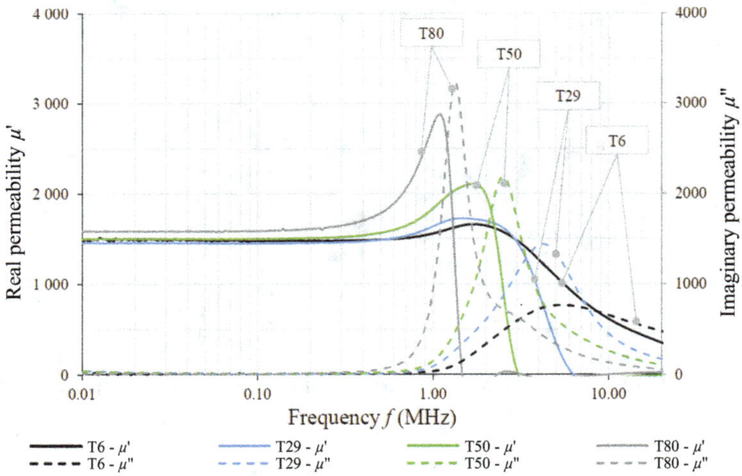

Figure 17.5 3F36 complex permeability vs. frequency

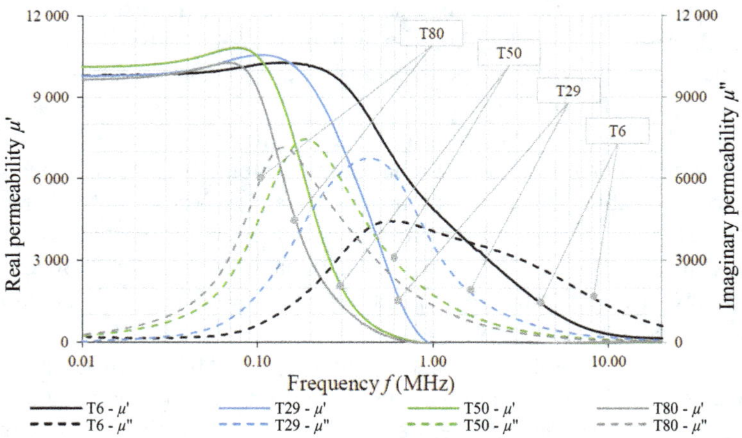

Figure 17.6 3E10 complex permeability vs. frequency

complex permittivity is essential to provide high-quality physical phenomena modeling and to improve the EMI filter design process. Complex permittivity describes the material interaction with an electric field. The real part of permittivity ε' defines how much energy from an external electric field is stored in a material. The imaginary part of permittivity ε'' is called the "loss factor," and it describes how dissipative or lossy the material is in the presence of an external electric field [13].

The complex admittance can be represented by a parallel model based on complex permittivity in the following manner:

$$\overline{Y} = j\omega C_P + G_P = j\omega\varepsilon_0\frac{A}{d}\left(\varepsilon_P' - j\varepsilon_P''\right) \tag{17.6}$$

$$\varepsilon_P' = \frac{C_P d}{\varepsilon_0 A} \tag{17.7}$$

$$\varepsilon_P'' = \frac{G_P d}{\omega\varepsilon_0 A} = \frac{\sigma}{\omega\varepsilon_0} \tag{17.8}$$

$$\tan\delta_d = \frac{\varepsilon_P''}{\varepsilon_P'} \tag{17.9}$$

where G_P is the measured conductance, C_P is the measured capacitance, ε_P' is the real component of the parallel relative complex permittivity, ε_P'' is the imaginary component of the parallel relative complex permittivity, ε_0 is the vacuum permittivity, d is the tested sample thickness, A is the test sample cross section, σ is the material conductivity, and $\tan\delta_d$ is the dielectric loss tangent.

The choice of the permittivity measurement techniques depends on many factors, such as the frequency, the expected value of complex permittivity, the tested material properties (homogeneous, isotropic), and the form (liquid, powder, solid), as well as the sample size and temperature. Figure 17.7 presents a summary of the available measurement techniques with reference to frequency and loss [21,22].

As the materials are low loss, the most suitable measurement technique for defining the complex permittivity of magnetic materials is the parallel-plate

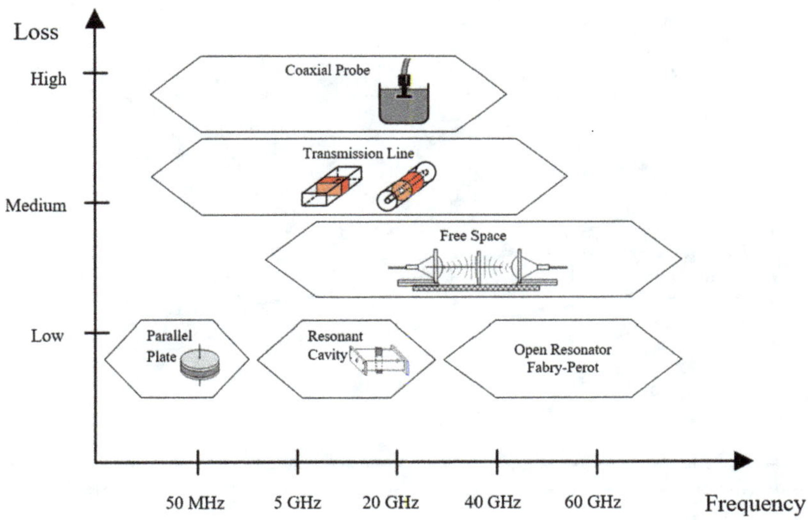

Figure 17.7 Summary of measurement techniques for material dielectric properties

method. This method uses a parallel-plate capacitor as a sample holder, with the tested material in between. The material is subjected to an AC source, while the actual voltage and current across the material are monitored by a LCR analyzer. As a result, the tested material parameters are derived by knowing the dimensions of the tested samples and by measuring their capacitance and dissipation factor. Physical sample size plays an important role in complex permittivity measurements. The size of the sample should be less than half of the electromagnetic wave wavelength for the frequencies of interest. Otherwise, the dimensional resonance will develop and affect or dominate the measured parameters. Samples with a small cross-section minimize these effects, providing high-quality material data at higher frequencies.

In order to demonstrate the permittivity variation in the Mn–Zn ferrite family, seven materials are selected to determine and compare electrical properties [21]. The real permittivity frequency characteristics of tested samples are shown in Figure 17.8. The real permittivity for all the tested samples drops as frequency increases. 3E10 material has the highest dielectric constant. The real permittivity of 3F36 ferrite is almost five times lower than 3E10. While the material datasheets provide generic parameters for all Mn–Zn materials [12–16], there is a significant parameter variance between materials, and without detailed characterization, the performance assessment is laden with substantial error.

The conductivity–frequency characteristics of the tested samples are shown in Figure 17.9. In the low-frequency region, conductivity is determined primarily by the conductivity of the filler, and the result is between 0.09 and 5 S/m. At higher frequencies, the conductivity is determined by the conductivity of the grain. The data show that the material's resultant conductivity changes strongly

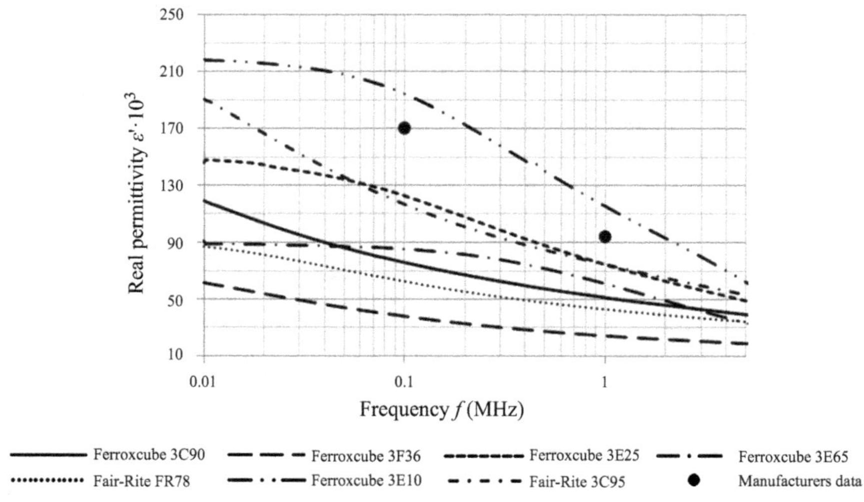

Figure 17.8 Real permittivity vs. frequency characteristic

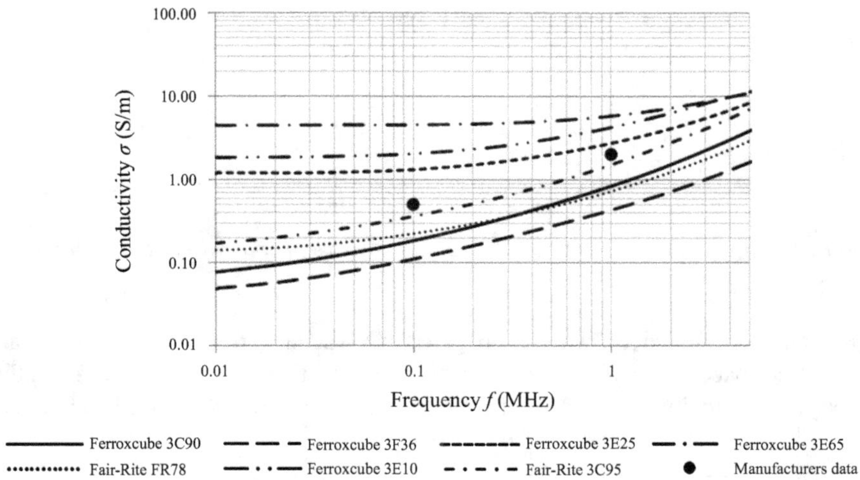

Figure 17.9 Conductivity vs. frequency characteristic

with the frequency. Similarly to the permittivity variation, the conductivity is usually provided by manufacturers as a generic figure that may lead to significant eddy current estimation error. The effect is magnified for large ferrite cores, where eddy-current losses become even more significant as they affect flux distribution in the core.

17.3 High-frequency effects in the magnetic core

Assuming that the electric field has one component while the propagation is in the z-direction, the electromagnetic wave equation yields the following expression in the time-domain [23]:

$$\overline{E}(z,t) = E_0 e^{-\alpha z} \cos(\omega t - \beta z) \tag{17.10}$$

A visualization of an electromagnetic wave propagating through a linear lossy material is shown in Figure 17.10. However, traditional electromagnetic wave visualization has certain drawbacks. As shown in Figure 17.10, the wave is limited to the z-axis, which is a significant simplification. The plane wave is extended in the x- and y-directions. Note that both electric and magnetic fields are related through material impedance; thus, the units on the two axes are different.

The propagation velocity v of the electromagnetic wave in the material, as defined by permeability μ and permittivity ε, is given by

$$v = \frac{1}{\sqrt{\mu\varepsilon}} \tag{17.11}$$

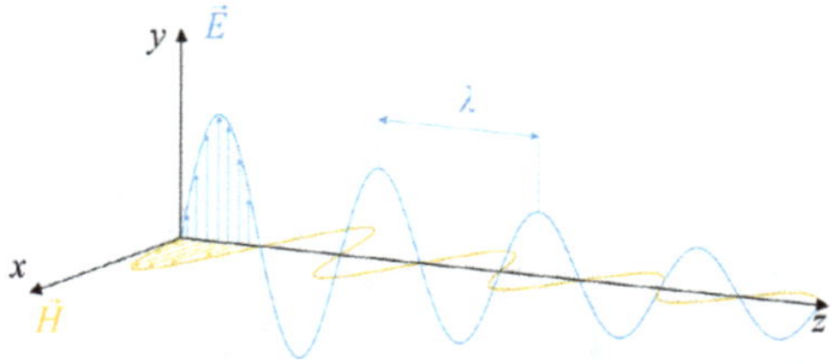

Figure 17.10 Visualization of electromagnetic wave propagation in lossy material

When an electromagnetic wave propagates in a vacuum, the permeability equals that of free space ($\mu = \mu_0$) and the permittivity equals that of free space ($\varepsilon = \varepsilon_0$), the propagation velocity becomes what we commonly refer to as the speed of light c_0:

$$c_0 = \frac{1}{\sqrt{\mu_0 \varepsilon_0}} = 299\,792\,458\ \frac{m}{s} \tag{17.12}$$

For small-signal excitation, magnetic material can be characterized by complex permeability and permittivity. The imaginary parts of the permeability and permittivity describe the losses within the ferrite material, while the real parts of the permeability and permittivity present the capability to store reactive energy. As both parameters $\overline{\mu}$ and $\overline{\varepsilon}$ are complex, the wave propagation velocity, $\overline{v}$, is also complex:

$$\overline{v} = \frac{1}{\sqrt{\overline{\mu}\overline{\varepsilon}}} = \frac{1}{\sqrt{(\mu' - j\mu'')(\varepsilon' - j\varepsilon'')}} \tag{17.13}$$

Magnetic materials have high permittivity and permeability. Therefore, the electromagnetic waves that enter the ferrite propagate slower than in free space, or equivalently, the wavelength in the ferrite is shorter than in free space. The wave speed $\overline{v}$, the frequency f, and wavelength $\overline{\lambda}$ are related by

$$\overline{\lambda} = \frac{\overline{v}}{f} = \frac{2\pi}{\omega\sqrt{(\mu' - j\mu'')(\varepsilon' - j\varepsilon'')}} \tag{17.14}$$

Frequency is the most fundamental attribute that describes the electromagnetic wave, as it never varies in linear materials. Wave number $\overline{k}$ combines wavelength, frequency, and material characteristics as follows:

$$\overline{k} = \frac{2\pi}{\overline{\lambda}} = \omega\sqrt{(\mu' - j\mu'')(\varepsilon' - j\varepsilon'')} = \beta - j\alpha \tag{17.15}$$

The imaginary part α of the wave number is an attenuation constant and defines the rate of extinction of the wave in a material. The real part β of the wave number is a phase constant, which is a measure of the phase shift per unit length. The wavelength is related to the phase constant. Therefore, an expression for the real part of the complex wave number provides an equation for the wavelength as an imaginary part of skin depth:

$$\alpha = -\mathrm{Im}\left[\bar{k}\right] = \frac{1}{\delta} \Rightarrow \delta = \frac{1}{\alpha} \tag{17.16}$$

$$\beta = \mathrm{Re}\left[\bar{k}\right] = \frac{2\pi}{\lambda} \Rightarrow \lambda = \frac{2\pi}{\beta} \tag{17.17}$$

17.3.1 Dimensional resonance and skin depth

In a low-loss magnetic material, the electromagnetic wave travels without significant energy loss. If the smallest cross-sectional dimension of the core (perpendicular to the magnetic and electric field) is equal to half a wavelength, then a standing wave will be formed across the cross section of the core. Under these specific conditions, the surface flux, which is in phase with the surface magnetic field, is canceled out by antiphase flux at the core center. This phenomenon is called dimensional resonance and it is characterized by the permeability dropping to zero.

A standing wave is a combination of two waves moving in opposite directions, each having the same amplitude and frequency. The phenomenon is a result of interference while the energy carried by the waves is either added or canceled. For opposite-moving waves, the interference produces an oscillating wave fixed in space. In relation to the core size, where the smallest cross-sectional dimension of the core is equal to half of the wavelength, the nodes and points where the standing wave is equal to zero are located at the core's outer edges. The antinode is a constructive interference with an amplitude equal to the sum of the amplitude of each of the waves. In the considered case, the antinode is situated at the core center where the flux amplitude reaches its peak. Flux amplitude decreases with increasing phase shift between the waves and is equal to zero when the waves are shifted 180°. This is called dimensional resonance and it directly affects apparent core permeability and, as a result, it also has an influence on inductance and impedance.

Magnetic materials with higher losses will not support standing waves as the electromagnetic wave is attenuated very quickly, and the dimensional resonance would not develop. Instead, the depth of the electromagnetic wave penetration is observed and is referred to as skin depth. The skin depth is inversely proportional to the wave attenuation constant. By analogy to the electrical conductor, the skin depth is defined as the equivalent depth where the electromagnetic wave can penetrate the material by a factor of *1/e*. The skin effect for electric current and magnetic flux is illustrated in Figure 17.11.

The alternating magnetic flux induces a counter-magnetic flux in a core made of conductive material whose dimensions are large enough to carry eddy currents. Eddy currents are closed loops of induced current circulating in planes perpendicular to the magnetic flux. The depth at which the eddy currents and electromagnetic waves penetrate into a material is affected by the frequency, conductivity, and magnetic permeability of the material. The penetration depth decreases with increasing frequency, conductivity, and magnetic permeability. Therefore, high

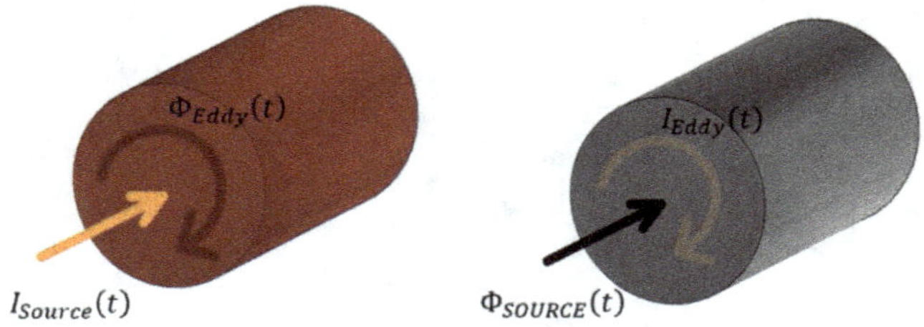

Figure 17.11 Skin effect in an electrical conductor carrying current and magnetic core carrying flux

frequency can be considered as the level at which the parasitics would have an effect on the performance of the core.

17.3.2 High-frequency effects in magnetic materials

Laminated structures, such as amorphous, nanocrystalline, and iron-nickel, by default, are conductive alloys where the eddy current effect is reduced by breaking the conduction path into many high-resistance paths. The practical minimum lamination thickness is the only limit for the design.

Bulky ferrite cores are subject to high-frequency effects such as dimensional resonance and skin effect within the core that result in a non-uniform, frequency-dependent magnetic flux distribution. The dimensional resonance and the skin effect are dependent on the size and shape of the core as well as the operating frequency. Electromagnetic wave parameters, the attenuation constant, and the wave number define and quantify high-frequency effects within the core. However, precise information on magnetic materials is mandatory for an in-depth analysis of high quality.

In reality, ferrite skin depth and dimensional resonance occur together and interact producing an effective resultant flux distribution in the magnetic core. Presenting both dependencies on one graph for each tested material allows for high-frequency effect identification within the magnetic core. The calculated wavelengths and skin depths for 3E10 and 3F36 material are shown in Figure 17.12. Let us take an example of a T90 ring core that has an outer diameter of 90 mm, an inner diameter of 60 mm, and is 20 mm high; the smallest cross-sectional dimension of 15 mm is indicated on the graph. For the T90 core, made of 3E10 material, the skin depth is smaller than half the wavelength for the entire frequency range. Magnetic cores made with 3E10 are dominated by skin depth caused by a combination of magnetic and electric losses. The electromagnetic wave is attenuated, and there is no dimensional resonance. For the T90 core, made of 3E10 material, the skin depth effect within the core is visible at 80 kHz. For the T90 core, made of 3F36 material, the wave propagation and attenuation must be considered together. At 1.7 MHz, the smallest core cross-sectional dimension is equal to half a wavelength, while the core is dominated by dimensional resonance. As the frequency increases, the skin effect becomes

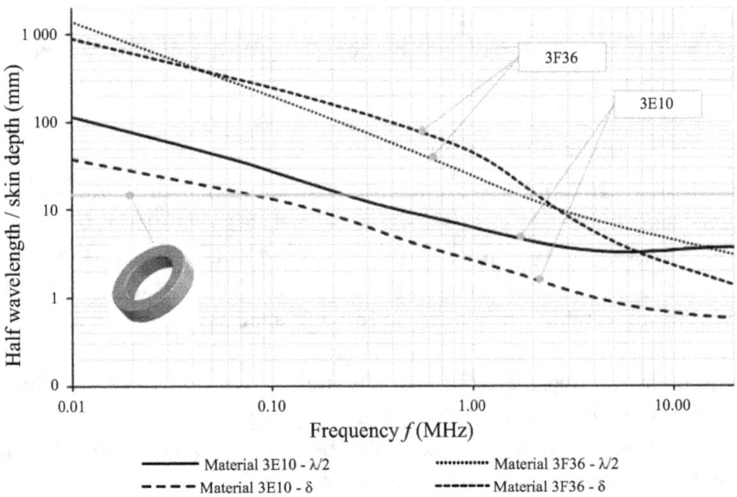

Figure 17.12 Electromagnetic wavelength and skin depth in ferrite 3E10 and 3F36

distinctive at 2.66 MHz, while above this frequency, the skin effect-related losses dominate.

The dimensional resonance and the skin effect are dependent on the size and shape of the core as well as the operating frequency. Dimensional resonance and skin depth in magnetic materials are defined by the electromagnetic wave and the properties of the core. Electromagnetic wave parameters, the attenuation constant, and the wave number define and quantify high-frequency effects within the core. However, precise information on magnetic materials is mandatory for an in-depth high-quality analysis.

17.4 Common-mode chokes

The EMI filter is a combination of passive components that can be represented as *R*, *L*, *C* networks. The largest and most customizable part of the EMI filter is the magnetic component, in this case, the CMC. The first choice material for CMC is a ceramic ferrite that provides a good balance between the manufacturing effort and performance. There are a number of publications on reducing the size and improving the performance of EMI filters. However, the effectiveness of the modeling depends on the model accuracy; therefore, a highly accurate model of the magnetic material is required for CMC size vs. performance analysis. As the application power increases, the performance is affected by the increased core size. The frequency effects developed in the core are linked to the size and, therefore, inductors do not scale well with frequency and power. As an example, a toroidal core with an external diameter of 25 mm is large enough to exhibit non-uniform flux distribution due to skin and dimensional resonance effects: for high permeability and permittivity materials, non-uniform flux distribution would be apparent at 1 MHz.

17.4.1 Filter impedance analysis

The fundamental principles of the magnetic design are valid across a range of applications, and trade-offs exist between the number of turns and the core size while being constrained by a number of physical effects that oppose ideal performance. The magnetic design must conform with industry standards and provide predictable and repeatable parameters within tolerance at acceptable manufacturing effort.

Magnetic design for an EMC filter, although similar to power inductor design, deals with high-frequency effects that are usually negligible for power inductors. A typical approach for the inductor design can be found in various publications. Key requirements are peak flux density, impedance, efficiency, dc resistance of the winding, and a parallel resistance that is inversely proportional to the ac power loss at a given frequency. In order to fulfill these requirements, a simple design flowchart can be applied that guides the designer through the material selection, the magnetic core sizing, the winding calculation, the power loss calculation, and the temperature rise estimation. The design flow is usually iterated until the design constraints are satisfied. The accuracy of the calculation flowchart depends on the model degrees of freedom and the quality of the input data.

The inductor for the EMC filter must provide attenuation across a wide frequency range as a countermeasure to the disturbance produced by the switching cell of the converter. The wide-bandgap semiconductor frequency spectrum is at a much higher level than what is known for silicon-based devices, as presented in Figure 17.2. Therefore, an accurate model shall be based on a complex permeability that allows for modeling across the frequency range.

The dynamic effects that are linked with the operating frequency are size-dependent and limit the performance of the component. Furthermore, the frequency-related eddy current and dimensional resonance depends on the material properties, and it is critical for the design algorithm to account for both phenomena. The correct identification of parameters boosts accuracy and reduces the development time. Modeling based on accurate material properties that account for dynamic phenomena overcomes the limits of the design process and helps mitigate the adverse dynamic effects caused by using bundled data for the entire ferrite range as is usually provided in the magnetic core datasheets.

An example plot of the impedance requirement for a CMC is shown in Figure 17.13. Typically, the requirement is provided as an impedance modulus vs. frequency, while the CMC design provides complex parameters of impedance based on a series model. Since the component is inductive, the impedance requirement is limited to the first quadrant of the complex plane. While the impedance absolute value is constant for a given frequency, there is an infinite number of solutions that form a quarter of the circle with a radius equal to the impedance

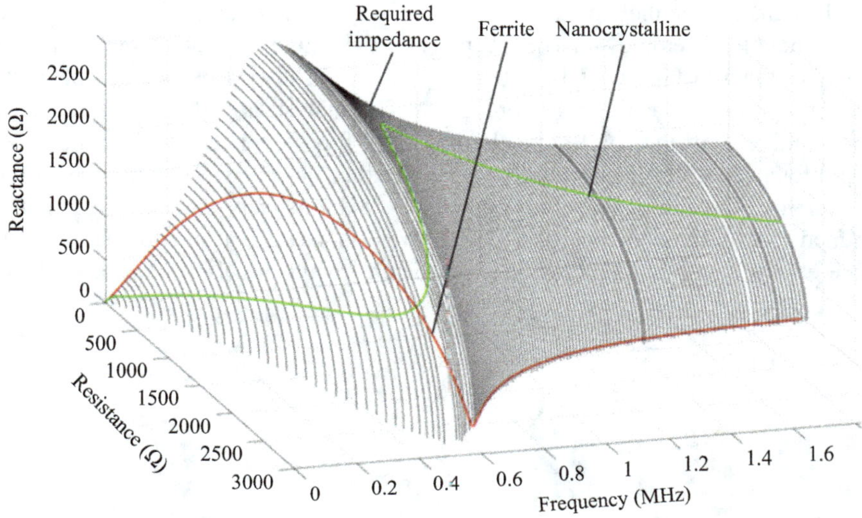

Figure 17.13 CMC impedance vs. frequency

absolute value. The ratio of the real and imaginary part sets the attenuation characteristic. The reactive energy may be a source of further oscillations at higher frequencies as it will add to the circulating energy envelope, while the real part is limited by the maximum power loss of the component. The imaginary part depends on the magnetic material properties, and it is proportional to the inductive part of complex permeability. The analyzed designs are mapped onto the impedance plane. The preferred design is based on the material, core geometry, winding turns, and associated parasitics.

The CMC filter provides an impedance characteristic that varies with frequency and changes its amplitude and angle to follow the desired frequency response. Detailed knowledge of the material properties is fundamental for material selection and prediction of the component characteristic [17].

17.4.2 Laminated ferrite core

Cores made of electrotechnical steel are often assembled by stacking insulated laminations, enabling the applied field to penetrate each lamination. The lamination thickness should be smaller than the skin depth for the required operating frequency. It has been known for decades that this method effectively controls eddy-current losses in iron-based cores. Eddy-current losses in laminated cores are reduced by a factor $1/n^2$, where n is the number of laminations, as the path of the induced electric field is closed in each lamination. In contrast to traditional laminations which are made of conducting material, the ferrite is perceived as a high-impedance bulk body. However, the ferrite bulk core at high frequencies also suffers from conduction and dimensional effects. Laminating the ferrite core can mitigate these effects and improve filter performance.

To demonstrate the lamination effects in a ferrite ring core, a 3D eddy-current FEA simulation is employed. The magnetic field distribution in solid and laminated cores is shown in Figure 17.14, while Figure 17.15 shows a plot of the magnetic flux density amplitude along the core radius. The analysis is performed at 500 kHz.

Intuitively, in the solid core, the magnetic flux is concentrated in the outer circumference, while the core center exhibits flux density weakening due to the skin effect. The laminated core, conversely, exhibits very little change in the magnetic field. Since the core is a closed toroid in shape, it is also prone to reluctance effects along its radius. At the considered frequency, high-frequency effects

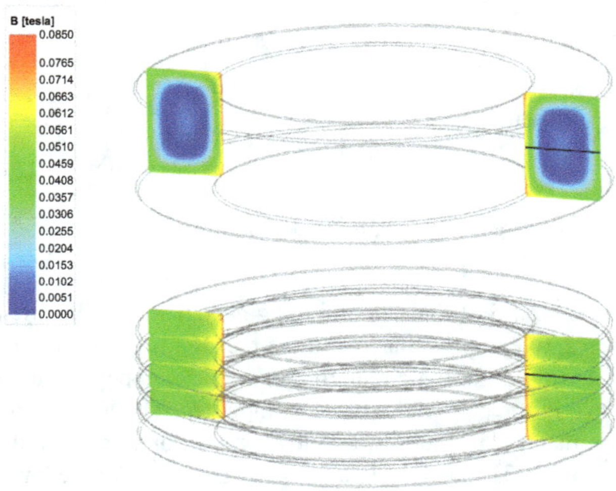

Figure 17.14 Flux density distribution in the solid and laminated ferrite ring core

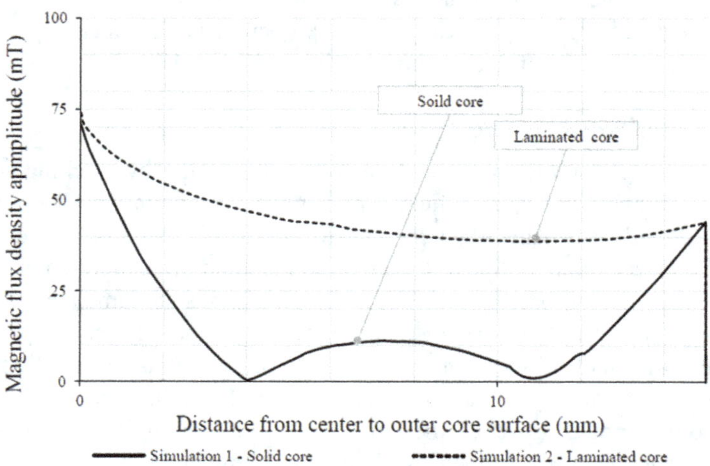

Figure 17.15 Flux density vs. core radius in solid and laminated ferrite ring cores

are developed in the solid core, while the laminated core flux distribution is simply dominated by the reluctance effect. At 500 kHz, the laminated core magnetic flux density gradually decreases along the cross section of the core, while the high-frequency effects are not significant.

The laminated core structure divides the conduction path into sub-regions that reduce the high-frequency effects. For such structures, the upper end of the frequency range is limited by the minimum lamination size. The lamination thickness impacts the core's complex characteristics at high frequencies. Thin laminations shift the skin and dimensional resonance effects to higher frequencies. The minimum lamination thickness is limited by the material grain size, shape, and manufacturing technology. The real permeability characteristic retains its stability at higher frequencies, while the eddy current losses are simultaneously reduced, which is visible in the reduced peak of the imaginary permeability characteristic. In order to visualize the impact of the lamination thickness on the core performance, the complex permeability for the T50/30/14 core, made of two materials, 3F36 and 3E10, is measured and contrasted with the core assemblies of two, four, and six laminations stacked together to form a magnetic core with the same size and volume. Figure 17.16 shows the examined cores. Core lamination thickness varies from 2.3 up to 7 mm, while the overall core thickness remains the same. Ferrite laminations are not insulated from each other in contrast to the state-of-the-art-laminated electrotechnical steel, which uses oxide film or insulating varnish to insulate each lamination. The ferrite core surface layer tends to oxidize during the sintering process and is considered a nonmagnetic surface layer. This oxidation layer is sufficient to provide insulation between ferrite laminations.

The effect of the core lamination thickness on the 3F36 ferrite real and imaginary permeability components is shown in Figures 17.17 and 17.18, respectively. As can be seen, the core with the smallest lamination shows the best performance. The real permeability characteristic is extended from 3 to 15 MHz, while the loss peak of the imaginary permeability characteristic is reduced by half. Lamination thickness does not affect the magnetic performance of the core at low frequencies.

The effect of lamination thickness on the 3E10 ferrite real and imaginary permeability is shown in Figures 17.19 and 17.20, respectively. The improvement is visible, and the real permeability characteristic is extended from 1 to 2 MHz, while the loss peak of the imaginary permeability characteristic is reduced by 30%. Since the 3E10 material is skin-effect driven, the performance gain is not as high as for 3F36.

17.4.3 Core size effect on lamination

Previous chapters show that the core's size has a strong impact on the solid core's high-frequency performance. In order to investigate this effect, the complex

Figure 17.16 T50 cores to test lamination thickness impact on core performance

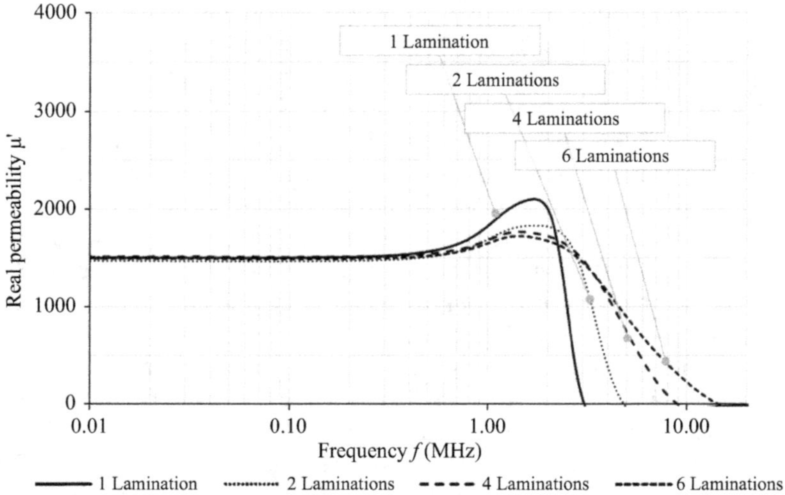

Figure 17.17 3F36 ferrite real permeability vs. frequency for various lamination thicknesses

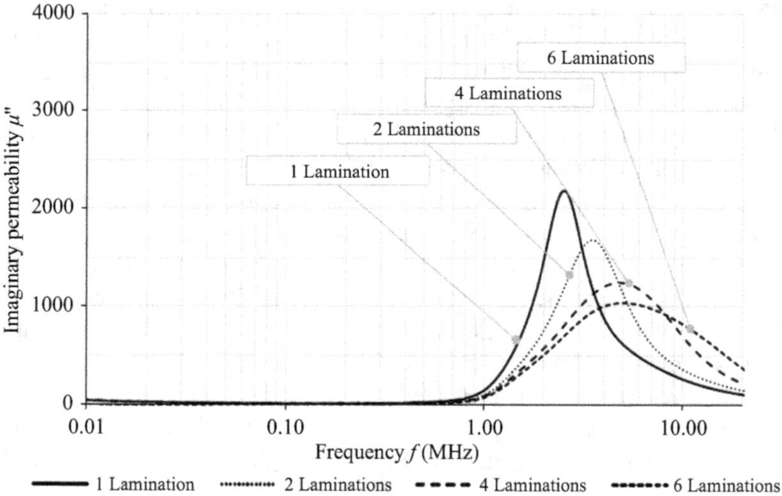

Figure 17.18 3F36 ferrite imaginary permeability vs. frequency for various lamination thicknesses

permeability characteristic for three bulk core sizes: T29, T50, and T80, made of two materials, 3F36 and 3E10, are compared with their laminated counterparts. The laminated cores are stacked together from smaller cores with a thickness of 3.5 mm. The T29-laminated core is stacked out of three laminations, while the T50 uses four and the T80 uses five laminations. Figure 17.21 shows the cores used for this test.

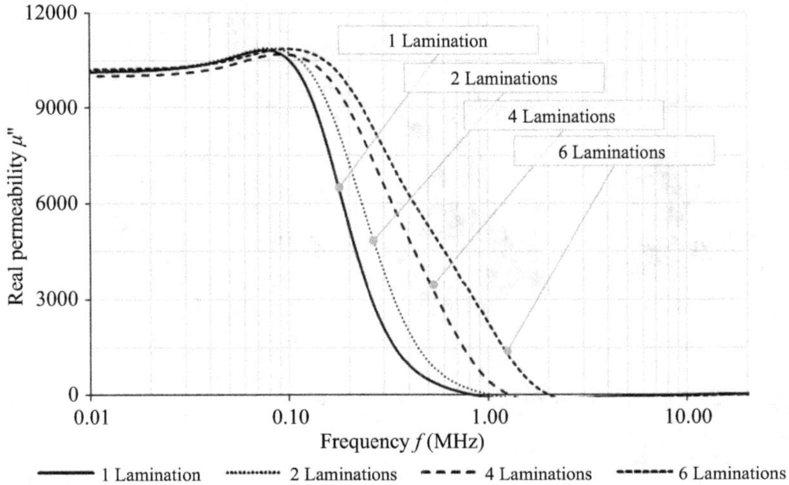

Figure 17.19 3E10 ferrite real permeability vs. frequency for various lamination thicknesses

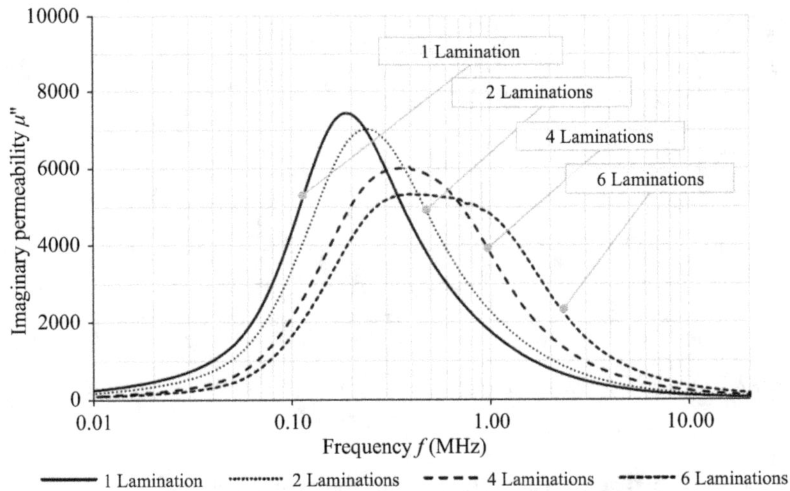

Figure 17.20 3E10 ferrite imaginary permeability vs. frequency for various lamination thicknesses

The complex frequency characteristics are presented in Figures 17.22 and 17.23 for the 3F36 material and in Figures 17.24 and 17.25 for the 3E10 material.

As can be seen, the laminated cores, regardless of their size and material, show significant improvement in the frequency characteristic. The laminated T80 core maintains its performance up to 6 MHz for the 3F36 material, exactly the same as

Figure 17.21 Cores used to test the core size effect on lamination performance

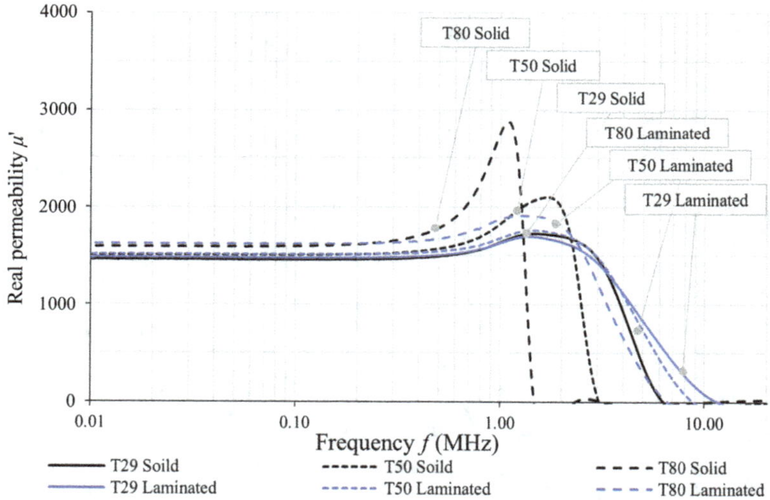

Figure 17.22 3F36 ferrite real permeability vs. frequency for various core sizes

the T29 solid core. The measured real permeability characteristic for the T29 bulk core, made of the 3F36 material, drops to zero at 6 MHz, while the laminated core keeps its characteristic up to 11 MHz. The biggest improvement in the imaginary permeability characteristic is visible in the T80 core. A significant loss increase, visible in the 3F36 material and caused by dimensional resonance, is reduced by half and shifted to higher frequencies.

17.4.4 Core impedance characteristics

The component characteristics can be shaped to the application requirement by combining materials of different characteristics. The potential of the solution is

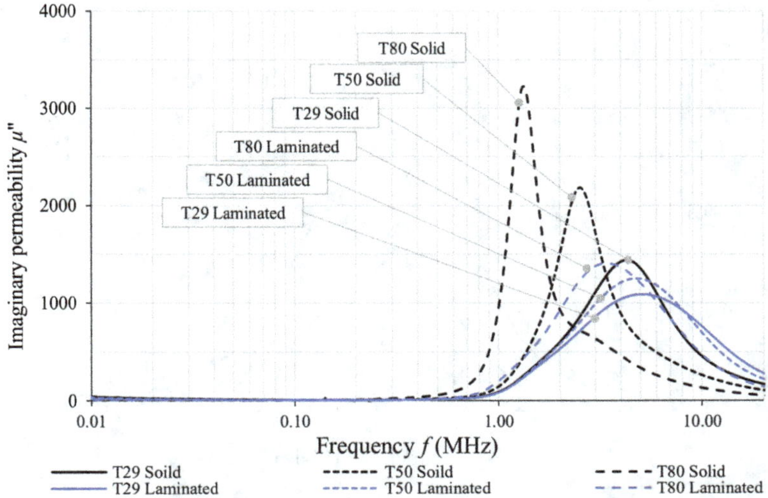

Figure 17.23 3F36 ferrite imaginary permeability vs. frequency for various core sizes

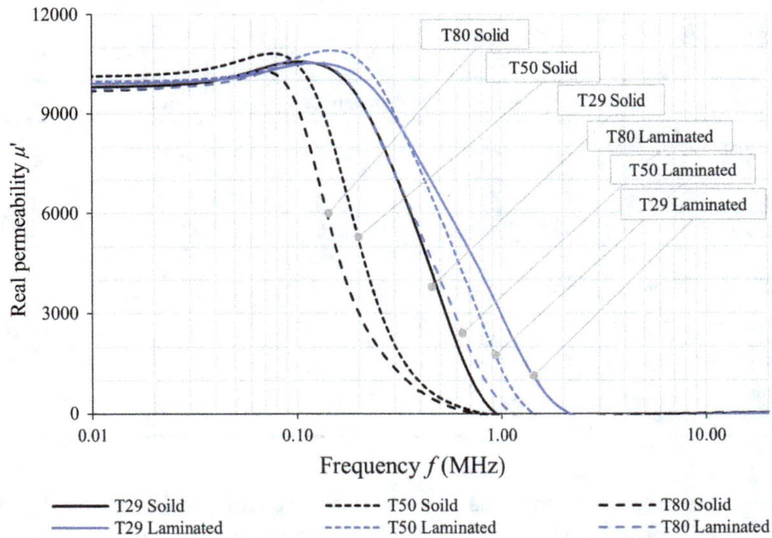

Figure 17.24 3E10 ferrite real permeability vs. frequency for various core sizes

shown in four examples based on the same T80 core size that compare and contrast the bulk ferrite cores with nanocrystalline and two ferrite-based structures, one made of the same material and the second being a hybrid structure combining two ferrite materials. The core details are shown in Table 17.3. In all cases, a one turn concentric winding is used to eliminate winding effects on the experiment.

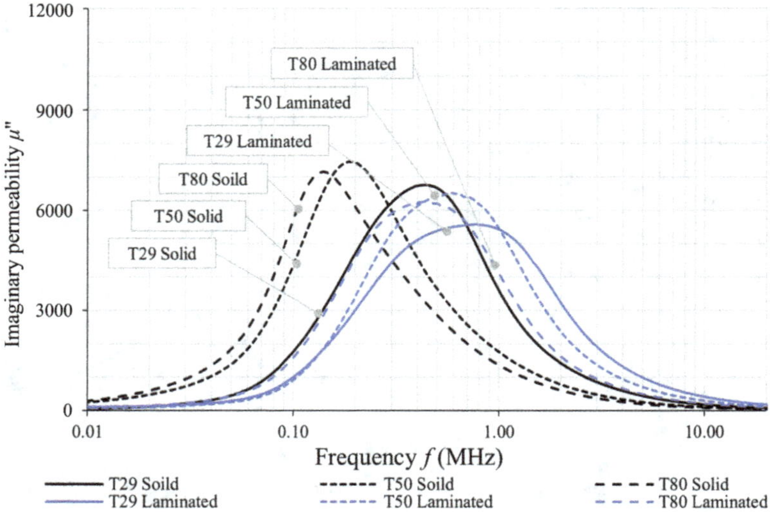

Figure 17.25 3E10 ferrite real permeability vs. frequency for various core sizes

Table 17.3 Tested core parameters

Parameter	Unit	Design 1.1	Design 1.2	Design 1.3	Design 1.4
Core material	–	Mn–Zn ferrite	Nanocrystalline	Mn–Zn ferrite	Mn–Zn ferrite + Ni–Zn ferrite
Core construction					
Core dimension OD × ID × H	mm	80×45×21	80×45×21	80×45×3.5 6 laminations	80×45×3.5 6 laminations
Material permeability	–	10 000	10 000	10 000	10 000/5200/ 250
Core cross section	mm^2	368	368	368	368

The measured core impedance frequency characteristics are shown in Figure 17.26. The solid Mn–Zn ferrite core, in comparison to the nanocrystalline core, has better performance in the lower frequency range. The laminated core Design 1.3, in comparison to the solid core, has significantly improved performance in the frequency range from 100 kHz, and up to 4 MHz impedance is higher than for nanocrystalline. A hybrid core with the same geometry as the Mn–Zn core combines properties of Mn–Zn and Ni–Zn ferrite material in one structure and shows significant improvement of impedance characteristic in the frequency range between 4 and 20 MHz that can challenge nanocrystalline material application. The core hybridization allows for shaping impedance characteristics.

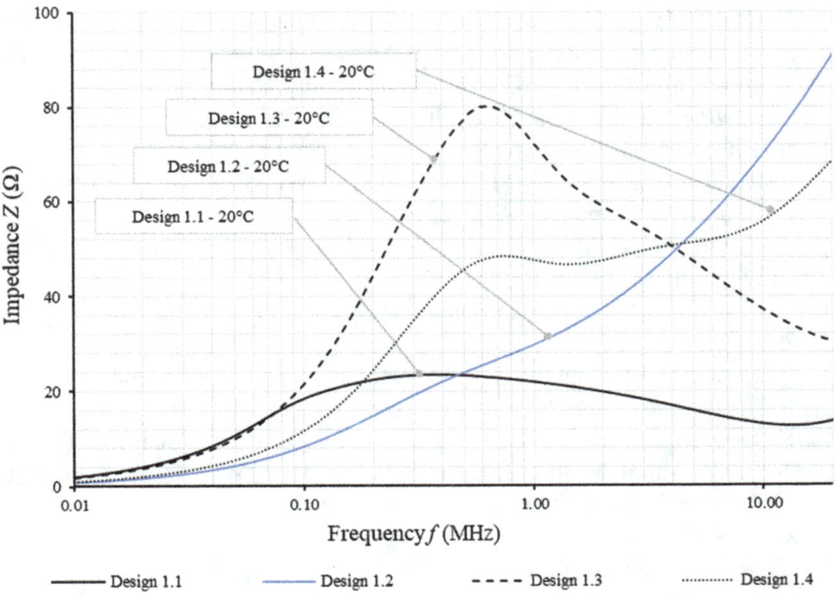

Figure 17.26 Tested cores impedance vs. frequency characteristic

The proposed laminated ferrite structure shows significant improvement of the impedance characteristic while maintaining the same core size and so allows for a smaller component with the same properties as the bulk one. The availability of laminated cores opens new horizons for EMI filter design.

References

[1] J. Millan, P. Godignon, X. Perpina, A. Perez-Tomas, and J. Rebollo, "A Survey of Wide Bandgap Power Semiconductors Devices," *IEEE Transactions on Power Electron.*, vol. 29, pp. 2155–2163, 2014.

[2] J. Biela, M. Schweizer, S. Waffler, B. Wrzecionko, and J.W. Kolar, "SiC vs. Si Evaluation of Potentials for Performance Improvement of Power Electronics Converter System by SiC Power Semiconductors," *IEEE Transactions on Industrial Electronics,* vol. 58, issue 7, pp. 2872–2882, 2011.

[3] Di Han, J. Noppakunkajorn, and B. Sarlioglu, "Comprehensive Efficiency, Weight, and Volume Comparison of SiC- and Si-Based Bidirectional DC–DC Converters for Hybrid Electric Vehicles," *IEEE Transactions on Vehicular Technology*, vol. 63, issue: 7, pp. 3001–3010, 2014.

[4] J.W. Kolar and J.E. Huber, "Future Power Electronics 4.0 3-Phase SiC/GaN Converter Systems," Tutorial at the *Applied Power Electronics Conference and Exposition (APEC2021)*, 2021

[5] H. Chung, S. Hui, and K. Tse, "Reduction of Power Converter EMI Emission Using Soft Switching Technique," *IEEE Trans. Electromagn. Compat.*, vol. 40, pp. 282–287, 1998.

[6] X. Wu, J. Yang, H. Chen, Z. Qian, “Analysis and Design Considerations for EMI and Losses of RCD Snubber in Flyback Converter,” in *Proceedings of Twenty-Fifth Annual IEEE Applied Power Electronics Conference and Exposition*, Palm Springs, CA, Murch 2010.

[7] A. Bhargava, D. Pommerenke, and K.W. Kam, “DC–DC Buck Converter EMI Reduction Using PCB Layout Modification” *IEEE Trans. Electromagn. Compat.*, vol. 53, pp. 806–813, 2011.

[8] H.W. Ott, *Noise Reduction Techniques in Electronic Systems*, John Wiley and Sons, 1988.

[9] H.W. Ott, *Electromagnetic Compatibility Engineering*, John Wiley and Sons, 2009.

[10] P. Bardos, “Predicting the EMC Performance of High-Frequency Inverters,” *IEEE Applied Power Electronics Conference*, pp. 213–219, 2001.

[11] N. Oswald, B.H. Stark, D. Holliday, and C. Hargis, “Analysis of Shaped Pulse Transitions in Power Electronics Switching Waveforms for Reduced EMI,” *IEEE Transactions on Industry Application*, vol. 47, issue: 5, pp. 2154—2165, 2011.

[12] www.ferroxcube.com

[13] www.chinadmegc.com

[14] www.hitachi.com

[15] www.vacuumschmelze.de

[16] www.mag-inc.com

[17] M. Kacki, M.S. Rylko, J.G. Hayes, and C.R. Sullivan, “Magnetic Material Selection for EMI Filters,” *IEEE Energy Conversion Congress and Exposition* (ECCE), Cincinnati, *OH*, 2017.

[18] E.C. Snelling, *Soft Ferrites: Properties and Applications*, *Newnes-Butterworth*; 1st Edition, 1969.

[19] Marcin Kacki, *Investigation of the High-frequency effects in Mn–Zn ferrites for EMI filter applications*, PhD dissertation, National University Of Ireland, Cork, Ireland, 2022.

[20] M. Kacki, M.S. Rylko, J.G. Hayes, and C.R. Sullivan, “Measurements Methods for High-Frequency Characterizations of Permeability, Permittivity, and Core loss of Mn–Zn Ferrite Cores,” *IEEE Transactions on Power Electronics*, 2022.

[21] O.V. Tereshchenko, F.J.K. Buesink, and F.B.J. Leferink, “An Overview of the Techniques for Measuring the Dielectric Properties of Materials.” *XXXth URSI General Assembly and Scientific Symposium*, Istanbul, Turkey, 2011.

[22] L. Brunetti, L. Oberto, and M. Sellone, “Material Measurements with Vector Network Analysers.” *Conference on Precision Electromagnetic Measurements (CPEM 2016)*, Ottawa, Canada, 2016.

[23] D.J. Griffiths, *Introduction to Electrodynamics, Prince Hall*, 1999.

Chapter 18

Common mode and differential mode suppression in power electronic systems

Peter Zacharias[1] *and Peter Wallmeier*[2]

The operation of power electronic systems is based on the medium-frequency switching of semiconductor switches, which control the transmitted current by means of the ratio of switch-on time to pulse period. High-voltage gradients of up to 250 V/ns are generated at the switches, depending on the semiconductor technology used. Therefore, parts of the power electronic system change their voltage level to the environment and ground extremely fast. Because each component is connected to its environment with a parasitic coupling capacity in addition to its regular wired integration into the system, unwanted currents also flow into this environment and to the reference mass. They propagate in the system and close either via the forward and return conductors or via the reference ground. These displacement currents can lead to significant voltage drops in line impedances and components, both inside and outside the system, which can ultimately result in interference and cause malfunctions in the system itself (internal electromagnetic compatibility) or in the environment (external electromagnetic compatibility) [1,2]. Rapidly changing currents also generate interference voltages via induced couplings to neighboring systems and thus disturbances in them. While induced interference is mainly countered by shielding measures, disturbances caused by capacitive couplings are attenuated by the introduction of impedances tuned to the disturbance. These impedances are generated by filters, which consist of the multi-stage interconnection of capacitances and inductors. Inductive components for these purposes are the focus of this chapter.

18.1 Problem description

The architecture depicted in Figure 18.1 should serve to describe the problem. A bidirectional buck converter with an ohmic-capacitive load is shown. The semiconductor switches T_1 and T_2 switch alternately. This changes the potential at the connection point of T_1 and T_2. If the parasitic capacitances C_1, C_2, and C_3 to the

[1]Department of Power Electronics, University of Kassel, Germany
[2]AEG Power Solutions, Germany

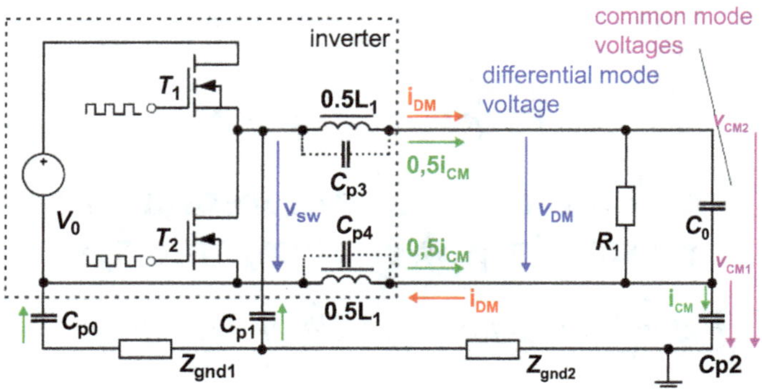

Figure 18.1 Equivalent circuit diagram to illustrate the formation and effect of differential and common mode voltages with symmetrically split choke L_1

environment did not exist, a current would only flow via the load connections. Its ripple depends on the switching frequency and the size of the inductance L_1. As this is caused by the voltage difference between the two outputs of the power converter, the associated signal is referred to as the differential mode signal. The rapid change in potential at the connection point of T_1 and T_2 results in a current via C_{p1}. The circuit closes via the virtual earth connection as well as via C_{p2} and V_0 and via C_0 and C_{p3}. This second channel for signal propagation is the common mode channel. The permissible values of the currents in both channels and the corresponding voltages are defined in the relevant standards depending on the application, e.g., commercial (FCC Part 15 class B, EN 61000-6-3, EN 61000-6-1), medical (IEC 60601-1-2), automotive (SAE, ISO7637, IEC CISPR-25, ISO1145-1, ISO1145-2), etc.

All inductive components as well as all electrical conductors through which current flows are surrounded by magnetic fields. If these fields also penetrate another component, voltages are induced there. These inductive couplings can be counteracted by electromagnetic shielding. Conducted interference can be minimized by inserting appropriate impedance networks. The corresponding measures are dealt with in this chapter.

18.2 Minimization of current ripple interference with filter chokes

The cutoff frequency of the load current is always considerably lower than the switching frequency. The ripple current resulting from the switching operations is the result of the voltage drop at L_1:

$$i_{L1}(t) = \frac{1}{L_1}\int v_{L1}(t)dt = \frac{1}{L_1}\int (v_1(t) - v_{out}(t))dt \tag{18.1}$$

This means that the amplitude of the switching frequency ripple current is proportional to $(L_1)^{-1}$. The larger the inductance L_1 and the larger the switching frequency, the smaller it becomes. The effective load current determines the required winding window A_w via the maximum permissible current density $J_{\max}$

$$A_a = \frac{N \cdot A_w}{k_f} = \frac{N \cdot I_{rms}}{k_f \cdot J_{\max}} \tag{18.2.}$$

with A_a being the available winding window of the core, A_w being the winding cross section, N being the number of windings, I_{rms} being the maximum RMS current, k_f being the conductor filling factor, and $J_{\max}$ being the maximum current density.

The magnetization cross section of the core results from the relationship for the maximum flux linkage that occurs

$$\Psi_{\max} = N \cdot B_{\max} \cdot A_e = L \cdot I_{\max} \Rightarrow A_e = \frac{L \cdot I_{\max}}{N \cdot B_{\max}} \tag{18.3}$$

where $\Psi_{\max}$ is the maximum flux linkage, $B_{\max}$ is the maximum magnetic flux density, A_e is the magnetization cross section, L is the inductance, $I_{\max}$ is the maximum current, and N is the number of turns.

If the two areas A_a and A_e are multiplied together to the area product G, a value independent of the number of turns N is obtained. This is a characteristic of certain core shapes and can be used for a selection

$$G = A_e \cdot A_a = \frac{L \cdot I_{\max}}{N \cdot B_{\max}} \cdot \frac{N \cdot I_{rms}}{k_f \cdot J_{\max}} = \frac{L \cdot I_{\max} \cdot I_{rms}}{B_{\max} \cdot k_f \cdot J_{\max}} \tag{18.4}$$

The area product of the core must be larger than the calculated value. However, this is only a rough approximation as, for example, re-magnetization losses and losses due to the proximity effect are not taken into account.

The inductance neglecting the stray field is given by

$$L = \mu_{r.eff} \mu_0 \frac{A_e}{l_e} N^2 = A_L \cdot N^2 \tag{18.5}$$

where $\mu_{r.eff}$ is the effective relative permeability of the core, l_e is the magnetic length of the core, and A_L is the inductance coefficient of the core).

In Figure 18.1, the design target is to maximize inductance L at the minimum size and loss of the filter or storage choke. Typically, high permeable cores are made from ferrite, laminated amorphous, or nanocrystalline ribbons with air gaps, or powder cores from iron alloy granulate are used. The magnetic field is stored in the air gaps. The winding material is, in most cases, enameled round wire or flat copper. If the current still has a significant ripple Litz wires are used to reduce eddy current loss due to skin and proximity effects.

An important property is the self-resonance of such a choke. As indicated in Figure 18.1, there is a parasitic capacitance C_{p3}, which is arranged parallel to L in terms of its effect. It forms a parallel resonant circuit. This causes the impedance of the component to increase until the resonant frequency is reached and then to change to a capacitive behavior. The inductance L can only be used below the resonant frequency. The parasitic capacitance depends on the number of turns and can be influenced within limits by the winding structure [3]. With MnZn ferrites, part of the capacitance is caused by the core, which is particularly noticeable with small numbers of turns [4]. The switching frequency should be a maximum of 10–20% of the resonant frequency.

The maximum current occurring during operation leads to maximum flux density. No saturation may occur under this condition, and the maximum flux density applied must therefore be below the saturation flux density. The alternating current ripple cause hysteresis losses in the core and eddy current loss in the core and winding. Taking these into account in the design requires calculation methods as described in Section 18.3. The area product can be used to select the smallest core as a starting point.

Powder cores are characterized by a high saturation flux density close to the one of iron (approx. 1.2 T), but at the same time, relatively high core losses. They also show a highly non-linear relation between inductance and current. The inductance drops with current and shows a soft characteristic. As the magnetic field is well guided by toroids inside the lower permeable core, fringing is very small, and the proximity effect in the conductors due to fringing fields is very low. It particularly does not occur due to the fine division of the air gaps by the powder grains and the shape of the core. Fringing losses do occur at the discontinuity introduced by air gaps in ferrite or nanocrystalline or amorphous tape cores. The magnetic field "escapes" from the core and passes through the winding or conductive materials in the close vicinity. This results in corresponding induced eddy current losses. The fringing radius can be significantly reduced by dividing the big air gap into several small air gaps. The inductivity of a gapped core choke is quite stable until the saturation point of the core material and then drops immediately "sharp" to an air choke.

Powder cores are characterized by a high saturation flux density (0.8–1.3 T), which is close to that of silicon steel (approx. 1.5–2.0 T). At the same time, powder cores exhibit relatively high core losses. They also show a strongly non-linear relationship between inductance and current. The inductance decreases with the current and exhibits a soft characteristic (Figure 18.2).

Due to the properties of the core material, the effective inductance changes once a certain value is exceeded. This means a "detuning" of the filter network. The effectiveness of a filter is therefore dependent on the main current. This is also clear from Figure 18.2.

Ferrites show significantly lower specific magnetization losses than powder cores. At the same time, the saturation flux density is much lower (250–500 mT), and they have a relatively high dependence on temperature.

At high power levels and frequencies multi-gapped ferrite cores are mainly used. A precise loss calculation with analytical models is difficult. In most cases, three-dimensional field simulations are used. Figure 18.3 shows examples of a

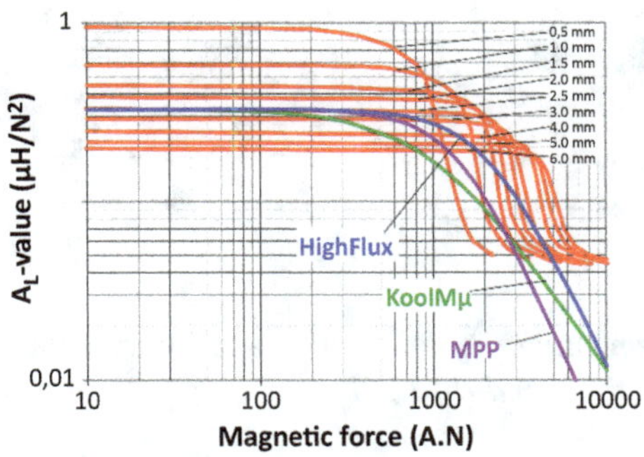

Figure 18.2 Comparison of the course of the A_L value of AMCC 160 cores (A_e = 623 mm², l_m = 292 mm) with different (total) air gaps with toroidal cores Magnetics 55337A2 (A_e = 678 mm², l_m = 324 mm, 125 µm, data: Magnetics, Honeywell) [5–7]

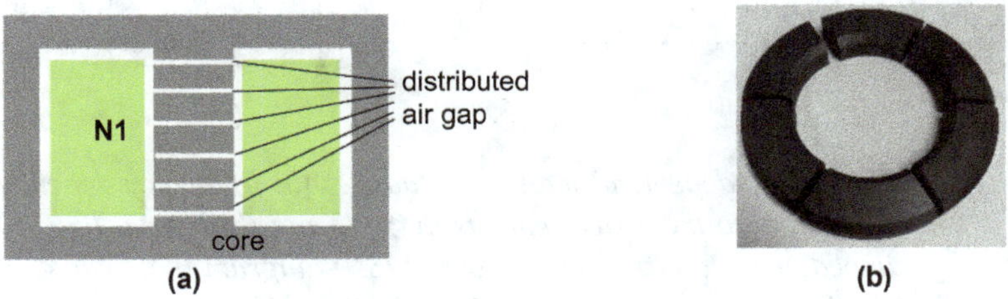

Figure 18.3 Examples of a multi-gapped core: (a) air gaps enclosed by a winding and (b) symmetrical division of an air gap in a toroidal core (photo: Zacharias)

multi-gapped core. If the ferrite multi-gapped toroidal core is wound as a toroid, the eddy current losses in the winding are kept low due to the low stray field of the winding itself. If the winding encloses the air gaps, this counteracts the "breaking out" of the magnetic field lines from the air gap.

18.3 Minimization of differential mode interference with LC and LCL filter

The effect of a simple inductive filter is enhanced if additional filter elements are used in the differential mode path (Figure 18.4). The smoothing inductance is divided in Figure 18.4(a) symmetrically between L_1 and L_2 as in Figure 18.1. If there is no symmetry here, some of the actual common mode signals also become differential mode signals. Effective suppression of these disturbance variables is then generally no longer possible.

High frequencies of differential mode signals are attenuated by C_1 during transmission as a multi-order filter is now achieved. The impedance of the

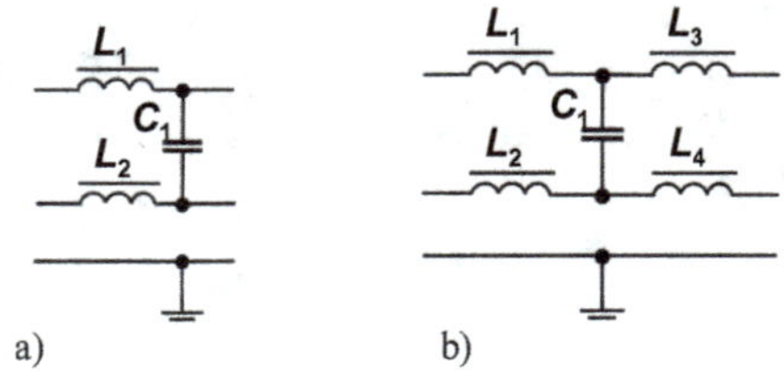

Figure 18.4 *Supplementing the differential mode filter with additional elements: (a) LC filter and (b) LCL filter*

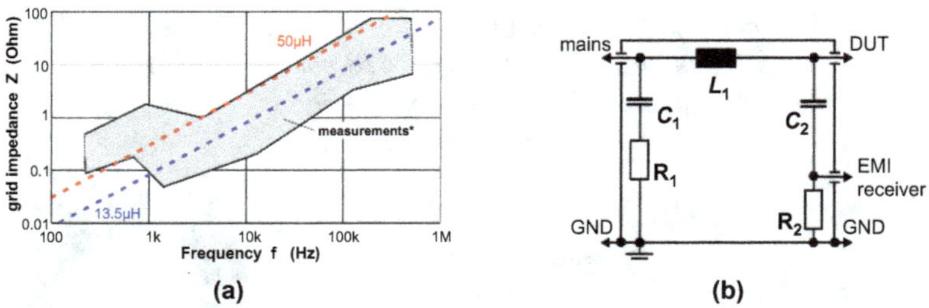

Figure 18.5 *Frequency-dependent mains impedance: (a) frequency dependence according to the measurements in * [8] and (b) typical line impedance stabilization network (LISN) diagram (part values for a particular standard (MIL-STD 461E): L_1 = 50 µH, C_1=8 µF, R_1=5Ω, C_2=250 nF, R_2 = 1 kΩ)*

inductance increases for harmonics with a higher frequency. The larger the impedance of the connected load, the larger the effect of this measure. In the case shown in Figure 18.1, the load itself contains a capacitance so that its high-frequency impedance is very low. The effect of C_1 is therefore comparatively small [19].

The effect can be increased if a further inductance is added symmetrically in the filter structure (Figure 18.4(b)). The impedance of the downstream inductors increases with the frequency. This results in a large suppression of the passed-on harmonics. However, the problem is that the current to be regulated by the converter must now be regulated in a dynamic system of a higher order. This may mean a higher measurement effort and more complex control.

For grid-connected power converters such as active front-end rectifiers or inverters for feeding solar and wind energy into the grid, the variants Figure 18.4 are very often used. The impedance of the connected grid varies depending on the location (Figure 18.5) and is assumed for measurements with a fixed inductance of 50 µH. The higher value is used because grid conformity is determined by measuring the voltage drops caused (Figure 18.4(b)). With the inductance of the grid as L_3 and L_4, Figure 18.4(a) produces a filter structure similar to Figure 18.4(b), which is effective at high frequencies and has little influence on the current controllers.

The higher the number and the smaller the distances to consumers and electricity suppliers, the lower the effective grid impedance [14].

18.4 Minimization of common mode (CM) interference

By definition, the CM component of the interference signals propagates in the signal channel between the power lines and earth. It can be assumed that the sum of all currents of the output connections of the power converter must equal zero. This also makes it possible to use an inductive coupling element with high permeability core to increase the impedance for the propagation in the wave channel between the connection lines of the inverter and GND. Figure 18.6 illustrates the principle. The currents i_{L1} and I_{L2} flow through the windings, each with the number of turns N, in such a way that the generated magnetic field strengths cancel each other out. Therefore, only the leakage inductance between the two windings is used for this signal path. This is proportional to N_2 and should be very small. For the common mode interference signal, however, the flux linkage windings are linked together with the same sign. This results in a very high impedance for signal propagation. This severely limits any CM currents. The voltage drops at the load impedance, or the connected mains are kept very low.

While the chokes for limiting differential interference are equipped with a low effective permeability, e.g., by inserting an air gap, in order to achieve sufficient storage capacity, the highest possible permeability values are advantageous for these so-called current-compensated chokes. The highest permeability values are achieved with amorphous and nanocrystalline metallic core materials. A few windings are sufficient to achieve high inductance values. At the same time, few turns result in a low winding capacitance and thus a high resonant frequency. Figure 18.7 shows the course of the impedance of a CM filter in comparison.

Due to the high permeability of the core made of amorphous material, the inductance at lower frequencies is higher than the inductance with a ferrite core. The values in the maximum impedance range are comparable. In general, amorphous and nanocrystalline materials can achieve a higher interference suppression effect with smaller sizes in the range of up to several 10 kHz. Saturation phenomena may be

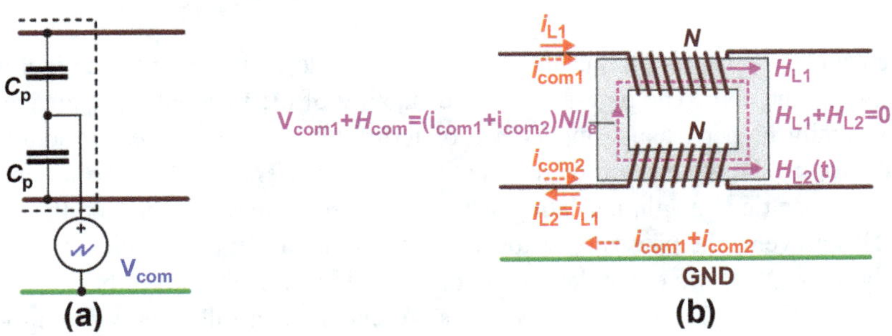

Figure 18.6 Basic approach against common mode interference: (a) model of common mode interference, design, and mode of operation of current-compensated chokes for use in common mode filters

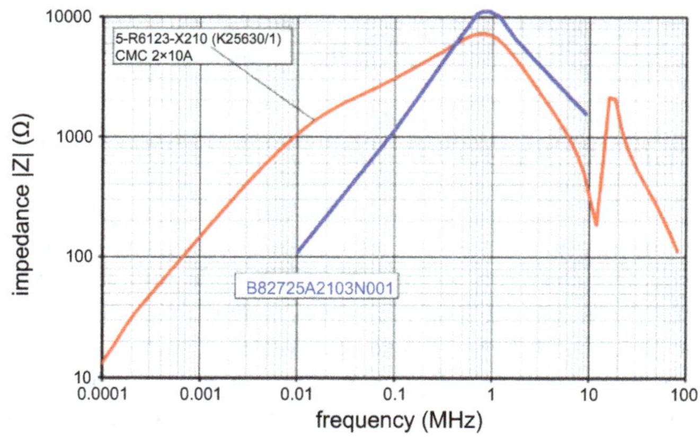

Figure 18.7 Comparison of the impedance curve of a winding of a current-compensated (2 × 10 A) ferrite choke (blue) with amorphous metallic core (T60405-R6123-X210, red) with that of a CM choke with ferrite core according to the manufacturer's datasheets (sources: TDK and VAC data sheets [13,20])

Figure 18.8 Examples of the basic design of chokes for differential and common mode chokes: (a) storage choke for current smoothing with a toroidal core, (b) current-compensated choke for 32 conductors, and (c) current-compensated choke for three conductor applications (photo: Zacharias).

problematic at higher amplitudes of the common mode current due to the high permeability of the core. The high-temperature stability of the material is advantageous. The generally higher leakage inductance of ferrite CM chokes makes an additional contribution to the attenuation of differential mode interference [3,9]. To ensure the required dielectric strength, insulating spacers are usually inserted between the windings. This prevents the phases from touching each other and adds isolation spacing.

Figure 18.8 shows examples of the basic design of such chokes based on toroidal cores. Figure 18.8(a) contains a storage choke with only one winding on a toroidal core of low permeability. Figures 18.8(b) and (c) contain images of current-compensated chokes for two or three current-carrying conductors with their characteristic structures. This means that with a largely symmetrical arrangement,

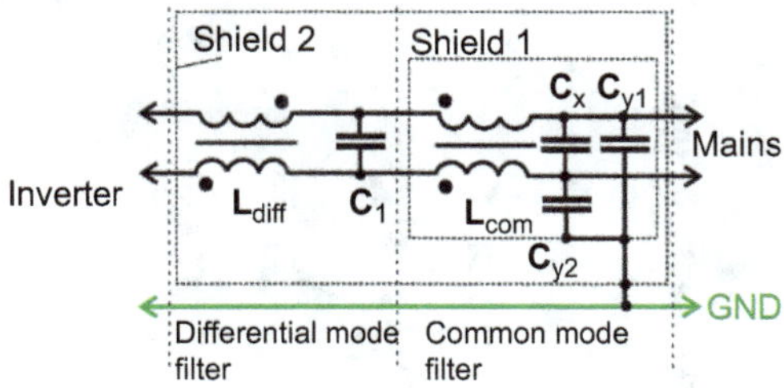

Figure 18.9 Basic structure of a filter as a countermeasure for differential mode and common mode interference

the insulation strength between the current-carrying conductors is increased with the aid of spacers.

The described interference suppression effect relates to two current-carrying output conductors of a power converter. However, the effect can be extended to several phases. To do this, several windings are applied to the core so that the resulting magnetic field strength approaches zero. This implies that the principle described can also be applied to three- and four-conductor systems.

Figure 18.9 shows the basic overall design for a filter against differential mode and common mode interference. The inductance of the DM filter is built up by two symmetrical windings on a core with two air gaps. The two air gaps are located in the windings to keep the stray field small. The low-inductance capacitor C_1 is part of the differential filter. The common mode filter is realized as a shielded module. L_{com} is the current-compensated choke as shown in this section. The capacitors C_{y1} and C_{y2} are used to divert as much as possible of the interference to GND. The remaining components cause the voltage drops at the mains or load impedance that are covered by the standard. Shielding measures are provided here to neutralize the effect of electrical and magnetic fields on the filter unit.

The shields "collect" the currents resulting from the parasitic capacitive couplings and discharge them to GND before they can leave the device. For this purpose, there should only be one electrical connection path if possible. Electrically conductive construction elements in the power converter should be connected to GND for this purpose.

Both L_{diff} and L_{com} are surrounded by stray magnetic fields. These can induce voltages in other components of the system. While very thin electrical shields (e.g., metalized paints) are sufficient to dissipate the effects of the electric field, the shielding of alternating magnetic fields requires thicker metallic shields—depending on the frequency.

18.5 Measures for shielding power electronic assemblies

The shielding of electrical assemblies against electric fields is known as Faraday cages [10–12]. A housing made of electrically conductive material forms an equipotential surface. Interaction between electric fields inside and outside the "cage" is thus largely prevented. Magnetic fields cannot be shielded in this way. Low-

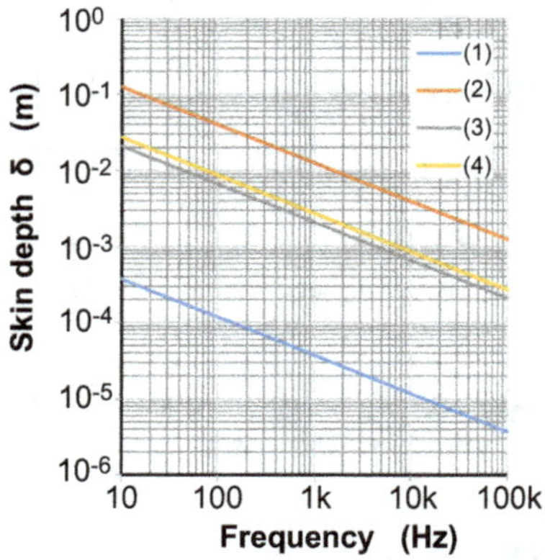

Figure 18.10 Calculated penetration depth of the magnetic field for the materials mu-metal ((1), $\mu_r = 10^5$, $\sigma = 1.8$ MS/m), austenitic stainless steel ((2) μ_r=1, 1.6 MS/m), aluminum ((3) μ_r=1, 34 MS/m), and copper 58 MS/m

frequency or DC fields are difficult to shield, as there are practically no non-conductors for magnetic fields. Similar to the electric field, low-frequency fields can be shielded by using a material that conducts the magnetic field well in these areas, so very high permeable materials like Mu-metal, permalloy.

However, the effect of induced eddy currents on electrically conductive materials is used to shield high-frequency magnetic fields. According to Lenz's rule, the eddy currents are directed in the opposite direction to the exciting current. Depending on the thickness and conductivity of the material used and the frequency, an opposing magnetic field is created. This compensates for the original field, except for a small amount. Behind the shield, the intensity of the original field can only be measured at a greatly reduced level. The larger the conductivity and permeability of the shielding material, the lower the "skin depth" δ of the magnetic field (Figure 18.10). The skin depth is calculated as follows:

$$H(x) = H(\delta) \cdot e^{-\delta x}, \text{ while } \delta = \frac{1}{\sqrt{\pi \mu_r \mu_0 \sigma f}}. \tag{18.6}$$

This means that with a shielding thickness of $d = \delta$, the magnetic field strength behind the shielding wall is

$$H(\delta) = \frac{H(\delta)}{e} \tag{18.7}$$

If, for example, a copper sheet 1 mm thick is used, the magnetic field at 10 kHz is attenuated to

$$\frac{H(\delta)}{H(0)} = \exp\left(-\frac{1 \text{ mm}}{\delta}\right) = \exp\left(-\frac{1 \text{ mm}}{661 \ \mu\text{m}}\right) = 0.22 \tag{18.8}$$

or 22%, respectively.

At 100 kHz, an attenuation of 0.825% can be expected. Pure aluminum is not very different from copper. However, it must be taken into account that many aluminum alloys do not achieve the high conductivity of pure aluminum. The penetration depth of the magnetic field than is correspondingly greater.

The permeability of mu-metal decreases at higher frequencies. This may make a combination of mu-metal with copper advisable. Combined sheets and foils are also available for this purpose. As explained, the shielding effect is based on eddy currents that act in the opposite direction to their cause. The residual magnetic field can still have a disturbing effect on the surroundings if the original fields are strong enough. Solutions can be found by means of multi-layered, nested shielding. When constructing such shielding arrangements, care must be taken to ensure that no electric circuits are formed between the shields. These could themselves become a source of interference due to voltage drops or magnetic fields.

18.6 The effect of insulating transformers

Transformers are often inserted between power converters and loads in order to isolate both areas from each other, to adapt the impedances/voltages to each other, and to prevent the transmission of interference signals. In an ideal transformer, only the common-mode signals can be influenced. The differential mode signals are transmitted via the magnetic field. If a leakage inductance is present, it makes a small contribution to reducing the ripple current.

Figure 18.10(a) shows a schematic structure of two windings wound on top of each other. The core of the transformer is earthed if it is made of metal. Figure 18.10(b) shows the effective parasitic capacitive elements. In addition to the capacitances to GND and the winding capacitances, a coupling capacitance is formed due to the insulation of the windings. The closer the coupling between the windings, the higher the coupling capacitance. This can reach the nF range in large transformers. The coupling capacitance is measured by short-circuiting the connections of windings N_1 and N_2. For calculations in simulations, it is advisable to divide the coupling capacitance as shown in Figure 18.11(c). Common-mode signals, i.e., signals to GND, can also be passed on via an insulating transformer.

If a non-short-circuited copper foil winding is inserted between N_1 and N_2 (Figure 18.12(a)), the common mode signals from the primary and secondary sides of the transformer can be diverted to GND [16–18]. Since the insulation of the windings must be ensured at the same time, a shield winding usually leads to a slightly higher leakage inductance between N_1 and N_2. A reduction in the insulation thickness results in an increase in the capacitance C_{GND} with a simultaneous reduction in the leakage inductance.

The measure described in Figure 18.12 provides the common mode signals with a path for derivation inside a power electronic structure. This means that, unlike differential mode signals, common mode signals cannot be passed on by magnetic induction.

Even close-coupled transformers have a stray field that extends into the environment. This can induce interference signals at other points. This effect can be significantly reduced by a simple measure. For this purpose, a short-circuited winding is implemented, which compensates the stray field as far as possible with its magnetic field. Figure 18.13 schematically illustrates the procedure and effect. Figure 18.13(a) schematically shows the propagation of the stray field around an

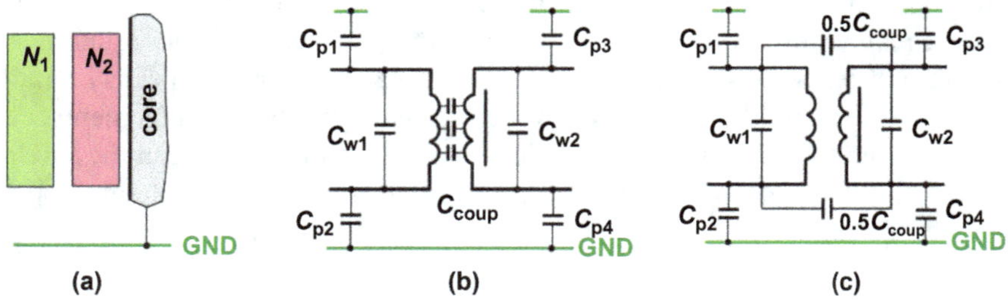

Figure 18.11 Mode of operation of parasitic capacitances in an insulating transformer: (a) schematic structure, (b) transformer with parasitic capacitances drawn in, and (c) simplified equivalent circuit diagram simulation purposes [15]

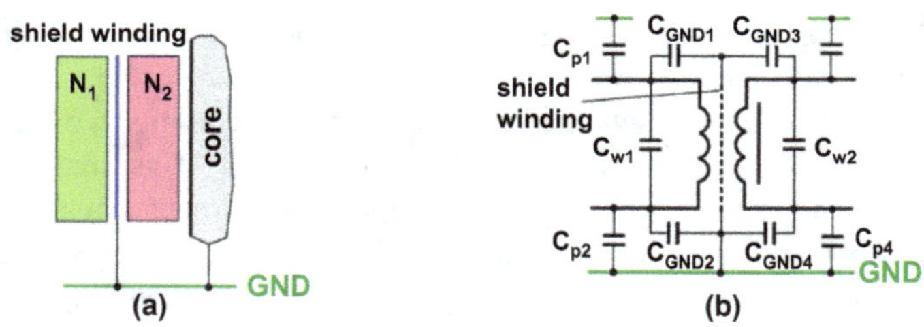

Figure 18.12 Mode of action of parasitic capacitances in an insulating transformer with shield winding: (a) schematic structure and (b) transformer with parasitic capacitances drawn in for simulation [15]

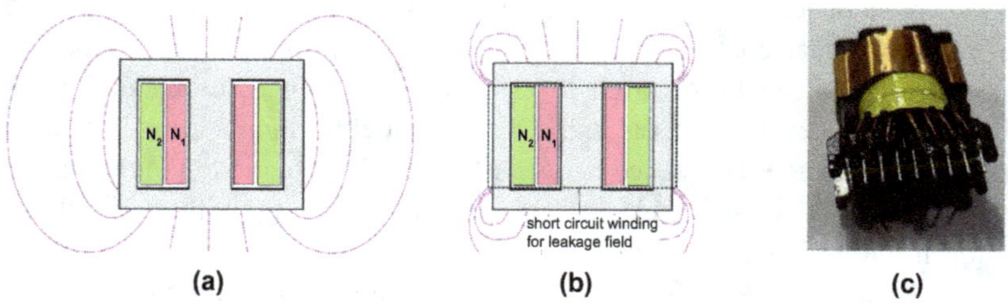

Figure 18.13 Winding for compensating the stray field of a transformer: (a) Yoke stray field of a transformer with EE section without compensation winding, (b) principle change in the stray field of a transformer when implementing a compensation winding for the stray field, and (c) design example, photo: Zacharias

EE transformer core. There is a stray field between the upper and lower yoke both when idling and under load. It spreads out in the vicinity of the transformer. If a copper foil is arranged as a short-circuit winding around the entire transformer, as in Figure 18.13(c), some of the outer field lines of the stray field are forced to close over the area within the short-circuit winding. This increases the effective magnetic resistance of the transformer environment and reduces the magnetic field strength in this area. The induction of interference voltages in the vicinity of the transformer is significantly reduced by this measure.

18.7 Integration of common mode and differential mode filter in one inductive component

The leakage inductance of a current-compensated choke also acts as a storage choke. It is usually very small and has only a minor influence on the behavior of the entire circuit. By adding an additional core between the upper and lower yoke, the stored energy can be increased to amplify the yoke leakage. Figure 18.14 schematically shows such an arrangement. In addition to the coupling core for the current-compensated chokes, a material area is inserted here, which has a lower effective permeability and is used to store magnetic energy. An EE core with an air gap could also simply serve this purpose. The windings N_1 and N_2 are connected in such a way that their magnetic field theoretically cancels each other out, as in a current-compensated choke. In practice, however, the magnetic field is forced into the scattering between the upper and lower yoke. As a result, the field lines mainly close over the middle leg of the core arrangement.

To illustrate this, Figure 18.15 shows a circuit that generates an exemplary situation of differential mode and common mode signals using the interleaved operation of two switches. The combined choke of Figure 18.15 is inserted here with its model in Figure 18.15(a). The yoke leakage leads to the formation of the inductance L_σ. The model allows the flux linkage in the individual areas to be viewed separately. The two flux linkage components Ψ_μ and Ψ_1 are the same. They therefore correspond to the flux linkage on a simple storage choke in a buck converter. This creates a DC bias, which may lead to saturation phenomena in the core.

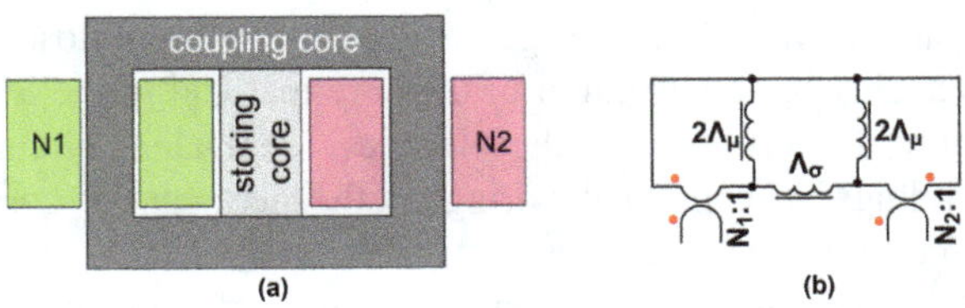

Figure 18.14 Current-compensated choke with increased leakage channel for combining common mode and differential mode choke in an EMC filter: (a) basic mechanical design and (b) electrical equivalent circuit diagram

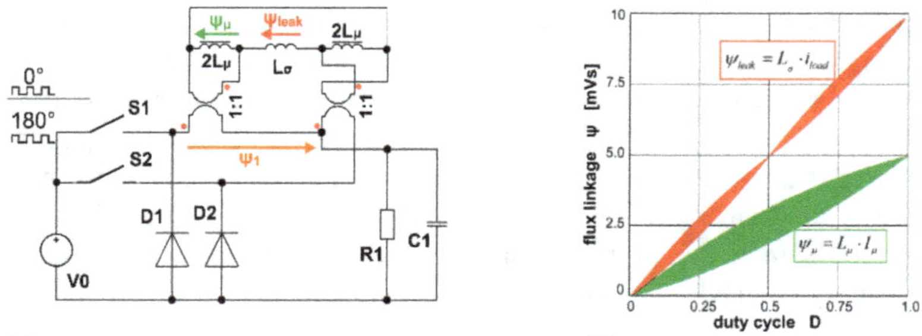

(a) (b)

Figure 18.15 Interleaved buck converter as an example for the treatment of differential mode and common mode signals by a choke according to Figure 18.14: (a) principle circuit (V_0 = 100 V, L_σ = 400 μH, L_μ = 50 mH and (b) flux linkage at the inductive components assuming ideal transformers with a transformation ratio of 1:1.

This arrangement is difficult to use as a coupling core for cut ribbon cores. The insertion of a center bar means that flux lines have a direction transversal to the tape material. This causes additional eddy current losses in the tape material of the coupling core. This is a design variant that is suitable for isotropic magnetic material, such as ferrite, in combination with powder material. Powder material ensures that hardly any fringing effects occur due to the fine distribution of the air gap in the material.

The two magnetizing inductances together with the yoke leakage inductance form a Δ circuit. The stress of the magnetizing inductances correspond to each other. The relatively high values of the flux linkage in the leakage channel must be taken into account when selecting the material for the center leg and determining the number of turns.

In the design shown in Figures 18.14 and 18.15, DC bias magnetization of the coupling core occurs depending on the duty cycle. Although the design measure is therefore effective, it can lead to saturation of the coupling cores. This becomes critical with the higher permeability of the coupling cores.

If a separate path is provided for the leakage flux, as in Figure 18.16, the problem of DC bias can be avoided. The windings receive additional storage cores for increased leakage flux. The leakage flux lines can now close within a core material. This enables good adaptation of the core material to the load parameters to be met. Figure 18.16(b) shows the electrical equivalent circuit diagram of the coupled windings. The internal wiring of the permeances corresponds to a *T*-equivalent circuit diagram of a transformer.

The mode of operation is also demonstrated here using a two-phase buck converter with interleaved operation. Figure 18.17(a) shows the schematic circuit structure with the inductor model. Figure 18.17(b) shows the calculated flux linkage under these conditions. The parameters were selected so that the resulting T-equivalent circuit is electrically identical to the Δ equivalent circuit. However, it

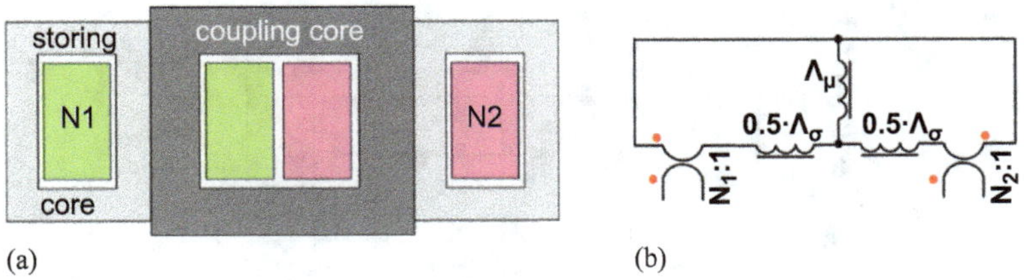

Figure 18.16 Combined choke for suppressing DC and differential mode signals with separate leakage channels: (a) basic mechanical design and (b) electrical equivalent circuit diagram

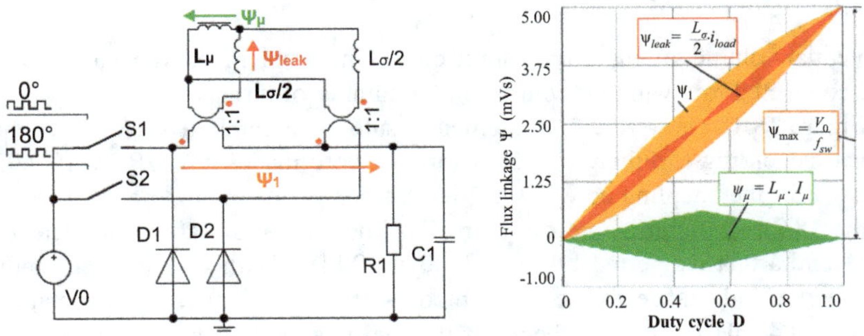

Figure 18.17 Interleaved buck converter as an example for the treatment of differential mode and common mode signals by a choke according to Figure 18.16: (a) principle circuit (V_0 = 100 V, L_σ = 400 μH, L_μ = 50 mH and (b) flux linking at the inductive components assuming ideal transformers with a transformation ratio of 1:1

is noticeable that the flux linkage of the magnetization inductance (green) does not contain a DC component. The saturation phenomena described above due to DC bias magnetization do not occur in this way by loading the circuit. Due to the closed-core shapes, this design form can also be used for cut-strip cores. The magnetic field does not transfer from one core material to another. The winding of toroidal cores is possible without further ado. The extra losses mentioned above therefore do not occur. It is also possible to combine completely different core materials.

There is no DC bias magnetization of the coupling core. The load is mainly caused by the AC magnetization of the coupling core. The storage cores must be designed for the maximum energy to be stored and the partial AC magnetization. In principle, cores with an air gap are also possible here. However, this results in fringing losses, which lead to additional heating of the component. The advantage

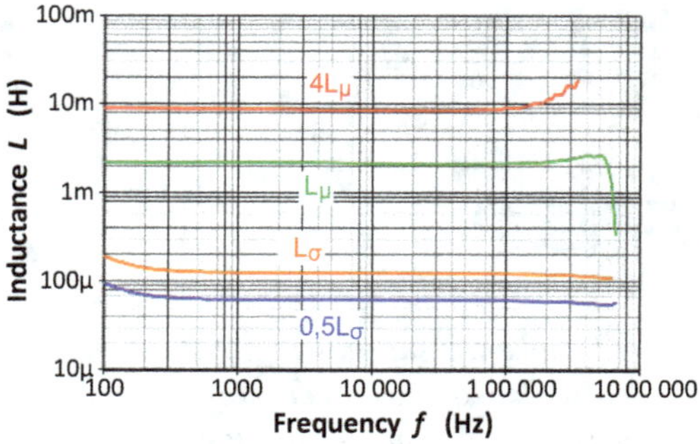

Figure 18.18 Measurement results on a combined choke designed according to Figure 18.16 consisting of a toroidal core VITROVAC 6030F (A_L = 3160 nH/T^2) and two toroidal cores made of iron powder (A_L = 100 nH/T^2), $N_1 = N_2 = 26$

of potential volume savings comes at the cost of the problem that the materials have to be connected to a winding with a certain number of turns.

Figure 18.18 shows the measurement results on a combined DM&CM choke realized according to Figure 18.16. A toroidal core made of VITROVAC 6030F was used for the coupling core [5,6]. With this core made of amorphous metal (see Section 18.2), a coupling core can be realized due to its saturation flux density of 0.82 T and a relative permeability of 3000 at 10 kHz. With 26 turns per winding, the effective inductance for common mode signals is 2.1 mH. An inductance of $4 \times L_\mu$= 8.4 mH is effective between the two phases. The additionally inserted toroidal cores made of metal powder have an A_L value of 100 nH/T^2. The storage inductance L_σ, which is effective for differential mode signals, has an inductance of 125 µH at 10 kHz. The frequency dependences shown were calculated from measured values (short-circuit and open-circuit tests). The impedance measurements were carried out with a Bode100 analyzer. At approximately 600 kHz, the choke has a parallel resonance point. Before this, the impedance increases due to parasitic capacitance. Above the resonance point, the choke has capacitive behavior. Other values can be set through combinations. The gain in installation space comes at the cost of losing one degree of freedom for dimensioning.

References

[1] T. Lazlo: Electromagnetic Compatibility in Power Electronics. IEEE Press 1995

[2] A. Charoy: Electromagnetic Compatibility of Power Converters. CAS-CERN Accelerator School: Power Converters, Baden, Switzerland, 7–14 May 2014, edited by R. Bailey, CERN-2015-003

[3] McLyman, W. T.: Transformer and Inductor Design Handbook. Vol. 121, 2004, ISBN: 0824753933

[4] FERROXCUBE Soft Ferrites and Accessories. Catalogue 2013

[5] Sekels GmbH: Magnetic Materials, Components and Systems, Amorphous C Cores, 2013

[6] Magnetics Inc.: Powder Cores. Catalogue 2017

[7] Honeywell Inc.: Metglas Powerlite – High Frequency C Cores. Technical Bulletin, 2001

[8] Höckel, M., and Meyer, J.: Measurement of Network Impedances (German). bulletin.ch 5 / 2019, pages 56–60

[9] Schliewe, J., Paulwitz, C., and Weber, S.: A Just Comparison of Ferrite and Nanocristalline Common Mode Chokes, PCIM Europe 2016 – 10–12 May 2016, Nuremberg, Germany, pp. 156–163

[10] Geetha, S., Satheesh Kumar K. K., Chepuri R. K. Rao, Vijayan, M., and Trivedi, D. C.: EMI Shielding: Methods and Materials—A Review. DOI 10. 1002/app.29812 (published online) 2009 *Journal of Applied Polymer Science.*

[11] X. C. Tong: *Advanced Materials and Design for Electromagnetic Interference Shielding.*

[12] Wallmeier, P.: Effects in Magnetic Components for Switched Mode Applications in the MHz Range, *IEEE APEC* 2004

[13] Vacuumschmelze: Common Mode Choke, specification item no. T60405-R6123-X210, March 2018.

[14] Smolenski, R.: *Conducted Electromagnetic Interference (EMI) in Smart Grids. eBook – Power Systems.* Springer, 2012, ISBN/EAN: 9781447129608

[15] Kovacevic, I., Müsing, A., Friedli, T., and Kolar, J. W.: Electromagnetic Modeling of EMI Input Filters. *International Conference of Integrated Power Electronics Systems (CIPS 2012)*, Nuremberg, Germany, March 6-8, 2012

[16] Wang, S.; Lee F. C.; and Odendaal W. G.: Improvement of EMI Filter Performance with Parasitic Coupling Cancellation. *IEEE Transactions on Power Electronics*, 2005, 20(5): 1221–1228

[17] Wang, S.; and F. C. Lee, F. C.; and van Wyk, J. D.: Design of Inductor Winding Capacitance Cancellation for EMI Suppression. *IEEE Transactions on Power Electronics*, 2006, (21), 1825–1832,

[18] Wang, S., van Wyk, J. D., and Lee, F. C.: "Effects of Interactions Between Filter Parasitics and Power Interconnects on EMI Filter Performance," in *IEEE Transactions on Industrial Electronics*, vol. 54, no. 6, pp. 3344–3352, Dec. 2007, doi: 10.1109/TIE.2007.906126.

[19] Hartmann, M.; Ertl, H.; and Kolar, J. W.: "EMI Filter Design for High Switching Frequency Three-Phase/Level PWM Rectifier Systems," *2010 Twenty-Fifth Annual IEEE Applied Power Electronics Conference and Exposition (APEC)*, USA, 2010, pp. 986–993, doi: 10.1109/APEC.2010. 5433382.

[20] Data sheet TDK/EPCOS: Power Line Chokes Current-compensated ring core double chokes, Series/Type: B82725A, January 2018.

Chapter 19

Active ripple cancelation

Peter Zacharias[1], Christian Noeding[1], Xiao Yu[1] and Alejandro Aganza-Torres[1]

19.1 The problem of ripple current

The majority of all power electronic applications are based on the principle of pulse-width modulation (PWM) [1]. A second power converter can be used for active ripple current compensation [2,3]. Since the following explanations refer to it, this principle is explained in Figure 19.1 using the example of a buck converter. A voltage (in the example $v_d(t)$) is applied by a converter INV1 to a series circuit of inductor L_1 and load. This increases the current $i_{L1}(t)$ in the inductor L_1 inserted to limit the current rise. When the switch changes the voltage, the current continues to be driven by the inductor L_1. The associated energy dissipation from L_1 decreases the current. This gives a triangular current waveform (Figure 19.1(b)), which oscillates around its mean value. A constant/stationary mean value of the current requires that the voltage average value at the choke L_1 is zero. By changing the duty cycle

$$D = \frac{t_{on}}{t_{on} + t_{off}} = \frac{t_{on}}{T} \tag{19.1}$$

the average voltage at the load can be changed. In order to obtain the smoothest possible curve, a filter is usually inserted between the choke and the load. In the example in Figure 19.1(a), this task is performed by capacitor C_1. Here, it bypasses the unwanted AC component of the current supplied by L_1 to the load or serves as a buffer for the energy required by the load.

According to Figure 19.1(a), the load R_1 is supplied at constant duty cycle D by an alternating current rms component, which results from the approximation equation

$$I_{AC.R1}(f_{sw}) \approx \frac{I_{AC.L1}(f_{sw})}{\sqrt{1 + (2\pi f_{sw} R_1 C_1)^2}} \approx \frac{U_0(D - D^2)}{2\sqrt{3} f_{sw} L_1 \sqrt{1 + (2\pi f_{sw} R_1 C_1)^2}} \tag{19.2}$$

If L_1 and/or C_1 can be made sufficiently large, the alternating component of the load current becomes very small. At the same time, the volume of L_1 and C_1

[1]University of Kassel, Department of Power Electronics, Germany

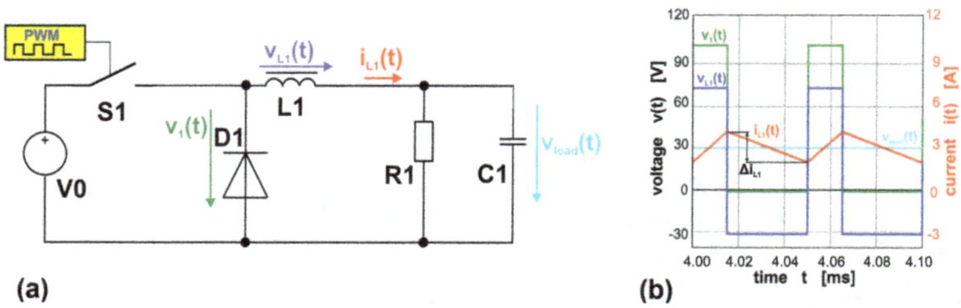

Figure 19.1 Origin of the current ripple by pulse width modulation: (a) Example circuit of a buck converter and (b) current and voltage waveforms for V_0 = 100 V, R_1 = 10 Ohm, L_1 = 0.5 mH, f_{sw} = 20 kHz, D = 0.3

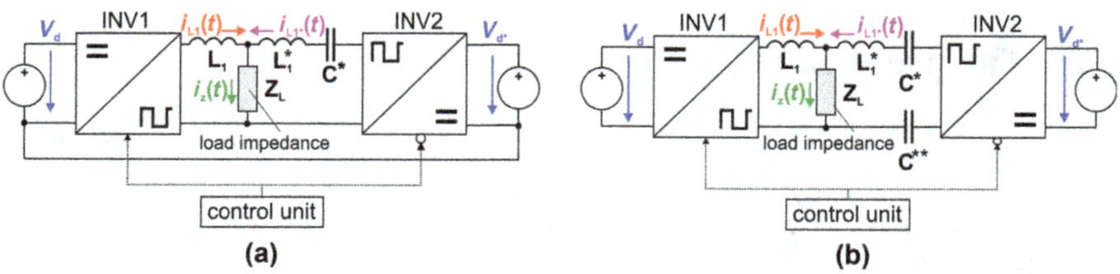

Figure 19.2 Principle of ripple compensation by impressing opposite-phase harmonics: (a) with galvanic coupling and (b) with purely capacitive coupling [18,23]

becomes very large. The system can also only react very slowly to changing duty cycles. In principle, the problem is alleviated if the switching frequency f_{sw} is increased. But even here there are limits. In the following, approaches are shown as to how the ripple current and thus the required storage elements can be reduced.

19.2 Active ripple cancellation by addition of an phase-inverted ripple current

A sufficiently low high-frequency impedance of the load is assumed. This means that the load impedance Z_L in the high-frequency range either exhibits capacitive behavior or can be represented by a voltage source.

The harmonics can be canceled, if a second power converter is constructed to a first PWM-controlled power converter in such a way that it produces exact current rises with the same Δi and inverse rise at the same duty cycle [6]. The average or low-frequency value of the current has to be eliminated in this case by an appropriate capacitor. In this case, the harmonic AC current components of power converter INV1 and INV2 cancel each other out (Figure 19.2(a)). This basic requirement results in the voltage ratio ($m_{INV1}=V_{load}/V_d$, with V_d = supply voltage of INV1) of converter INV2. The DC component or the low-frequency component with the actual useful power is supplied in this way by power converter INV1.

INV2 only supplies the power to compensate the high-frequency harmonics for the reactive power and must therefore be designed for a much lower power. This possibility has already been described in [2,4,5]. How well the suppression/elimination of the ripple current works depends on the ratio of the switching frequency to the highest frequency content of the active power to be transmitted. For grid-connected power converters in active front ends of industrial plants or inverters in photovoltaic systems or converters in wind turbines, there are many possible applications. At mains frequencies of 50/60 Hz, the switching frequencies are far above this.

Since the compensation current is a pure alternating current, there is also the possibility of capacitive (Figure 19.2(b) coupling).

The effect of the presented method shall be demonstrated by a simple example. Figure 19.3 shows a single-phase inverter in a half-bridge circuit with a split voltage source. The partial capacitive load (R_z, C_z) is located at the DC-link mid point of the bridge. The current through L_1 is modulated as a sinusoidal current. The ripple current would produce a high switching frequency component in the load. However, subtracting the current i_L* leaves almost only the line-frequency power for the load. The current i_L* is a pure reactive current. The generated high-frequent AC voltage of both half-bridges is equal. However, the low-frequency part of the voltage is carried by C* in the compensating converter. The rated reactive power of the two chokes of the same value $L_1 = L$* are therefore very different

$$Q_1 \approx 2\pi 50 \text{ Hz} \cdot L_1 \cdot I_{L1}^2 + 2\pi f_{sw} \cdot L_1 \cdot I_{ripple}^2 \tag{19.3}$$

with $I_{L1.sat} < \hat{I}_{L1} + \max\left(\hat{I}_{ripple}\right)$ and

$$Q^* \approx 2\pi f_{sw} \cdot L^* \cdot I_{ripple}^2 \tag{19.4}$$

with $I_{L*.sat} < \max\left(\hat{I}_{ripple}\right)$

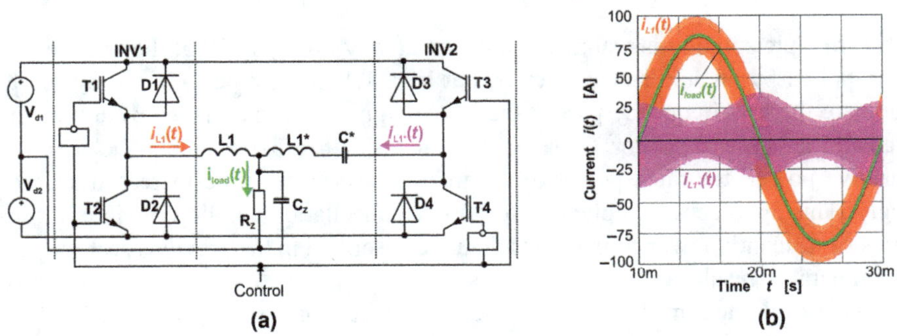

Figure 19.3 Example circuit to explain the effect of the presented compensation method with auxiliary converter: (a) circuit of a single-phase inverter in a half-bridge circuit and (b) diagram of the calculated currents.

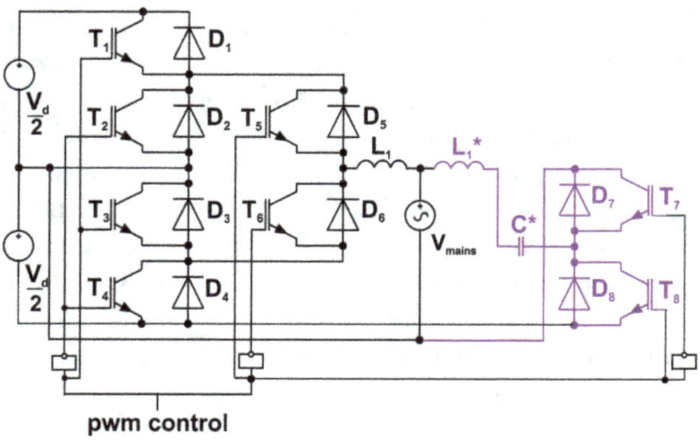

Figure 19.4 Three-level circuit with auxiliary converter for compensation of the current ripple

L_1 will be designed so that the saturation limit is not exceeded even at maximum useful current. L^* must essentially supply the compensation current for L_1 at the corresponding switching frequency. As a result, both the rated power for L^*, but also for the INV2 converter, is significantly smaller than for INV1. If necessary, it is also possible to "upgrade" an existing design if the required connection points of INV1 can be reached.

In principle, the method can also be extended to multilevel circuits, as Figure 19.4 shows. The high-frequency ripple current is generated by the switch pair T_5/T_6 and derived by the switch pair T_7/T_8 via L_1* and C*. Switch pairs T_1/T_3 and T_2/T_4 are used for low-frequency switching of the potentials for the positive and negative DC terminals of T_1/T_2.

19.3 Use of inductive components for current ripple cancelation

The principle solution is realized in a different way in Figure 19.5: the inversion is not applied by an auxiliary converter, but by a 1:1 transformer. C_1, C_2 is used to separate the low-frequency part of the PWM signal from the magnetization of the transformer. The capacitor C_3 has a similar task. In order to apply the required reactive power, the rated power of the main converter must be increased accordingly. This approach to implement the ripple cancellation solution obviously contains significantly fewer components than the described in the previous section with a comparable result.

Due to the assumed low impedance of the network or the load, the left side with L_1 and the right side with L_2 are practically decoupled completely. The load of L_1 thus remains unaffected by the function of L_2. However, one can achieve much smaller ripple values of the load current by combining L_1 and L_2. Figure 19.6 helps to visualize the task. Here, a generalized model calculation for $A_e = A_w$ has been realized. The weight minimum of L_1 corresponds to high THDi values. In the

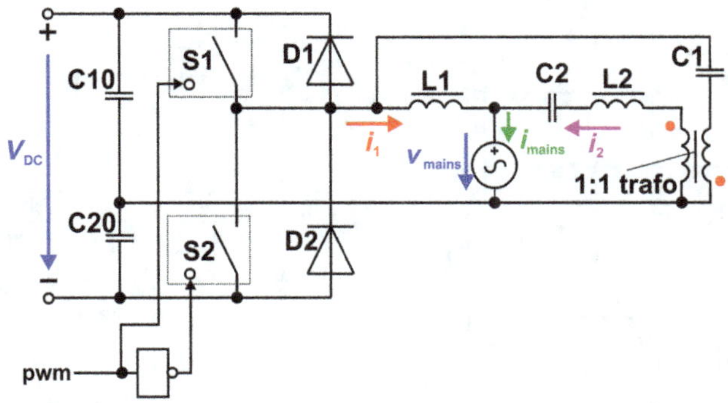

Figure 19.5 Transformer for generating an inverted voltage + capacitors for separating the low-frequency voltage and current components [18]

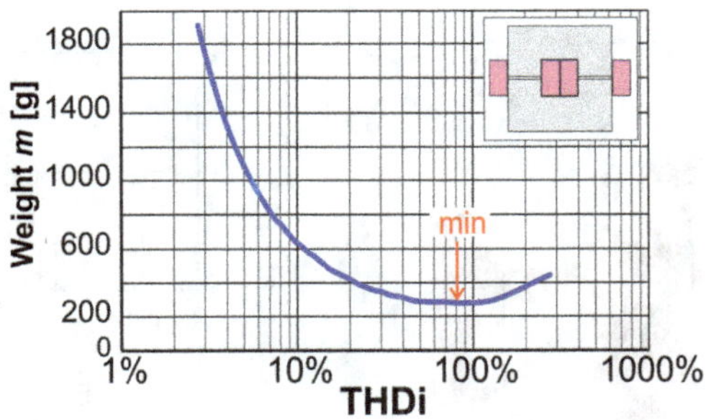

Figure 19.6 Calculated effect of the required THDi at $A_e = A_w$ and S = 1 kVA, U_{mains} = 230 V/f_{mains} = 50 Hz on the weight of the choke L_1 for a single-stage filter [18]

model case used, a THDi = 2% is required, which would correspond to a calculated weight of about 1.8 kg. If it were now possible to select L_1 close to the minimum for the weight and to combine it with a further inductive component L_2, which would have approximately the same size, a weight saving could be achieved for a low THDi.

Under the conditions $S = 1$ kVA, $V_{mains} = 230$ V, $f_{sw} = 32$ kHz, $f_{mains} = 50$ Hz, and THDi = 2%, the dependencies of maximum flux linkage and RMS current of L_1 shown in Figure 19.7(a) are obtained for the effect of the inductance on the size. The product of maximum flux linkage and maximum current is a measure of the energy to be stored in L_1 (Figure 19.7(b)). If one chooses a value belonging to 110% of the minimum, one obtains approx. 500 μH = $L_1 = L_2$ as a target value for the design of corresponding components.

When this tuning is done, at maximum current (1 kW at 230 V_{rms}), the curves for currents $i_{L1}(t)$, $i_{L2}(t)$, and i_{mains} shown in Figure 19.8 are obtained. The ripple component of i_{L1} is almost completely compensated by the current i_{comp}. The

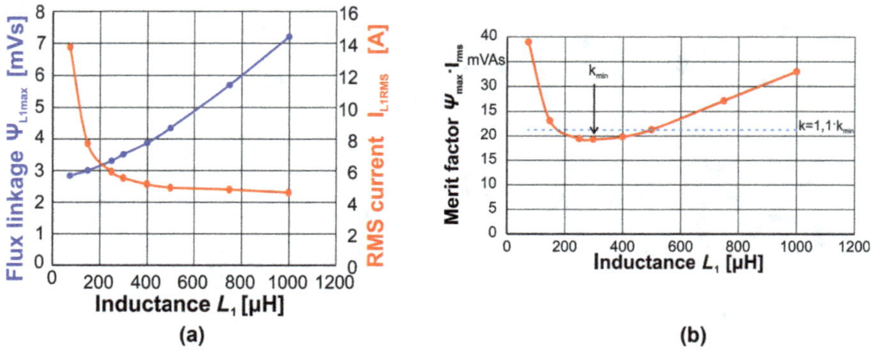

Figure 19.7 Influence of the inductance L_1 on the size of this component: (a) flux linkage and RMS current as a function of the size of L_1 and (b) merit of order factor for the size as a function of L_1 [18]

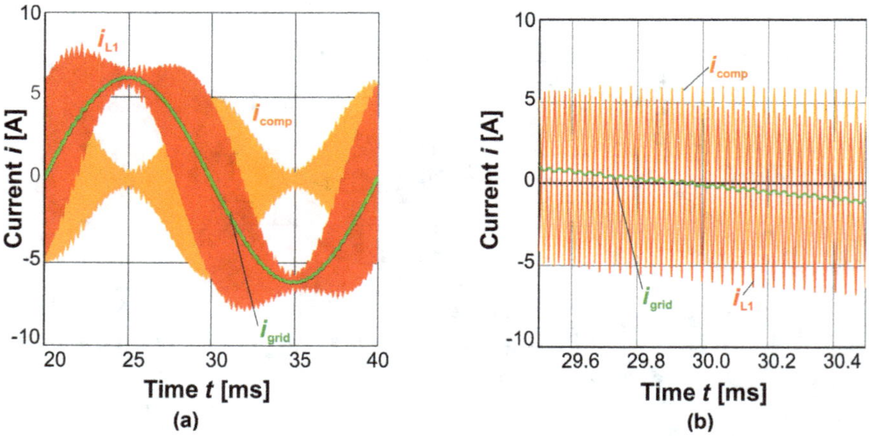

Figure 19.8 Reactor current $i_{L1}(t)$, compensation current $i_{L2}(t)$, and load current i_{grid}: (a) full mains period 20 ms and (b) zoom near zero crossing at 30 ms

remaining ripple is due to device tolerances. Deviations from the ideal are also due to the reactive current requirement of the transformer.

Thus, the ripple current cancelation method works for both solution approaches. The additional effort is still problematic. The statements made above show that there is a minimum size L_1. Different design approaches with varying degrees of integration for the selected example are shown below.

(A) Separate design of the inductive components.

First, a realization with L_1, L_2, and transformer as separate components was carried out. The result of this design is shown in Figure 19.9. The cores EE42 and PQ40 were used. The total weight for these inductive components is in this case 495 g. This value should not be taken as an absolute minimum of the weight, since only cores were used, which were available at during the experimental evaluation. However, it is clear that for each component separately the maximum permissible flux density and the permissible current density determined from the simulation must be observed as limit values. This results in a relatively high space requirement.

With the inductive components shown in Figure 19.9 and additional capacitors $C_1 = C_2 = 4.7\ \mu$F, the results in Figure 19.10 have been achieved. One can see the drastic reduction of the current ripple in both cases shown:

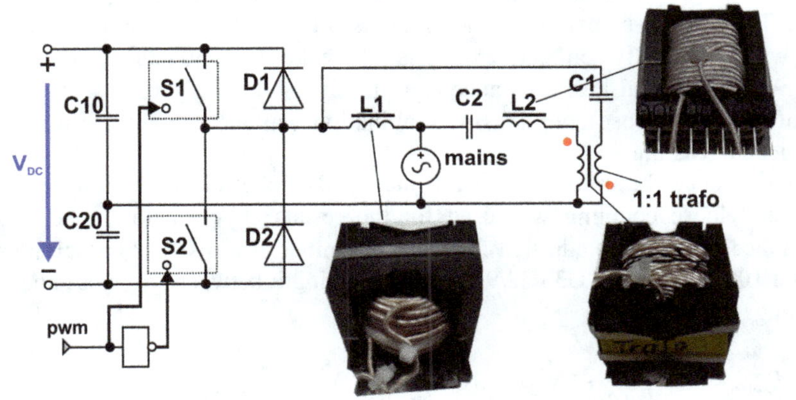

Figure 19.9 Realization of the magnetic components for a filter with ripple cancellation according to Figure 19.5: L_1 (PQ40; $R_{DC.L1}$ = 0.12 Ohm), L_2 (E42/21/20; 0.17 Ohm), transformer (PQ40; $R_{DC.N1} = R_{DC.N2}$ = 0.13 Ohm) based on [18]

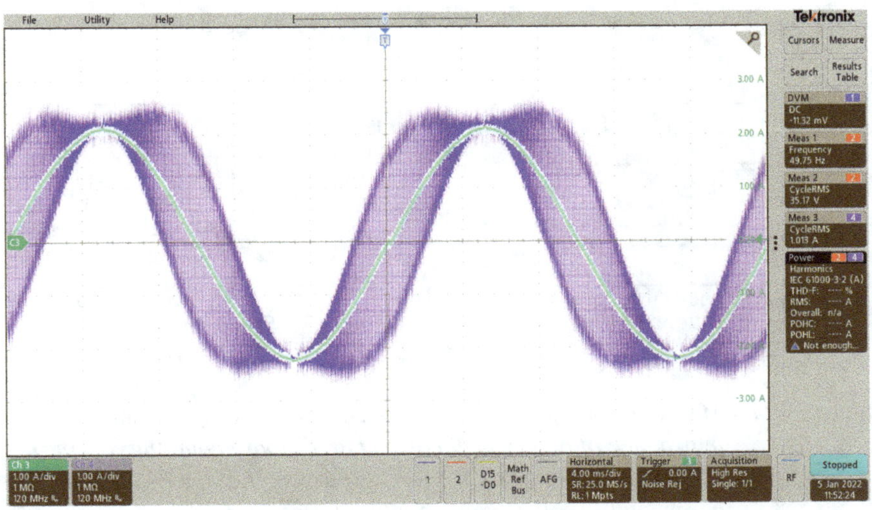

Figure 19.10 Current $i_{L1}(t)$ (dark blue) and load current (light blue) with the filter implemented according to Figure 19.9 with modulation index m = 0.8

(B) Combining L_2 and transformer into one component.

L_2 can be designed as a separate element or as a component of a transformer. The latter then leads to a magnetic integration of L_2 into the transformer. One can establish a fixed or a loose coupling. The measured AL values for the dispersion between 2 windings of a transformer with a U33/22/9 are shown in Figure 19.11. The tight coupling of the windings is shown in blue (AL′ = 4.3 nH/T^2 @ 32 kHz). The simple loose coupling (AL″ = 10^2 nH/T^2 @ 32 kHz) can be modified by "blocking" the path of the magnetic field lines between upper and lower yoke by electrically conductive materials. Counter fields due to eddy currents then constrict the magnetic leakage flux and increase the magnetic leakage field resistance. The price to be paid is slightly increased losses. By shaping the yoke stray field with electrically conductive material, the yoke stray inductance is somewhat lower, and the serial loss resistance is a little bit higher. In the present case, the use of tight and loose coupling in combination was more favorable and easily technically feasible.

To produce the transformer with integrated choke, only the combination of close and loose coupling was used for the results shown below. The leakage inductance L_{leak} can be adjusted for a given number of turns NL2 (which must be at least 100 for the core U33/22/9 used) by splitting it between N_{L2} close and loose coupling:

$$L_{\text{leak}} \approx 4,3 \frac{nH}{Turns^2} \cdot N'_{L2}2 + 103 \frac{nH}{Turns^2} \cdot N''_{L2}2 \tag{19.5}$$

The leakage conductance values were determined experimentally from test measurements. Using 100 turns per winding results in a leakage inductance of 1 mH for the loose coupling. 500 μH is achieved at $N' = 30$ and $N'' = 70$. The schematic design of the compensated filter is then shown in Figure 19.12(b). The notations and transformation rules in [7,8] were used.

Then, for $L_1 = L_2 = 0.5$ mH, the following curves are obtained:

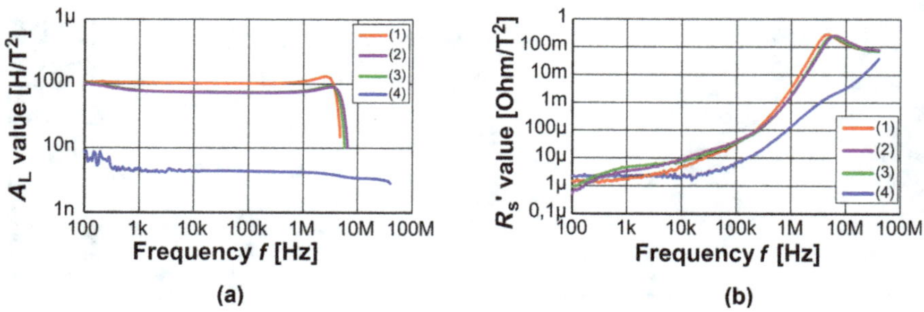

Figure 19.11 Measured AL value and serial resistance R_s of the leakage impedance of a U33/22/9 core: (1) red: loose coupling, (2) purple: 1.5 mm Cu sheet as yoke cover, (3) green: 1.5 mm Al sheet as yoke cover, and (4) blue: tight coupling based on [18].

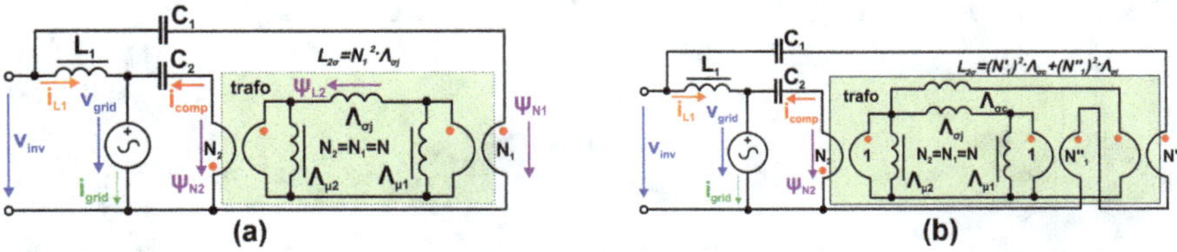

Figure 19.12 Unification of the functions L_2 and transformer: (a) electrical equivalent circuit with only loose coupling and (b) with combined loose and close coupling of N_1 and N_2 based on [18].

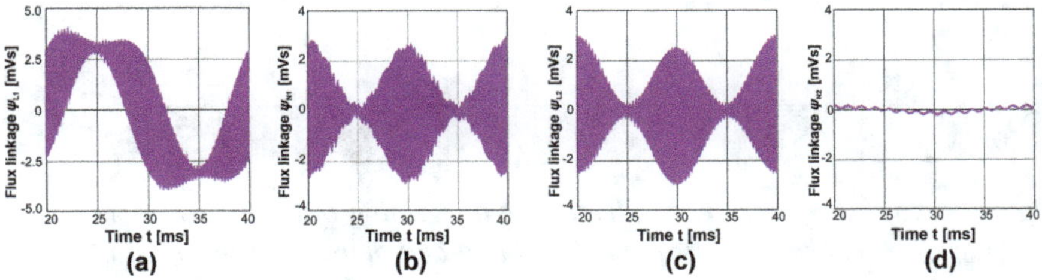

Figure 19.13 Calculated flux linkage at the three windings of the arrangement and in the leakage inductance of the transformer: (a) at the winding of L_1, (b) at N_1 of the transformer, (c) at the leakage inductance L_2, and (d) at the winding N_2 [18]

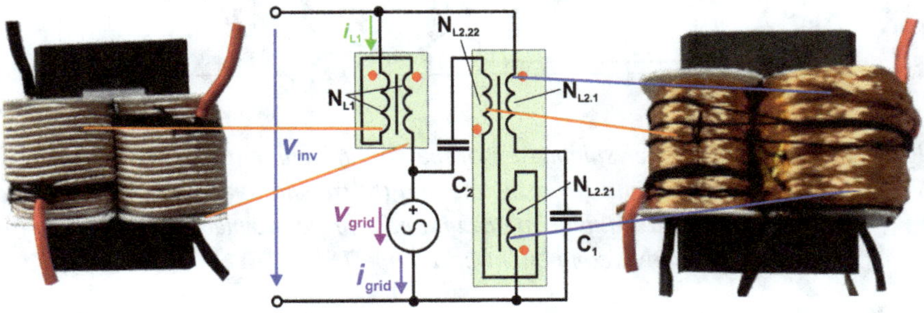

Figure 19.14 Example of the structure of the inductive components when combining choke L_2 and transformer (U33/22/9) choke (510 μH, 0.14 Ohm) L_1, and transformer ($R_{DC.L2.1}$ = 0.17 Ohm; $R_{DC.L2.2}$ = 0.31 Ohm) with windings for fixed and for loose coupling (c) electric equivalent circuit according to [23]

The last picture for N_2 is interesting because it means that only about half of the core contributes to the core losses, and the output of N_2 is actually in short-circuit operation. A core with a larger winding width and thus lower yoke leakage permeance would also allow a design without distributed windings. Here, the core U33/22/9 has been used.

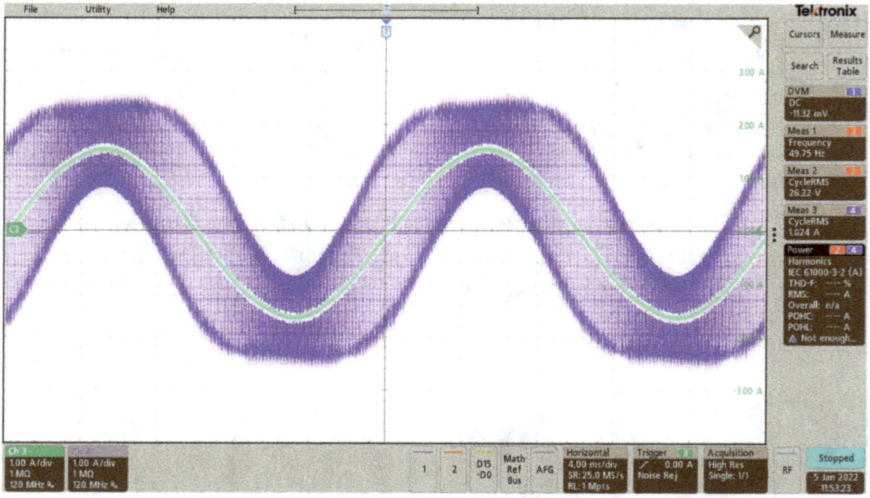

Figure 19.15 Measurement results for the integration of L_2 and transformer: choke current $i_{L1}(t)$ (dark blue) and output current (light blue)

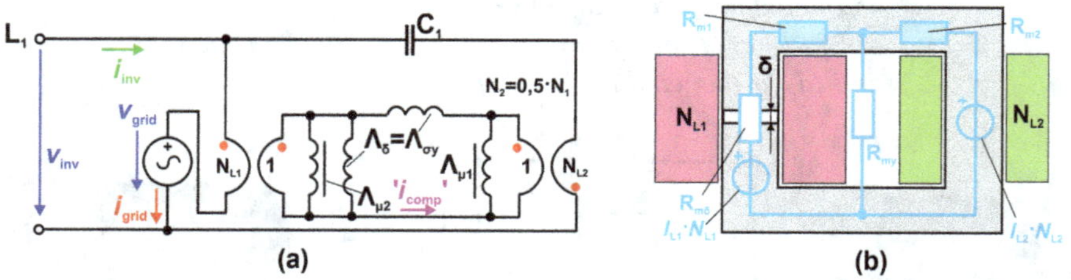

Figure 19.16 Transformer summation of the "main" current and the inverted ripple current: (a) principle and (b) translation of the principle into a magnetic equivalent circuit and a principle mechanical design for the inductive component based on [18]

(C) Magnetic integration of the three components L_1, L_2, and transformer

When one reaches at variant B, the question arises whether it is possible to move the summation of inverter current i_{inv} and compensation current i_{comp} into the transformer. Evaluating the previously observed and measured conditions, we reach the arrangement shown in Figure 19.16(a). The winding N_{L1} forms the inductor L_1 mainly with the air gap δ. To achieve the desired compensation effect, L_1 and L_2 (Figure 19.16(b)) must be of equal size. Because the summation of the currents has now been moved to the magnetic part of the arrangement, this means that $\Lambda_\delta = \Lambda_{\sigma y}$ must be. $\Lambda_{\sigma y}$ is essentially fixed by the core used and its proportions. There are possibilities of limited "tuning" of this value (see below). Taking this

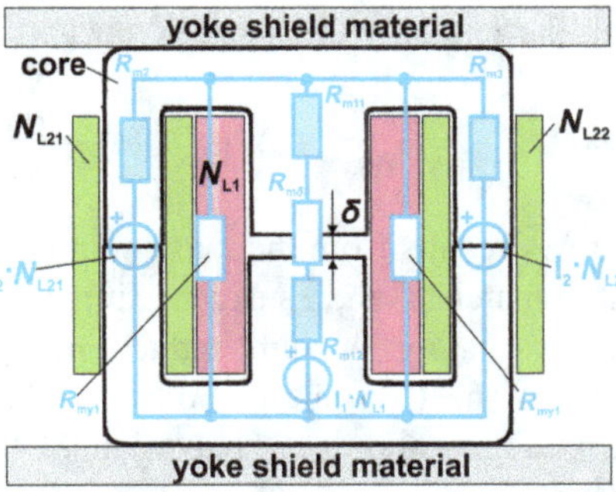

Figure 19.17 *Mechanical design and magnetic equivalent circuit with optional shielding materials for leakage field tuning ($N_{21} = N_{22}$) as an example of a fully integrated ripple filter based on [18]*

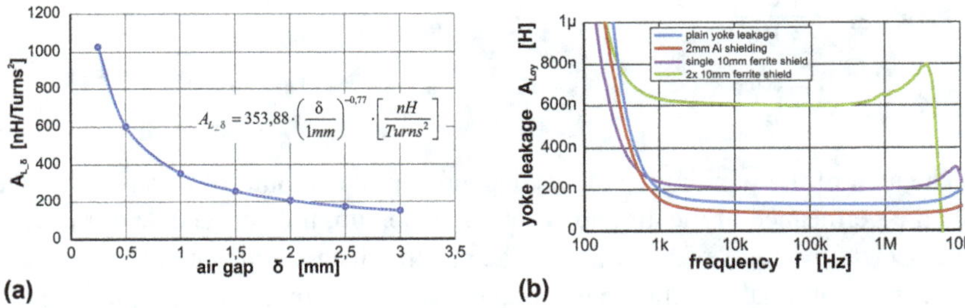

Figure 19.18 *Characteristic properties of the E42/21/29 core: (a) AL value of the core as a function of the air gap and (b) measured change in the yoke leakage conductance $A_{L\sigma y}$ using different shielding materials [18]*

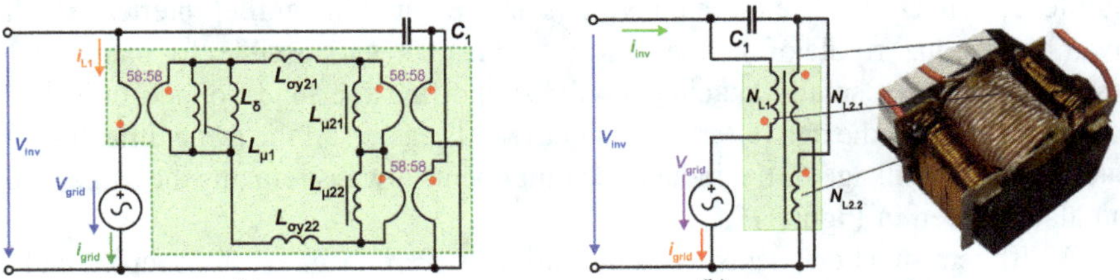

(a) (b)

Figure 19.19 *Realization of a completely magnetically integrated filter: (a) Realization example of the inductive component (core E42/21/20, $C_1 = 4.7\ \mu F$, $R_{N1.DC} = 0.125$ Ohm, $R_{N21.DC} = 0.162$ Ohm; $R_{N22.DC} = 0.166$ Ohm) and (b) circuit diagram with integrated inductive component [18]* relationship into account, there is only one suitable air gap δ for each core

$$L_1 = L_2 = N_{L1}^2 \cdot \Lambda_\delta \approx N_{L1}^2 \cdot \frac{\mu_0 \cdot A_e}{\delta} = N_{L1}^2 \cdot \Lambda_{\sigma y}\ \delta = \frac{\mu_0 \cdot A_e}{\Lambda_{\sigma y}} \tag{19.6}$$

The number of turns N_{L1} results from the maximum current and the maximum flux linkage. If you want to match L_1 and L_2 exactly, it makes sense to select the air gap so that it applies approximately. A slight reduction is possible by inserting a matched shield of aluminum sheet on one or both yoke sections. In any case, it should be noted that the air gap and core shape cannot be selected completely independently of each other.

Central conditions for the function of the fully integrated variant are approximately equal permeances of air gap and yoke leakage region and $N_1 = 2N_2$. At the same time, the saturation induction (or the maximum induction given by the losses) of the core material must not be exceeded. This results in the following conditions for the core cross sections

$$\begin{aligned} A_{Fe1} &\geq \frac{\Psi_{\max .N1}}{N_1 \cdot B_{\max}} \\ A_{Fe2} &\geq \frac{\Psi_{\max .N2}}{N_2 \cdot B_{\max}} \end{aligned} \tag{19.7}$$

The area of the winding windows depends on the amount of dissipated power. While the permeance of the air gap is set via its length, the permeance of the yokes can be set via the ratio of the yoke surface to the distance between the yokes. To achieve truly minimal weights, special cores have to be made for this purpose. The solution is therefore more of an option for series production.

19.4 Interleaved operation of PWM converter branches

When PWM converters are operated interleaved, individual harmonics of the load current are partially canceled at the load. A similar effect can be achieved with multilevel converters. The more converters are operated in parallel interleaved, the lower the residual ripple of the load current. This makes it possible, for example, to design particularly small capacitances for supporting the load voltage or for filtering. This makes the converter stage significantly faster. At the same time, the AC load of the flux linkage of the inductive components always remains the same. This can also be seen in Figure 19.20.

A different situation occurs when the two phases are magnetically coupled with a high coupling factor (Figure 19.21) [14–16]. The transformer with a high coupling factor acts as a common mode choke and takes over a large part of the AC flux linkage. In this way, there is a strong (decoupled) division of DC and AC flux linkage between the storage chokes and the coupling elements. The diagrams in Figure 19.21(b) are obtained by observing the flux linkage at the storage chokes and the coupling transformers while changing the duty cycle D in the range $D = 0...1$. It becomes clear that

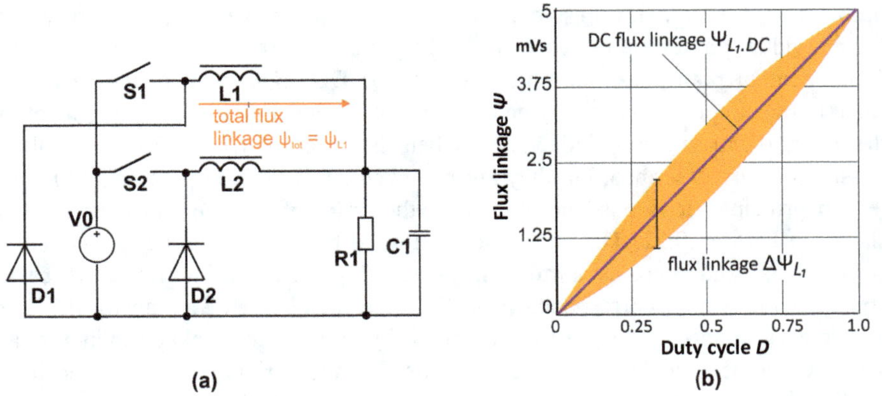

Figure 19.20 Example of a two-phase interleaved operated buck converter: (a) circuit and (b) simulated DC and AC flux linkage at the inductors and ripple current in the load at V_{d1} = 100 V, L_x = 0.1 mH, R_{load} = 10 Ohm, f_{sw} = 20 kHz

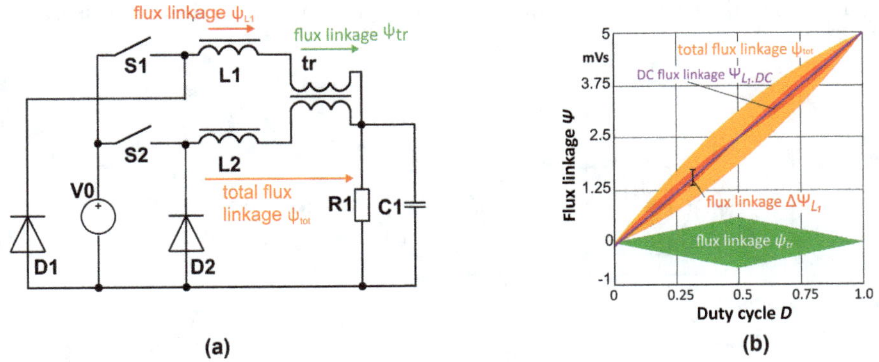

Figure 19.21 Example of a two-phase interleaved operated buck converter: (a) circuit and (b) distribution of the flux linkages to the inductive components

already with a coupling of only two phases, a strong reduction of the AC flux linkage in the storage reactors occurs. In this case, it is expedient to always couple only two phases and to balance the average values of the phase currents by means of a control algorithm, so that the situation in the 1:1 transformer Tr1 is the same as for a current-compensated reactor. In the case of high permeability core material of the transformers, the problem of saturation can be alleviated by inserting a small air gap. For the ratio of magnetizing inductance L_m of the transformer to the storage inductance L_x, $L_m > 4L_x$ should be set for the example Figure 19.20.

The number of coupled phases can be further increased. In the process, the fraction of AC flux linkage in the storage reactors is further reduced. To avoid the problem of saturation, it is best to always couple only two phases in such a way that the individual phases are driven with a voltage that is 180° out of phase. If the DC component of both phase currents is equal, the DC component of the magnetic force in the coupling elements cancels itself out. As a result, the DC component of the flux in the magnetizing inductance is then also zero. In [9,14,17], other coupling possibilities are described, but they are inconvenient for the purpose sought here.

In principle, it is possible to reverse the order of storage and coupling, as shown in Figure 19.22. This makes it possible to reduce the number of inductive components, and thus, as a rule, the space required. Starting from the basic approach (a), which corresponds to a T equivalent circuit of a transformer, the equivalent variants (b) and (c) are obtained by shifting the leakage inductors as separate components Ls to the outside. From (c), the variant (d) is obtained by a $\Delta - Y$ reshaping. The stored magnetic energy is always the same for the same external conditions.

For closely coupled windings N_1:N_2 (Figure 19.23), the structure of toroidal cores can be easily derived. Actually, closely coupled windings (coupling factor k close to 1) are used to achieve low leakage inductances. Leakage flux is the flux that is not linked to the other winding. In order to significantly increase the leakage inductance in the basic setup shown, one must resort to suitable design measures. If

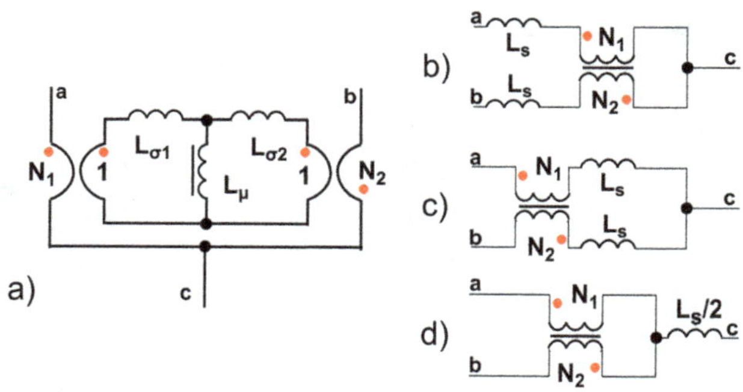

Figure 19.22 Principle possibilities of coupling two phases for the construction according to Figure 19.21: (a) T equivalent circuit of a transformer, (b) and (c) simulation of the T equivalent circuit with separate storage chokes, and (d) combination of the storage chokes in one component 10

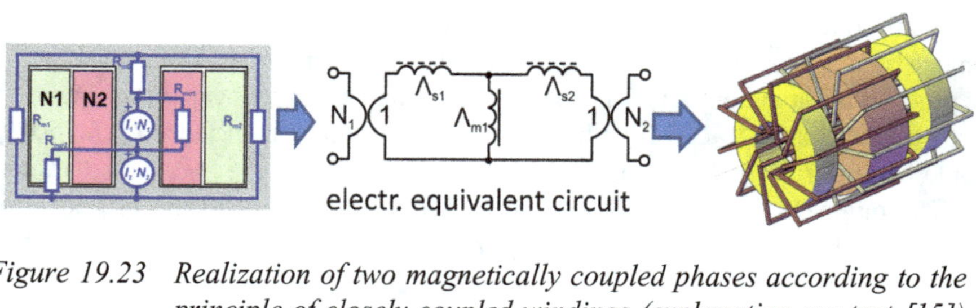

Figure 19.23 Realization of two magnetically coupled phases according to the principle of closely coupled windings (explanation see text [15])

the permeability in the leakage channels is increased, the leakage inductances/storage inductances are consequently increased, as the example in Figure 19.23 shows. The winding design is rather awkward to manufacture. If we now refer back to the equivalent circuit Figure 19.25(b) shown in Figure 19.22(b), we finally obtain the arrangement shown in the photo with three separately wound toroidal cores. The two outer toroidal cores carry the DC flux and a small portion of the AC flux. When the phase currents are equal, the center toroid is loaded with AC flux only.

However, the concept of loosely coupled windings (Figure 19.24) is also suitable for the setup in principle. In the basic design, the two windings are mounted on different core sections. The core has the task of coupling, while the increased stray field of the yoke takes over the storage function. Various

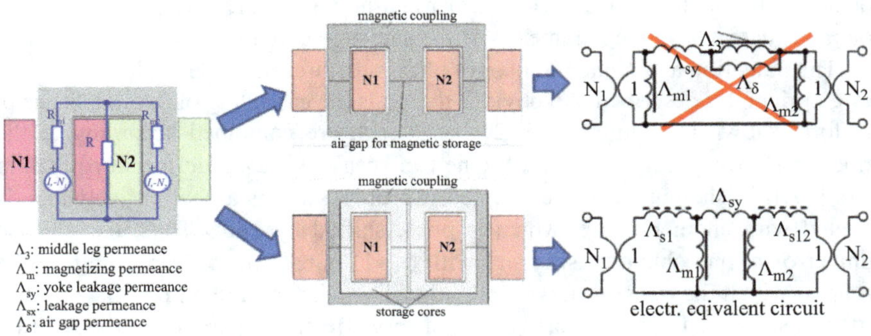

Figure 19.24 Use of loosely coupled windings for magnetic coupling: the upper path leads to a strong DC flux loading of the core material. The lower approach provides the DC flux with a material particularly suitable for this purpose (see also [10, 21]).

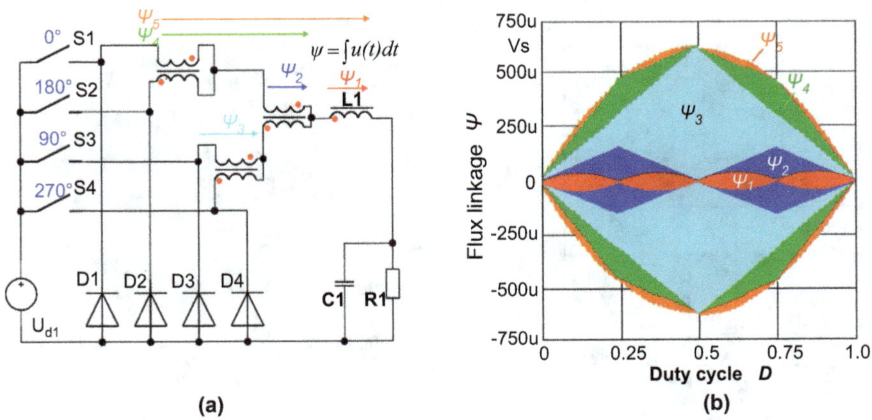

Figure 19.25 Cascaded coupling of four phases of a buck converter: (a) circuit diagram and (b) loading of the individual inductive components with AC flux linkage [15]

approaches are known from the literature to realize the energy of the stray channel by a center leg with air gap or by a center leg with low permeability. However, from the electrical equivalent circuit for this variant, shown in Figure 19.24, it is clear that a relatively high DC flux loading of the core sections occurs. This limits the possible volume savings. The variant shown below, as in Figure 19.23, carries the non-concatenated (leakage) flux in separate core sections made of a material with high saturation induction and low permeability. The yoke permeance Λ_{sy} remains, which may result in an asymmetry of the flux partitioning between the right and left parts of the coupling core. One will choose the largest possible yoke spacing in this design principle, which is a magnetic integration with only two windings, to keep Λ_{sy} as small as possible. The equivalent electrical circuit has a strong resemblance to that in Figure 19.23 and resembles it when $\Lambda_{sy} \to 0$. The presence of only two windings potentially results in a small size. Whether and how large the saving in weight and volume is depends on the parameters of the available core materials.

The transition to a higher number of phases requires multiple application of the design principles described. An obvious design is the cascading of several stages in the form shown in Figure 19.25. 2 phases each are combined according to the principle shown in Figure 19.22(d). One can speak of a cascade form or pyramid form [17,18]. The output of the coupled two phases forms a new phase, which is coupled with another phase, which is phase-shifted by 180°. Thus, for a phase number of p, one obtains a structure with $(p-1)$ transformers. Depending on the group to which the transformers belong, they have different rated powers and thus different sizes. As Figure 19.25(b) shows, a very effective separation of the AC flux linkage from the storage inductors occurs. The AC component becomes proportionally smaller $\sim(p-2)$ for a phase number p at the storage inductors. This type of coupling is obviously only possible for a phase number $p = 2^n$.

Another possibility is the circular coupling [11–13] of the phases, as shown in Figure 19.26. This type of coupling requires a total of as many transformers as

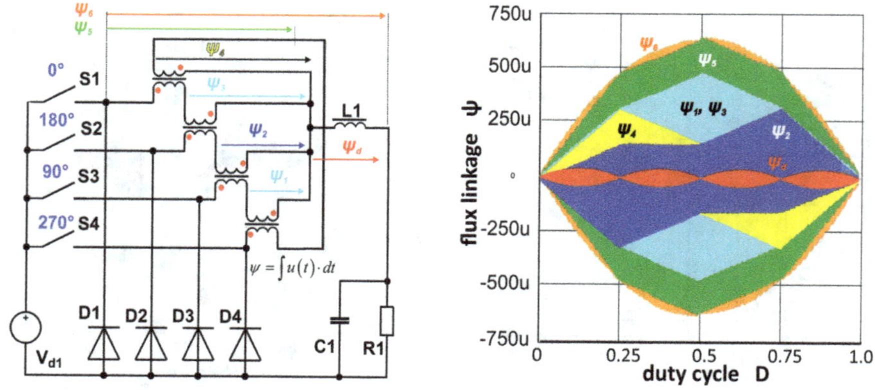

Figure 19.26 Circular coupling of four phases of a buck converter: (a) circuit diagram and (b) loading of the individual inductive components with AC flux linkage [15]

phases. The last phase of the group is coupled to the first one. This makes it possible to use transformers of the same size. The ± symmetry of the load with AC flux linkage is not perfect with this type of coupling, as Ψ_2 shows.

In principle, core materials with particularly high permeability are suitable. However, too high permeability leads to high sensitivity to unbalances of the coupled phase currents. To keep these low, a small air gap in the core is recommended. Otherwise, care must be taken to ensure that the total current is distributed evenly among the individual phases. This can be achieved with controllers that correct the specified duty cycle slightly according to the difference in the average current values. The higher the clock frequency, the smaller the steps in changing the duty cycle must be possible.

Although actually only even phase numbers should be possible to fulfill the requirement for a phase shift of 180°, Figure 19.27 shows that odd phase numbers are also possible. In this example, five phases are used. The phase intervals $\Delta\phi$ between two phases are divided in such a way that one obtains $\Delta\phi = 360°/p$. The coupled phases are selected in such a way that one comes as close as possible to a phase shift of 180°.

Figure 19.27 shows that the desired effect can also be achieved with five phases. The effect of the coupling becomes particularly clear when the operation of the circuit as an inverter is examined (Figure 19.28). The ripple of the output current follows the AC flux linkage of the smoothing reactor L_d. The total phase current shown then resembles that of a five-level circuit of an inverter, although these are only simple two-point circuits. This points to an interesting application possibility: At high currents, very low ripple currents can be achieved even at relatively low switching frequencies if interleaved operation of partial phases is combined with magnetic coupling. In this way, only partial currents have to be commutated. The commutating inductances therefore do not have to be as small as with directly parallel-connected transistors or corresponding semiconductor chips. Magnetic coupling also permits particularly small current ripples, as can be seen in Figure 19.28.

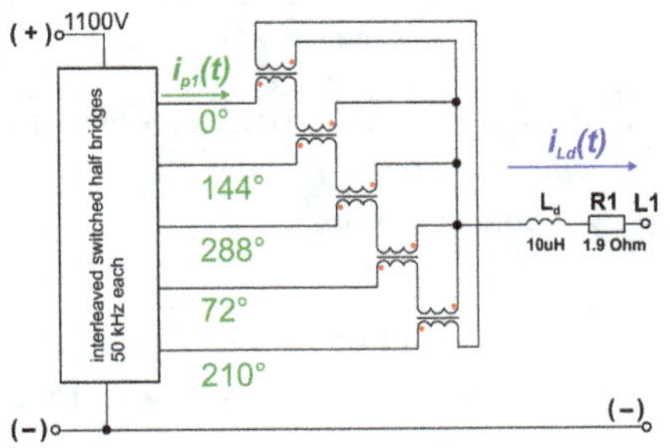

Figure 19.27 Application of circular coupling to an odd number of phases p = 5 using the example of a half-bridge for the production of a sinusoidal alternating current [15]

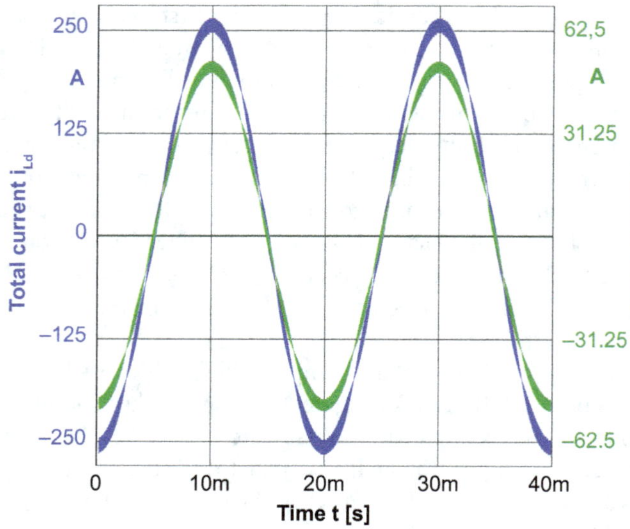

Figure. 19.28 Current ripple formation for an inverter constructed from five single phases in a half-bridge circuit for the 3ph–690V network with grounded neutral point [15]

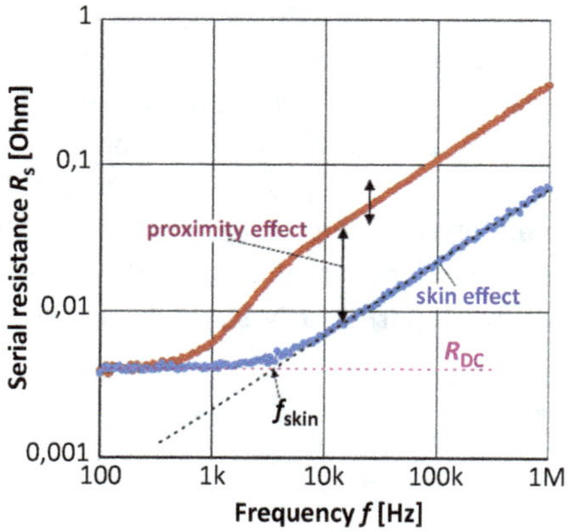

Figure 19.29 Measurement of the serial resistance content at the same conductor (copper, 3 mm × 4 mm = 12 mm²): as single winding (blue) or 15 windings Ø45 mm × 65 mm (red)

The reduction of the ripple current of the output choke L_1 affects the losses of the component in two ways. In principle, the losses can be approximated by the equation

$$P_{loss.tot} \approx k^*(f)\hat{\psi}^a V_e + R_{DC}I_{DC}^2 + (R_{AC}(f) - R_{DC})I_{AC}^2 \tag{19.8}$$

Here,

R_{DC}: DC resistance of the winding, R_{AC}: AC resistance of the winding, $\hat{\psi}$: peak value of the flux linkage, and α: flux density exponent of the Steinmetz formula.

Because of the triangular current, the following can be approximated:

$$\hat{\psi} = L_1 \hat{I}_{AC} \approx \sqrt{3} L_1 I_{AC} \tag{19.9}$$

This results in the equation

$$P_{loss.tot} \approx k^*(f)\hat{\psi}^{\alpha} V_e + R_{DC} I_{DC}^2 + (R_{AC}(f) - R_{DC})\left(\frac{\hat{\psi}}{\sqrt{3} L_1}\right)^2 \tag{19.10}$$

This means that the current ripple can have a considerable effect on the core losses but also on the winding losses. This plays a role especially for large conductor cross sections. The AC resistance is determined by the proximity effect, fringing effect, and skin effect and is therefore very dependent on the structure of the winding. Figure 19.29 illustrates this. The impedance of the same conductor was measured without a core with one turn or wound up with 15 turns. In the latter case, the proximity effect dominates already at significantly lower frequencies than the skin effect. The AC components of the electrical resistance give the associated losses a high weighting factor. Fringing losses, e.g., in the vicinity of air gaps in the core, must be added here.

19.5 Multiple interleaved operation of PWM converter branches

Using Figure 19.24 (lower variant) as a starting point, it is easy to extend the arrangement to more than two phases for coupling with integrated storage. Such an extension is shown in Figure 19.30. The principle in this design is to use a common

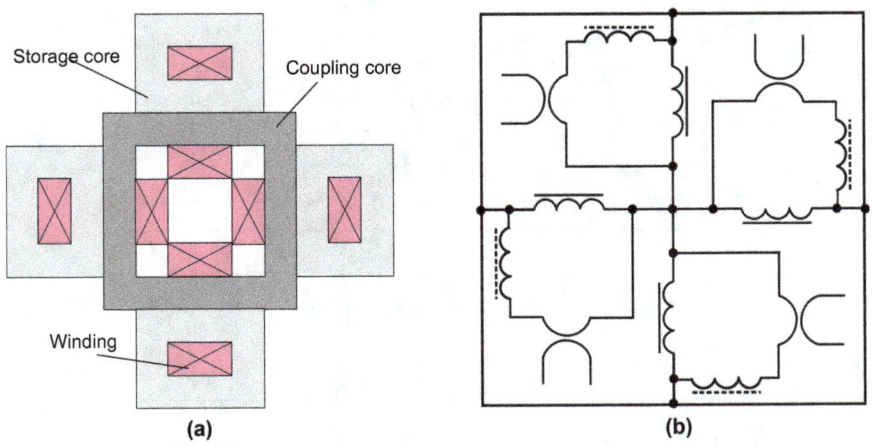

Figure 19.30 Arrangement for coupling of four phases with integrated storage: (a) principle arrangements of the cores for coupling and magnetic storage and (b) schematic of the electrical equivalent circuit for the magnetic arrangement according to (a)

core for coupling for all windings. The corresponding flux of these windings is then chained to each other. For the non-concatenated flux for windings, a separate path is provided using core materials with high saturation induction. This is then the place for the storage function of the resulting four-phase inductive component.

In the literature on the subject of multiphase magnetic coupling, one often finds the arrangement shown in Figure 19.31(a). The storage function is assigned to the air gaps in the cores of the individual windings. Another part of the energy is stored in the field of the yoke leakage. The core sections carrying the windings are penetrated by both DC flux and AC flux. This makes material selection problematic. The air gap permeance and the yoke leakage field permeances are of the same order of magnitude. This limits its effectiveness as a common mode reactor of multiphase systems.

The coupling of the individual fluxes is somewhat confusing. This becomes clear from the equivalent electrical circuit (Figure 19.31(b)). The voltage drops generated directly by the windings are connected to each other by a series circuit via the yoke leakage inductors. If the cores of the winding carriers are arranged in the ring by means of the yokes, the voltage drops of the coreless inductors are also arranged in the ring. The effect is then that of a multiphase common mode reactor with low magnetizing inductance. A disadvantage is that the DC fluxes are only partially compensated in the yokes. This limits the amount of component volume that can be saved.

The arrangement in Figure 19.32 corresponds electrically to an integrated coupling element of the variant for pair-wise coupling of windings according to Figure 19.27.

The coupling arrangement in Figure 19.32 is implemented in such a way that the DC fluxes in each core cancel each other out. Thus, the cores ideally carry only AC flux components. If the individual cores are arranged in a ring, all induced voltages are connected in series in the electrical equivalent circuit. The coreless

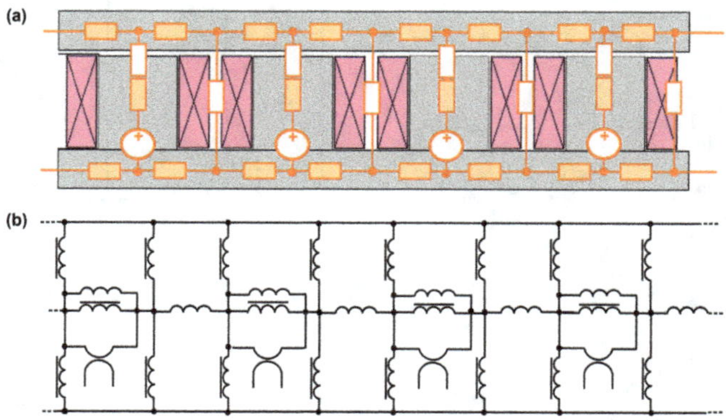

Figure 19.31 Multiphase coupling of windings with air gap via common yokes [19]: (a) sketch of the principle structure and (b) equivalent electrical circuit

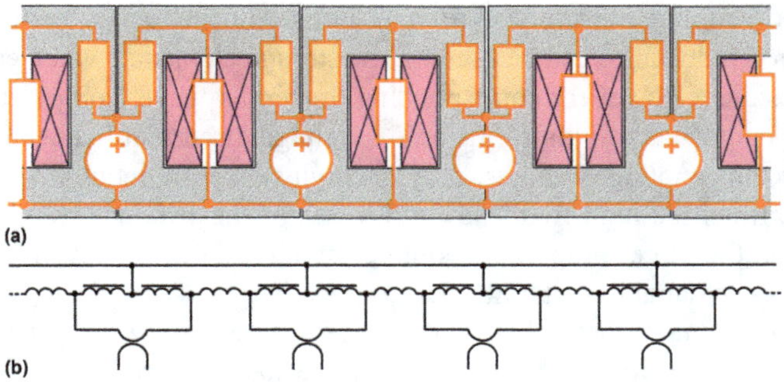

Figure 19.32 Pair-wise current-compensated winding coupling in a magnetically integrated approach: (a) principle structure and (b) electrical equivalent circuit

inductors represent the permeances of the yoke stray field, which provide a slight decoupling of the individual phases. Ideally, the permeances of the core are much lower than those of the yoke stray field. The magnetic storage effect of the array is therefore small and must be implemented separately, as described above.

19.6 Summary

The standard solution to the problem of reducing the ripple current of pwm power electronic converters is to use a sufficiently large inductance for the choke L_1 smoothing the current. However, large chokes also mean large weight and/or volume and thus low power density. There are various measures that can be taken to achieve more favorable solutions in terms of circuitry and design.

One variant is to inject a ripple current of the opposite phase into the load. This requires a synchronous current generator that produces the 180° shifted ripple current at the load. Ideally, the ripple components of the main current and the compensation current add up to zero. The second converter required for this must be designed only for the reactive power of the harmonic harmonics. The effect of this design approach is to emulate a very large reactor.

The same effect can also be achieved by a purely passive compensation network. The phase inversion is achieved by a transformer whose winding is connected accordingly. Here, too, the effect is comparable to that of a very large choke. This means that the measures described here can be used when large inductance values would actually be required. Since less energy has to be stored, the system is only marginally slower. In principle, it is possible to integrate the inductive components. Whether this makes sense depends on the technical parameter combinations.

The ripple current causes losses in both the core and the winding of an inductive storage element. Magnetic coupling of interleaved phases allows extensive separation of DC and AC components of the magnetic flux to inductive storage elements and coupling elements. With appropriate dimensioning, materials with a high saturation inductance can be used for the storage elements, which enables very compact designs. However, corresponding materials have comparatively high specific AC magnetization losses. If the AC voltage or current load can essentially be shifted to the coupling elements, only the high saturation induction plays a role. At the same time, the high-frequency AC losses of the winding are also reduced. Depending on the target properties, volume savings can be achieved by assigning load quantities to suitable materials.

The ratios of multiphase DC/DC converters were mainly illustrated using the example of buck converters. However, they are just as applicable to boost converters [9,20]. A higher number of phases leads to a higher effective switching frequency for EMI. This reduces the required filters. It is possible to combine buck and boost converters to form a half-bridge branch [21,22]. This means that the method of several interleaved phases can also be used for inverters that are composed of half bridges.

References

[1] Piasecki, S.; Jasinski, M.; Rafał, K.; Korzeniew-ski, M.; and Milicua, A.: Higher Harmonics Compensation in Grid-Connected PWM Converters for Renewable Energy Interface and Active Filtering. *Przeglad Elektrotechniczny (Electrical Review), R.* 87 NR 6/2011.

[2] F. Z. Peng, H. Akagi and A. Nabae, "A Study of Active Power Filters Using Quad-Series Voltage-Source PWM Converters for Harmonic Compensation," in *IEEE Transactions on Power Electronics*, vol. 5, no. 1, pp. 9–15, 1990

[3] H. Akagi, Active Harmonic Filters. in *Proceedings of the IEEE*, vol. 93, no. 12, pp. 2128–2141, 2005

[4] Bruce Carsten: US000005929692A, (11.07.1997), 27. 07.1999; Ripple cancellation circuit with fast load response for switch mode voltage regulators with synchronous rectification.

[5] Ching-Tsai Pan, Ching-Ming Lai and Yu-Lin Juan: Output Current Ripple-Free PWM Inverters. IEEE Transactions on Circuits and Systems—II: Express Briefs, Vol. 57, no. 10, 2010.

[6] M. A. Alharbi, M. S. A. Dahidah, S. A. Ali, S. A. E. Ethni and V. Pickert, "Ripple-Free Multiphase Interleaved Stacked Converter for High-Power Applications," in *IEEE Transactions on Power Electronics*, vol. 37, no. 12, pp. 14770–14780, 2022, doi: 10.1109/TPEL.2022.3194979.

[7] Peter Zacharias: *Magnetic Components*. Springer Nature. 2022, ISBN 978-3-658- 24742-3

[8] P. Sernia, D. Finn and G. Walker, "The Effects of Coupled Inductors in Parallel Interleaved Buck," in *Australasian Universities Power Engineering Conference*, vol. 1, pp. 1–5, 2003.

[9] H. Shin., J. Park, S. Chung, H. Lee, and T. Lipo, "Generalized Steadystate Analysis of Multiphase Interleaved Boost Converter with Coupled Inductors," *IEEE Proceedings – Electric Power Applications*, 152, (3), pp. 584–594, 2005

[10] Chuanyun Wang: Investigation on Interleaved Boost Converters and Applications. Dissertation, Virginia Polytechnic Institute and State. 21st July 2009.

[11] G. J. Capella, J. Pou, S. Ceballos, J. Zaragoza and V. G. Agelidis, "Current-Balancing Technique for Interleaved Voltage Source Inverters with Magnetically Coupled Legs Connected in Parallel," in *IEEE Transactions on Industrial Electronics*, vol. 62, no. 3, pp. 1335–1344, 2015, doi:10.1109/TIE.2014.2345345.

[12] J. Salmon, S. Singh: Multilevel Voltage Source Parallel Inverters and Coupled Inductors. *WO2020248078A1.*

[13] J. Park, H. Byun, B. Kim, S. Kim and C. Won, "Analysis and Design of Coupled Inductor for Interleaved Buck-Type Voltage Balancer in Bipolar DC Microgrid," in *Energies*, vol.13, no.11, pp. 2775, 2020.

[14] Jie Liu: Investigation of Multiphase Power Converter Using Integrated Coupled Inductor Regarding Electric Vehicle Application. Dissertation, University of Kassel, 2013.

[15] P. Zacharias and X. Yu: Separation of AC and DC Flux Linkage Using the Example of Buck Converter. *EUROCON2023*, Turin.

[16] J. Lee, H. Cha, D. Shin, K. Lee, D. Yoo, and J. Yoo, "Analysis and Design of Coupled Inductors for Two-Phase Interleaved DC-DC Converters," in *Journal of Power Electronics*, vol. 13, No.3, pp. 339– 348, 2013.

[17] Dipankar De, Christian Klumpner, Chintanbhai Patel, Kulsangcharoen Ponggorn, Mohamed Rashed, Greg Asher: Modelling and Control of a Multi-Stage Interleaved DC–DC Converter with Coupled Inductors for Super-Capacitor Energy Storage System. *IET Power Electronics* 2013 doi:10.1049/iet-pel.2012.0529

[18] P. Zacharias, C. Nöding, A. Aganza-Torres: Ripple Cancellation of PWM Power Converters by Magnetically Integrated Filter Technology. *25th European Conference on Power Electronics and Applications (EPE'23 ECCE Europe)*, Aalborg, Denmark, 2023, 1-10. 10.23919/EPE23ECCEEurope58414.2023.10264376.

[19] Chen, Minjie. (2021). On Ripple Reduction of PWM Coupled Inductors. 10.36227/techrxiv.14773515.v1.

[20] Chia-Chou Chang, Chen-Chan Lee, Jyun-Chun Huang and Yaow-Ming Chen: "Ripple Current Cancellation for Boost Converters," *2017 IEEE 3rd International Future Energy Electronics Conference and ECCE Asia (IFEEC 2017-ECCE Asia)*, Kaohsiung, Taiwan, 2017, pp. 671–676, doi: 10.1109/IFEEC.2017.7992119.

[21] X. Huang, X. Yang, J. Wu, P. Ren and W. Chen, "Optimal Design of Coupled Inductors for Multiphase Buck Converter," 2022 IEEE PEAC, Guangzhou, Guangdong, China, 2022, pp. 734–738, doi:10.1109/PEAC56338.2022.9959239.

[22] G. Aulagnier, K. Abouda, E. Rolland, M. Cousineau and T. Meynard, "Benefits of Multiphase Buck Converters in Reducing EME (Electromagnetic Emissions) Analysis and Application to On-Chip Converters for Automotive Applications," 2015 IEEE EMC, Dresden, Germany, 2015, pp. 102–107, doi: 10.1109/ISEMC.2015.7256140.

[23] P. Zacharias: What Challenges are Near-Ideal Switches to Inductive Components in Power Electronics? (Engl.), *National Conference "Bauelemente der Leistungselektronik und ihre Anwendungen 2023"*, ETG-Fb. 171

Conclusion

Wrap-up and future directions

The book summarizes the most important developments from the authors' point of view. All authors are active in the field of power electronic applications of inductive components. The main tasks of inductive components in power electronic converters are temporary energy storage, current delay/smoothing, and coupling/transformation. These tasks require material properties that vary greatly in their combination in different materials. The first part of this book takes this into account. The basics of material properties as well as the possibilities of modeling and calculation are presented here.

The past developments described in this book have already enabled the rapid development of electrical energy converters. At the same time, this is being stimulated by the extremely innovative development of power electronic semiconductor switches in the wide bandgap range, but also on a Si basis. However, the chapters make it clear that it is not the properties of the semiconductors that are limiting.

In addition to the ever-increasing penetration of social life by technologies that require specific energy converters, there is a trend toward increasing power. One example of this is solid-state transformers. Due to the limited blocking voltage of the semiconductor switches used, higher and higher currents are being used in applications.

The combination of high currents with fast switches in power supply technology leads to problems in complying with the permissible limit values of the components. It has been shown how these problems, for example, can be reduced by splitting the current over several inverter legs.

The generation of electromagnetic interferences also always plays an important role in power electronic circuits. Other applications could otherwise be impaired in their use. Filters, shielding, and circuitry design measures can help here. Attention was therefore also paid to this topic.

In contrast to electrical components, magnetic components are often heavy and bulky. The reason for this is the saturation induction of the core materials and high specific losses in the core and windings, which have to be dissipated from the component. Several articles deal with the combination of several tasks in one component and explain the necessary considerations. The examples presented should be seen as a suggestion to take up and apply these considerations to own work.

It is to be expected that the development toward integrated magnetic components will continue in the future. The combination of the construction of power electronic converters with additive manufacturing techniques is likely to offer completely new, as yet unexplored approaches. Magnetic components are characterized by comparatively high robustness, resistance to interference, and durability. It can therefore be assumed that the problem of limited material resources will be a driving force in the future to further develop optimized inductive components and improve power electronic converters based on knowledge of the internal connections.

Index

www.ingramcontent.com/pod-product-compliance
Lightning Source LLC
LaVergne TN
LVHW011557200926
848484LV00001B/327